ALAMEDA CREEK WATERSHED **Historical Ecology Study**

FEBRUARY 2013
SAN FRANCISCO ESTUARY INSTITUTE

Prepared for and funded by

THE SAN FRANCISCO PUBLIC UTILITY COMMISSION

and the

ALAMEDA COUNTY FLOOD CONTROL & WATER CONSERVATION DISTRICT

Prepared by

SAN FRANCISCO ESTUARY INSTITUTE

Bronwen Stanford
Robin Grossinger
Julie Beagle
Ruth Askevold
Robert Leidy[1]
Erin Beller
Micha Salomon
Chuck Striplen
Alison Whipple

[1]U.S. Environmental Protection Agency, San Francisco

In Cooperation with

ALAMEDA COUNTY RESOURCE CONSERVATION DISTRICT

Additional funding from

ZONE 7 WATER AGENCY AND THE ALAMEDA COUNTY WATER DISTRICT

HISTORICAL LAND COVER, EARLY 1800s
(see figure 9.2b for land cover ca. 2000)

- Tidal Marsh, Pond, Slough
- Tidal Flat
- Bay/Deep Channel

- Valley Freshwater Marsh
- Perennial Freshwater Pond
- Wet Meadow
- Willow Thicket

- Alkali Meadow
- Alkali Sink Scrub-Vernal Pool Complex
- Alkali Vernal Pool Complex
- Alkali Playa Complex

- Sparsely Vegetated Braided Channel
- Confined Riparian Woodland/Savanna
- Sycamore Alluvial Woodland

- Oak Savanna
- Grassland

- Perennial Stream
- Intermittent Stream
- Slough/Overflow Channel
- Distributary
- Spring

N
1 mile
1:135,000

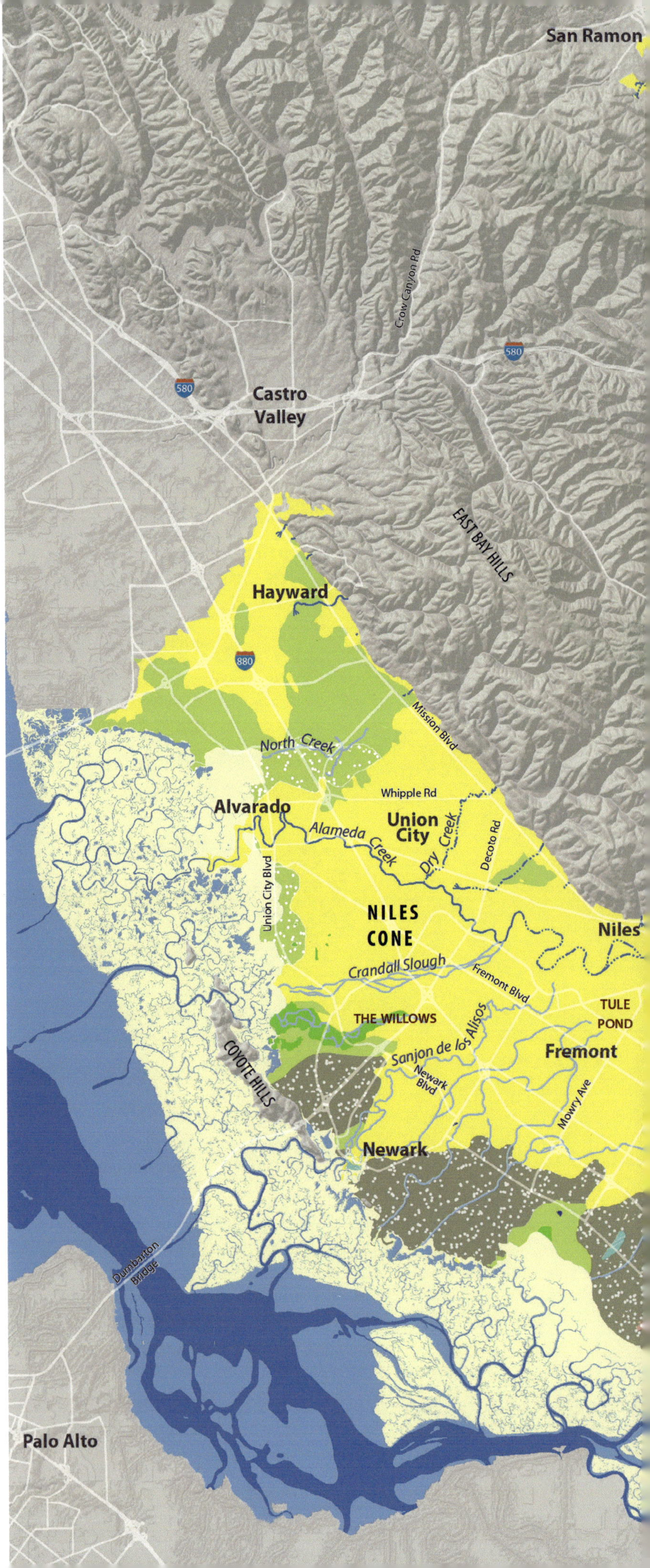

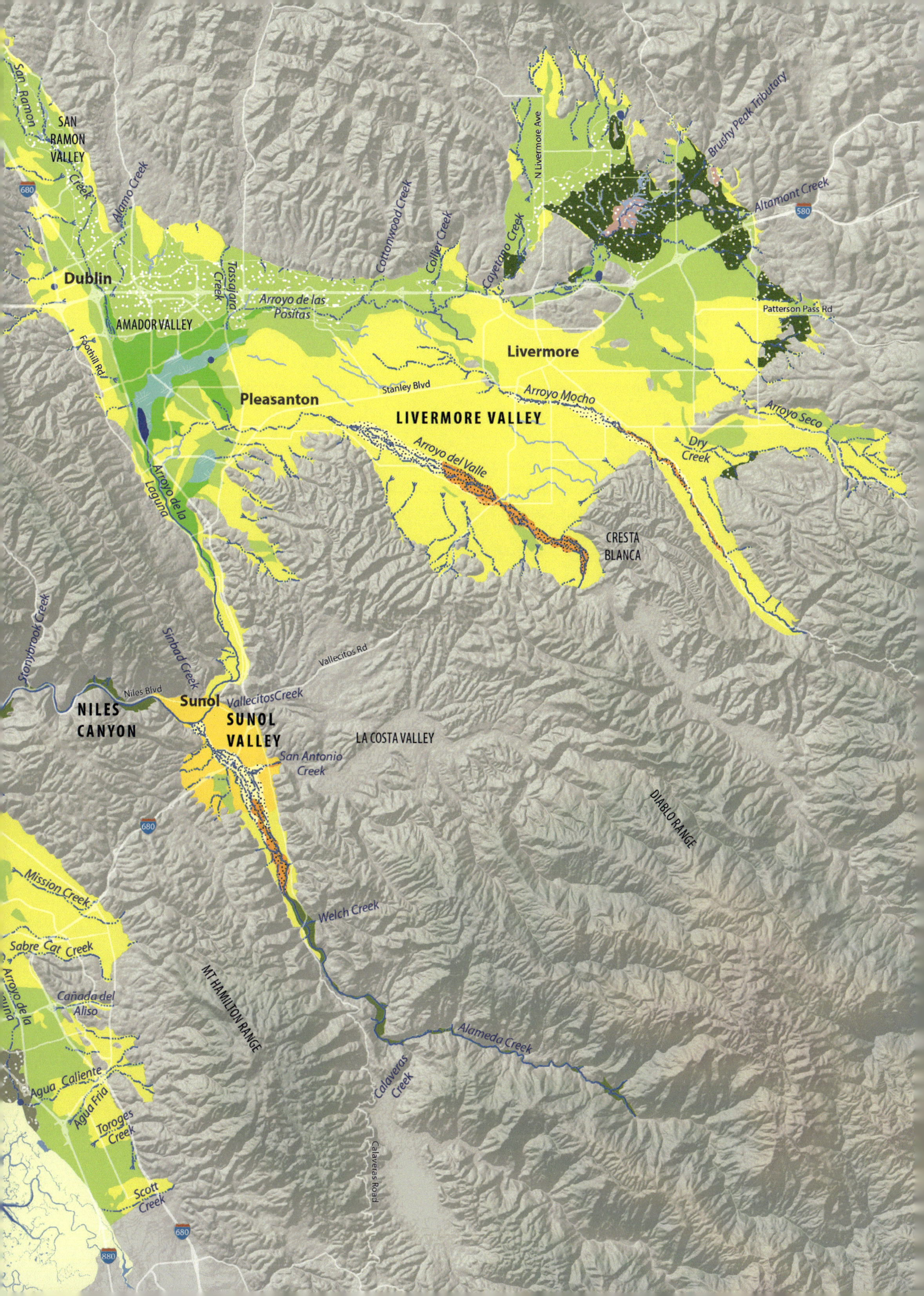
San Ramon Creek
SAN RAMON VALLEY
Alamo Creek
680
Dublin
AMADOR VALLEY
Foothill Rd
Tassajara Creek
Cottonwood Creek
Collier Creek
Cayetano Creek
N Livermore Ave
Brushy Peak Tributary
Altamont Creek
580
Arroyo de las Positas
Patterson Pass Rd
Livermore
Pleasanton
Stanley Blvd
Arroyo Mocho
LIVERMORE VALLEY
Arroyo Seco
Arroyo del Valle
Dry Creek
Arroyo de la Laguna
CRESTA BLANCA
Stonybrook Creek
Sinbad Creek
Vallecitos Rd
Niles Blvd
Sunol
VallecitosCreek
NILES CANYON
SUNOL VALLEY
LA COSTA VALLEY
San Antonio Creek
680
DIABLO RANGE
Mission Creek
Welch Creek
Sabre Cat Creek
MT HAMILTON RANGE
Arroyo de la Laguna
Cañada del Aliso
Alameda Creek
Agua Caliente
Agua Fria
Toroges Creek
Calaveras Creek
Calaveras Road
Scott Creek
680
880

Suggested citation:

Stanford B, RM Grossinger, J Beagle, RA Askevold, RA Leidy, EE Beller, M Salomon, C Striplen, AA Whipple. 2013. Alameda Creek Watershed Historical Ecology Study. SFEI Publication #679, San Francisco Estuary Institute, Richmond, CA.

Report and GIS layers are available on SFEI's website, at www.sfei.org/projects/AlamedaCreekHE

Front cover, top: Sunol Dam 1888, #158383, © San Francisco Public Utilities Commission; bottom: Livermore Valley in Thompson and West (1878), courtesy David Rumsey Map Collection.

Back cover, from left to right and top to bottom: Thompson and West 1878, courtesy David Rumsey Map Collection; USDA 1939; photo by Bronwen Stanford; photo by Amy Wolitzer; photo by Alison Whipple; Lewis 1860, courtesy The Bancroft Library, UC Berkeley; AD-866, © San Francisco Public Utilities Commission; Unknown ca. 1890, © San Francisco Public Utilities Commission; photo by Robin Grossinger.

CONTENTS

FIGURES

TABLES

BOXED TEXT

ACKNOWLEDGMENTS

We would like to express our gratitude to the volunteers and staff at the historical societies, libraries, archives, and agencies where we gathered the data that form the backbone of this report. Particular thanks are due to Phil Holmes of the Museum of Local History, Alison Moore and Katherine du Tiel at the San Francisco Public Utilities Commission (SFPUC) archives, and Amy Kasameyer at the Jepson Herbarium. James Thorne at UC Davis provided access to the Wieslander dataset. Thanks to J. Fong and D. Catania at the California Academy of Sciences for providing access to fish records (Catania) and for taking photographs of fishes (Fong) used in Chapter 8. Other institutions whose staff generously assisted us in accessing their materials include The Bancroft Library, Water Resources Center Archives, and Earth Sciences and Map Library at UC Berkeley; Alameda County Water District; Alameda County Department of Public Works; Alameda County Recorder; Museum of the San Ramon Valley; Niles Silent Film Museum; Union City Historical Museum; Museum on Main; California Historical Society; Society of California Pioneers; Livermore Heritage Guild; Bureau of Land Management; California State Library; and the Fremont, Oakland, and San Francisco public libraries. Without their help and careful curation of historical materials this project would not have been possible.

Staff from the Alameda County Resource Conservation District assisted with data collection, interviews, and the acquisition and orthorectification of historical aerial photographs. We would particularly like to thank Leslie Koenig, Morpheus Anima, and Ling He. Assistance was also provided by a number of SFEI staff, including Kristen Cayce and Sean Baumgarten. In addition, three interns from the Bill Lane Center for the American West at Stanford contributed to this project: Alec Norton, Melissa Runsten, and Jenny Rempel.

Laurel Collins of Watershed Sciences assisted with many phases of this project, including data collection, analysis, and review. She also provided a number of historical photographs. Jeff Miller of the Alameda Creek Alliance guided us to valuable historical data and local experts.

A number of long-time residents of the Livermore Valley area agreed to do interviews for the upland pilot portion of this project. We would like to extend a special thanks to Paul Banke, Rich Fletcher, Tim Koopman, and Dan Sacchau for sharing their invaluable knowledge of the area. Field trips, including one with Heath Bartosh of Nomad Ecology and several with members of the Alameda Creek Watershed Council, contributed greatly to our understanding of the watershed.

Several scientists helped with the fisheries chapter, including Pete Alexander of the East Bay Regional Park District, Ken Gobalet of California State University at Bakersfield, Brian Sak of SFPUC, Bill Trush of McBain and Trush, and Kristina Cervantes-Yoshida of UC Berkeley. Local resident Tony Logan also provided his expertise.

We would like to thank our technical review team for contributing their expertise to the interpretation and discussion of historical conditions and providing comments on the report. The team included Heath Bartosh, Peter Baye, Scott McBain, Janet Sowers, and Jim Thorne. Lester McKee and Sarah Pearce of SFEI provided additional internal technical review.

This project was funded by the San Francisco Public Utilities Commission and the Alameda County Flood Control and Water Conservation District (ACFCWCD), with additional contributions by Zone 7 Water Agency and the Alameda County Water District (ACWD). The Alameda County Resource Conservation District (ACRCD) served as fiscal sponsor. We benefited from the knowledge, guidance, and support of many staff of the sponsoring agencies, including Carla Schultheis, Ellen Natesan, and Brian Sak at SFPUC; Hank Ackerman, Sharon Gosselin, Manny DaCosta, and Kwablah Attiogbe at ACFCWCD; Carol Mahoney and Joe Seto at Zone 7; Eric Cartwright and Evan Buckland at ACWD; and Leslie Koenig and Amy Evans at ACRCD. In particular, Tim Ramirez of SFPUC and Ralph Johnson of ACFCWCD were crucial to the initiation and success of this project.

1 • PROJECT SUMMARY

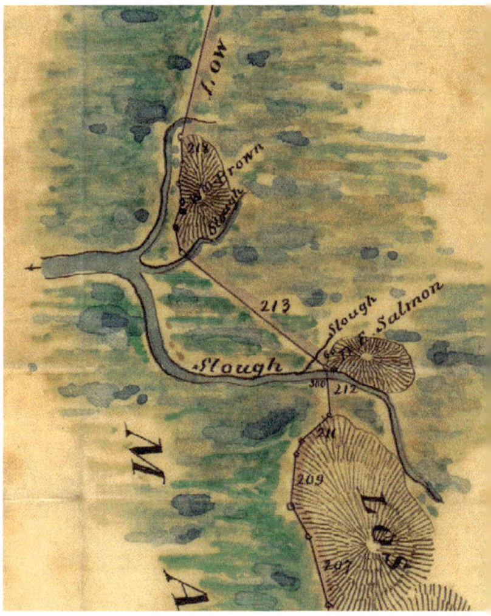

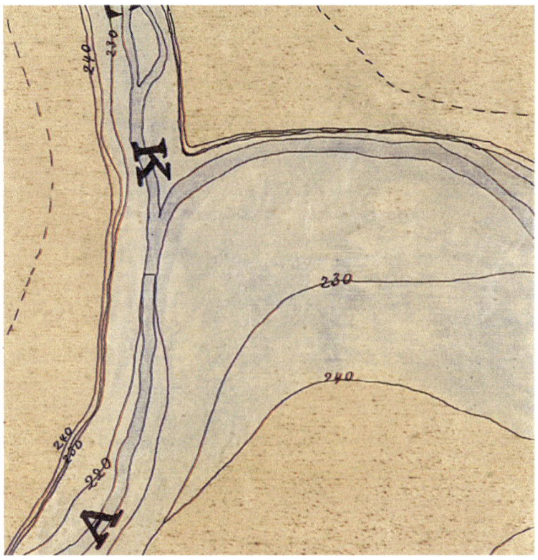

Two hundred years ago the valleys and plains of the Alameda Creek watershed were dominated by grasslands and seasonal wetlands, and shaped by streams that overflowed annually. While substantial ecological resources still remain, the region has been subject to extensive modification: today over half of the valley floor is covered by urban development and many creeks flow through artificial channels. A number of important natural features have been lost in the course of this transformation. How did the Alameda Creek watershed look and function before development? What processes controlled the distribution of habitats and native species? Which physical controls are relatively intact, and which have shifted? How can we enhance the ecological health of the region in the future?

Historical ecology is the study of landscapes and ecosystems in the historical period, defined here as the period of written human record. In California this corresponds to a time of rapid change following the arrival of the Spanish, making an understanding of conditions at the start of this period particularly valuable. This report depicts conditions in the early 1800s—shortly after the arrival of Europeans but before large-scale agriculture and major modifications to the landscape. Data used in this report extend from 1769 through the 21st century, and range from travel diaries and family photographs to parcel maps and General Land Office surveys.

Report overview

The Alameda Creek Watershed Historical Ecology Project synthesized hundreds of historical data sources to create a picture of the historical landscape and explore the implications for contemporary management (fig. 1.1). The goals of this report are to describe the stream network and habitat patterns across the study area in the early 1800s to inform present-day management decisions, and to provide managers with new tools to assess ecosystem functions. Along with the associated project geo-database, the report provides a spatially comprehensive assessment of the historical distribution and abundance of habitat types of the Livermore-Amador Valley, Sunol Valley, and Niles Cone/Fremont Plain. It also explores fish assemblages and habitat patterns of the surrounding uplands. This information can help us understand the changes of the recent past and identify new strategies to work towards healthy watersheds that support both people and natural ecosystems. This report does not provide a comprehensive land and water use history or attempt to closely document change over time. Rather, it provides a view of historical conditions in the early 1800s and broadly explores the impact of land use trends on the historical system.

Without a detailed understanding of the former characteristics of the region—and how these characteristics changed in response to human alterations to the landscape—appropriate ecological and hydrological restoration targets can be difficult to determine (fig. 1.2). Understanding the nature of the historical landscape is not a trivial task, however: early

Figure 1.1. Views of the Alameda Creek watershed through time. These images represent a selection of sources used in the report. (from left to right and top to bottom: Thompson and West 1878, courtesy David Rumsey Map Collection; USDA 1939; photo by Bronwen Stanford; photo by Amy Wolitzer; photo by Alison Whipple; Lewis 1860, courtesy The Bancroft Library, UC Berkeley; AD-866, © San Francisco Public Utilities Commission; Unknown ca. 1890, © San Francisco Public Utilities Commission; photo by Robin Grossinger) data are idiosyncratic, challenging to interpret, and scattered in archives across the region. Historical ecology is the tool we use to interpret early data within the context of the present-day landscape, and it is an essential component of crafting sound, site-specific environmental restoration objectives, which demand detailed data as the basis for management strategies (Collins and Montgomery 2001, Grossinger 2005).

Historical ecology has particular relevance in the context of global climate change. Climate change is expected to cause overall temperature increases and changes in precipitation patterns in California, as well as a potential increase in the frequency of disturbance events such as fires and flooding (ICF International 2010a). The native California landscape was well adapted to a highly variable, episodic climatic regime, and buffered the effects of environmental extremes while providing diverse ecological functions. As we anticipate a more variable climate in the future, we can learn from the ways in which dynamic historical ecosystems were able to respond and adapt to extreme conditions in the past (Harris et al. 2006). An understanding of the underlying drivers influencing landscape patterns can help us design more resilient landscapes for the future (Grossinger 2012, Safford 2012, Wiens et al. 2012).

This study is designed to advance public engagement in the watershed and directly support several significant planning efforts, including the Alameda

Figure 1.2. Sycamores along Alameda Creek through Sunol Valley. Sycamore alluvial woodland is an example of a habitat type specifically adapted to the summer-dry creeks found in the watershed. Sycamore alluvial woodland in Sunol Valley historically produced large woody debris and supported the development of pools. (photo by Alison Whipple, February 2012)

Watershed Habitat Conservation Plan, Alameda County flood control planning, the Alameda Creek Watershed Council watershed planning, and the South Bay Salt Pond Restoration Project. This study was funded by the San Francisco Public Utilities Commission (SFPUC) and the Alameda County Flood Control and Water Conservation District (ACFCWCD), in partnership with the Alameda County Resource Conservation District (ACRCD). Additional funding was provided by Zone 7 Water Agency and Alameda County Water District (ACWD).

A host of restoration and monitoring activities are being pursued through the watershed, including tidal marsh restoration, stream and riparian habitat restoration, and efforts to control erosion and sedimentation. Increasingly, scientists and managers are appreciating the important role that pockets of habitat within highly developed landscapes can play (Leidy et al. 2011, Sanderson and Huron 2011). Historical ecology can enable managers to develop a set of tools with which to direct restoration of increasingly functional systems within these urbanized landscapes.

Regional context

The Alameda Creek watershed is the third largest watershed draining into the San Francisco Bay (after the Sacramento and San Joaquin rivers), covering about 700 square miles. It has experienced substantial uplift and lateral displacement due to the number of faults that cross the watershed, including the Hayward Fault and the Calaveras Fault. Three distinct alluvial plains or lowlands occur within the watershed—Livermore-Amador Valley, Sunol Valley, and the Niles Cone. Rising between these lowlands are the Diablo, Altamont, and East Bay hills. These hills enclose the Sunol and Livermore valleys and border the Niles Cone. The region experiences a two-season Mediterranean climate, characterized by high inter- and intra-annual variability. Virtually all precipitation occurs between October and April.

Today the region is home to over 500,000 people, a third of the population of Alameda County (U.S. Census Bureau 2010). Livermore, historically known for its valuable springs, was initially settled by Robert Livermore, and was an important stop for travelers journeying east from the Bay to mine gold. Pleasanton was centered on slightly higher ground at the edge of a vast marsh complex, but has today spread over the full extent of the now-drained marshland. Dublin grew rapidly from sparsely settled rangeland in the mid-20th century. Fremont, a conglomeration of five historical cities, today spreads across 87 square miles, covering much of the Niles Cone. It includes Mission San José, one of the most prosperous of the Spanish missions, and was a valuable agricultural region. Union City and Newark, positioned on large tidal sloughs along San Francisco Bay, were important landings and centers for transport of agricultural goods. All of these cities experienced a population explosion beginning in the 1950s; the population in Livermore-Amador Valley more than tripled from 1950 to 1960 and again from 1960 to 1970. Today the study area is a mix of highly urbanized areas and open rangelands and agricultural lands, some of which still support native land cover types.

The Alameda Creek watershed still supports a number of distinct native communities, including sycamore alluvial woodland, alkali meadows, and vernal pools and associated species (Keeler-Wolf et al. 1996, Keeler-Wolf et al. 1998, Holland 2009). Alameda Creek has been identified as one of the best examples of potential steelhead habitat in the Bay Area as well as an anchor watershed for steelhead (Leidy et al. 2005a, Becker et al. 2007). The creek has been the subject of recent attention focused on removing barriers to steelhead runs and improving steelhead habitat through the watershed.

Transformation of the historical watershed

The Alameda Creek watershed has been modified to serve human needs from the earliest human occupation (fig. 1.3). Costanoan Ohlone people built shell mounds along Alameda Creek and harvested salt from the salt ponds in the tidal marsh. Spanish and American settlers used deep subtidal channels as access points to the Bay and built landings there. Farmers learned the flood patterns of Alameda Creek, and used the overflow across the Niles Cone as a valuable deposit of fresh soil. Brick makers mined the fine clay deposits along the lower reaches of streams to make their bricks; gravel miners extracted gravel from the coarse gravelly deposits further upstream.

Figure 1.3. "Construction of Niles Stone Dam, fish ladder and aqueduct inlet." Construction of dams and channels in the late 1800s and early 1900s rapidly transformed the watershed. This dam was located near the bottom of Niles Canyon. (#62759, © San Francisco Public Utilities Commission)

Alameda Creek watershed was the site of focused hydromodification from the beginning of the Euro-American period. Farmers developed wells in artesian zones to access the groundwater and directed the path of overflow from Alameda Creek such that sediment would fill low points and deposit over the tidal marsh, converting it to farmland. Farmers also cleared wetlands in the Amador Valley and on the Niles Cone to access land irrigated by groundwater. One of the more large-scale modifications to natural processes was the water system developed by the Spring Valley Water Company (SVWC) to provide water from Alameda Creek to San Francisco. Engineers harnessed natural processes of recharge, transport, and groundwater storage to make water available for consumption in the city. Canyon channels were used for transport, and water was directed across gravels so that it would percolate into groundwater aquifers (fig. 1.4). The Pleasanton marsh complex, a large wetland that historically occupied the area now covered by Pleasanton, was viewed as a giant underground reservoir, and engineers worked to maximize water output from the marsh by reducing evapotranspiration.

Residential development brought a further series of modifications. Creeks (including Alameda Creek) were channelized to control flooding and maximize accessible land. Riparian vegetation was cleared to make space for farmland. Reservoirs and dams altered the movement of sediment and water through the watershed. Grazing caused local incision and increased erosion. Pavement sped runoff. Introduced species (particularly eucalyptus) replaced native riparian trees.

Among the most significant modifications to the landscape is the transformation to confined streams, incised channels, drained wetlands, reservoirs, and diked tidal marshlands. In the Livermore-Amador Valley, total channel length has increased by 30%, from 197 miles in 1800 to 256 miles in 2010. This is due largely to the construction of channels through former wetlands. On the Niles Cone, overall channel length has also increased by one-third, and only 22 of the 133 miles of channel match the

Figure 1.4. "Alameda Creek at the filter galleries." Engineers slowed and spread the flow of Alameda Creek through Sunol Valley to allow percolation into the filter galleries by damming Alameda Creek at Sunol Dam, directly downstream. This slowed the movement of water from the valley into Niles Canyon. (image from SVWC n.d., courtesy San Francisco History Center, San Francisco Public Library)

historical channel network. Even many of the channels that do match their historical alignment function differently than they did historically, due to armoring and lack of access to the historical floodplain for flood storage.

Overall, 90,000 acres, or 78% of the study area, is covered with agriculture, development, or salt ponds—all habitat types not historically present. Even those native habitat types that remain, such as tidal marshes and seasonal wetlands, have been impacted by introduced species and modifications to the hydrologic regime. However, looked at another way, the fact that almost 40% of the study area has not been developed and is used for agriculture, rangeland, or open space provides managers with an unusual amount of flexibility to consider a watershed plan for the future.

Report structure

This study focuses on the historical habitat types and drainage patterns of the alluvial plains of the Alameda Creek watershed, southern Alameda County (fig. 1.5). The study centers on Alameda Creek, following the creek from the Alameda-Santa Clara county boundary downstream to the San Francisco Bay near Union City. The northern sub-watershed in Livermore-Amador Valley is also included (arroyos Mocho, del Valle, las Positas, and de la Laguna, as well as a number of smaller tributaries—Alamo, South San Ramon, Tassajara). The study also covers the surrounding lowland alluvial plains. Throughout this report "Alameda Creek" will be used to refer to the historical path of Alameda Creek. The new, modified path through Niles Cone will be referred to as the Flood Control Channel.

We divide the study area into three regions. "Livermore-Amador Valley" includes all of the study area upstream of where Arroyo de la Laguna enters Sunol Valley. In analyses we lump Niles Canyon, Sunol Valley, and the upper Alameda Canyon into "Sunol Valley-Niles Canyon." The "Niles Cone" includes everything downstream of Niles Canyon, extending north and south of the true cone of Alameda Creek (see fig. 1.5b).The study area encompasses 120,885 acres, including the tidal marshlands (but not all of the adjacent tidal flat). In addition, a pilot study (Chapter 7) investigates the application of historical ecology tools to a 30,000 acre portion of upland adjacent to the study area.

This chapter opens the report with an overview and introduction to the historical watershed. Chapters 2 and 3 provide a review of our methods and a brief land and water use history. Chapters 4 through 6 provide detailed descriptions of the historical habitats and drainage networks in each of the three distinct regions of the study area: Livermore-Amador Valley, Sunol Valley-Niles Canyon, and the Niles Cone. The upland pilot portion of this report (Chapter 7) provides details on the historical habitats and modification in the uplands. Chapter 8 presents data on historical native fish assemblages for the entire watershed, including upper stream reaches. Chapter 9 presents a conceptual landscape synthesis, comparison to present-day conditions, and some of the management implications of this study.

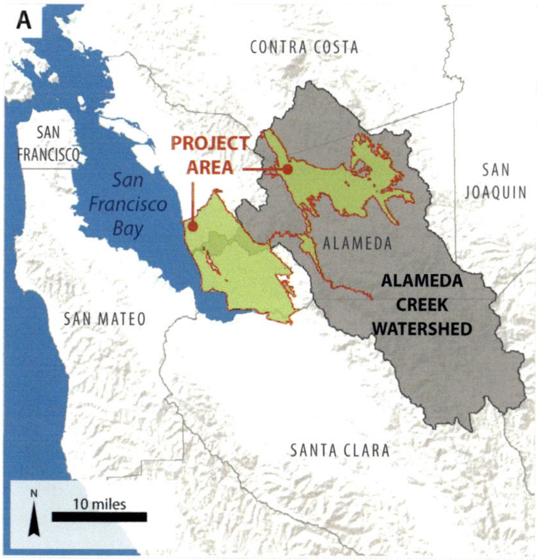

Figure 1.5. Project area locator maps. The project area encompasses the alluvial plains of the Alameda Creek watershed, covering more than 120,000 acres from Livermore Valley through Sunol Valley to the tidal marshes along the San Francisco Bay. Major towns included in the study area are Fremont, Union City, Newark, Livermore, Dublin, southern San Ramon, and Pleasanton. The Alameda Creek watershed extends over 400,000 acres, far beyond the study area, from Contra Costa through Alameda and Santa Clara counties (A; watershed boundary shown in dark grey). For analysis we divided the study area into three regions: Livermore-Amador Valley, Sunol Valley and Niles Canyon, and Niles Cone (B). The upland pilot covers an area in the hills south of Pleasanton and east of Sunol Valley. The contemporary stream network for the project area is displayed in B.

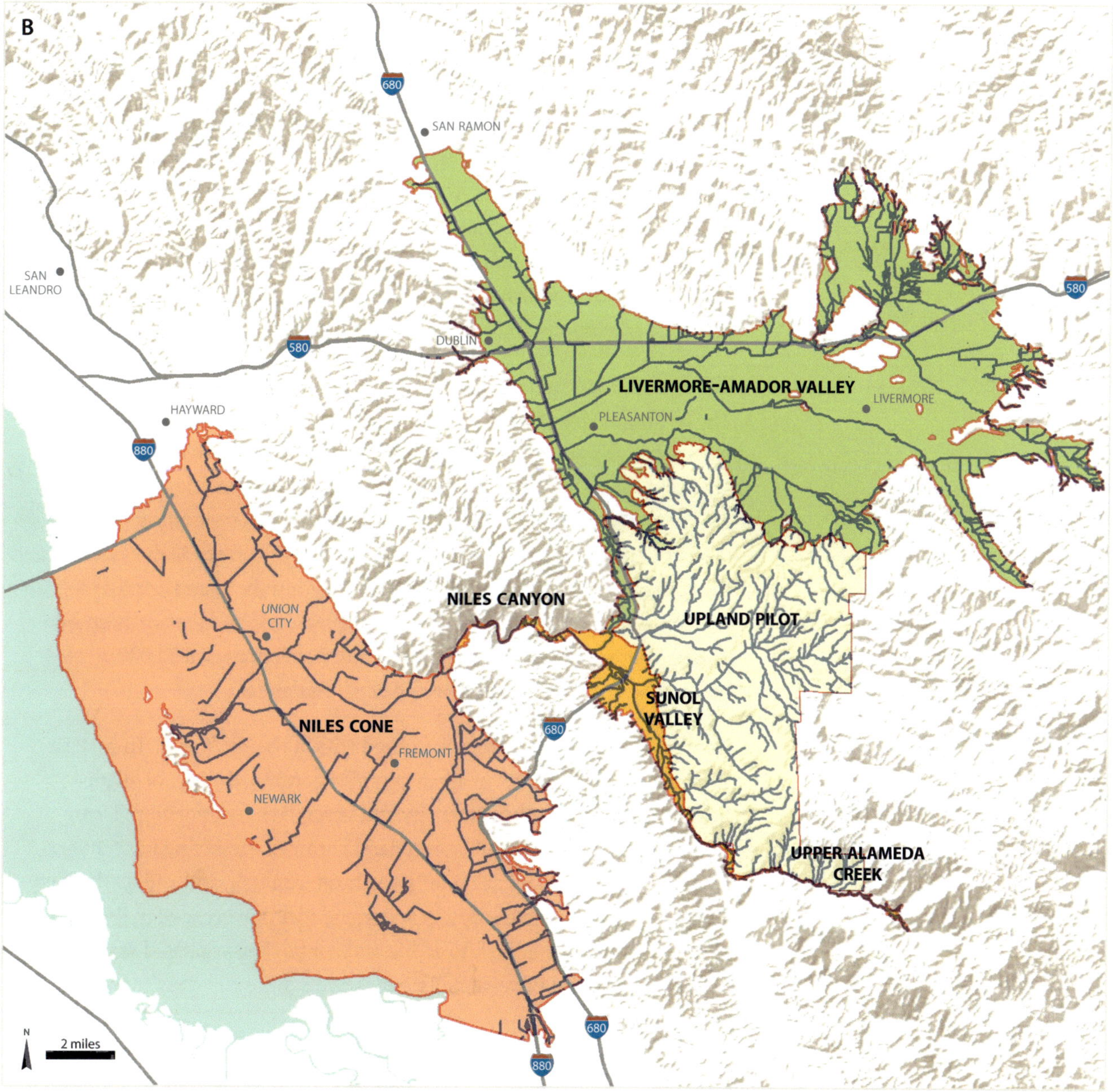

Overview of findings: historical habitats and functions

This section reviews key findings from the historical ecology analysis. Further description and analysis is provided throughout the report. Box 1.1 on the following pages summarizes many of these concepts.

Landscape-scale patterns

The historical landscape provided ecosystem services in distinct patterns related to underlying physical processes. Broad, braided streams provided efficient groundwater recharge through their porous gravel beds, as well as coarse sediment storage. Natural basins and overflow zones forced water to slow and sediment to settle, providing fine-sediment storage and reducing downstream flood peaks. The diversity of habitat types supported a broad array of native species in close proximity, from vernal pool fairy shrimp (*Branchinecta lynchi*) and Livermore tarplant (*Deinandra bacigalupii*) to willow thickets, oaks, and sycamores.

Habitat patterns in the Alameda Creek watershed followed fairly steep and heterogeneous physical gradients, resulting in a diversity of habitat types within a small area. Stream form and riparian (streamside) vegetation were shaped largely by the alluvial fans of the three large, sediment-rich streams—Arroyo del Valle, Arroyo Mocho, and Alameda Creek (see box 1.1). These streams all drained from hills that supply coarse sediment and spread into broad stream beds supporting sycamore alluvial woodland. In contrast, streams characterized by finer materials, such as Arroyo las Positas, had sinuous, narrow channels with sparse tree cover.

The stream patterns in turn controlled the distribution of valley floor habitats. The sediment-rich streams emerged from their canyons to deposit coarse gravels broadly across the valley floor. These areas were well drained and tended to support grasslands. As slope decreased further downstream, the ground surface intercepted groundwater, sometimes supporting wetlands. At the same time, braided, gravelly channels often re-formed in one channel and the stream deposited finer sediments both within the channel and as overflow (see box 1.1). At the western edge of the Niles Cone and the northern side of Livermore-Amador Valley, this fine sediment supported 34,000 acres of seasonal wetlands (wet meadow and alkali wetlands).

Three large wetland complexes formed within the study area. In eastern Livermore, the Springtown alkali complex contained a mix of alkali habitats, varying in alkali intensity and vegetation composition. Near present-day Pleasanton, a large wetland complex maintained water year-round, supporting a mix of ponds, freshwater marsh, and willow thickets. The tidal marsh along the bay shore provided a third type of major wetland system, grading from fresh to saline and containing vegetated marsh plains, tidal channels, and unvegetated pannes and salinas.

The three alluvial plains within the study area had very distinct habitat patterns. (Throughout this report, we use "habitat" to refer to land cover types, rather than only to the environment for a specific species (Lindenmayer et al. 2008).) Livermore-Amador Valley was split almost evenly between wetland and dryland habitats, and 37% of the wetlands (7,500 acres) were alkali wetlands. Sunol Valley was dominated by grassland and oak savanna, with significant cover by in-channel habitats (including sycamore alluvial woodland). The Niles Cone region, by far the largest, had an almost equal area of grassland and tidal marshland (27,000 and 25,800 acres, respectively). The remaining 20% of the cone (15,700 acres) was covered with willow thicket and seasonal wetlands, including vernal pool complex.

Application of historical findings

Understanding the landscape patterns and processes of the recent past can help us identify impacted areas with the potential to contribute to more functional, resilient systems and improve overall ecological health. Although most historical features can never be fully restored, many lost functions could be significantly recovered within the watershed. The recently released East Alameda County Conservation Strategy calls for the restoration of natural patterns of flow, runoff, and overflow where possible, which requires a knowledge of the historical characteristics of the system (ICF International 2010a).

Historical information is not predictive of the future, but rather is a useful tool for developing practical restoration targets and designs. Some physical controls, including land use and climate, can change. Other controls, however, such as topography and geology, may be relatively stable. By showing what types of habitats persisted where, regional historical analysis can help us understand the relative importance of these processes, how they have changed, and how these processes may affect the sustainability or maintenance needs of proposed designs. When integrated with contemporary data and future projections, historical information helps managers identify restoration opportunities and develop realistic strategies.

The following paragraphs outline a few features of the historical landscape that have been lost and discuss ways that the ecological functions they provided could be restored in the watershed. While the rest of the report provides a more comprehensive treatment of the study area, the paragraphs below serve as introductory examples, describing some of the most striking features of the historical landscape and considering the effects of their absence.

PLEASANTON MARSH COMPLEX Beneath present-day Pleasanton are the remains of a 2,600 acre marsh complex, extending as wide as two miles in places (fig. 1.6). Surrounded by seasonal wetlands and supporting a mix of open water ponds, freshwater marsh, and dense willow thickets, the marsh complex would have provided habitat for a wide range of species, including the yellow billed cuckoo (*Coccyzus americanus*), native fishes such as thicktail chub (*Gila crassicauda*), even grizzly bears (*Ursus arctos horribilis*; Treultein and Fages 1972). The streams and groundwater of the Tri-Valley area all drained towards the marsh, which provided an

CANYON REACH Within the bedrock canyon, the stream is confined to a single channel along most of its length and flows across a coarse substrate composed of boulders and cobble. Water is perched within the bedrock canyon and cannot easily sink into the ground. Combined with high groundwater, this typically results in perennial surface flow. Consistently available water supports a riparian forest, with a mix of oaks (*Quercus*), sycamores (*Platanus racemosa*), and more hydrophilic species such as alders (*Alnus*) and willows (*Salix*).

MULTI-THREAD REACH As the high energy stream exits the canyon, it is no longer confined. Due to a combination of the lack of confinement, decreased slope, and a change in bed materials, the stream loses transport capacity and begins to deposit coarse sediment (gravels and cobbles) to build a conical alluvial fan, often forming a multi-thread channel with braiding and distributaries among small islands and bars. At different stages of fan formation, this pattern can vary. Water begins to sink through the gravels to the underlying aquifers, and the stream enters a losing reach, which typically has surface flow for only a portion of the year. The typical riparian vegetation is sycamore alluvial woodland, or a mix of sycamores and oaks, which are able to thrive with intermittent surface flow and accessible groundwater, and help to stabilize the bars. Towards the bottom of this reach, as depth to groundwater increases, riparian trees typically become sparser.

BOX 1.1. CONCEPTUAL STREAM TRANSECT

This drawing illustrates some of the habitat diversity historically found on local streams, and the physical controls that supported that diversity. Four reach types with distinct stream morphology, substrate, groundwater levels, surface flow, and riparian (or streamside) vegetation patterns can be identified. The reaches varied in length depending on slope and depth to groundwater, as well as the presence or absence of other creeks or geographic features. While this diagram does not represent any specific stream in the Alameda Creek watershed, components of these patterns are found here and in other nearby watersheds (e.g., Coyote Creek, upper Pajaro River), and the model can provide a helpful way of conceptualizing the connections between groundwater, substrate, and vegetation (Grossinger et al. 2006, 2008). See the detailed descriptions of Alameda Creek (both in Sunol Valley and on the Niles Cone), Arroyo Mocho, and Arroyo del Valle for more information on variations to this pattern.

One of the major factors shaping channel and riparian form on large streams in the study area is the alluvial fan pattern. Alluvial fans are formed as streams emerge from canyons and spread across an alluvial plain (Blair and McPherson 1994, Knighton 1998). Streams with a large source of sediment deposit a cone of this coarse sediment over time across the surrounding plain, creating a fan shape. Stream flows sink into the porous fan substrate, causing intermittent flow and distinct patterns of groundwater, riparian vegetation, and stream morphology. Although the dynamic processes of fans are not captured in the diagram at left, it shows an example of the reach types that can be created by a fan system. Understanding these patterns is critical to understanding how stream habitats were distributed along Arroyo del Valle, Arroyo Mocho, and Alameda Creek. (illustration by Jennifer Natali)

SINGLE-THREAD REACH Further downstream, the stream reforms into a single stem channel as it loses power and sediment supply and experiences a decrease in slope. The stream begins to flow over finer gravels and sands. The stream bed and the ground surface slope to intersect groundwater levels, and the stream enters a gaining reach, which typically has water year round. Riparian vegetation is often mixed riparian forest.

DELTA DISTRIBUTARY REACH Finally, as the stream continues to decrease in slope and stream power, the substrate shifts to silts and clays. Many of these streams intercept groundwater as they flow through distributary channels into wetlands. Riparian vegetation typically consists of willows, transitioning to marsh species.

important freshwater resource in an otherwise largely summer-dry valley. Organic matter from the marsh was transported downstream, likely supporting fisheries, and steelhead may have fed in the marsh during their outmigration. The marsh stored fine sediment that washed from the hills, storing some of the fine materials that now move downstream to the lower reaches of Alameda Creek and out to San Francisco Bay. It absorbed flood water as well, dampening peak flows, and slowly released water downstream through the summer.

Early settlers began to clear and drain the edges of the marsh in the mid-1800s, but it wasn't until the 1890s and early 1900s that people made a concerted effort to drain the entire marsh (see box 4.5). By 1912, only small remnant patches remained, and much of the former marshland had been converted to agriculture and hop fields. Traces remain—the portion of the wetland complex that sits under the Valley Trails neighborhood in Pleasanton was only recently removed from the FEMA floodplain (Zone 7 Water Agency 2006).

With the loss of the Pleasanton marsh complex, the Alameda Creek watershed lost a host of ecosystem support functions (e.g., organic matter production, sediment storage, denitrification). Some types of functions provided by the marsh can potentially be partially restored. Stream re-design to slow the passage of water and create places to store fine sediment can help relieve erosion and sedimentation problems in Arroyo de la Laguna and further downstream. Riparian re-vegetation along canals can re-establish riparian habitat for birds and other species. Other functions could be sought in other parts of the watershed. For example, organic matter to provide food for steelhead may need to be addressed

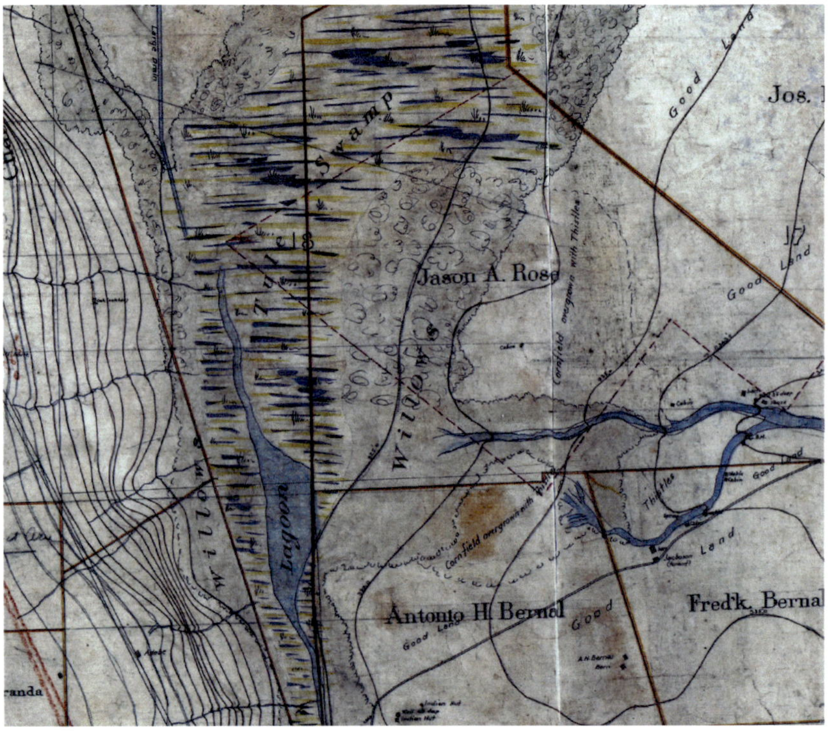

N
1,000 feet

Figure 1.6. Pleasanton marsh complex. This early map shows the variety of habitats present in the Pleasanton marsh complex, including willows, an open water lagoon, and a marsh or "tule swamp." (Allardt [1880]1907, courtesy The Bancroft Library, UC Berkeley)

through wetlands and off-channel habitat downstream in the tidal marshes or other reaches of the creek.

SEASONAL WETLANDS AND DISCONTINUOUS STREAMS A vast amount (34,000 acres) of seasonal wetlands occurred across the study area, compared to fewer than 5,000 acres today, many of which are artificial or degraded (fig. 1.7). The diversity of seasonal wetland types (wet meadow, alkali meadow, valley sink scrub, alkali playa, vernal pool complex) created an unusual number of distinct plant communities, many of which are rare today (Keeler-Wolf et al. 1998, ICF International 2010a). Associated wildlife included California tiger salamander (*Ambystoma californiense*), red-legged frog (*Rana aurora draytonii*), longhorn fairy shrimp (*Branchinecta longiantenna*), and vernal pool fairy shrimp (*Branchinecta lynchi*), all of which are federally listed, as well as a host of additional rare plants, including Livermore tarplant (*Deinandra bacigalupii*) and palmate-bracted bird's beak (*Cordylanthus palmatus*). Rich plant matter produced in seasonal wetlands helped support fish and other aquatic life downstream. Seasonal wetlands stored fine sediment and helped disperse flood waters and dissipate stream energy (Sedell et al. 1990, ICF International 2010a). These seasonal wetlands were typically fed by spreading streams, which supplied fine sediment and water as they spread into the wetlands and underlying groundwater.

Today, impervious surfaces over much of the valley floor speed runoff rather than allowing it to sink into the ground or spread into wetlands (e.g, White and Greer 2006). The present-day stream network is highly connected in comparison to the historical network, moving water efficiently across the land, bypassing wetlands, and creating high flood peaks.

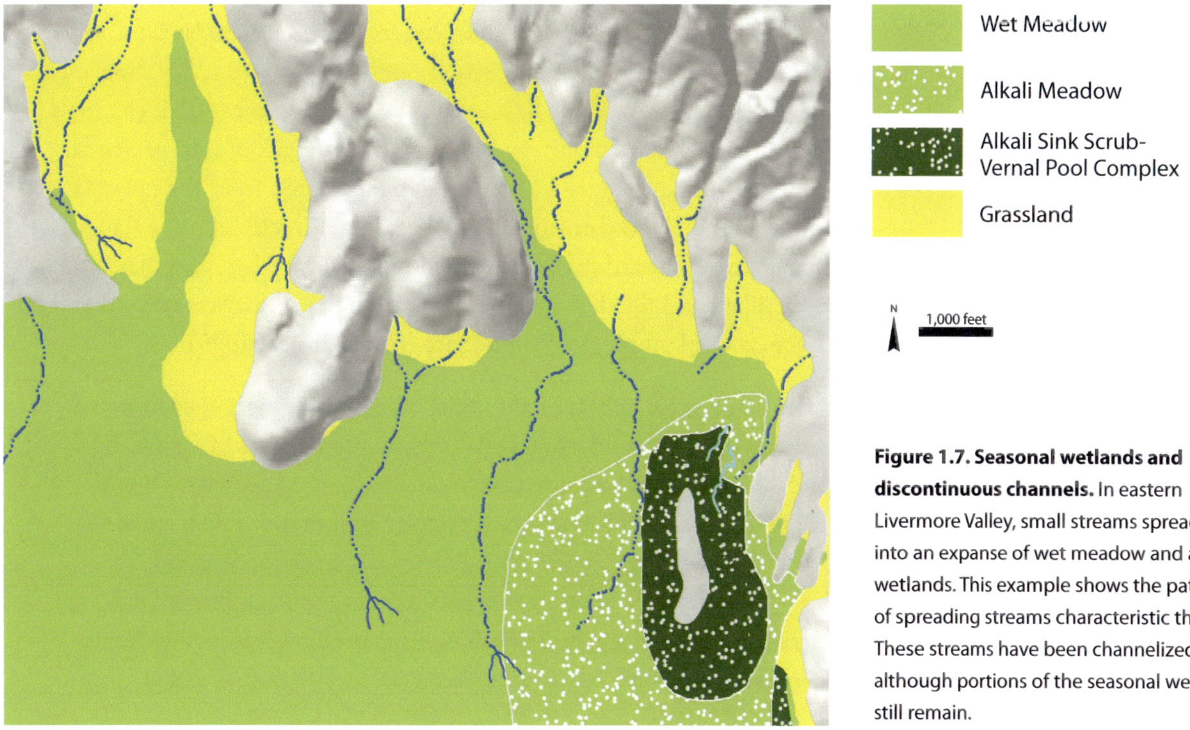

Figure 1.7. Seasonal wetlands and discontinuous channels. In eastern Livermore Valley, small streams spread into an expanse of wet meadow and alkali wetlands. This example shows the pattern of spreading streams characteristic the area. These streams have been channelized today, although portions of the seasonal wetland still remain.

However, some significant remnants of seasonal wetland persist. Large areas of undeveloped rangeland in the eastern portion of the county still support alkali wetlands, and the Don Edwards National Wildlife Refuge in Fremont contains remnant vernal pools. The hydrology in both cases has been altered, but many of the native species persist, particularly in places where grazing controls annual grasses.

Although the flooding associated with discontinuous channels is not desirable in an urbanized landscape like that present today in Livermore-Amador Valley and the Niles Cone, restoration projects and Low Impact Development (LID) strategies can be designed to slow water movement, increase recharge, and support wetlands in contained areas, which may help restore some of the functions provided by streams with formerly discontinuous channels. Protection of the watersheds surrounding remnant seasonal wetlands to ensure continued water inputs to these areas will be essential to help support these wetlands into the future (Coats et al. 1988, Keeler-Wolf et al. 1998, Holland 2009, ICF International 2010a).

TIDAL MARSHLAND Along the edge of the San Francisco Bay were 25,800 acres of tidal marshland, representing a bayside buffer from one to over three miles wide. The marshland contained a dense network of tidal channels, along with salinas and pannes. An artesian zone beneath the landward side of the marsh provided freshwater inputs, and marsh habitats transitioned from saline to brackish to freshwater conditions. The marshlands provided a wide range of ecosystem functions, including sediment storage, flood water storage, water filtration, protection from wave erosion, and a range of habitats along the fresh-brackish transition. The marshes likely supported many fishes, including steelhead, as well as a host of plant and bird species.

Although a few areas of tidal marsh exist today within the study area, most of them have been diked and converted to salt ponds (fig. 1.8). The historical Alameda Creek estuary no longer exists; today the Flood Control Channel receives little tidal influence, and most tidal channels have been obliterated or have silted up through lack of tidal action (Collins and Grossinger 2004). Sediment delivered from the uplands settles in the Flood Control Channel or flows directly into the Bay rather than being delivered more directly to the tidal marshes. Marshes have been diked and filled, greatly reducing both the area of marshland and the diversity of plant species found there (Goals Project 1999).

Attention has recently turned to tidal marsh restoration. Restoration projects currently underway seek to reopen some areas of former tidal marshland within the study area to tidal influence and restore many of the historical tidal channels. Particular attention should be paid to reestablishing a fresh-brackish-saline transition zone to provide the broadest array of habitat types possible, and to protect the wetland-terrestrial ecotone, most of which was occupied by seasonal wetlands (Goals Project 1999, ICF International 2010a, Beller et al. 2013). As sea

A

B

Figure 1.8. Diked former tidal marsh. (A) Former tidal marsh, now diked ca. 1905, looking west through the Coyote Hills. Salt ponds are visible in the distance. (B) This restoration project at Eden Landing seeks to reopen many former salt ponds to tidal action and restore tidal marshlands. (A: courtesy California Historical Society, CHS2012.935; B: by Bronwen Stanford, 2011)

levels rise, tidal marshes will need to maximize sediment supply to help them maintain elevation. Restoration projects could direct sediment from Alameda Creek into the tidal marshes.

CHANNEL COMPLEXITY Historically channels were more varied in form than today, ranging from broad swales to well defined meandering streams and wide braided channels. The channel beds were often complex and varied, with pools, riffles, bars, and side channels. This range of channel types supported a variety of different aquatic species, and channel complexity translated to more in-stream refugia. Complex stream beds also provide distinct types of habitat that may support unique species assemblages and provide critical refuge during dry periods, floods, human modifications, or other disturbances (Sedell et al. 1990).

Stream restoration projects can help introduce some complexity back into the system. Riparian cover contributes both shade and woody debris. Restoration projects can consider historical stream patterns and present-day hydrology to determine the most appropriate channel type to construct. Efforts can also seek to protect those areas that have been less modified and may still provide some refugia and variety of channel form.

VARIATIONS IN FLOW Historically most streams in the lowlands of the Alameda Creek watershed flowed for only a portion of the year. Much of the time the creek beds were dry. Summer-dry streams provide a different set of functions than streams that flow year-round. In many cases, native fish and plant species are more adapted to summer-dry conditions and high levels of environmental stressors than introduced species, and can out-compete them in these harsh conditions (Leidy et al. 2011). Some native plant species, such as the sycamore, rely on intermittent flood flows for germination (Keeler-Wolf et al. 1996).

Flow in several of the large streams in the Alameda Creek watershed is managed today through dam releases or inputs from the Delta or Hetch Hetchy (including Del Valle, Mocho, Alameda Creek, San Antonio, and Las Positas). The East Alameda County Conservation Strategy (EACCS) emphasizes the importance of maintaining natural variation in stream flows wherever possible to maintain the natural sediment transport patterns and support native species that may depend on seasonal flows (ICF International 2010a). To support some of these intermittent stream values, managers can consider mimicking the natural hydrologic regime whenever possible.

RIPARIAN DIVERSITY The varied flow patterns and channel morphology of the historical stream network resulted in an impressive variety of distinct riparian habitat types. Riparian habitats in the Alameda Creek watershed varied in width, species assemblage, and density, from sparse oak and sycamore trees (*Quercus* spp. and *Platanus racemosa*) along meandering, intermittent streams, to dense willow thickets (*Salix* spp.) along Arroyo de la Laguna and the Pleasanton marsh complex. In the eastern alkali wetlands, alkali meadow with iodine bush (*Allenrolfea occidentalis*) was found along the streams, while along braided Arroyo del Valle and Alameda Creek through Sunol Valley sycamore alluvial woodland grew within the broad bed of the creek. Perennial reaches along Alameda Creek supported mixed riparian forest, including alders and cottonwoods. This complexity and diversity is typical of river systems in dry Mediterranean climates, due to the variations in stream power, depth to groundwater, elevation, channel form, and substrate that are characteristic of these systems (Sandercock et al. 2007).

Although the watershed has been modified, many of the natural hydrologic patterns persist. Healthy riparian ecosystems can help reduce flooding and erosion and improve water quality (Tiegs et al. 2005, ICF International 2010a). Riparian vegetation can also provide movement corridors for a host of species and provide nutrients and woody debris for aquatic ecosystems (ICF International 2010a). In the Alameda Creek watershed, riparian restoration would benefit from considering reach-scale targets rather than assuming the same approach will be suited to the entire length of the creek. Because of both natural controls and management effects, adjacent reaches may be best suited for different riparian species; for example, the sycamore alluvial woodland found along upper Arroyo Mocho is sustained by processes no longer present on nearby lower Mocho or Las Positas.

2 • METHODOLOGY

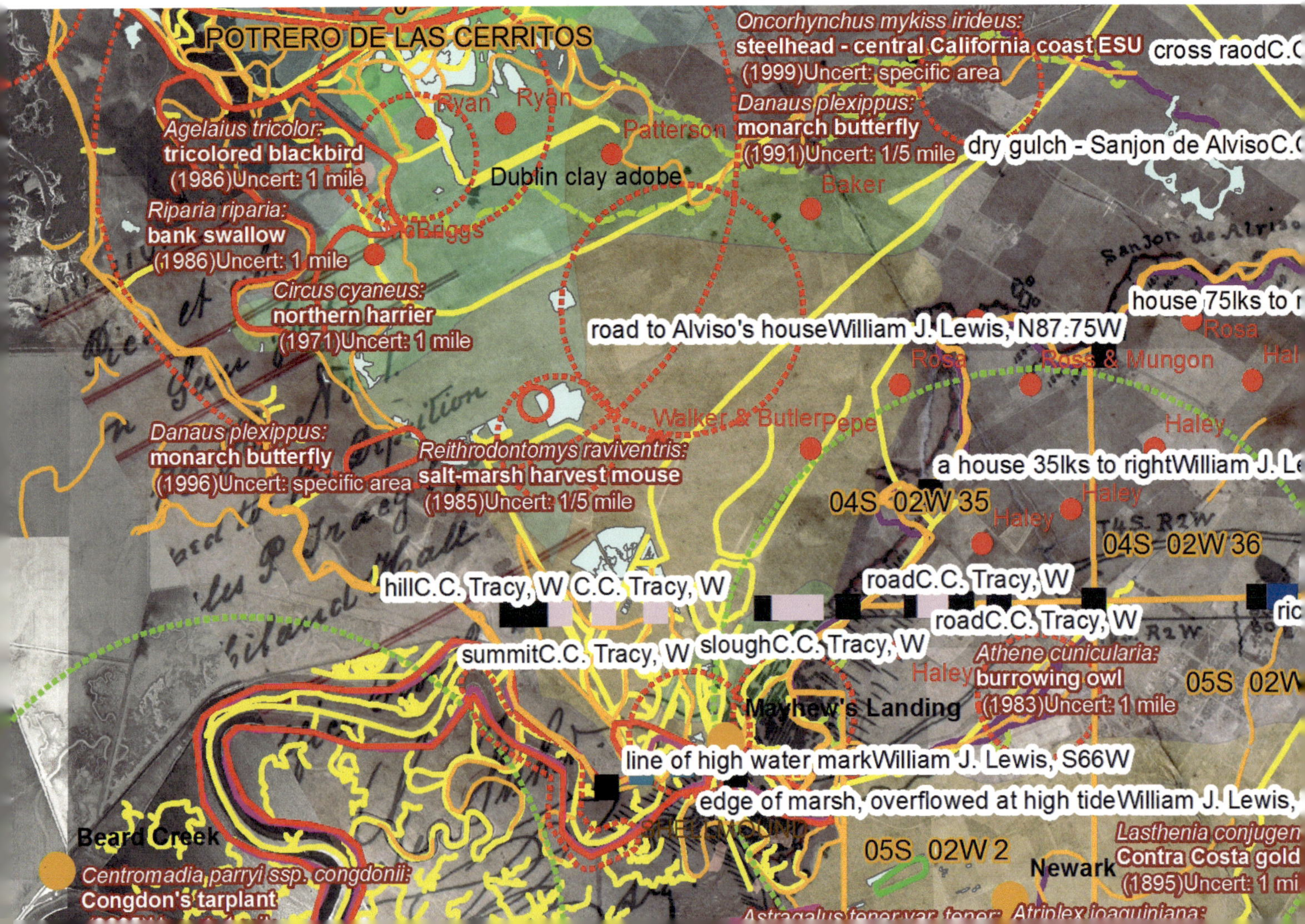

This chapter and the following historical context chapter contain the background information on which the rest of the report builds. The chapter describes the methods we used to map historical habitat types across southern Alameda County. The discovery, organization, and interpretation of historical data form the foundation of this project. Outlined in this section is the intensive process through which data spanning disparate places and eras were synthesized for this study (fig. 2.1).

Briefly, we divide a historical ecology research project into five distinct steps. The first two steps are data collection and data compilation—the process of building and organizing a large historical dataset. The third and fourth steps are data synthesis and analysis, in which we map historical habitats in a geographic information system (GIS), using our compiled datasets to compare this habitat map against present-day land cover patterns. The fifth step is reporting and technical review.

The following paragraphs provide detail on data collection, data compilation, and mapping methodology. This section also includes an introduction to historical sources and definition of habitat type classifications.

Data collection

A substantial variety and quantity of historical data are needed for accurate assessment of the historical landscape (Grossinger 2005). With this in mind, we assembled a diverse range of historical records spanning two centuries and compiled these data into a map of historical landscape patterns prior to substantial Euro-American modifications.

Assembled materials include: (1) textual data (e.g., Spanish explorers' accounts, Mexican land grant case court testimonies, General Land Office records, early travelogues, court cases, herbarium records, and county histories and reports); (2) maps (e.g., Mexican land grant maps, early city and county maps and surveys, USDA soil surveys, and US Geological Survey maps); and (3) paintings and photography (both ground-based and aerial).

To acquire these sources, we visited local historical archives, public libraries, county offices, and regional archives. In total, we visited 22 source institutions to collect data (table 2.1). We also reviewed material available online and conducted searches of over 30 electronic sites and databases.

We acquired full or partial copies of approximately (roughly estimated) 900 maps, 700 documents, and 1,200 photographs (box 2.1). These represent a small fraction of the documents reviewed at the archives themselves. While we reviewed thousands of documents for this study, historical research is never completely exhaustive, and the local historical record is extensive. Additional sources will undoubtedly surface

Figure 2.1. Historical data. (top) A surveyor gathering data on a stream cross section along Arroyo de la Laguna. (bottom) This selection of the data compiled in GIS emphasizes the complexity and variety of the dataset. These data sources were synthesized to create the habitat map. (Top: AD-862, June 1916, © San Francisco Public Utilities Commission) containing ecological information that will enrich the descriptions and information incorporated in this report.

Data compilation

Data compilation is the process of organizing the large volume of heterogeneous data used in this study into more accessible formats for interpretation, comparison, and integration at the local and landscape scale. As part of this process, we georeferenced, or spatially placed in the GIS, both maps and spatially locatable quotes; we read textual sources and transcribed relevant quotes into one comprehensive document; and we created large-scale maps, or "base maps," displaying compiled data onto which we transferred non-georeferenced data. High-priority maps for georeferencing were chosen based on the quality and quantity of features shown. They were georeferenced to contemporary orthorectified aerial imagery (USDA 2009), using ESRI's ArcGIS 9.3.1 and 10 software. This allowed us to compare historical data to each other and to contemporary aerial photography and maps. Approximately 120 maps were georeferenced.

Relevant quotes were extracted from textual material and transcribed into a text document. Quotes were organized by broad geographic area (Livermore-Amador Valley, Sunol Valley, Niles Cone) and by subject (e.g., riparian vegetation, wetlands, channel geometry). In addition, quotes pertaining to land use history, irrigation history, and climate were

Table 2.1. Data collection agencies, libraries, and archives visited. This list represents the physical data repositories we visited over the course of this project.

Institution Visited	Location
Alameda County Library	Fremont
Alameda County Recorder	Oakland
Alameda County Surveyor	Hayward
Alameda County Water District	Fremont
The Bancroft Library, UC Berkeley	Berkeley
California Academy of Sciences, Ichthyology Department	San Francisco
California Historical Society	San Francisco
California Society of Pioneers	San Francisco
California State Lands Commission	Sacramento
California State Library	Sacramento
Earth Sciences and Map Library, UC Berkeley	Berkeley
Livermore Heritage Guild	Livermore
Museum of Local History	Fremont
Museum of the San Ramon Valley	Danville
Museum on Main	Pleasanton
Niles Silent Film Museum	Fremont
Oakland History Room, Oakland Main Library	Oakland
San Francisco Public Library	San Francisco
San Francisco Public Utilities Commission Archives	San Francisco
Union City Historical Museum	Union City
University and Jepson Herbaria Archives	Berkeley
Water Resources Center Archives	Berkeley (now Riverside)

transcribed. Over 130 pages of quotes were transcribed, and were used to assist interpretation and reporting. In addition, about 100 of these quotes were spatially specific enough to be locatable on our base maps. These were mapped and included in our GIS as an independent data layer. We also relied heavily on two specialized sources, described below.

Aerial photos

One heavily used source was historical aerial photography, which required the orthorectification and mosaicking of 211 aerial photographs into a comprehensive, continuous coverage of the entire study area. All aerials covering the western half of the study area were from 1939, and were acquired from the Earth Sciences and Map Library at UC Berkeley. Aerials covering the eastern half of the project area were taken in 1940 and were mostly provided (and orthorectified) by the RCD/NRCS, although a few of these also came from the Earth Sciences and Map Library. The photomosaic was particularly useful for identifying wetlands, upland habitats, and former creek alignments within the pre-urban, agricultural setting. Aerial photos were interpreted based on intercalibration between visible signatures and descriptions in contemporaneous sources. Although taken after many significant landscape changes had occurred, these photographs show strong traces of earlier landscape features that have since been lost. The spatial consistency, accuracy, and high level of detail made them an invaluable source for the project.

General Land Office survey data

Initiated by the U.S. Continental Congress's Land Ordinance of 1785, the Public Land Survey (PLS) field notes of the General Land Office (GLO) provide some of the most detailed descriptions of landscape and vegetation prior to the extensive environmental changes that followed European contact (Buordo 1956). The GLO surveys have allowed researchers to place these ecological data with a level of accuracy, consistency, and spatial extent rarely available from other sources of this era. Progressing from Ohio to the West Coast, the GLO survey reached Alameda County in 1851 and continued in the area until 1879. The survey established townships of 36 square miles divided into square-mile sections. The section and township corners ideally form a square grid across the landscape at a resolution as fine as the quarter-section. However, many areas in California, including parts of the study area, lack a complete network of inner township section lines as a result of private Mexican land grant holdings (White 1991, Grossinger et al. 2007).

We adapted methods developed by the Forest Landscape Ecology Lab at the University of Wisconsin-Madison to store, display, and analyze the GLO data within a GIS environment (Manies 1997, Radeloff et al. 1998, Sickley et al. 2000). One of the primary benefits of the ArcMap (ESRI) form developed by the Wisconsin group is the ability to place the survey points efficiently and accurately within a contemporary spatial coordinate system. In addition,

BOX 2.1. HISTORICAL DATA FOR ALAMEDA CREEK WATERSHED

This study involved the collection and compilation of a wide array of historical sources, spanning multiple centuries, languages, and formats. Historical documents form the backbone of our historical mapping and analysis, from Spanish-language explorers' journals and 1850s land grant cases to soils mapping and aerial photography of the mid-20th century.

Since each source was produced by individuals in different social contexts and with variable goals, understanding the provenance of the sources we draw on is a fundamental starting point for understanding our findings. Shown below are examples and brief descriptions of some of the primary sources used in this study.

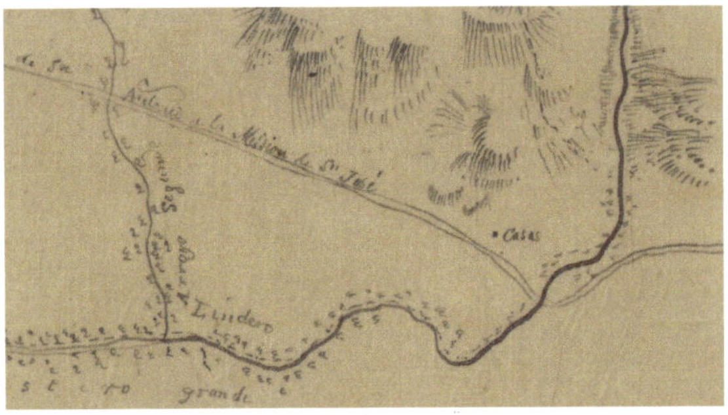

Mexican land grant sketches (1840s-1860s). As the Mission system disintegrated, influential Mexican citizens submitted claims to the government for land grants. A *diseño*, or rough sketch of the solicited property, was included with each claim. *Diseños* often show notable physical landmarks which would have served as boundaries or natural resources, such as creeks, wetlands, springs, and forests. While *diseños* are not as spatially accurate as subsequent surveys, they provide extremely early glimpses of former landscape features and patterns. (USDC 1840c, courtesy The Bancroft Library, UC Berkeley)

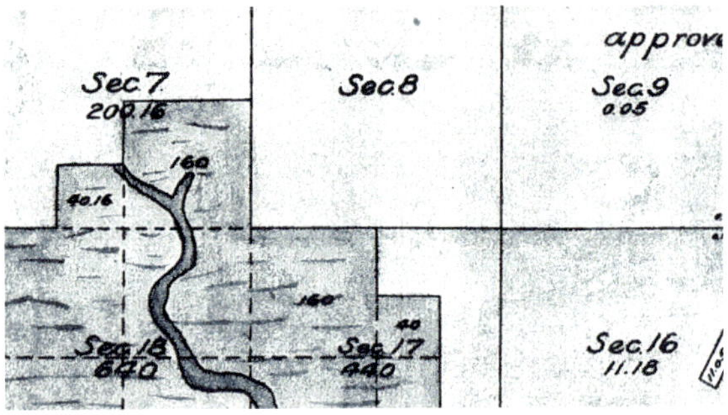

General Land Office public land surveys (1851-1879). In areas not claimed through the land grant system, the U.S. Public Land Survey imposed a grid of straight lines on the landscape, dividing property into six-mile square townships. Each township was further subdivided into 36 one-mile sections, each section containing 640 acres. Surveyors methodically surveyed section lines along these transects, noting cultural and natural features they encountered along the way. Survey notes and plat maps from these surveys are useful for their ecological information. (courtesy Bureau of Land Management)

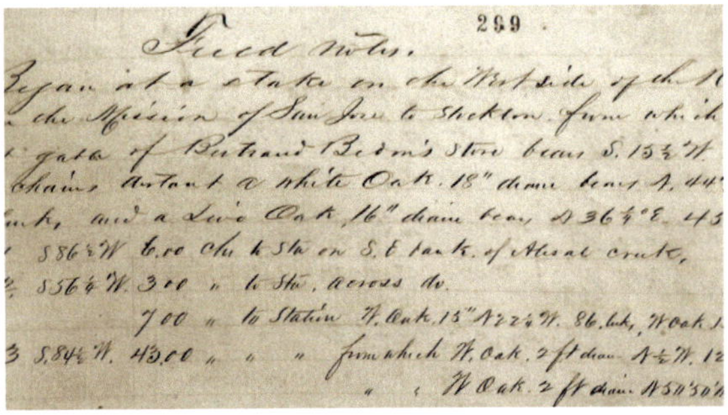

Textual accounts (1769-2012). Written accounts can provide a wealth of detailed information with nuance about landscape dynamics not available on maps. Spanish expeditions provide the earliest accounts; later sources such as land grant case testimonies, newspaper articles, ornithological records, county histories, and travelogues give rich perspectives from early visitors and residents. (text courtesy Alameda Department of Public Works)

U.S. Coast and Geodetic Survey maps or T-sheets (1857-1896). The U.S. Coast and Geodetic Survey was established in 1807 by Thomas Jefferson to create navigation maps. Though the maps only cover the coastline and immediately adjacent areas, they are a highly valuable source because of their impressive detail and accuracy, scientific rigor, and relatively early survey dates. The maps covering the landward portion of the coastline are known as "topographic sheets" or "T-sheets." We refer frequently to these "T-sheets" throughout the report. (Kerr 1857, courtesy NOAA)

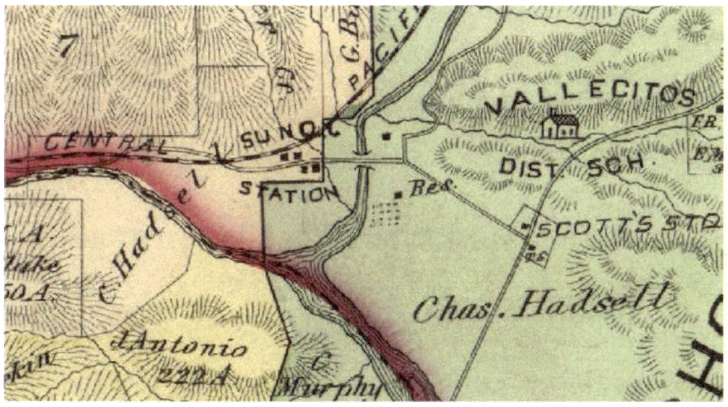

City and county surveys (1850-1940s). Local surveyors produced numerous maps, including many surveys of individual parcels. These maps, often surveyed at a large scale, contain details not included in other regional mapping efforts such as sloughs and side channels, smaller ponds and wetlands, or clusters of trees. Though coverage is inconsistent, these maps are invaluable in constructing an understanding of local ecosystem dynamics. (Thompson and West 1878, courtesy David Rumsey Map Collection)

U.S. Geological Survey topographic maps (1899-1973). Around 1900, the USGS (established in 1879) began producing topographic quadrangles at 1:62,500 for Alameda County. Though the maps are relatively coarse, they provide some of the earliest consistent, comprehensive coverage for the entire region. (USGS 1906)

Historical aerial photography (1939-40). A Depression-era program to ensure crop stability and soil conservation practices resulted in extensive aerial photographic coverage for much of the county. The historical aerial imagery used in this study is from 1939 and 1940, and represents the earliest complete coverage. While the photographs were taken after substantial modification, the photos nevertheless reveal relict ecological features, traces of which are often still present in the landscape. (USDA 1939-40, courtesy Earth Sciences and Map Library, UC Berkeley and ACRCD/NRCS)

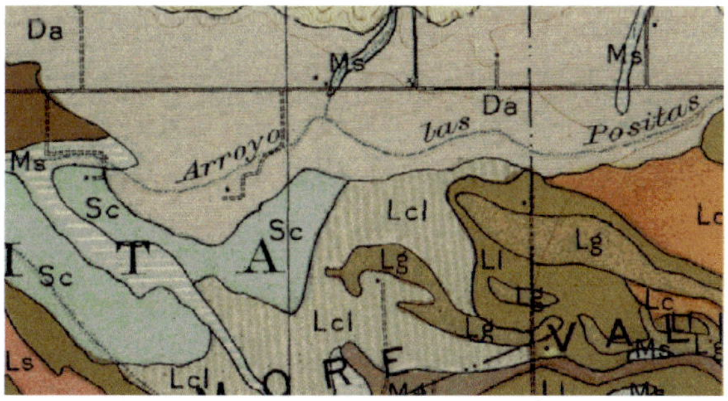

U.S. Department of Agriculture soil surveys (1910-1914). Early soil surveys were developed to describe variability in the agricultural viability of regional soils. These maps, and their accompanying reports, are a key source in the inference of historical habitat extent and location. Descriptions of soil properties and agricultural use can provide insight into former habitats, in particular providing spatially accurate detail on the extent of wet meadows and alkaline habitats. (Westover and Van Duyne 1910)

Landscape photography (1860s-1950s). Historical photographs represent a category of diverse historical data that can provide extremely localized, accurate information. Photographs can capture the conditions of a given place and time in a manner that provides substantial detail about specific species presence and landscape structure. (AD-863, 1916, © San Francisco Public Utilities Commission)

the resulting database can be easily manipulated for subsequent analyses. These data have been used most often in reconstructions of historical forest structure and composition in the Midwest and Northwest and only rarely in California systems (Radeloff et al. 1999, Collins and Montgomery 2001, Bloom and Bahre 2005, Brown 2005, Whipple et al. 2011).

The GLO survey contains rich information about local ecology and hydrology, including descriptions of vegetation composition, creek seasonality and channel geometry, and ponds and marshes. Where surveyors noted entering timber or crossing a stream, we used this information to delineate habitat boundaries. The descriptive field notes also provided more general narrative accounts of the landscape, including details on wetlands, oak groves, and open plains.

Mapping methodology and analysis

We used a GIS to interpret and synthesize our information into digitized data layers representing the historical landscape characteristics of the Alameda Creek watershed. The GIS was used to collect, catalog, compile, digitize, analyze, and display our sources. By spatially relating sources from many time periods, we were able to examine habitats through time (fig. 2.2). The relational database component of GIS allows for storage of many attributes about a single feature, which we used to integrate our disparate sources and document the provenance of our interpretation of the historical landscape. Using GIS, we were able to synthesize complex arrays of sources by assembling maps and narrative information from different periods, allowing us to assess each data source, more accurately map each feature, and better understand change over time. We used ESRI's ArcGIS 9.3.1 and 10 software.

Accurate interpretation of documents produced during different eras, using different methods or techniques, for differing purposes, and with different authors, surveyors, or artists can be challenging (Harley 1989, Grossinger and Askevold 2005). Our dataset of disparate sources (each representing a different scale, time period, and level of accuracy) prevents mapping each area at the same level of detail. Undoubtedly many features were undocumented in the historical record and thus are not depicted on our map.

To address these issues, we interpreted our data through an iterative process of source inter-calibration using GIS and other techniques. Our dataset of overlapping sources allowed us to compare an array of complementary documents. This approach provided independent verification of the accuracy of original documents and our interpretation of them (Grossinger 2005, Grossinger and Askevold 2005). In some cases, a high density of sources documenting a particular feature allowed for high mapping confidence of both presence and extent. However, many features are documented by only one source or simply may not have any specific early source that describes the habitat. In these cases extrapolation based on soil types, topography, hydrology, and general descriptions was necessary. These varying suites of sources are recorded on a feature-by-feature basis in the GIS.

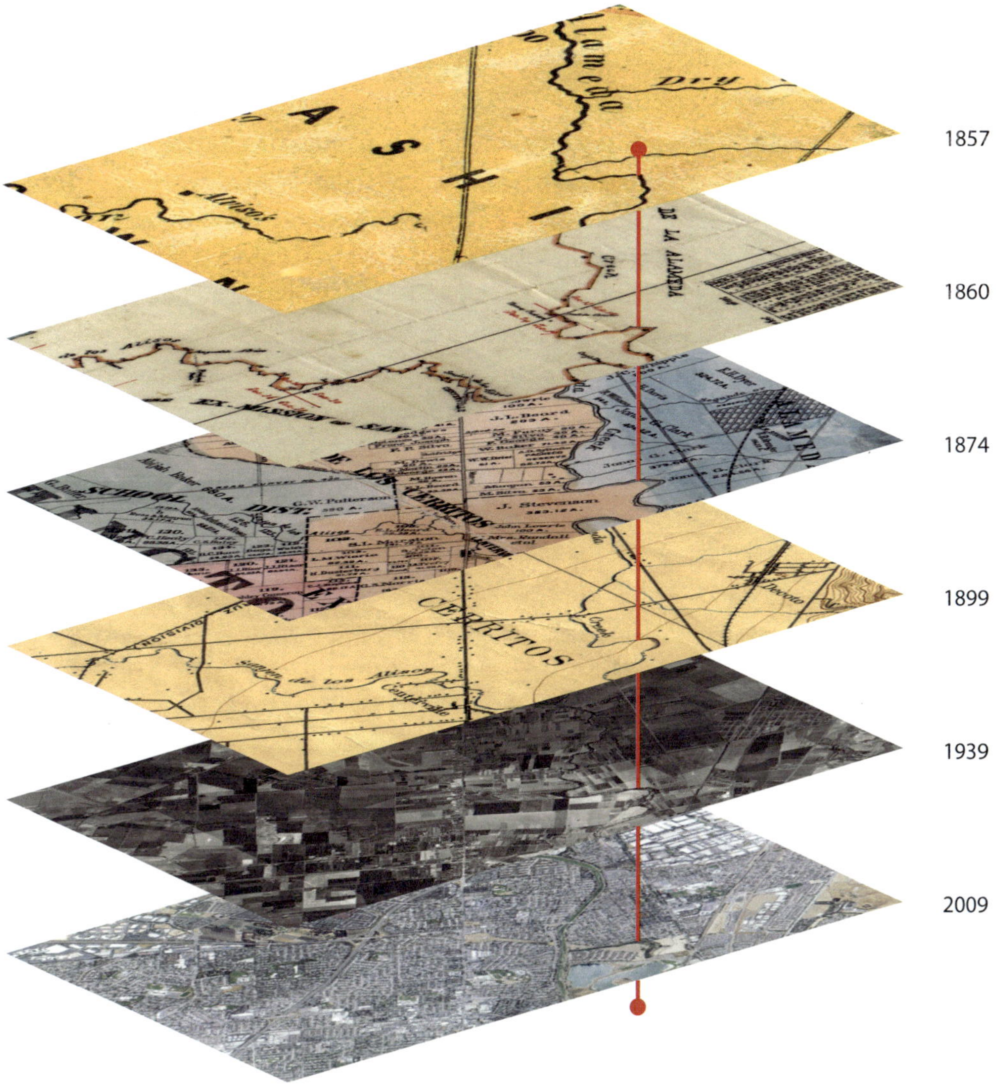

We examined historical data for evidence of landscape characteristics prior to significant Euro-American modification (referred to as early 1800s for simplicity). Our goal was to map landscape features at the watershed scale as they existed, on average, prior to and during the early decades of Euro-American settlement (1770s-1850s). Despite inter-annual and decadal variability, mean climatic characteristics during the period for which historical data were obtained were relatively stable (Dettinger et al. 1998; see climate history p. 62). We incorporated many later sources (i.e., outside of the target time period) that were found to record features that clearly corresponded to features documented by earlier sources, and thus provided more accurate mapping of these features. For example, a feature shown on an early source (e.g., a *diseño)* that confirmed the general presence of the feature but not its location, could be mapped from a later source that showed only remnants but with greater spatial precision (e.g., a historical aerial photo).

Figure 2.2. Stacked historical data. This graphic shows how different data sources can be layered to show change over time. By comparing across sources in this way, we can cross-check information to build a more robust depiction of historical conditions. These images reveal the consistent placement of a now-lost bend of Alameda Creek in Fremont. (top to bottom: Higley 1857, courtesy Bureau of Land Management; Lewis 1860, courtesy The Bancroft Library, UC Berkeley; Allardt 1874, courtesy The Bancroft Library, UC Berkeley; USGS 1899a; USDA 1939; USDA 2009)

To document the mapping sources used and the classification and mapping accuracy certainty associated with individual features, we assigned each feature a set of attributes including a primary "digitizing source" and supporting sources ("interpretation sources"), as well as estimated certainty levels (table 2.2). In the example above, the *diseño* would become the interpretation source (confirming the presence of the feature), and the later source would be the digitizing source (showing the exact outline of the feature). Our confidence in a feature's interpretation (classification), size, and location was assigned as a set of three certainty levels based upon the number and quality of sources and our experience with the particular aspects of each data source (following standards discussed in Grossinger et al. 2007). The application of attributes on a feature-by-feature basis allows users to assess the accuracy of different map elements and identify the original data, serving as a catalog of information sources (Grossinger 2005).

A significant part of the analysis process involved comparing historical conditions to present-day conditions where analogous contemporary data were readily available. We used the land cover mapping recently completed by ICF International (2010a) where available (in the Livermore-Amador Valley and Sunol Valley), and CalVeg for the Niles Cone. For stream networks and tidal marshlands we used the recently completed Bay Area Aquatic Resource Inventory (BAARI) mapping (SFEI 2011). We developed a crosswalk to modern habitat classes (see table 2.3) and we assessed change over time through comparisons such as the difference in overall length and connectivity of streams, size and position of wetlands, and extent of sycamore and oak woodland.

The following sections describe our historical habitat classes and then outline the methods used to integrate and synthesize data in the GIS to depict broad classes of habitats on the map, both for the purpose of visual representation of historical habitats and channels and for analysis of the historical landscape. For more information on the accuracy of a particular habitat polygon, please refer to the GIS metadata.

Table 2.2. Certainty levels. Each mapped feature was assigned a certainty level of high, medium, or low for each of three characteristics. Interpretation describes our certainty that the habitat type assigned to the feature is accurate and that the feature existed historically. Size describes our certainty that the feature's spatial extent is accurately depicted. Location is our certainty that it existed in exactly that spot. Together these certainty levels help us record the uncertainties inherent in the mapping process.

Certainty Level	Interpretation	Size	Location
High/ "Definite"	Feature definitely present before Euro-American modification	Mapped feature expected to be 90%-110% of actual feature size	Expected maximum horizontal displacement less than 50 meters (150 ft)
Medium/ "Probable"	Feature probably present before Euro-American modification	Mapped feature expected to be 50%-200% of actual feature size	Expected maximum horizontal displacement less than 150 meters (500 ft)
Low/ "Possible"	Feature possibly present before Euro-American modification	Mapped feature expected to be 25%-400% of actual feature size	Expected maximum horizontal displacement less than 500 meters (1600 ft)

Habitat classification

We developed 13 non-tidal habitat types based on historical evidence and modern classification systems (table 2.3). These classes balance a desire to preserve the detail often available in the historical record while creating meaningful classes that are comparable to contemporary classification systems and applicable across the entire study area. They are an attempt to capture broad-scale patterns. Within each of these classes there would have been complex fine-scale patterns and considerable variation in species assemblages. In some cases, the character of historical data makes direct translation to a single contemporary vegetation class impossible. Riparian habitats were not mapped, with three exceptions: sycamore alluvial woodland, sparsely vegetated braided channel, and confined riparian woodland/savanna (see riparian methods p. 33 for details). Six riparian classes were conceptually mapped for major streams in the study area in figure 9.21.

The following definitions provide brief explanation of the habitat types outlined in table 2.3. They are in large part derived from contemporary descriptions and classification systems outlined elsewhere (e.g., Holland 1986, Goals Project 1999, Holstein 2000, Collins and Grossinger 2004, Barbour et al. 2007, Grossinger et al. 2007, Sawyer et al. 2009, ICF International 2010a). For more detailed descriptions of each type, please refer to these documents. Where possible, photos from the study of each habitat are included to illustrate the habitat type.

San Antonio Creek

Springtown, Livermore

HABITAT TYPE DEFINITIONS

Creek. Creeks were mapped as a line feature that was shown as dashed for reaches that were typically summer-dry, and as a solid line for reaches that maintained surface flow through the dry season. Creeks that spread into multiple distributary channels are shown with a forked crow's foot symbol. Creeks were surrounded by riparian vegetation, not shown on the habitat map.

Slough. Drainages that were not directly connected to an upland watershed (i.e., were emergent on a valley bottom) were classified as sloughs. Sloughs are broad, shallow, and less well-defined than creeks—they tend to carry water only during high flows and are not typically mapped by historical sources. Many of them may be remnant former creek channels or intermittently connected distributary channels.

San Antonio Creek

Alameda Creek, Sunol Valley

Upper Alameda Creek

Tule Pond, Fremont

Sycamore Alluvial Woodland. Sycamore alluvial woodland is a sycamore-dominated (more than 50% of relative cover in tree layer of *Platanus racemosa*) riparian woodland type that grows along the alluvial benches of braided streams (Keeler-Wolf et al. 1996). Understory includes California buckeye, blue elderberry, and mule fat, but the habitat type may also include areas of unvegetated channel (Holland 1986). Our mapping of sycamore alluvial woodland may include some areas that would not meet the classification guidelines applied to present-day mapping as we were limited by the detail available in historical sources.

Sparsely Vegetated Braided Channel. Braided stream beds with low density tree cover were mapped as gravel bed. These areas would have been covered with largely unvegetated gravel and riverwash and occasional sycamores (*Platanus racemosa*), willows (*Salix* spp.), and shrubs (i.e., mule fat (*Baccharis salicifolia*)).

Confined Riparian Woodland/Savanna. Streams flowing through confined bedrock canyons (Niles Canyon, upper Alameda Creek) were mapped with confined riparian woodland/savanna. In some places vegetation was mixed riparian forest, but in others the canyon broadened slightly and vegetation became more sparse and savanna-like. This confined riparian forest type contained a mix of species, including willows (*Salix* spp.), alders (*Alnus* spp.), sycamores (*Platanus racemosa*), and oaks (*Quercus* spp.).

Perennial Freshwater Pond. Perennial ponds are non-vegetated depressional areas containing standing water throughout the year. They often occurred within larger complexes of marshland, willows, and seasonal wetlands and likely contained pondweeds (*Potamogeton* spp.). Springs and high groundwater often helped sustain these features through the dry season.

Valley Freshwater Marsh. Valley freshwater marshes are persistent emergent freshwater wetlands typically dominated by bulrushes (*Bolboschoenus* and *Schoenoplectus* spp.), cattails (*Typhus* spp.), sedges (*Carex* spp.), and rushes (*Juncus* spp.). These wetlands are seasonally flooded (Cowardin et al. 1979); soils generally have a high organic content and are usually saturated (Holland 1986).

Willow Thicket. Willow thickets (or willow swamps) are palustrine forested wetlands that occur in large stands, rather than as riparian vegetation along a creek, and are associated with areas of emergent groundwater (Cowardin et al. 1979, Goals Project 1999, Collins and Grossinger 2004, Beller et al. 2011). They were often referred to as *sausals* in early Spanish documents and are largely absent from the landscape today (Collins and Grossinger 2004). They contained a mix of willow species *(Salix)*, and included dense thickets dominated by shrub-sized willows with occasional larger trees, in addition to willow "groves" which tended to include more established trees. Willow thickets in this project area typically bordered perennial wetland types.

Don Edwards Refuge, Fremont

Wet Meadow. Wet meadows are temporarily or seasonally flooded herbaceous communities characterized by poorly drained, clay-rich soils. They can be flooded for days or weeks depending on precipitation and topography, and stay moist longer than adjacent, better-drained areas. Vegetation included grasses and a significant component of obligate and facultative wetland species such as wire rush *(Juncus balticus)*, irisleaf rush *(Juncus xiphiodes)*, buttercup *(Ranunculus californicus)*, and blue eyed grass *(Sisyrinchium bellum)* (Holstein 2001).

near Frick Lake, Livermore

Alkali Meadow. Alkali meadows are temporarily or seasonally flooded herbaceous communities characterized by poorly drained, clay-rich soils that have a high residual salt content (over 0.2% in the first six feet; Westover and Van Duyne 1911). They support a distinctive, salt-tolerant plant community, including some species characteristic of salt marshes and/or vernal pools (Baye et al. 2000, Holstein 2000). These habitats are typically subject to seasonal soil saturation and temporary to seasonal flooding, with subsequent drying through the summer (Holland 1986, Elmore et al. 2006). Dominant plant species can include salt grass (*Distichlis spicata*), wild barley (*Hordeum* spp.), orachs (*Atriplex* spp.), tidy tips (*Layia* spp.), goldfields (*Lasthenia* spp.), alkali heath (*Frankenia salina*), tarplants (*Centromadia* spp.), and alkali weed (*Cressa truxillensis*).

Springtown, Livermore

Alkali Sink Scrub. Alkali sink scrub (also known as valley sink scrub) is similar to alkali meadow, but is dominated by a scrub cover typically including iodine bush (*Allenrolfea occidentalis*), seep weed (*Suaeda* spp.), and other *Chenopodiaceae* species (Holland 1986). Alkali sink scrub was restricted to eastern Livermore Valley in this study area, and is mapped as part of a complex with vernal pool complex and alkali playas.

Don Edwards Refuge, Fremont

Alkali Vernal Pool Complex. Vernal pools are seasonally flooded depressional wetlands underlain by a hardpan or claypan with hummocky microtopography (Keeler-Wolf et al. 1998, Holland 2009, SFEI 2011). These wetlands are associated with a specialized set of plants, including *Navarretia* spp., popcornflower (*Plagiobothrys* spp.), downingia (*Downingia* spp.), and goldfields (*Lasthenia* spp.) (Goals Project 1999). Vernal pool complex includes the pools themselves and the surrounding matrix of grassland or seasonal wetland.

Alkali Sink Scrub-Vernal Pool Complex. A habitat type developed for our habitat map containing both sink scrub and vernal pools, along with patches of alkali meadow. These habitat types were combined because they often occurred as a complex and could not be distinguished.

Springtown, Livermore

Alkali Playa Complex. Alkali playas are alkali wetlands with particularly high levels of soil alkalinity (over 1% in the first six feet; Westover and Van Duyne 1911). These areas are composed of a mosaic of alkali meadows and more sparsely vegetated alkali playas or flats. Salt grass (*Distichlis spicata*) and iodine bush (*Allenrolfea occidentalis*) are still significant components, but this habitat class also includes large expanses of non-vegetated, seasonally flooded areas (<10% plant cover) with local alkaline concentrations too high to support substantial vegetation (Holland 1986).

Oak Savanna. Oak savannas contain a low density of oaks spread over open, low herbaceous cover. The oaks are irregularly spaced, forming occasional clusters and treeless areas. Oak savanna tree density is variously defined in the literature, ranging from less than 10% to as much as 25% tree cover (see Sawyer and Keeler-Wolf 1995, FGDC 1997, Allen-Diaz et al. 1999, Davis et al. 2000, ICF International 2010a). Oak savanna in this study area occurred at very low density, with valley oaks (*Quercus lobata*) as the dominant tree.

Grassland. Grassland is a mix of low herbaceous cover, including grasses and forbs, with occasional oaks (Holstein 2000, Minnich 2008). Oak savanna and grassland are distinguished by oak density: the oaks in grassland occur at densities lower than in savanna. Grassland was the default habitat type for this study, which may have resulted in overmapping of grassland area. (Note: The grassland pictured includes a mix of species different from those present historically.)

Tidal Marsh. Tidal marshes are defined as intertidal areas that support at least 10% cover of vascular vegetation adapted to intertidal conditions (Goals Project 1999). Plant species distribution within the tidal marsh is determined largely by elevation and salinity (and thus inundation frequency); dominant plant species include Pacific cordgrass (*Spartina foliosa*), pickleweed (*Sarcocornia pacifica*), alkali heath (*Frankenia salina*), and high marsh species such as salt grass (*Distichlis spicata*) (Goals Project 1999, Baye et al. 2000, Collins and Grossinger 2004). Areas of freshwater influence support different species assemblages, including *Typha* and *Schoenoplectus* spp. (Baye et al. 2000).

Salina/ Marsh Panne. Salinas and marsh pannes are largely unvegetated, open-water depressions within tidal marshes. Salinas occur as large bodies at the backshore edge of the tidal marsh, often producing natural seasonal salt deposits, while marsh pannes occur as smaller bodies scattered through the marsh (Goals Project 1999, Collins and Grossinger 2004). They are irregularly or seasonally flooded. Their hydrology (alternately flooded and dry) and high salinity prevents most plants from establishing, but they may support submerged vegetation such as widgeongrass (*Ruppia maritima)* and algae (Baye et al. 2000).

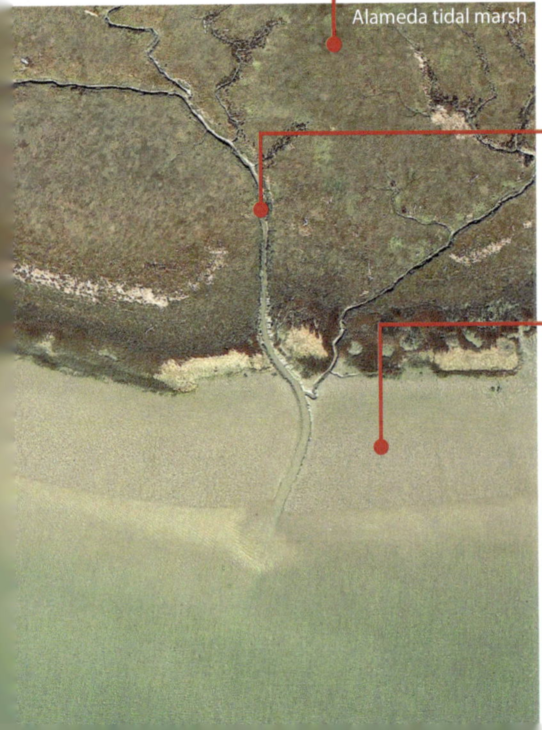

Tidal Channel. We classified sloughs within the tidal marsh as tidal channels, including both deep, subtidal and shallower, intertidally exposed channels (Goals Project 1999, Collins and Grossinger 2004). Some tidal channels had fluvial connections and provided sediment storage and fish habitat.

Tidal Flat. Tidal flats are areas with less than 10% vegetative cover, and are typically covered with clay, silt, sand, or shell. These areas are inundated frequently enough to prevent most plant growth and occur directly bayward of vegetated tidal marshland (Collins and Grossinger 2004).

Table 2.3. Crosswalk between historical ecology and contemporary habitat classification systems. The CNDDB types listed represent a sampling of the contemporary habitat types that could crosswalk to the historical types—the list is not exhaustive. We include the more general Holland (1986) communities as well as CNDDB alliances, and Cowardin (1979) classification where we were unable to link the historical land cover to a specific alliance.

	Habitat Type	California Terrestrial Natural Communities (CNDDB 2010), NDDB (Holland 1986)	Cowardin et al. (1979)
Riparian cover	Sycamore Alluvial Woodland	California sycamore woodlands (61.310.00)	
	Sparsely Vegetated Braided Channel	Mulefat thickets (63.510.00), California sycamore woodlands (61.310.00), Sandbar willow thickets (61.209.00)	
	Confined Riparian Savanna-Woodland	California sycamore woodlands (61.310.00), Fremont cottonwood forest (61.130.00), Sandbar willow thickets (61.209.00), Arroyo willow thickets (61.201.00), Valley oak woodland (71.040.00)	
Wetland	Perennial Pond	Pondweed mats (52.107.00), Cattail marshes (51.050.00)	Lacustrine / palustrine freshwater wetland. Permanently flooded.
	Valley Freshwater Marsh	Coastal and valley freshwater marsh (CTT52410CA), California bulrush marsh (52.114.00), Baltic and Mexican rush marshes (45.562.00), Soft rush marshes (45.561.00), Cattail marshes (51.050.00), Pale spike rush marshes (43.230.00)	Palustrine persistent emergent freshwater wetland. Temporarily to permanently flooded, permanently saturated.
	Willow Thicket	Arroyo willow thickets (61.201.00), Red willow thicket (61.205.00)	Palustrine forested (scrub/shrub) wetland. Temporarily flooded, temporarily to permanently saturated.
	Wet Meadow	California goldfields-dwarf plantain-six-weeks fescue flower fields (44.108.00), Soft rush marshes (45.561.00), Baltic and Mexican rush marshes (45.562.00), Creeping rye grass turfs (41.080.00), Pale spike rush marshes (43.230.00), Western ragweed meadows (33.065.00)	Palustrine emergent wetland. Temporarily flooded, seasonally saturated.
	Alkali Meadow	Wildflower field (CTT42300CA), Alkali meadow (CTT45310CA), Alkali seep (CTT45320CA), Salt grass flats (41.200.00), Creeping rye grass turfs (41.080.00), Alkali heath marsh (52.500.00), Tar plant fields (44.160.00),	Palustrine emergent saline wetland. Temporarily flooded, seasonally saturated.
	Alkali Sink Scrub	Valley sink scrub (CTT36210CA), Iodine bush scrub (36.120.00), Alkali weed - Salt grass playas and sinks (46.100.00), Salt grass flats (41.200.00), Creeping rye grass turfs (41.080.00) Alkali heath marsh (52.500.00), Tar plant fields (44.160.00), Alkali sacaton grassland (41.010.00)	Palustrine emergent saline wetland. Temporarily flooded, seasonally to permanently saturated.
	Alkali Vernal Pool Complex	Northern hardpan vernal pool (CTT44110CA), Northern claypan vernal pool (CTT44120CA), Fremont's goldfields - Downingia vernal pools (42.007.00), Fremont's goldfields - saltgrass alkaline vernal pools (44.119.00), Alkali weed- salt grass playas and sinks (46.100.00), Alkali sacaton grassland (41.010.00), Alkali heath marsh (52.500.00)	Palustrine emergent wetland. Temporarily flooded, seasonally saturated.
	Alkali Playa Complex	Alkali playa (CTT460000CA), Alkali weed- salt grass playas and sinks (46.100.00), Alkali sacaton grassland (41.010.00), Salt grass flats (41.200.00), Alkali heath marsh (52.500.00)	Palustrine saline wetland. Temporarily to seasonally flooded, seasonally to permanently saturated.
Dryland	Oak Savanna	Valley oak woodland (71.040.00), Popcorn flower fields (43.300.00), Creeping rye grass turfs (41.080.00), Purple needle grass grassland (41.150.00)	
	Grassland	Native grassland (CTT42100CA), Wildflower field (CTT42399CA), California poppy fields (43.200.00), Popcorn flower fields (43.300.00), Creeping rye grass turfs (41.080.00), Purple needle grass grassland (41.150.00), Fiddleneck fields (42.110.00)	
Tidal Marsh	Tidal Marsh	Pickleweed mats (52.215.12), California cordgrass marsh (52.020.00), Salt grass flats (41.200.00), Alkali heath marsh (52.500.00), American/ California/Hardstem bulrush marshes	Estuarine haline intertidal persistent emergent wetland. Irregularly flooded, permanently saturated.
	Salina/ Marsh Panne	n/a	Estuarine haline intertidal wetland. Irregularly flooded, unconsolidated bottom.
	Tidal Channel	n/a	Subtidal.

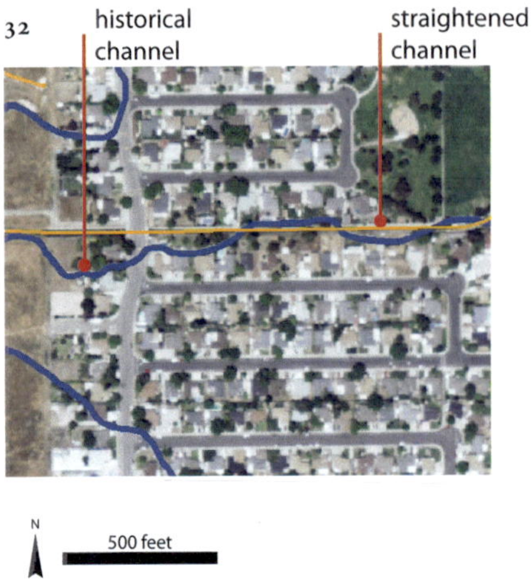

Figure 2.3. Change in sinuosity. The historical course of Arroyo las Positas, shown in blue, was much more sinuous than the present-day course, shown in orange. However, as the two lines fall within 50 feet over much of their length, these segments would be recorded as no change in our analysis of positional change, although we did capture the historical course in our historical stream network. (USDA 2009)

Stream and riparian habitats

CHANNEL NETWORK All watercourses were mapped as line features in ArcMap. The historical stream network is based on the BAARI mapping (SFEI 2011), modified where the historical stream course clearly differed from the contemporary alignment. We were also able to compare our mapping to other recent mapping efforts, including Sowers (1999) and the National Hydrography Dataset (NHD; USGS). Where the historical position was within 50 feet (15 m) of the mapped contemporary position and the shape of the creek did not appear to have been modified, we maintained the contemporary line feature. This accounts for slight mapping errors due to differences in mapping scale or georeferencing errors. However, if the contemporary and historical channel shapes appeared different in the aerial photography (e.g., more sinuous historical line), we modified the channel network to match the historical shape, even if the historical channel position did not deviate by more than 50 feet from the contemporary line (fig. 2.3). This allowed us to accurately reflect changes in shape and sinuosity.

To map the historical drainage network, we first compared early aerial imagery (USDA 1939 and 1940) to contemporary imagery (USDA 2005 and 2009) and contemporary channel mapping to identify post-World War II modifications. To evaluate earlier change we compared the 1939-40 network with earlier maps depicting channel plan form and information from GLO surveys. Soils, topography, and landscape features contributed to our understanding and depiction of creeks. The final historical creek layer represents our best understanding of hydrography prior to significant Euro-American modification. All contributing sources and associated certainty levels for each creek reach are recorded in the GIS attributes.

To maintain a consistent depiction of channel density over time, we attempted to match the level of detail shown in the contemporary layers. The extremely detailed BAARI mapping captured many small and ephemeral drainages, particularly in the uplands, and in some areas we were not able to match the level of detail shown by BAARI, which may result in a slight undermapping of historical stream courses. However, most channels shown across the valley floor are large enough that they were captured at similar levels of detail by BAARI and our historical mapping, so this was not a significant problem within the study area.

We were not able to map channel courses to equal detail throughout the project area. Most historical maps showed the general path of a channel rather than capturing all of the historical bends and turns. As a result, streams that were still relatively intact by the time of the 1939-40 aerial photography tend to be mapped with a higher level of detail than streams that were mapped from map sources or had been channelized by the time of aerial photography. In particular, very small drainages may be over- or under-mapped in some areas due to differing levels of detail and the challenge of distinguishing pre-European ephemeral stream courses from the gullies resulting from early grazing.

Historical channels were also coded with intermittent or perennial flow. To assess flow patterns, we began with early USGS flow designations, and modified this picture with additional information from GLO surveys, maps, and textual accounts. On lower Alameda Creek, in particular, a large body of textual evidence helped establish changes in flow patterns. In general, in the absence of information we assumed a creek to be intermittent, as this was the more common pattern, and so the historical record would be more likely to note the presence of a creek with perennial flow.

Many channels historically lost definition on the alluvial plain rather than directly connecting to another channel. We represented the terminus, or distributary, of defined channels with a forked symbol on the habitat map. If we had information about the historical location of distributary channels, we mapped those rather than using the symbol. This symbol was not applied to sloughs or small disconnected drainages, many of which functioned themselves as distributaries. We also did not apply the distributary symbol to very small first-order drainages.

RIPARIAN HABITATS In general we did not map riparian (streamside) habitats. Riparian habitats (particularly riparian habitat widths) are inconsistently documented in the historical record and have been dramatically altered today, making it impossible to meaningfully and consistently map historical riparian habitats across the study area. However, we were able to use those data that were available to develop a conceptual graphic showing patterns of riparian vegetation (see Chapter 9). We created tables to summarize riparian characteristics across the watershed, using broad vegetation and width classes to reflect the lack of precision in our measurements.

Width classes were based on prior work defining riparian corridors at SFEI (Collins et al. 2006, Williams et al. 2012). Reaches were assigned to a width class using a combination of textual accounts, maps, and remnants visible in the historical aerials. In many places riparian vegetation had been removed by the time of the historical aerials. In these cases we estimated width based on those remnants that persisted and extrapolated from this for the rest of the reach.

We did map three types of well-documented riparian habitats in the GIS. Along braided reaches we mapped a sparsely vegetated braided channel type to describe a broad zone of creek influence and to show a markedly different type of creek process. Within some of these braided reaches we also mapped sycamore alluvial woodland, a class of riparian vegetation. In canyons, we mapped the narrow canyon floor as confined riparian woodland/savanna to capture the mixture of sparsely wooded areas and mixed riparian forest which bounded the stream.

Mapping outer banks of braided channels. Braided stream reaches are supported by factors including abundant bed load, erodible banks, high variability in discharge, and steep slope. This pattern often occurs downstream of a large slope break, change in channel confinement, or across an alluvial fan (Knighton 1998; see box 1.1). Individual channels are separated by bars and islands and are often unstable, although they may also persist for decades or centuries (Knighton 1998). To identify braided reaches, we looked for streams that were depicted in historical maps with a broad channel and distinct islands and bars (e.g., Thompson and West 1878, Allardt 1874, USGS mapping ca. 1900), and a scoured multi-thread pattern on the historical aerials (USDA 1939-40). We also looked for creeks with a bed that reached at least 200 feet wide. This allowed us to focus on those broad systems that were distinctly different from nearby single-thread channels. In the Alameda Creek watershed, these criteria for a broad, braided system were met along major reaches of three creeks: Alameda Creek through Sunol Valley, and Arroyos Mocho and del Valle through Livermore-Amador Valley.

To map outer banks along these three reaches we relied largely on historical aerial photography (USDA 1939-40), with calibration from earlier historical sources (e.g., large-scale county maps, GLO survey notes) and from the 2007 county LiDAR survey (fig. 2.4). This corridor spans the area between the tops of the outer banks and represents the zone that was under direct influence of the creek such that it supported distinct channel scour patterns and often distinct riparian vegetation. Significant modifications had already occurred by 1939, when this imagery was captured, so in many places the banks we mapped were substantially narrower than the likely historical extent. Our mapped riparian corridor provides a conservative estimate of the extent of active creek influence.

Sycamore alluvial woodland. We mapped sycamore alluvial woodland within some of the braided channels. This class captures swaths of sycamore alluvial woodland rather than distinguishing individual clusters or trees and is not as detailed as the contemporary mapping due to the limited nature of the historical data. We mapped sycamore alluvial woodland primarily from the historical aerials (USDA 1939-40) with some calibration from contemporary sources (USDA 2009, CNDDB 2010) and historical textual accounts.

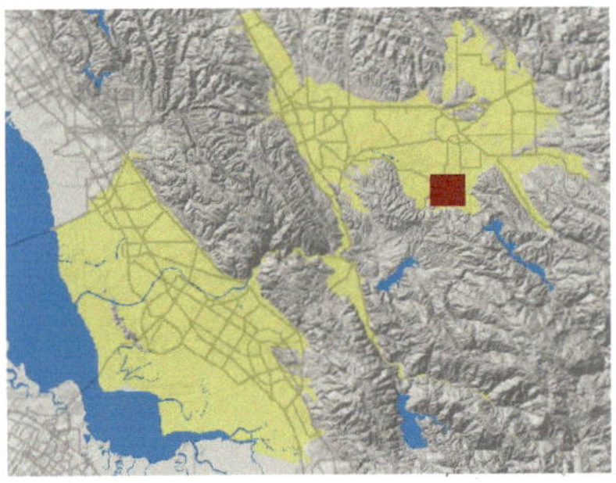

Figure 2.4. Terraces in the historical aerials. The same stretch of Arroyo del Valle is shown in 2007 LiDAR and the 1940 aerials. In both images the abandoned terraces of the historical floodplain are visible as lines to either side of the creek. See also figure 4.39. (USDA 1940)

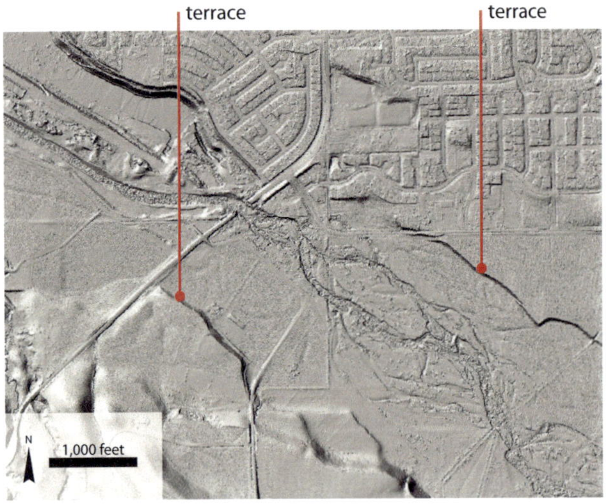

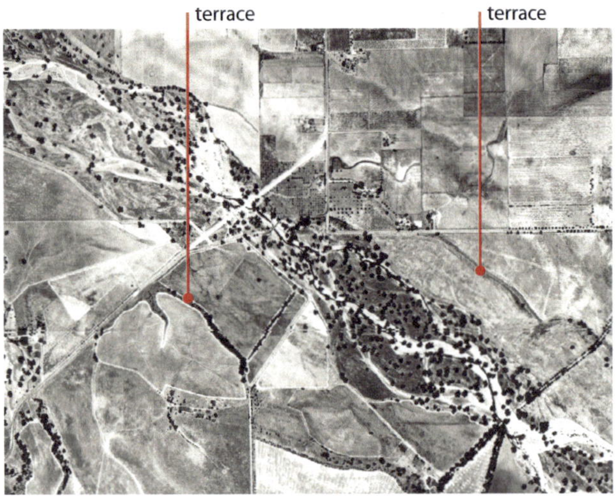

Sycamore alluvial woodland is a riparian habitat type of special concern. Of the 17 significant stands of Central California sycamore alluvial woodland today, three occur within the study area, on streams with managed flows (Keeler-Wolf et al. 1996). Sycamore alluvial woodland can help stabilize braided channels. For regeneration, sycamores prefer intermittent flooding with a high stable groundwater table (Keeler-Wolf et al. 1996). Throughout California sycamore regeneration is low, likely due to grazing of young shoots and a lack of flood flows due to reservoirs.

Confined riparian woodland/savanna. The upland vegetation class surrounding confined canyons such as Niles Canyon and the canyon surrounding upper Alameda Creek was often historically characterized by woodland, a habitat type that we did not map within the study area. The narrow canyon bottoms themselves were occupied by a mix of riparian forest and woodland/savanna where the canyon bottoms widened (fig. 2.5). In most cases, riparian habitats were folded into the surrounding habitat class. To capture this upland surrounding habitat, we mapped the canyon floors as confined riparian woodland/savanna. This class represents the surrounding hillside vegetation extending into the canyons.

Within the canyons, we mapped the entire canyon floor width as this habitat class. At the mouths of the canyons, we mapped to the edge of the remnant secondary terrace visible in the LiDAR as a proxy for the zone of higher stream influence.

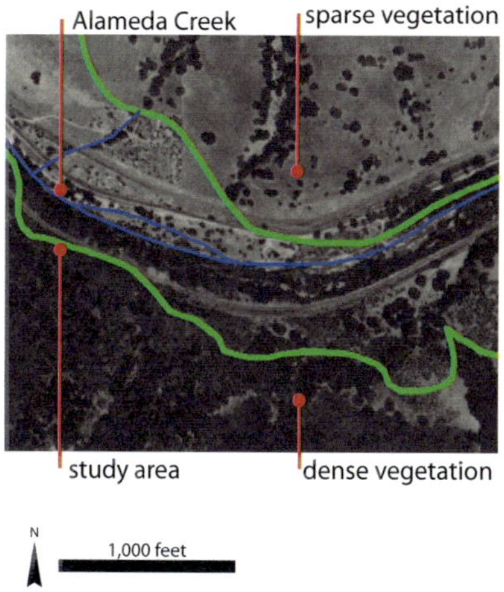

Figure 2.5. Confined riparian woodland/savanna. Note the open savanna along the northern side of Alameda Creek (shown in blue), which contrasts with the more densely vegetated southern slope. The green lines represent the boundaries of the study area. The hillside vegetation north of the channel would be classified as savanna, while that to the south is scrub or woodland. (USDA 1939)

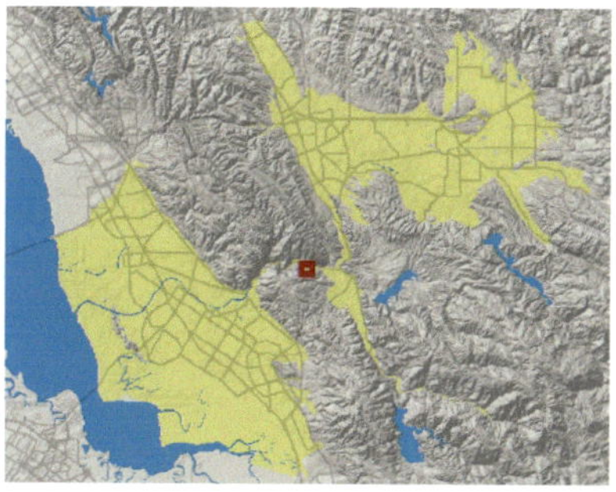

Wetland habitats

The wetland types we mapped included seasonal wetlands (wet meadow, alkali meadow, alkali sink scrub-vernal pool complex, alkali playa complex, alkali vernal pool complex), tidal wetlands (tidal marsh, salina, marsh panne, tidal channel), and perennial freshwater wetlands (valley freshwater marsh, pond, willow thicket). Methods for each class differed and are described below.

WET MEADOWS To map wet meadows we used a combination of the historical and modern soil survey data, along with the quaternary geology mapping (Knudsen et al. 2000, Witter et al. 2006). These sources were calibrated with images, maps, and GLO survey data where available.

Soil surveys were our most detailed source of information for wet meadows. Two historical soil surveys and two contemporary soil surveys together cover the study area (Westover and Van Duyne 1910, Holmes and Nelson 1914, Welch et al. 1966, Welch 1981). These surveys carefully assess the grain size and drainage characteristics of each soil type to inform agriculture, essentially mapping wetland areas. By evaluating descriptions in these surveys, along with the appearance of soils on the historical aerial photographs and information about location and topography, we were able to identify seasonally wet areas (Grossinger et al. 2006, 2007). In particular, we looked for descriptions of clay soils including phrases such as "sticky," "water stands for days," "poor drainage," and "tendency to puddle." Wet meadows were mapped from a combination of historical and contemporary sources. Small patches (less than 5 acres in area) within wet meadows were merged with the surrounding habitat.

ALKALI HABITATS We mapped alkali habitats (including vernal pool complex) using a three-step process. First, we used the historical soil survey to map the extent of alkali-influenced soils (Homles and Nelson 1914; Westover and Van Duyne 1911). Then we used these maps and additional data to delineate the extent of vernal pool complexes. Once this preliminary map was completed, we used additional information to develop and map alkali-influenced habitat classes, modifying the original map of alkali extent as appropriate.

Mapping alkali extent. To map the extent of alkali-affected habitats, we relied primarily on the historical soil surveys. These surveys included descriptions of the quality of the land for agriculture with an assessment of the concentration of alkali salts in the soil, providing the earliest systematic analysis of soil quality and alkali concentrations.

On the Niles Cone, the historical soil survey identified two soils with high alkali content (Yolo clays and Laguna loam; Holmes and Nelson 1914). Although the soil survey covers the entire Bay Area, these soils were reported to have strong alkali influence specifically in the Niles Cone: Yolo clays contain "much alkali" near Newark, and Laguna loam "carries alkali" near Alvarado (Holmes and Nelson 1914:91, 97).

In the Livermore-Amador Valley, the report accompanying the 1910 soil survey included a map documenting alkaline soils across the valley with salt concentrations in the first six feet ranging from below 0.2% to over 1% (Westover and Van Duyne 1911). Using these data, we mapped a zone of alkali-influenced soils in all areas where alkali concentrations were above 0.2%, adjusting for topography in places where error was likely introduced due to the large scale of the map.

Mapping vernal pool complex extent. To map the extent of vernal pool complexes we relied on five source types: historical soil surveys, contemporary soil surveys, historical aerial imagery, contemporary aerial imagery, and contemporary vernal pool mapping (Holland and Placer Land Trust 2009, ICF International 2010a).

The soil surveys identified three historical soils and two modern soils as having "hog wallow" (mounded) topography over much of their surface, a common descriptor for vernal pools (Holland 1978; table 2.4). We restricted our draft map of vernal pool extent to the alkali boundary: vernal pools and alkali soils both develop in regions underlain by a claypan or a hardpan, with poor drainage and seasonal inundation (Holland 1986, Keeler-Wolf et al. 1998). The draft vernal pool extent map was the intersection of the vernal pool soils with the alkali boundary.

In eastern Livermore Valley we were able to use aerial photography and contemporary mapping to refine this initial map of vernal pool complexes. In some places, particularly north of Springtown, we extended our mapping to include areas where the contemporary and historical aerials showed vernal pool complexes (fig. 2.6). We used Holland's 2005 map of vernal pool extent to add some areas that had fallen outside of our alkali boundary but were still on vernal pool soils. In the Niles Cone region, vernal pools have largely been covered with development, although there are some remnant areas and historical species records support the vernal pool presence indicated by the soil survey.

Our mapping of vernal pool complexes is fairly conservative. Additional areas in eastern Livermore Valley likely had vernal pool complexes, but we were limited to mapping areas with clearly visible vernal pool patterns in 1940 aerial imagery, or areas with other evidence supporting the presence of vernal pools.

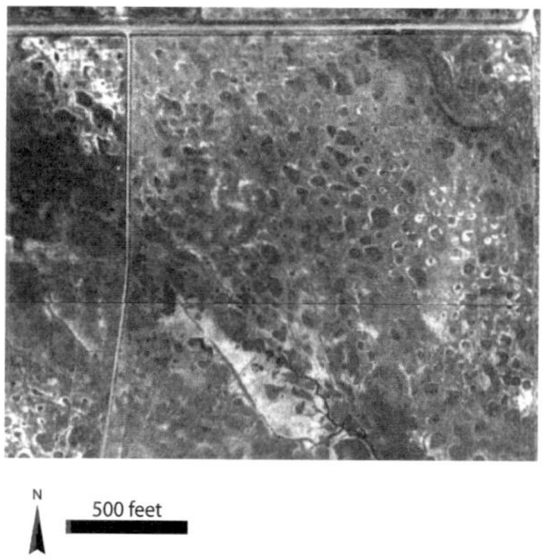

Figure 2.6. Vernal pool complex in historical aerial photography. This imagery, from an area near Springtown just south of Hartford Avenue, clearly shows the circular pools and hummocky pattern characteristic of vernal pool complexes. (USDA 1940)

Mapping alkali-influenced habitat classes. Distinct alkali land cover types in the project area included alkali meadow, alkali sink scrub, alkali playa, and vernal pools, all of which are still present today in some limited extent (Coats et al. 1988, Holland 2009, ICF International 2010a). These alkali habitats were interspersed. The distribution of each habitat type was controlled by local microtopography and small differences in hydrology, which were difficult to capture or distinguish at the landscape scale. Since our data did not support a parcel-scale mapping of individual pockets of each habitat type, we combined habitat types to create broad alkali habitat complexes. For example, the alkali vernal pool complex class is a complex containing both vernal pools and alkali meadow. The alkali sink scrub-vernal pool complex class contains interspersed alkali sink scrub, vernal pool complex, and alkali meadow. These habitat complexes more accurately represent the interspersed patches of different land cover types.

Alkali meadow was mapped on wet meadow soils with alkali presence between 0.2% and 1%, as mapped by Westover and Van Duyne (1911). Alkali meadow was present in each of the four alkali habitat classes, but was most dominant in the alkali meadow type. Soil surveys report that observable crop effects begin at concentrations above 0.2% (Westover

Table 2.4. Soil type descriptions indicating vernal pools. The soil types listed in the table were used to map vernal pool complexes. These were assembled from three distinct soil surveys, two historical and one more recent. The excerpts from the soil descriptions indicate the presence of hummocky mounds, or "hog wallows."

Soil Type	Excerpt	Source
Ulmar loam	"some areas show a tendency to the hog wallow topography"	Westover and Van Duyne 1911
Ulmar fine sandy loam	"the greater part of the type is more or less marked by hog wallows"	Westover and Van Duyne 1911
Yolo clay	"a great part of these areas has a hog-wallow surface, carries much alkali, and is of little agricultural value"	Holmes and Nelson 1914
Pescadero clay	"most of the soil is in large, nearly level areas that have hog-wallowed, or hummocky, microrelief"	Welch et al. 1966
Solano fine sandy loam	"most of this soil is in small, nearly level areas that have hog-wallowed microrelief"	Welch et al. 1966

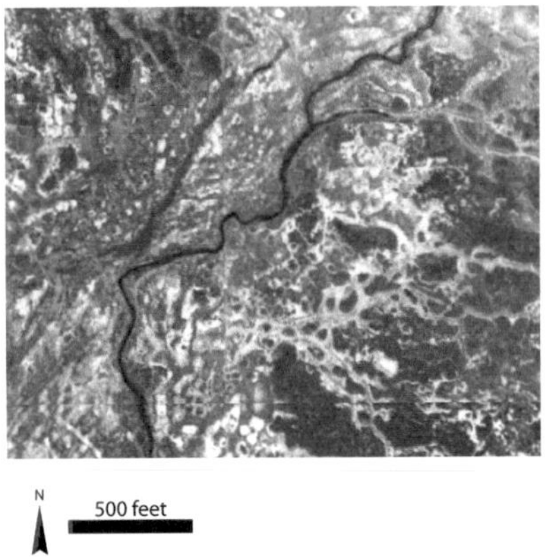

Figure 2.7. Playas and vernal pools in eastern Livermore. Playas are distinguishable by their irregular white scalded pattern. This contrasts with the more regular, round pattern of vernal pool complexes. (USDA 1940)

and Van Duyne 1911, Carpenter and Cosby 1939), and previous research has indicated that this level of soil salinity resulted in observable plant community effects (e.g., salt grass and other salt-tolerant species; Beller et al. 2011, Stanford et al. 2011). Alkali meadow generally includes smaller patches of grasses, scalds, and vernal pools.

Vernal pool complex was mapped in areas that contained vernal pools surrounded by alkali meadow (see above for details on vernal pool complex and alkali boundary).

In eastern Livermore Valley, alkali sink scrub and vernal pool complex often occurred together. In areas with alkali concentrations between 0.2% and 1% and evidence of these habitat types, we mapped alkali sink scrub-vernal pool complex. The complex is a general classification that contains more sink scrub in some areas and more vernal pools in others.

Alkali playa complex covers the highest intensity alkali region. This class was mapped in areas with a high concentration of alkali salts in the historical soil survey (above 1% in the first six feet; Westover and Van Duyne 1911). The area has a visibly different appearance in the historical aerial photography: in place of vernal pools, large scalds spread across much of the region (fig. 2.7). These scalds or playas are intermixed with other alkali habitats, including alkali meadow and particularly alkali sink scrub.

TIDAL MARSHLAND Mapping of the tidal marshland of the Alameda Creek watershed was based on the EcoAtlas (SFEI 1998) and the South Bay T-sheets project (http://maps.sfei.org/tsheets), corrected in some places for this project. For most of the study area, the tidal marsh composition and extent was derived from the U.S. Coast Survey and U.S. Coast and Geodetic Survey large-scale maps (see the following topographic maps, or T-sheets: Kerr 1857b, Westdahl 1896a, Westdahl 1896b, Kerr 1857a, Morse and Westdahl 1896, Rodgers and Kerr 1857, Morse and Westdahl 1896-7). See Grossinger et al. 2011 for detailed methods.

In the area adjacent to the historical course of Alameda Creek where the channel connects with the tidal marsh, southward to Patterson's Landing (approximately the intersection of Newark Boulevard and Lowry Road in Fremont), the eastern extent was less obvious because the earliest T-sheet stopped short of the tidal edge (Kerr 1857b), and the later T-sheet showed an altered tidal edge from early development (Westdahl 1896a). To map the eastern tidal edge in this area we combined the later T-sheet (Westdahl 1896a) with a number of additional sources, including the land grant boundary (roughly representing the tidal edge) found on the contemporary USGS quadrangles and other maps (Whitney 1873, Allardt 1874, Thompson and West 1878); the five-foot contour from the contemporary USGS quadrangle; and the tidal edge depicted on the historical USGS quadrangle (USGS 1899a).

OTHER WETLAND FEATURES Other wetland habitats were mapped from spatially explicit historical sources documenting the presence and extent of depressional marshes, ponds, and willow groves on the alluvial plains. Where multiple sources showed the same feature, sources were synthesized to produce the most likely representation of the historical feature. Topography, historical soils maps, and early aerial imagery were used to refine the shape and extent of wetland features in the absence of other available documentation.

This process undoubtedly under-represents the historical extent and distribution of wetland features. Some known wetland features are documented in the textual record, but were ultimately not recorded with enough accuracy to render them mappable in the GIS (fig. 2.8). Other wetlands were likely left undocumented by the available historical record. Subsequent research may reveal more information about the presence or location of additional wetland features.

Dryland habitats

The dominant dryland habitat types within this project area were grassland and low-density oak savanna (here we define "dryland" habitats to be well-drained terrestrial habitats without regular cycles of flooding). Historical sources in general contained much less spatially explicit documentation of dryland habitat features as compared with wetlands. Although some GLO surveyors distinguished between them, these habitat types were not depicted or distinguished on early maps, which complicated our attempts to map them. By the time of early photography and the 1939-40 aerials, many oaks had already been cleared. As a result, we focused on producing a meaningful representation of patterns of dryland vegetation cover at the landscape scale rather than attempting to depict small details. We relied on the few sources that did address dryland habitats, including GLO surveys,

Figure 2.8. Willows on the Niles Cone. The forest of trees to the right in this image may represent riparian forest along Alameda Creek, or a willow thicket on the plain. This property adjoined Alameda Creek, along Crandall Slough. We lack adequate information to map this feature, but it is an example showing that additional small groves likely existed across the plain. (Thompson and West 1878, courtesy David Rumsey Map Collection)

textual descriptions, historical soil surveys, landscape photography, and historical aerial photography.

The grassland habitat type covers the vast majority of the dryland habitat area and was the default habitat type (i.e., in the absence of other evidence, we mapped grassland). In this report we use grassland to describe a broad range of low herbaceous cover types, including forbs, bunchgrasses, and rhizomatous ryegrasses (see grassland treatment in chapters 4-7). This habitat type may also include areas of sparse tree cover or places with small groves that were below the minimum mapping unit, not locatable, or not documented in the historical record. In many cases, riparian corridors were lumped with the surrounding grassland.

We mapped oak savanna where there was spatially precise documentation of low density oaks as an important landscape feature in textual descriptions and GLO survey data. Areas with documentation of oaks that lacked adequate spatial data to permit mapping are discussed in the text. Additional patches of oak savanna almost certainly existed within the mapped grassland areas.

Inevitably, historical documents reveal more detail about the historical landscape than is represented in the habitat map. For example, not all features noted by the GLO surveyors were subsequently mapped in the habitat layer, as some information was more detailed than our mapping standards allowed (e.g., a small grove of oaks without any details to describe the size or shape was mapped as part of the grassland class). As is the case with most ecosystems, transitions between oak savanna and low herbaceous cover were gradual and diffuse, extending over broad areas. The habitat mapping represents regional-scale transitions as opposed to local-level detail (such as small groves, or narrow zones of riparian oaks along ravines). Many of these local characteristics, though not mapped, are explored further in this report.

Crosswalk between historical and present-day land cover classes

To perform a comparison between historical and contemporary land cover for the study area, we used four contemporary land cover layers: BAARI wetlands (SFEI 2011), BAARI baylands (SFEI 2011), EACCS (ICF International 2010a), and CalVeg (2004). Due to differences between classification systems, we grouped many specific habitat types together to form more general classes. In particular, we combined a number of terrestrial wetland types to find common groupings. Table 2.5 summarizes the crosswalk that we created to compare acreages for each land cover type.

We used different datasets to calculate land cover on the Niles Cone, the baylands, and Livermore-Amador and Sunol valleys. In each region we used the best available data. As a result, the level of detail in present-day mapping is not consistent across the study area.

Table 2.5. Crosswalk between Alameda County land cover classification systems. We developed this crosswalk to allow for comparisons between our historical mapping and contemporary mapping. In the table, the column "crosswalk class" lists the classes we developed to allow comparisons. The land cover classes for each classification system that were included in each "crosswalk class" are listed to the right. This crosswalk includes both one-to-many and many-to-one comparisons. On the Niles Cone we combined several types of wetlands into the class "wetland" to allow us to compare historical and present-day acreages. Riparian types, with the exception of sycamore alluvial woodland, were not included in this crosswalk due to inconsistencies between historical and present-day mapping.

<table>
<tr><th>Crosswalk Class</th><th>HE Mapping Class</th><th>BAARI Wetland Mapping (2011) (Niles Cone)</th><th>CalVeg (2004) WHR Class (Niles Cone)</th><th>ICF International 2010a (Sunol-Niles, Livermore)</th></tr>
<tr><td>Tidal</td><td></td><td></td><td></td><td></td></tr>
<tr><td>Bay Associated Tidal Flat</td><td>Bay Associated Tidal Flat</td><td>Tidal Bay Flat, Tidal Nascent Vegetation</td><td></td><td></td></tr>
<tr><td>Marsh Associated Tidal Flat</td><td>Marsh Associated Tidal Flat</td><td>Tidal Marsh Flat</td><td></td><td></td></tr>
<tr><td>Panne</td><td>Panne, Salina</td><td>Tidal Panne</td><td></td><td></td></tr>
<tr><td>Shallow Bay/ Channel</td><td>Shallow Bay, Shallow Channel</td><td>Shallow Bay</td><td></td><td></td></tr>
<tr><td>Vegetated Tidal Marsh</td><td>Tidal Marsh</td><td>Tidal Vegetation</td><td></td><td></td></tr>
<tr><td>Artificial Salt Pond/ Flat</td><td></td><td>Lagoon Perennial Open Water Natural, Lagoon Perennial Unvegetated Flat Unnatural, Lagoon Perennial Vegetated Flat Unnatural</td><td></td><td></td></tr>
<tr><td>Tidal Ditch</td><td></td><td>Tidal Ditch, Tidal Engineered Channel</td><td></td><td></td></tr>
<tr><td>Non-tidal</td><td></td><td></td><td></td><td></td></tr>
<tr><td>Grassland/ Savanna</td><td>Grassland, Savanna</td><td></td><td>Annual Grass</td><td>California Annual Grassland</td></tr>
<tr><td>Urban</td><td></td><td></td><td>Urban</td><td>Golf Course/ Urban park, Rural- Residential, Urban-Suburban, Ruderal</td></tr>
<tr><td>Agriculture</td><td></td><td></td><td>Cropland</td><td>Cropland, Developed Agriculture, Orchard, Ornamental Woodland, Vineyard</td></tr>
<tr><td>Seasonal Freshwater Wetland</td><td>Wet Meadow</td><td rowspan="2">Depressional Vegetated Natural, Depressional Vegetated Unnatural, Seep or Spring Natural, Seep or Spring Unnatural</td><td rowspan="5">Fresh Emergent Wetland</td><td>Seasonal Wetland</td></tr>
<tr><td>Perennial Freshwater Marsh</td><td>Valley Freshwater Marsh</td><td>Perennial Freshwater Marsh</td></tr>
<tr><td>Willow Thicket</td><td>Willow Thicket</td><td></td><td></td></tr>
<tr><td>Alkali Wetland</td><td>Alkali Meadow, Alkali Playa Complex</td><td>Playa Unvegetated Flat Unnatural, Playa Open Water Unnatural, Playa Unvegetated Flat Natural, Playa Vegetated Unnatural</td><td rowspan="2">Alkali Meadow and Scalds, Valley Sink Scrub, Alkali Wetland</td></tr>
<tr><td>Alkali Vernal Pool Complex</td><td>Vernal Pool Complex, Alkali Sink Scrub-Vernal Pool Complex</td><td>Vernal Pool, Vernal Pool Complex</td></tr>
<tr><td>Open Water</td><td>Pond</td><td>Depressional Open Water Natural, Lacustrine Open Water Unnatural</td><td>Lacustrine</td><td>Pond, Quarry Pond, Reservoir, Riverine</td></tr>
<tr><td>Sycamore Alluvial Woodland</td><td>Sycamore Alluvial Woodland</td><td></td><td></td><td>Sycamore Alluvial Woodland</td></tr>
</table>

To calculate land cover change for the Niles Cone region we used a combination of the BAARI mapping (SFEI 2011) for wetlands and the 2004 CalVeg layer for both drylands and wetlands. The CalVeg layer has a lower resolution and level of detail than our historical mapping or the ICF layer we used for Livermore-Amador Valley. BAARI mapping is detailed but only includes wetlands. To capture the most comprehensive mapping of the Niles Cone, we combined the two layers. Our final calculation of contemporary wetland cover for the Niles Cone includes all of the wetland areas mapped by each layer, so it may slightly overestimate wetland area.

To calculate change in the baylands, we used the detailed BAARI (SFEI 2011) baylands layer. Salt pond levees (not mapped by the BAARI layer) were coded "developed" even where they were covered by some vegetation.

To calculate land cover change for Sunol Valley and Livermore-Amador Valley we used the 2010 layer completed for the East Alameda County Conservation Strategy (EACCS) by ICF International. For analyses we excluded the portion of the study area that extends into Contra Costa County because this was not included in the ICF layer.

The crosswalk does not indicate precise correspondence. For example, present-day tidal vegetation may not have the same habitat quality as historical tidal marsh habitat. Some areas mapped as historical seasonal wetlands may not satisfy the definition of a present-day wetland, and some areas mapped as wetlands today may perform few of the functions that historical wetlands did. The crosswalk is intended only to allow a rough comparison and highlight areas of dramatic change.

A few habitat types were excluded from the crosswalk. All riparian types (with the exception of sycamore alluvial woodland) were excluded because we did not map historical riparian vegetation extent. Instead, we describe change in riparian cover in Chapter 9. Many areas mapped as woodland by CalVeg or ICF International were in fact riparian areas, and were excluded. As a result, acreages generated using this crosswalk do not add up to the total acreage of the study area. Over the entire study area, the calculated acreages for historical and present-day land cover differ by 1,120 acres, approximately 1%.

Salt Co
UNION
E.B.Granger
J.L.Beard
School
Hanson
J.G.Kilpark
T.Pacheco
NILES
Scott
W.Brown
Potrero de los Cerritos
WASHINGTON T.
Kelsey
Shinn
P.Morrison
R.Bo
School
H.Overacker
Tyson
Indian Mound
Ind. Mound
CENTERVILLE
Overacker
RED HILL
Sanjon de los Alisos
Niehaus
A.Alviso
School
R.Blacow
Forbes
Chadman
Palmer
Tule Swamp
Indian Mound
J.J.Riser
Baker
Eggers
A.Cameron
Mission
Beard's Embarcadero
San Jose
Brophy
WASHINGTON CORNE
Mayhew's Landing
A.D.Forbes
O.Morgan
Ryan
Babb
James
Call
I.T.Walker
Hagan
Salt Works
Mowry's Landing
Ind. Mounds
Swamp
Mowry's Creek
Mayhew &
Mathews

An understanding of the land use context is critical to appropriate interpretation of historical landscape patterns. Native land use shaped tree, grass, and brush distribution long before the historical period, and then with the arrival of the Europeans, agriculture and water diversions began to reshape and transform the county into the largely urbanized landscape we see today (fig. 3.1). Historical data represent a variety of time periods, depicting the landscape under a number of land use practices. Human and physical influences on the landscape work together: the ecology and physical features of the landscape controlled where and how people settled, and settlement patterns in turn dictated the patterns of human influence. Climate is important to interpretation as well—years of extreme rainfall or drought caused flooding and crop failures, influencing land use, the historical record, stream flows, and vegetation patterns.

The following sections provide a brief summary of major land use, water use, and rainfall trends in southern Alameda County. We focus on 19th century trends in agriculture, irrigation, settlement, and water use that provide context for the landscape-scale ecological and hydrologic changes occurring in the county.

Land use

Native land management era

Records of human presence in the region date to at least 14,000 years ago (Milliken et al. 2009). At the time of European arrival, the study area was inhabited by people of the Chochenyo Ohlone/Costanoan language group. A number of distinct tribelets, or smaller autonomous groups, were represented, including the Irgin, Tuibun, and Alson along the Bay; the Causen in Sunol Valley; and the Pelnen, Seunen, and Ssouyen in Livermore-Amador Valley, with the Ssaoam and Yulien along the eastern edge of the Valley (Milliken 2008).

There were Rancherias of Digger Indians near the then open lake, below the present town of Pleasanton, and also in the canyon in the vicinity of the present town of Sunol.

—OAKLAND TRIBUNE 1898:134

The Ohlone left no written records, so our knowledge of them is derived largely from archeological records and European accounts. Early explorers recorded a series of villages both across the Niles Cone and through San Ramon and Sunol valleys. When Pedro Fages traveled across the Niles Cone in 1772, he noted to the east "five villages with about six houses in each and quite a number of heathen inhabiting them," and additional villages to the west (Treutlein and Fages 1972). In the southern San Ramon Valley a few days later he encountered three villages in short succession, one of them "an extremely big village of heathen who perhaps numbered in excess of two hundred" (Treutlein and Fages 1972). Population density at the time of contact was an estimated 2.7 to 5.6 people per square mile, based on Mission records (Milliken et al. 2009).

Figure 3.1. Land use in the Alameda Creek watershed. The photograph shows the construction of the Sunol Dam at the top of Niles Canyon, an important step in the transformation of the watershed. The dam was completed in 1901. The map shows the mix of land uses near the Coyote Hills in 1873. Note the salt works, "Indian Mounds," and growing town of Centerville. (Top: #158383, © San Francisco Public Utilities Commission; Bottom: Whitney 1873, courtesy David Rumsey Map Collection)

Shellmounds along the bayshore represent some of the most prominent traces of Ohlone presence. Within the project area, several mounds were mapped along Sanjón de los Alisos and east of Coyote Hills in 1873

(Whitney 1873). These mounds date to ca. 500 BC, and appear to have been largely abandoned 1,000 years ago (Lightfoot and Luby 2002).

The Ohlone made use of a wide variety of plant species. Products harvested included acorns, grasses and forbs, intertidal and nearshore marine products, and tule to construct rafts and innumerable other goods (Kroeber [1925]1976). Font noted in 1776 that, "they eat grass and herbs and some roots like medium-sized onions, which they call amole, and in which those plains greatly abound" (Font and Bolton 1933:356; *amole* may refer to soaproot or cacomite, Clarke 1952).

Although Native people in California did not practice agriculture, they did modify the landscape in a variety of important ways. Tribal groups throughout the region managed lands under their influence with practices such as seed beating, burning of scrub and grasslands, harvest of grasses, and use of digging sticks to turn the soil (see Lewis 1985, Stewart et al. 2002, Anderson 2005, Nabhan and Martinez 2012).

Of particular interest to land managers as well as tribes today is the use of fire to shape local ecosystems. Native groups used fire to control the distribution of chaparral, maintain grassland cover and forage for wildlife, control pathogens, improve access to acorns, and aid in hunting rabbits and other small game (Kroeber [1925]1976, Keeley 2002, Stewart et al. 2002, Anderson 2005). Records of Native burning for Alameda County are limited and incomplete, but burning almost certainly shaped the distribution of vegetation in the landscape ca. 1800 (Keeley 2002, Keeley 2005). Two early sources claimed that no timber remained on the Niles Cone due to excessive burning. Trapper Langsdorff described burning as follows:

> The native Indians have now and then, thoughtlessly, and simply to make a bonfire, set fire to the forests, and burned down large tracts, leaving few trees standing. (Langsdorff 1806, in McCarthy 1958:116)

A later quote complained of "all traces of vegetation having been burned up by the Indians, who sometimes adopted this method of annoying the Yankee traveller" (Ryan 1850). Although we do not know the extent of Native burning in the study area, it is clear that the landscape the Europeans encountered had already been shaped by millennia of human management.

Mission era (1776-1836)

In 1769, Fray Juan Crespí led the first European expedition across the Niles Cone, describing a "large creek with a large flow of running water and a great many trees on its bed" which followed a course through a "large plain with very good soil" (Crespí and Bolton 1927). Over the next few years, Pedro Fages, Crespí, Juan Bautista de Anza, and Pedro Font made additional preliminary explorations of the East Bay. Significant European impacts in the study area began when neighboring Mission Santa Clara was founded in 1777. By 1797, one-third of inhabitants of the Fremont area had moved to Mission Santa Clara (Milliken 1995).

In 1795 Mission San José was founded within present-day Fremont (fig. 3.2). The Mission San José site was chosen for the fact that it had "much fine land and easily worked," and that several small creeks were able to provide water for irrigation (Dantí and Sal 1795 in Cook 1957). The rich soil and temperate climate as well as the proximity to San Francisco via water, helped the productivity of Mission San José to outpace that of its predecessors, and the mission achieved considerable wealth (Oakland Tribune 1898:134). Historian Halley described how the natural resources available influenced perceptions and use of the mission site:

> A fine site, a healthy climate, abundance of the purest water (which ran perennially from unfailing springs through the Mission garden), with the Calaveras and Alameda Creeks close by. Wood was near and abundant. Game was ever within shot. The pasturage was all that could be desired. The soil was as rich and mellow as a ripe apricot. The belt on which it was situated was warm and ever free from killing frosts. An embarcadero was only a few miles distant, and within an hour's walk were warm mineral springs, possessed of potent healing qualities. (Halley 1876:15)

The influence of the mission reached inward to Livermore-Amador Valley as well as north along the East Bay. Eventually over 55 tribelets and nine distinct languages were represented in the mission registry, with a maximum population of almost 2,000 Native people (Milliken 2008). People from western Livermore began moving to Mission San José in 1801, and the last independent tribelets had disappeared from both Livermore-Amador Valley and the Niles Cone by 1808 (Milliken 2008). People from

Figure 3.2. Mission San José, 1881. This image shows Mission San José during the early American period. (Ford drawing, courtesy Library of Congress online)

within the study area made up a large proportion of the total population at the mission—in 1809, 80% of Mission San José's native Californian residents spoke Chochenyo Ohlone (Milliken 2008).

The inhabitants of the mission built a reservoir and aqueduct system using water supplied by Mission Creek to irrigate gardens, orchards, vineyards and olive groves near the mission buildings (Sandoval n.d.). Mission San José grew wheat, barley, corn, and beans (Bancroft [1886] 1966), and the first grist-flour mill was constructed to mill wheat from mission lands (Sandoval n.d.). Cattle and sheep grazed widely over the study area. At its peak, 35,602 head of stock were reported (Bowman 1947).

This mission for the whole decade [1830s] was probably the most prosperous in California both before and after secularization. Its highest population of 1,866 souls was reached in 1831 … crops were uniformly good, the yield being larger in proportion to the seed sown than elsewhere, and live-stock increased steadily to the end … 18,000 cattle, 15,000 sheep, 1,900 horses.

—BANCROFT [1886]1966:724

In 1833 the Mexican Congress passed an act calling for the secularization of the California missions, and confiscated mission properties. However, Mission San José maintained its holdings until 1836 (Milliken 2008). As the mission finally dissolved, its landholdings were distributed among the Mexican families in the area, and many of the people who had lived at the mission worked at these ranchos. Although mission property was intended to support the Native converts, only a single piece of land was granted to one group among the 1,900 Indians living at the mission at this time (Milliken 2008).

Ranching and the rancho era (1836-1860s)

After the dissolution of Mission San José, the Mexican government distributed several land grants to private individuals in Alameda County (fig. 3.3). The first land grant was made in 1836 when Fulgencio Higuera received land south of the Mission at Agua Caliente. Additional land grants were made in 1839 (Las Positas, El Valle de San Jose, and Santa Rita) and 1842 (Potrero de los Cerritos and Arroyo de la Alameda).

The owners of the ranchos raised cattle, probably of Andalusian stock, and a type of small sheep called the churro (Stokle 1968:70). Their cattle grazed on "wild oats and clover and nutritious grasses from valley to hill tops and even the mountain sides" (Oakland Tribune 1898:134).

Although ranching dominated much of the landscape in the mid-1800s, some grain production existed on Niles Cone. Don José de Jesus Vallejo built a flour mill at Niles on Alameda Creek in 1841 (Kern 1983), and J.M. Horner built a mill at Alvarado in 1853 that used steam instead of water power (Halley 1876:109). For the most part, landowners did not practice commercial agriculture and cultivated only small gardens.

As a wave of settlers arrived in the 1850s in the aftermath of the Gold Rush, the rancho system began to collapse and smaller plots were created out of subdivisions of the ranchos. Agriculture began to push cattle off of the valley floors and into the hills (Daily Alta California 1863). Climate also played a role—as in much of the state, the major flood and drought of 1861-1864 reduced cattle populations and hastened a shift towards agriculture. In

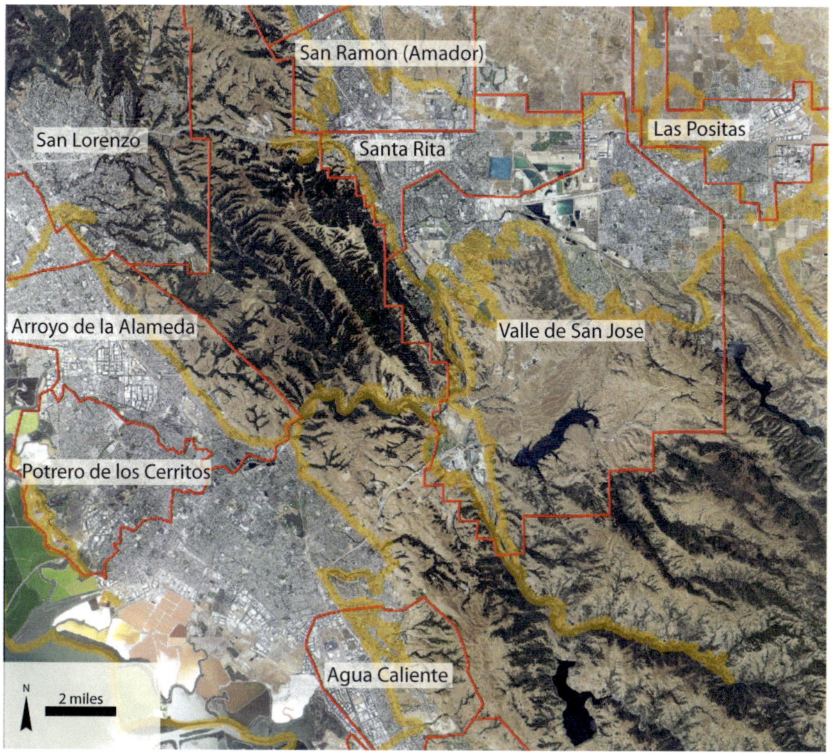

Figure 3.3. Mexican land grants of southern Alameda County. Within the study area (outlined in orange) there were a number of Mexican land grants. These red outlines represent the boundaries finally approved by the U.S. government in the mid- to late 1800s. (USDA 2009)

the 1860s, the number of cattle on rangelands in Alameda County dropped from 51,321 head to just 8,156 (Burcham 1982).

The impact of early cattle and sheep grazing in the Alameda Creek watershed is unknown. With the establishment of Mission San José, cattle and other stock spread across the Niles Cone, up Niles Canyon and Sunol Valley, and through Amador Valley north to present-day Walnut Creek (Durán 1827 in McCarthy 1958, Bowman 1947; fig. 3.4). At the peak of Mission stock holdings in 1826, stock ranging densities reached an estimated one head for every six acres of grazing land (45% cattle, 30% sheep, 25% hogs; Bowman 1947). In the late 19th century sustainable stocking densities for California were considered to be five to ten acres per head for cattle in "valley land" (Bancroft et al. [1890]1970). Stocking densities for the uplands (a significant part of the grazed area) would have been lower. However, these densities were calculated for American cattle, which were only introduced to California in the 1850s (Adams 1946) and were much larger and required more forage than their Mexican counterparts (Burcham 1956), so the equivalent Mexican cattle densities would have been less impactful.

Five years ago the valley [near Pleasanton] was nearly all open and occupied by cattle ... Many of these cattle, shut out from the fertile valleys which were open a few years ago, are now restricted to the hills, and are in imminent danger of suffering for want of pasturage.

—DAILY ALTA CALIFORNIA 1863

By the 1850s and 1860s, with the introduction of larger American cattle at higher densities, it is likely that cattle and sheep grazing began to have increasing effects on ecological and morphological processes in the county. Grazing can cause compaction of the soil, erosion, sedimentation, and incision, as well as potentially precipitate a wholesale change in vegetative cover, and it may have shaped the landscape depicted in late 19th century sources. In particular, the introduction and spread of invasive European

Figure 3.4. Cattle in Livermore Valley, 1917 and 2011. Cattle were an intimidating presence across the plains. Historian Halley described that "the wild cattle roamed about threateningly. In the morning the vast herd would be at the bay, at noon at the Alameda creek, and at night they would move back to the bay" (Halley 1876:484). These images show the persistence of cattle in Livermore Valley. (A: Anderson Ranch, Finley Road Tassajara, courtesy Museum of San Ramon Valley, Edmonston binder vol. 2 p. 126; B: Springtown Preserve, photo by Ruth Askevold, May 2011)

grasses had widespread impacts on soil stability and hydrology (through different root structure) as well as on grassland community composition. As we were unable to quantify these impacts, they are treated only generally here. However, we did attempt to consider the influence of cattle as we interpreted historical sources.

The old timers used to talk about Livermore Valley, and Tassajara, [that] we were both a grainery and a hay shed for all the agricultural business [such as the] dairies in Fremont and Newark ... [During] the time of the gold rush, after SF proper became more and more built up, basically the East Bay supplied food for San Francisco.

—PAUL BANKE, LONGTIME RESIDENT, PERS. COMM.

Agriculture and the early American era (1850s-1930s)

With the rise of agriculture, patterns in Livermore-Amador Valley and the Niles Cone began to diverge. Both were initially used as cattle rangeland, but Livermore-Amador Valley became the source of grain for the region, while Niles Cone was used to grow a diverse array of fruits and vegetables.

During the first half of the 19th century, ranchers in Livermore-Amador Valley widely believed that the gravelly soils could only be used for stock-grazing (Likins 1874, *Daily Alta California* 1860c, Davis 1868, Wood 1883, *Oakland Tribune* 1898). In 1856, Joseph Livermore sowed 160 acres to the first field of grain in Livermore-Amador Valley; he was followed by several landowners in the western part of the valley the next year (fig. 3.5; DeNier 1927). Rapid conversion to grain followed. In his 1883 history of the county, Myron Wood noted that "Amador Valley, formerly the valley of San José,

where the padres of that old mission pastured their cattle, is now the great grain district of this county … Less than twenty-five years ago this valley was a cattle ranch" (Wood 1883:35). Grain was farmed across the valley and up hillsides (Koopman pers. comm.).

Figure 3.5. Robert Livermore's Farm, 1878. Robert Livermore's farm was situated along Arroyo las Positas, near natural springs and a willow grove. (Thompson and West 1878, courtesy David Rumsey Map Collection)

While wheat was the dominant crop in Livermore-Amador Valley through the late 1800s, settlers also planted grapes, orchard trees, and some row crops. As the Pleasanton marsh was drained in the early 20th century, this formerly flooded area opened for cultivation. The groundwater provided sub-irrigation and a hops industry flourished in the Pleasanton area at the turn of the 19th century (McGown 1902). At its height, there were 400 acres devoted to hops just north of town; low prices caused a decrease in acreage by 1911 (Westover and Van Duyne 1911). Grapes also spread, particularly in the eastern portion of the valley. By 1898, 5,000 acres had been planted with grape vines (Oakland Tribune 1898). Testimony from the early 1900s described the crops grown near Pleasanton:

> Properties immediately surrounding Pleasanton are being used for beet raising, alfalfa, dairying, hay and grain, truck gardens … They were used for hops at one time. The hop fields were removed some three years ago … Most of your Pleasanton land is alfalfa, that will raise alfalfa without irrigation. It is good for beets, potatoes, corn and beans. (Gale 1915:173)

At the same time that wheat spread through Livermore-Amador Valley, farmers on the Niles Cone began to shift their energies from grain to orchards and row crops (fig. 3.6). In 1855 Henry C. Smith planted the first substantial orchard on the banks of Alameda Creek in what is now Union City (Shinn [1889]1991). Fresh soil deposited by overflows from Alameda Creek and access to groundwater for irrigation created fertile farmland, and fruit orchards began to replace wheat and barley fields (Brown 1986). By 1884, as grain production declined, the flour mills at Niles and Alvarado closed (McCann and Hinkel 1937, Kern 1983). However, just as some row crops were planted in Livermore-Amador Valley, some cattle remained on the Niles Cone, particularly dairy cattle:

Figure 3.6. Niles Cone agriculture. This postcard shows horses moving through plum orchards on a road between Irvington and San Jose. (Irvington-San Jose Road, Irvington, CA, Fisher Collection, courtesy Museum of Local History, Fremont)

> Dairying in Washington Township … The climatic conditions and character of the land are peculiarly suited to the dairying industry. The moisture allows large crops of volunteer as well as sown feed to follow in quick rotation. Many places cut the alfalfa the year round. The wild grasses and sweet scented herbs that also spring up and cover the uncultivated lands with a carpet of green furnish rich grazing and give to the product of this section a delicacy of flavor that is not excelled. (*Township Register* 1910)

> *Shipments to the city were made by way of the embarcaderos … Every fruit grower who lived in Washington township during the early sixties, will remember the system; the four miles haul over the dusty road, the cool marsh, the tules and cattails, the blue water, the schooners at the wharf, the beauty and mystery of it all.*
>
> —SHINN [1889]1991:25

Through the early agricultural era, Alameda Creek was used to transport agricultural products and other goods to the wider market. Farmers from as far away as Livermore brought their grain, hides, and vegetables to landings to be shipped to San Francisco (fig. 3.7; Akers 1931). Early orchard owners planted near the Bay and shipped fruit by schooner (Shinn [1889]1991). The arrival of the railroad transformed land use through the study area. When the Western Pacific and Central Pacific railroads arrived in Alameda County in 1869, more farmers had access to rapid transportation, and could grow perishable foods further from the Bay (fig. 3.8). This allowed for mass conversion to high-value crops (Shinn [1889]1991, Sandoval 1985, Corbett 2005). By 1903, farmers on the Niles Cone had planted 5,000 acres with fruit (Carruthers 2000).

Distinct regions of the Niles Cone were used for different crops. Early landowners grew barley and oats on the lowlands and wheat on the uplands (Halley 1876). Farmers planted orchards and vineyards inland, up to the

Figure 3.7. Landings along San Francisco Bay. (top) A series of landings along the Bay provided farmers with access to markets in San Francisco. Landings changed names over time, but included (south to north) Warm Springs Landing, Mowry's Landing, Plummer's Landing, Jarvis/Mayhew's Landing, Patterson Landing, and Barron's Landing, as well as Alvarado along Alameda Creek. (Thompson and West 1878, courtesy David Rumsey Map Collection)

Figure 3.8. Railroad at Hall's Station. (bottom) Railroads transformed the transportation of goods across Niles Cone and even from Livermore-Amador Valley to San Francisco. (Thompson and West 1878, courtesy David Rumsey Map Collection) "very summit" of the hills, and vegetables along the bayshore (*San Francisco Chronicle* 1890, Lee 1916b). Overflow lands near the tidal marsh were used for pasture (Morse and Westdahl 1896-7), and beets became particularly important just inland of the tidal marsh edge (fig. 3.9). Beets are brackish-tolerant crops with wild ancestors that were native to European salt marsh edges, and were planted at the marsh edge in Sonoma and Suisun as well (Baye pers. comm.).

E. H. Dyer planted 150 acres with beet seed in 1869, and the following year a sugar beet mill was constructed nearby, becoming the first successful sugar beet mill in the United States (Halley 1876, Oakland Tribune 1898, Sandoval 1985). Beet production extended to the Livermore-Amador Valley and by 1911, 2,000 acres were devoted to the production of sugar beets; half of this land was owned by a single company (Westover and Van Duyne 1911:12). In 1895 a local newspaper described the fertility of the land near Alameda Creek:

> The great body of water [Alameda Creek], which must have an outlet, overflows its banks nearly every winter and boats are quite a convenient means of communication ... The best of winter potatoes, onions and other vegetables are raised on this river-bottom land, and no better land can be found in the State for sugar-beets. (*The Morning Call* 1895)

Much of the tidal marsh was converted to salt ponds through the later 19th century (fig. 3.10). The natural marsh salinas had been used by native

Figure 3.9. Crops across the Niles Cone. This lithograph shows the variety of crops grown across Niles Cone in the late 1800s. The mouth of Niles Canyon is at the right. Row crops abut a young orchard, while grain fields and homesteads stretch into the distance. Note the two trains and the riparian trees along Alameda Creek through the canyon. This image looks south from the mouth of the canyon. (Thompson and West 1878, courtesy David Rumsey Map Collection)

people as a source of salt long before the arrival of Europeans, and the Mission collected salt from the marsh in great quantities (Sullivan 1934), but the artificial management of these ponds began in 1852 (Brown 1960). In the latter half of the 19th century, salt production emerged as a great source of wealth: by 1893, the industry's total output was 37,850 tons per year (Coquhoun 1893:22). The Union Pacific Salt Company, the largest in the state, produced on average 5,000 to 7,000 tons of salt per year at the mouth of Alameda Creek by the end of the 19th century (Oakland

Figure 3.10. Alvarado salt works at Newark. Note the extensive salt ponds already in place by 1878 (A). (A: Thompson and West 1878, courtesy David Rumsey Map Collection; B: courtesy Robert Fisher, Museum of Local History, Fremont)

B

Figure 3.11. Rapid development in Pleasanton. These paired images from 1939 and 2009 show the massive transformation that has occurred in Amador Valley in the 20th century. In 1939, Pleasanton was a small development on the higher ground to the southeast of the former marsh complex, and agriculture covered the former marshland. In 2009, to the right, the entire area has been transformed to residential and industrial development. (USDA 1939; USDA 2009)

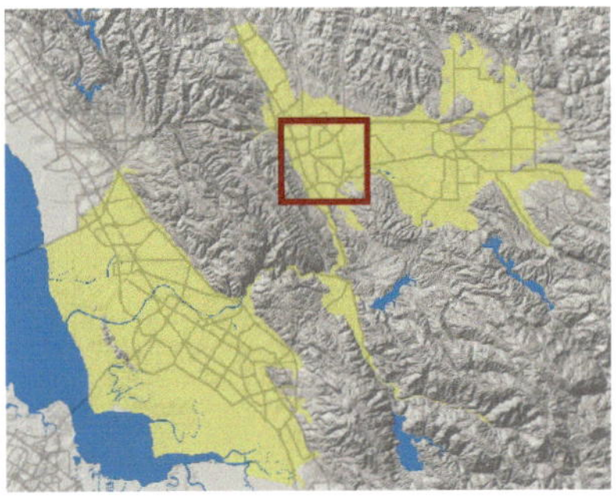

I can remember when you could drive through the valley and find roses, cut flowers, sugar beets, alfalfa, hops, lots of grain, now all of that is gone. All of the best land, went into housing. Because they were flat and level and easy to access.

—PAUL BANKE, PERS. COMM.

Figure 3.12. Land use timeline. (opposite page) A summary of major events and trends in land use from European contact to the present. Population data drawn from U. S. Census Bureau.

Tribune 1898:134). Today, Cargill still produces salt on land in the San Francisco Bay, but in 2003 federal and state government purchased 16,500 acres of salt ponds with the goal of restoration (Rodgers 2010).

Residential development and the 20th century

Through the 20th century, urbanization has been a major shaping force. The 1950s marked the beginning of a period of rapid growth, as the population began to grow by tens of thousands of people per decade. Population on the Niles Cone jumped from around 10,000 in 1950 to over 60,000 in 1960, and is over 300,000 today. In Livermore-Amador Valley, the population in Livermore and Pleasanton rose from 6,600 in 1950 to 20,260 in 1960, and has reached almost 200,000 (U.S. Census Bureau 2010). This rapid growth in population required conversion of agricultural lands to residential developments. Lands that had been used for agriculture and ranching were converted to subdivisions of homes to accommodate the influx of people (fig. 3.11, fig. 3.12).

The agriculture that did continue was increasingly supported by irrigation (Clark 1915). Dry years, fewer floods, and declining groundwater levels due to over-pumping all helped to make irrigation a higher priority (Lee 1916b). By 1915 there was "very little grain" on the Niles Cone, and 85% of the agricultural land was irrigated (West 1937). By the mid-1900s, the main crops even in Livermore-Amador Valley had shifted from dry-farmed wheat to orchards and tomatoes (McCann and Hinkel 1937). However, even today some grazing continues, particularly in eastern Livermore Valley (fig. 3.13).

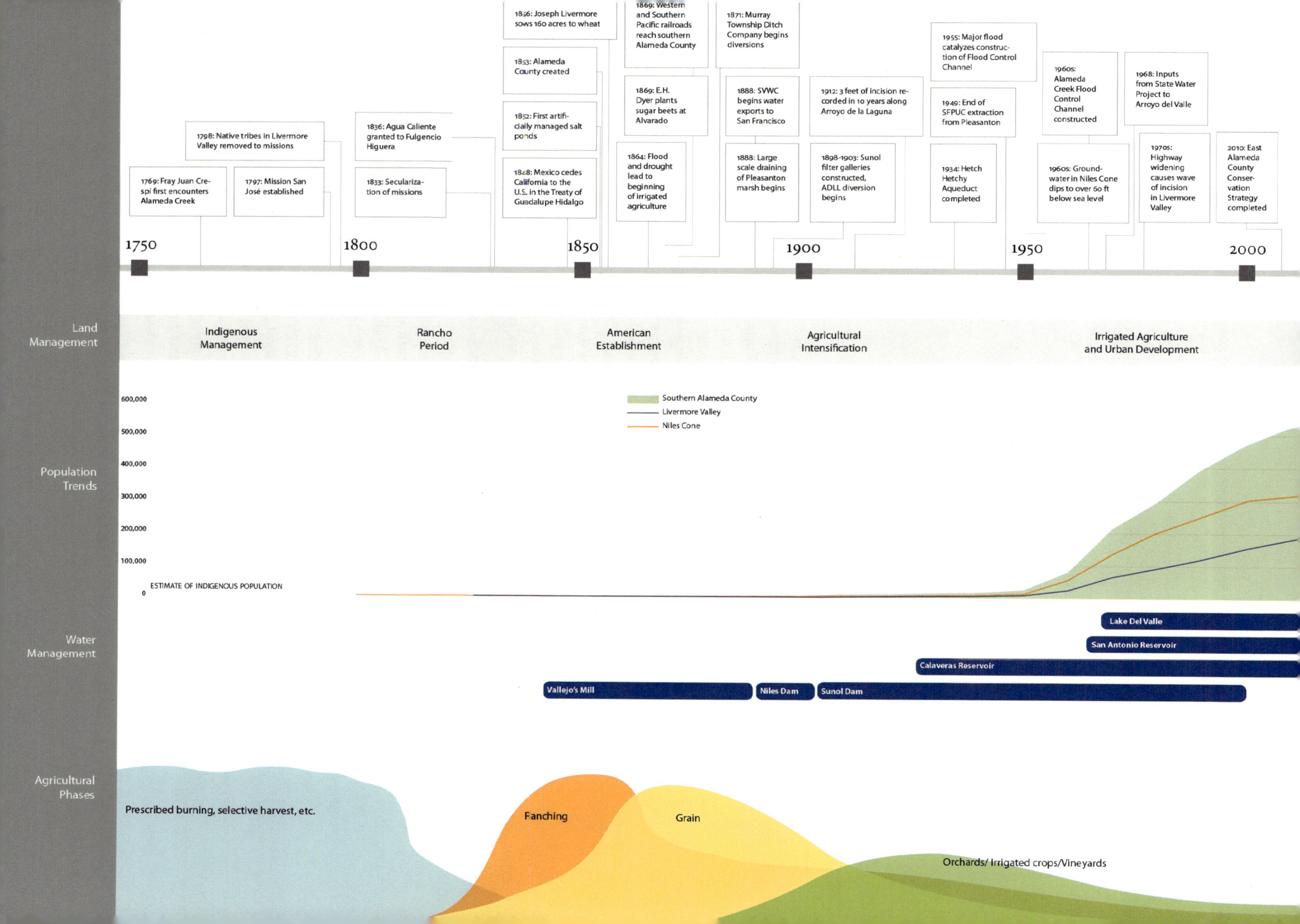

1769: Fray Juan Crespí first encounters Alameda Creek
1798: Native tribes in Livermore Valley removed to missions
1797: Mission San José established
1836: Agua Caliente granted to Fulgencio Higuera
1833: Secularization of missions
1856: Joseph Livermore sows 160 acres to wheat
1853: Alameda County created
1852: First artificially managed salt ponds
1848: Mexico cedes California to the U.S. in the Treaty of Guadalupe Hidalgo
1869: Western and Southern Pacific railroads reach southern Alameda County
1869: E.H. Dyer plants sugar beets at Alvarado
1864: Flood and drought lead to beginning of irrigated agriculture
1871: Murray Township Ditch Company begins diversions
1888: SVWC begins water exports to San Francisco
1888: Large scale draining of Pleasanton marsh begins
1912: 3 feet of incision recorded in 10 years along Arroyo de la Laguna
1898-1903: Sunol filter galleries constructed, ADLL diversion begins
1955: Major flood catalyzes construction of Flood Control Channel
1949: End of SFPUC extraction from Pleasanton
1934: Hetch Hetchy Aqueduct completed
1960s: Alameda Creek Flood Control Channel constructed
1960s: Groundwater in Niles Cone dips to over 60 ft below sea level
1968: Inputs from State Water Project to Arroyo del Valle
1970s: Highway widening causes wave of incision in Livermore Valley
2010: East Alameda County Conservation Strategy completed
1750
1800
1850
1900
1950
2000
Land Management
Indigenous Management
Rancho Period
American Establishment
Agricultural Intensification
Irrigated Agriculture and Urban Development
Population Trends
600,000
500,000
400,000
300,000
200,000
100,000
0
ESTIMATE OF INDIGENOUS POPULATION
Southern Alameda County
Livermore Valley
Niles Cone
Water Management
Lake Del Valle
San Antonio Reservoir
Calaveras Reservoir
Vallejo's Mill
Niles Dam
Sunol Dam
Agricultural Phases
Prescribed burning, selective harvest, etc.
Fanching
Grain
Orchards/ Irrigated crops/Vineyards

Figure 3.13. "Pastoral farm scene," 1950. This image shows pasture lands in eastern Livermore Valley. This photograph was taken in 1950, and shows the continuing presence of grazing in the eastern valley. (courtesy Livermore Heritage Guild)

The channelization of creeks and construction of dams helped promote development. In the 1960s the Alameda Creek Flood Control Channel ended annual flooding by Alameda Creek on the Niles Cone, opening land to development that would have historically been periodically inundated (McKeown 1975b, Goold 1986, Furtado 1987). The channelization of numerous small streams through the Livermore-Amador Valley reduced flooding across the valley floor, and helped contain the increased runoff from paved surfaces in the developing cities (Fletcher pers. comm.). This resulted in higher peak flows downstream (and more flooding), channel incision, and increased sediment delivery to Fremont. Dams on Calaveras Creek (1925), San Antonio Creek (1965), and Arroyo del Valle (1968) helped further control flows. The dam on Calaveras Creek, in particular, removed a large source of coarse sediment from the watershed, and now captures flows from both Alameda Creek itself (through the diversion dam) and the large tributaries Calaveras Creek and Arroyo Hondo.

Once urbanization began, land use changes rippled through the local economy, helping drive the shift to more urbanization. Longtime resident Paul Banke described how cattle raised in the Livermore-Amador Valley were fattened on beets near Newark and shipped to slaughterhouses in San Francisco (Banke pers. comm.). As the slaughterhouses and sugar beet mills closed, ranching became less profitable.

In 1897 Dr. W.H. Cope discovered near the center of Pleasanton a large natural well which was the beginning of a water-works for the town. It was found that 42,000 gallons flowed from the well every twenty-four hours.

—MCCANN AND HINKEL 1937:215

The gravel quarries have also changed the shape of the landscape in the 20th century. The coarse gravelly substrate in Livermore and Sunol valleys lent itself to gravel mining, and local extraction of gravel from stream beds for road construction was documented beginning in the late 1800s (*Echo* 1894a, c), as was the development of a limestone quarry near the mission (Morse and Westdahl 1896-7). Large-scale mining developed through the 20th century, with gravel extraction along Arroyo del Valle and then Alameda Creek in Sunol Valley and the upper Niles Cone (Davis 1976, Whitfield 1986). Today, as gravel resources are depleted, many of the mines have been turned over to the regional parks service and water agencies.

Groundwater use

The two largest groundwater basins of the Alameda Creek watershed lie beneath the Niles Cone and Livermore-Amador Valley, each fed by its respective subwatershed, and each containing multiple distinct aquifers and aquitards (clay layers separating aquifers). Both historically supported artesian springs. Both basins also experienced heavy use in the early 1900s, leading to a drawdown of water levels to the point that artesian water sources dried up and export was halted. The extensive water use contributed to the drying of seasonal wetlands, marshes, seeps, and springs. Today both groundwater basins have recovered and are once more used as a water resource. Water use trends for these two subwatersheds are covered briefly here; for more details see one of the many water reports for this area (Williams 1912, Dockweiler 1912b, Clark 1924, West 1937, Morris et al. 1960, Fisher et al. 1966, Figuers 1998, CA Department of Transportation 1998).

Livermore-Amador Valley

Due to the relatively dry climate and lack of perennial streams in Livermore-Amador Valley, those water sources that were available were noted and prized in the earliest records. Travelers described "a pool of good water located in the western portion of the valley" (Father Viader 1810 in Cook 1960:258) and remarked on the "excellent water" that could be found near Livermore, likely at Las Positas (Moerenhout [1849]1935). The springs associated with the Pleasanton marsh complex were so valuable that they led to a property dispute, resolved by drawing the land grant line through the center of one particularly valuable spring (Amador 1861, Dyer 1862a, Healey 1863). In the eastern portion of the valley, Robert Livermore used a ditch to direct water from Las Positas springs to his wheat field for irrigation as early as 1844 (Wood 1883). Local landowners, including Livermore, also began digging artesian wells near Las Positas in the 1850s and 1860s (Wood 1883:464), and in the mid-1870s water from Arroyo Mocho and Las Positas springs was supplied to the City of Livermore by the Livermore Spring Water Company (Wood 1883).

Efforts to contain the Pleasanton marsh complex were likely undertaken by the very earliest settlers, and there are records of some canals being built in the late 1870s and 1880s to reclaim marsh land for agriculture (Figuers 1998, Corbett 2005). However, the substantial push to drain the marsh and develop it as a water source occurred in the 1890s and 1900s, through a combination of wells tapping the underground supply and canals directing water to Arroyo de la Laguna and thence to the Sunol Filter Beds (Baker 1914, Espy 1950). By 1898, 53 artesian wells discharged into Arroyo de la Laguna (Lee 1916a), and by 1912 several engineering reports were produced to estimate the historical extent of the marsh (Tibbetts 1907a, SVWC 1912b; for more details see box 4.5). These reports used the historical outlines of the marsh to help determine the depth to groundwater that would minimize loss to evaporation and to determine places that water could be rerouted across porous gravels to maximize recharge (SVWC 1912b).

From Santa Rita westerly to Dublin ... the ground is observed to be in a state of almost complete saturation, and the nature of the plant life indicates that this is a permanent condition. This condition also extends southerly almost to the producing wells of the Spring Valley Water Company, situated southwest of Pleasanton, and has every appearance of being due to true artesian forces.

—MULHOLLAND AND LIPPINCOTT 1912:17

The extraction of water impacted water supply in the valley. In 1902 some artesian wells near Pleasanton still rose to a height of four feet above ground (McGown 1902), and water stood less than three feet from the surface over the area of the marsh (Mulholland and Lippincott 1912). However, by 1916 wells that had formerly been purely artesian needed to be pumped at some times of the year, and the lowering of the water table was noticeable (Lee 1916a; fig. 3.14). Well records showed that west of Livermore water fell as much as 50 feet in the dry year of 1913 (Lee 1916b).

A transformation of the stream network accompanied the extraction of groundwater resources. Streams with discontinuous channels such as Arroyo Mocho and Arroyo las Positas were channelized to form a continuous, water-removing network beginning in the 1880s (Whitney 1873, Nusbaumer 1889a, Tibbetts 1907b). By the time of the 1906 USGS mapping, the stream network looked much like the contemporary network.

The Pleasanton sub-basin in Amador Valley continued to be used as a water source for the SFPUC through the 1940s, when low groundwater levels caused pumping to cease (Fisher et al. 1966, CA Department of Transportation 1998; fig. 3.15). With the creation of the Del Valle Reservoir in 1968, flood flows to the system were greatly reduced. Today recharge through abandoned quarry ponds and releases of State Water Project water through Arroyo Mocho and Aroyo del Valle help to maintain groundwater levels (Department of Water Resources 2006). Groundwater is pumped for water supply for the valley, and water levels are closely monitored to ensure that they do not interfere with mining in the remaining quarry ponds.

Figure 3.14. Pleasanton Well Rig. Pumping stations such as this one were used to extract groundwater in the area that once supported the Pleasanton marsh complex. This water was piped to San Francisco by the Spring Valley Water Company. (#82211, © San Francisco Public Utilities Commission)

Alameda Creek

Water diversions began even earlier on Alameda Creek. From the earliest European settlement on the Niles Cone, people used shallow wells and ditches to take water from Alameda Creek, Mission Creek, and small hill drainages (Bryant [1848]1985). As early as 1841, Vallejo's Mill at Niles diverted water from Alameda Creek, disrupting flow patterns (Kern 1983, Sandoval 1985).

In the 1860s water companies began to develop artesian wells near the mouth of Alameda Creek, and the San Francisco-based Spring Valley Water Company (SVWC) began working to acquire water rights in the area (Williams 1912, CA Department of Transportation 1998). Local landowners felt sufficiently threatened by water export to band together and incorporate to form Washington and Murray Township Water Company in 1871. This water company created a ditch to irrigate lands south of the creek (Lee 1916b:2823, CA Department of Transportation 1998). In 1888 SVWC began piping Alameda Creek water from Niles to San Francisco (Clark 1901, Dockweiler 1912b). Some engineers were not convinced that diversions and groundwater pumping would have an effect on groundwater levels. As late as 1889, the *San Francsico Chronicle* (1889a) reported Professor Hilgard stating, "The artesian wells do not derive their waters from the creek, nor, indeed from any common source."

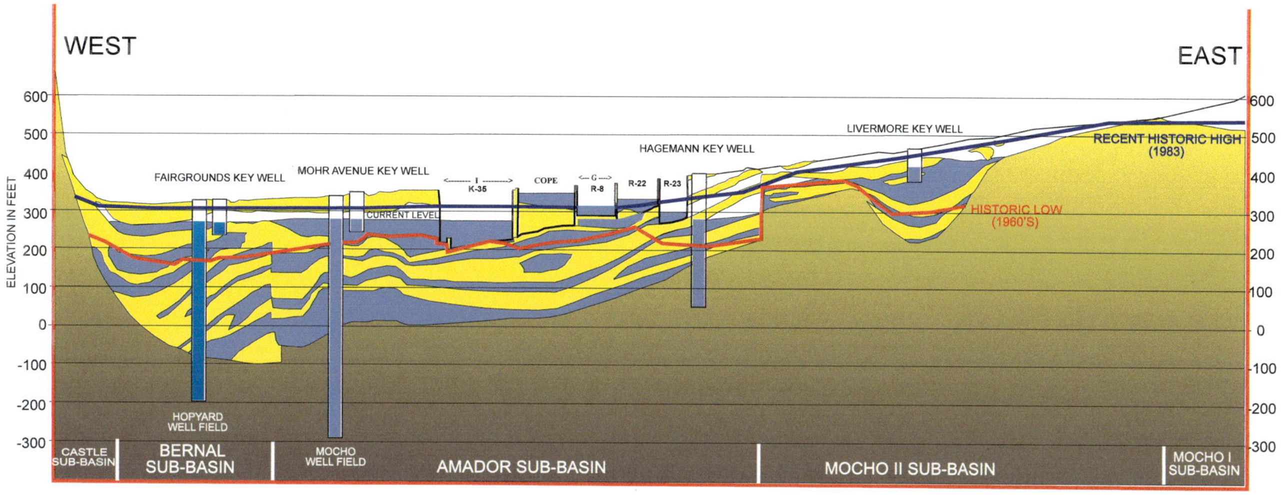

Figure 3.15. Groundwater in Livermore-Amador Valley. This graphic shows the groundwater (blue) and groundwater sub-basins of Livermore-Amador Valley and the historical low for groundwater levels, reached in 1960. (graphic from Zone 7 Water Agency 2011)

By 1900, the diversion point for SVWC water from Alameda Creek had been moved upstream from Niles to Sunol Dam at the head of Niles Canyon. The Sunol Dam allowed for more complete capture of water in an extensive set of underground filter galleries in Sunol Valley that collected water from both Alameda Creek and Arroyo de la Laguna (see box 5.1). In addition, SVWC was beginning to export water from the Pleasanton wells, which was piped downstream to the Sunol filter galleries (Williams 1912, Lee 1916b). Calaveras Reservoir was completed in 1925, further limiting flows downstream. The Report of the Special Committee of Merchants Exchange reported in 1905 that "the entire flow of Alameda creek, both above and below the surface, was annually stopped from the latter part of June until November, inclusive" (in Dockweiler 1912b:39). A 1937 report stated that "in 10 years out of 19 [from 1919 to 1937] the consumption of water exceeded the entire runoff of Alameda Creek" (West 1937).

I remember particularly that there was an artesian well behind the little house, which ran continuously. This ran into a ditch, and down on into the marsh. As the pumping in the valley get heavier, water level dropped, and I would think that the artesian flow stopped sometime around, perhaps, 1915. These wells have never flowed since.

—DONALD P. PATTERSON 1977, DESCRIBING WELL NEAR OLD DUCK CLUB ON THE NILES CONE

At the same time that upstream flows were diverted, water extraction from the aquifers of Niles Cone increased, further draining water supply. Water was removed both for local agriculture, which came to rely increasingly on irrigation, and for export to other regions of the East Bay and to San Francisco (Lee 1916b, Figuers 1998). Areas that had been used to grow dry-farmed crops such as grains were instead cultivated with water-hungry row crops and alfalfa (Whitfield 1986). The dropping water table meant that orchards that previously had been able to tap into groundwater increasingly needed irrigation as well (Patterson 1977). A surge in well creation was driven by a series of dry years:

> Prior to the dry years of 1912 and 1913 there were very few irrigation pumping plants, and such as there were had been installed for the irrigation of special crops such as nursery stock or because of the owners' individual opinion in the matter. The dry years, however, have taught the value of a pumping plant on an orchard tract as insurance against drought and during the past three or four years a number of plants have been installed. (Lee 1916b:2865)

The saltwater intrusion, as I remember, started about in 1920. That was before centrifugal pumps came in. The water was primarily used for agriculture. As time went on, the water tables started to go down. You could only pump a certain height with the pumps they had. Then they developed the centrifugal pumps, so then they pumped down deeper.

—WHITFIELD 1986

Sandy Figuers (1998) reported that in 1910, SVWC was pumping 15-20 million gallons per day (mgd), and an additional 15-25 mgd were being pumped by local farmers and the People's Water Company (supplying water to the East Bay). By 1920, Bailey reported 1,450 wells in the district, "about 350 of which are fair producing wells" (Bailey 1920:5). By 1930, Oakland's water rights had been bought out in an effort to reduce the pressure on the aquifer (Whitfield 1986).

The extensive use of groundwater and diversion of surface flows combined to draw down groundwater levels. The Hayward Fault acts as a groundwater barrier, such that levels recorded upstream of the fault are often much higher than those directly below the fault (Clark 1915, Pierce 1948, Morris et al. 1960). Most water extraction occurred downstream of the fault. By the early 1900s, water companies began deepening local farmers' wells to forestall complaints about falling groundwater (Report of Special Committee of Merchants Exchange 1905 in Dockweiler 1912b), and formerly artesian wells "in the marsh or lower area ... ceased to flow. The drying up of springs is further evidence, likewise the greater depth to water in wells" (Dockweiler 1912c). Beginning in 1913, groundwater levels in portions of the Niles Cone fell below sea level (Morris et al. 1960), and in 1920 levels dipped to seven feet below sea level around Alvarado (West 1937, Figuers 1998). By 1928, portions of the Newark aquifer "contained water that was unsuitable for irrigation use" due to salt water intrusion (Morris et al. 1960:23). Thin aquitards (clay layers) and deep wells allowed water to percolate down into overdrawn deeper aquifers. By 1950, a combination of improperly sealed wells and lowered groundwater levels had allowed salt water to enter the underlying Centerville aquifer, where water levels were 50-80 feet below sea level (Figuers 1998:44). Water levels at a well drawing from the Newark aquifer reached 67 feet below sea level in October of 1961 (fig. 3.16).

Today the groundwater basin has recovered through active recharge, and is used again as water storage by the Alameda County Water District (Figuers 1998, Alameda County Water District 2001). Water from Alameda Creek and the State Water Project are used to recharge the groundwater basin and to continue the process of extracting brackish groundwater from contaminated aquifers (Alameda County Water District 2001). Private wells are also still operated across the Niles Cone (Alameda County Water District 2001).

Climate history

Climate and weather patterns in Alameda County are driven by a combination of regional physiographic characteristics, local topography and land use, the interaction of maritime and continental air masses, and global climate patterns. Over long time scales, climate changes cause shifts in habitat distribution and abundance. Shorter-term climatic variation also influences native habitat patterns indirectly by affecting land use: droughts can instigate greater reliance on groundwater, new irrigation practices, or the failure and abandonment of a crop; extreme winter floods can catalyze stream channelization or alter riparian habitat. Historical landscapes and land use change should be interpreted within the context of climate history.

Alameda County has a Mediterranean climate, characterized by high inter- and intra-annual variability in rainfall. Virtually all precipitation occurs between the months of October and April. Topography controls rainfall patterns—the lowlands within the study area receive an average of less than 16 inches of rain per year, compared with 20-30 inches in the surrounding hills (Williams 1912, DWR 2006; fig. 3.17).

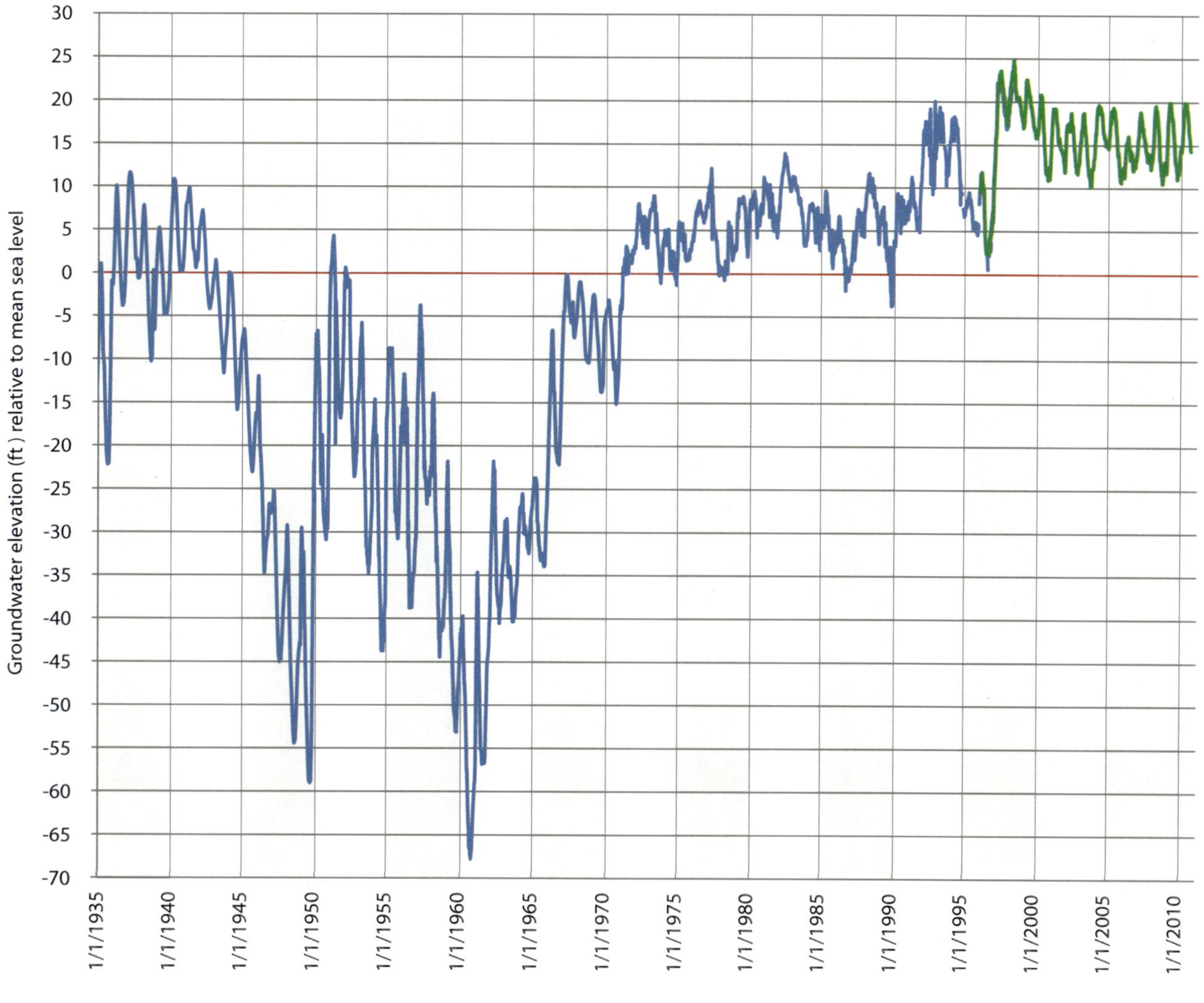

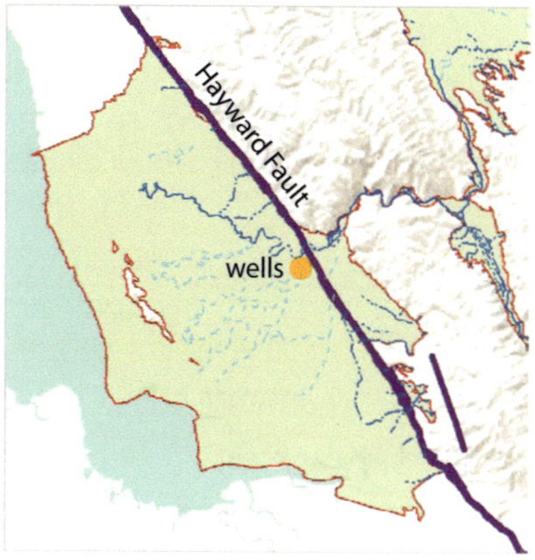

Figure 3.16. Groundwater levels in Niles Cone. These data, drawn from the longest well of record on the Niles Cone, show the pattern of falling and then recovering groundwater levels observed across the Cone. The well 28D002 (shown as a blue line) was recorded from 1935 through 1999, and well 29A006 (shown as a green line) provides the record from 1999 to 2011. Both wells draw from the Newark aquifer just below the Hayward Fault, where the ground surface is 60 feet above mean sea level. The locator map at right shows the location of the wells (both wells are represented by the orange dot). Wells were sampled once per month until 1950; since then they have been sampled once per week. Groundwater levels plunged rapidly in the late 1940s and then again around 1960. Today, through active recharge and monitoring, groundwater levels have recovered and stabilized, although the natural seasonal rise and fall of the water table is apparent. Threats of returning saltwater intrusion require the Alameda County Water District to maintain relatively consistent groundwater levels. (data courtesy ACWD)

Figure 3.17. Rainfall across Alameda Creek watershed. This map shows the distribution of rainfall across the study area, displayed as inches of average annual rainfall. Annual rainfall is highest in the hills and then decreases across the valley floors. The study area is outlined in light green. (data from Alameda County Public Works Department)

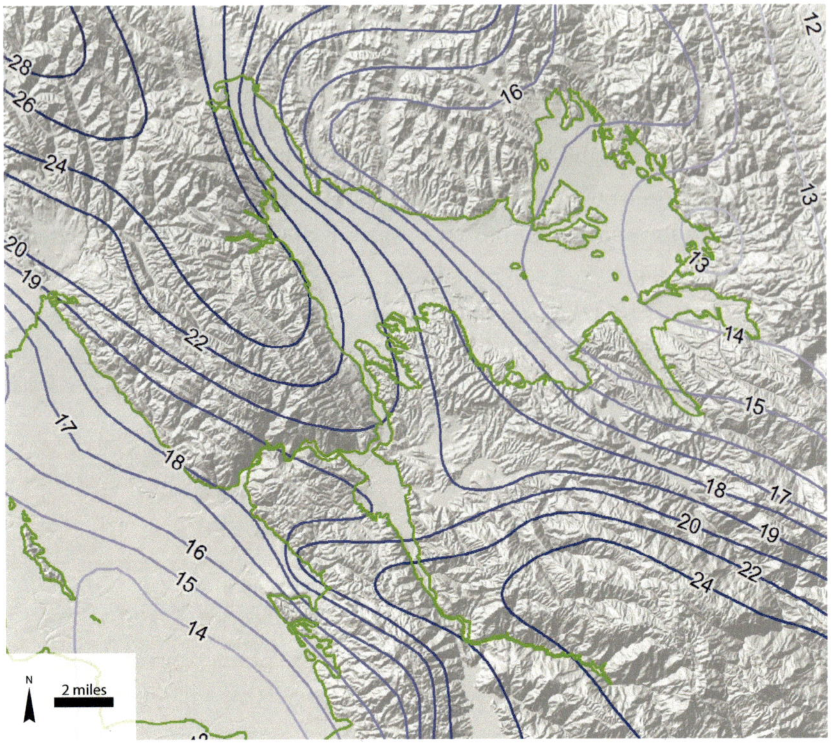

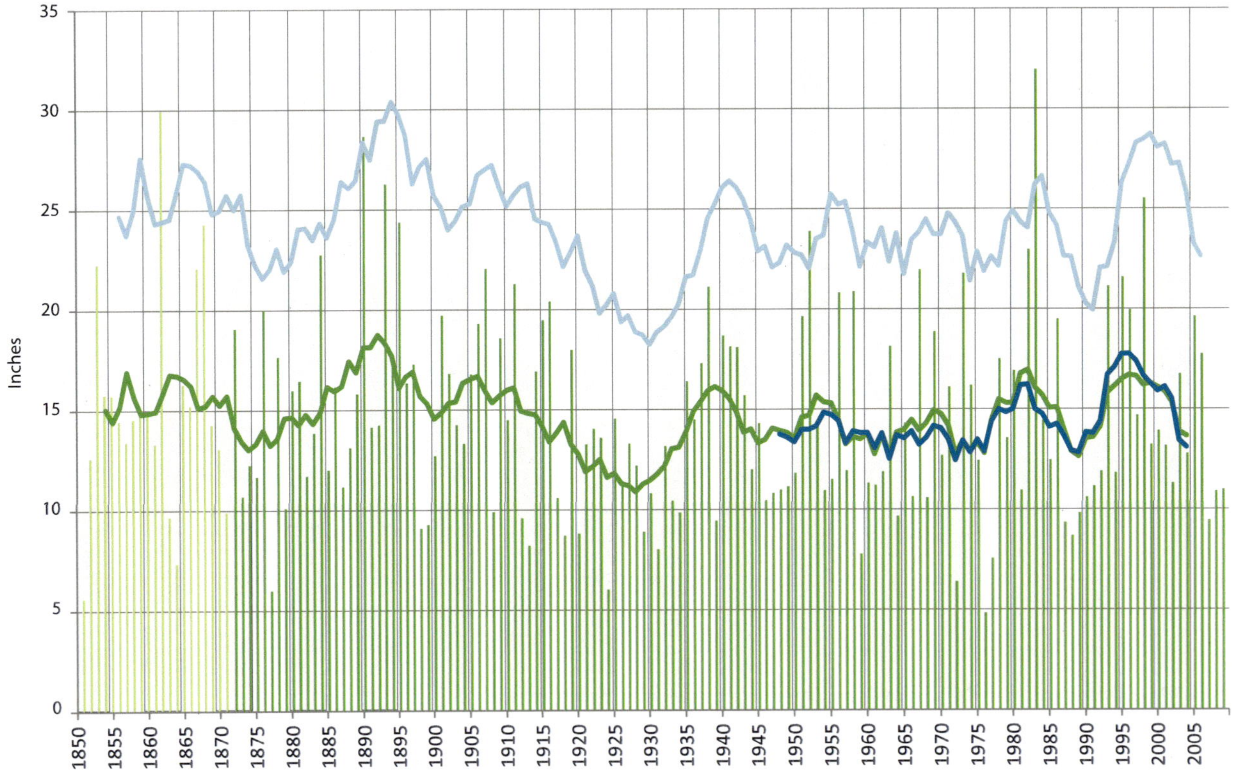

Figure 3.18. Historical rainfall data for Alameda County. Data for Livermore were extrapolated from San Francisco data for the years 1850 to 1871. The rainfall gauge in Newark began in 1942 and matches the Livermore record fairly closely, so only the mean is shown. The average rainfall at Livermore between 1872 and 2009 was 14.56 inches. The Mt. Diablo gauge is included as a proxy for the upper watershed, and is approximately 10 inches greater than Livermore and Newark (see fig. 3.17). Data are grouped by water-year, which runs from October to September, and are labeled with the year in which the water-year ends (e.g., rainfall in December 1955 shows up in water year 1956).

A U.S. Weather Bureau rainfall gauge was installed in the town of Livermore in 1871, providing an unusually long record (Williams 1912). To extend the record further back, we used the San Francisco record (beginning in 1849) and regression methods to estimate rainfall in Livermore. The resulting rainfall record is displayed in figure 3.18. For the period of record, the average rainfall in Livermore was 14.56 inches/year. We also collected rainfall data for the Newark station, established in 1942. These data tracked the Livermore data closely, indicating that within the lowlands of the study area, annual precipitation is fairly consistent. Rainfall in the upper watersheds can be an average of 10 inches/year higher.

Prior to the American period, rainfall patterns can be estimated using other sources. Paleoclimate records (including tree rings and accumulated sediment deposits) indicate that 4,000-2,000 years ago, the Bay Area experienced unusually cool and wet conditions compared to prior millennia, resulting in rapid sediment deposition and tidal marsh development (Malamud-Roam et al. 2006). In response to these conditions and the end of rapid sea level rise, tidal marshes in the Bay expanded through the period from 3,500 years ago to the 1800s (Malamud-Roam et al. 2006). More recently, reconstructions for California show that the years 900-1350 CE were a relatively dry period, followed by a "Little Ice Age" from 1400-1800 CE during which there were cooler temperatures and higher average annual rainfall (Raab and Jones 2004, Stahle et al. 2001). For at least the last 270 years, precipitation patterns have been relatively stable in western North America, despite inter-annual and decadal variability (Dettinger et al. 1998, Malamud-Roam et al. 2007), although other analyses of tree ring data suggest that the early historical period (1760-1820) had below average rainfall in California (Fritts and Gordon 1980).

Neither these long-range climate trends nor more recent historical data capture the intensity of individual storm events, which may have more impact on flooding and sediment sourcing than the total annual rainfall, but they do point to trends and abnormally wet or dry years which can affect wetland size/duration and dry season stream flow. A richer picture can be gained from the addition of textual accounts. For example, damaging floods were noted in December 1889, November 1892, January 1895, March 1907, January 1909, March 1911, January 1914, February 1938, November 1951, January 1952, December 1955, and April 1958; all but two of these years had rainfall over 20 inches (*San Francisco Chronicle* 1889b, Branner 1912, USACE 1961). Textual accounts help highlight particularly notable climatic events, including the flood of 1862 and following drought, which catalyzed land use changes through California. The relatively dry period in the early 1900s coincided with (and exacerbated) intensive groundwater pumping in Alameda County, contributing to groundwater overdraft. An understanding of rainfall data, along with land use patterns, provides important context to aid the interpretation of historical documents.

Highways and railroad beds have been washed out in the county, and streams and marshes are overflowed. The Alameda creek is full from bank to bank, while the water is very high at Alvarado. The tidal canal, or such part as has been excavated, has several feet of water in it, the banks have caved in, and several of the dredges wrecked and the railroad track torn up.

—*SAN FRANCISCO CHRONICLE* 1889B, DECEMBER 12

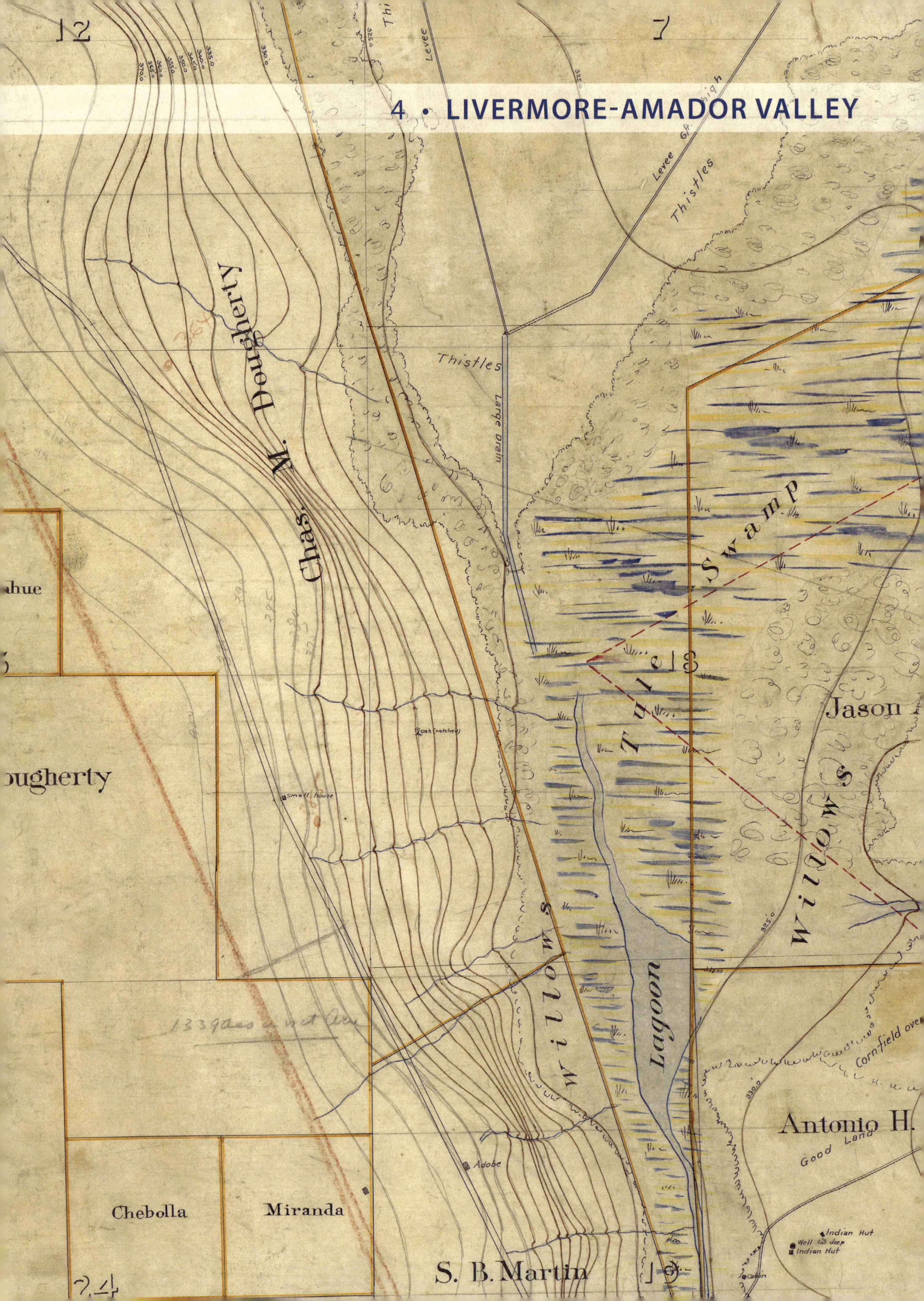
12
7
Levee
Thistles
Chas. M. Dougherty
Thistles
Large Drain
Tule Swamp
18
Jason
Willows
Dougherty
Small house
Lagoon
Willows
Cornfield
Antonio H.
Good Land
Adobe
Chebolla
Miranda
S. B. Martin
Indian Hut
Well to deep
Indian Hut
24
19

Livermore-Amador Valley has developed rapidly in recent decades, today containing a complex mix of urban and suburban development, agriculture, and rangelands. Two centuries ago, the valley was dominated by vast open expanses of grasslands and seasonally flooded wet meadows. Two large wetland complexes occurred at low points in the valley, one east of present day Livermore and one west of Pleasanton (fig. 4.1). Two broad braided creeks, Arroyo del Valle and Arroyo Mocho, drained from the Diablo range into the southern portion of the valley, while a series of smaller creeks with narrower channels drained into the valley, particularly from the north. This chapter explores the historical ecology of the Livermore Valley area, including Livermore and Amador valleys and southern San Ramon Valley (fig. 4.2). (The western portion of the valley, including Pleasanton, is known as Amador Valley, while the eastern portion, including Livermore, is Livermore Valley. Together they are known as the Livermore-Amador Valley.) Today this area contains the cities of Livermore, Pleasanton, and Dublin, as well as the southern edge of San Ramon.

The soil of the [Livermore] valley is generally good on the borders, but towards the center it is either wet or heavy and withal somewhat saline, on the higher parts dry and gravelley...On the hills that surround this basin are to be found fossiliferous sand stones and among the alluvium in some localities are to be found considerable quantities of fragmentary shells.

—TRUSK 1854 IN WILLIAMS 1912:40

Distinct habitats divided the valley both from north to south and from east to west, responding to persistent physical gradients. The northern valley was characterized by fine-grained clay soils, which flooded each year to form extensive wet meadows, or seasonal wetlands. In contrast, the southern valley was made of more well-drained soils, which supported dry grasslands. Moving from east to west, high alkali concentrations in the wetlands east of Livermore contrasted with relatively fresh wetlands in the western valley, where perennial water was available. The following paragraphs describe these gradients and the habitats they supported in more detail.

Geology and soils largely controlled the distribution of habitats across the valley (see fig. 9.2c). The fine-grained Great Valley Sequence composes the hills to the north of Livermore-Amador Valley, while the coarse and erosive Franciscan Complex makes up the hills to the south (Graymer et al. 1996). Formations to the north and east contain salty marine deposits, which erode and wash downstream to create high levels of salinity in the eastern valley floor (Westover and Van Duyne 1911, Carpenter et al. 1984, Edwards and Thayer 2008, Mikesell et al. 2010). These differences in source geology were reflected in the soils of the valley floor. Clay and silty soils with some alkali influence were found across the northern half of the valley. Coarse loams and gravelly loams covered the southern valley (Westover and Van Duyne 1911; fig. 4.3)

Figure 4.1. Pleasanton marsh complex. One of the most prominent historical features of Livermore-Amador Valley was the 2,600 acre Pleasanton marsh complex, shown here in a highly detailed map based on an 1880 survey. The marsh complex supported a rich variety of habitats and wetland types. By the early 1900s it had been cleared and drained. (Allardt [1880]1907, courtesy The Bancroft Library, UC Berkeley)

The topography of Livermore-Amador Valley is controlled by the movement of faults and uplift of hills (box 4.1). To the west the valley is bounded by the East Bay hills, flanked by the Hayward and Calaveras faults; to the east the valley is bounded by the Altamont Hills and the Greenville Fault; and to the north and south the valley is bounded by the Diablo Range. A number of smaller faults and thrusts cross the valley,

HISTORICAL LAND COVER, EARLY-1800s

- Valley Freshwater Marsh
- Pond
- Wet Meadow
- Willow Thicket

- Alkali Meadow
- Alkali Sink Scrub-Vernal Pool Complex
- Alkali Playa Complex

- Sparsely Vegetated Braided Channel
- Confined Riparian Woodland/Savanna
- Sycamore Alluvial Woodland

- Oak Savanna
- Grassland

- Perennial Stream
- Intermittent Stream
- Slough/ Overflow Channel
- Distributary
- Spring

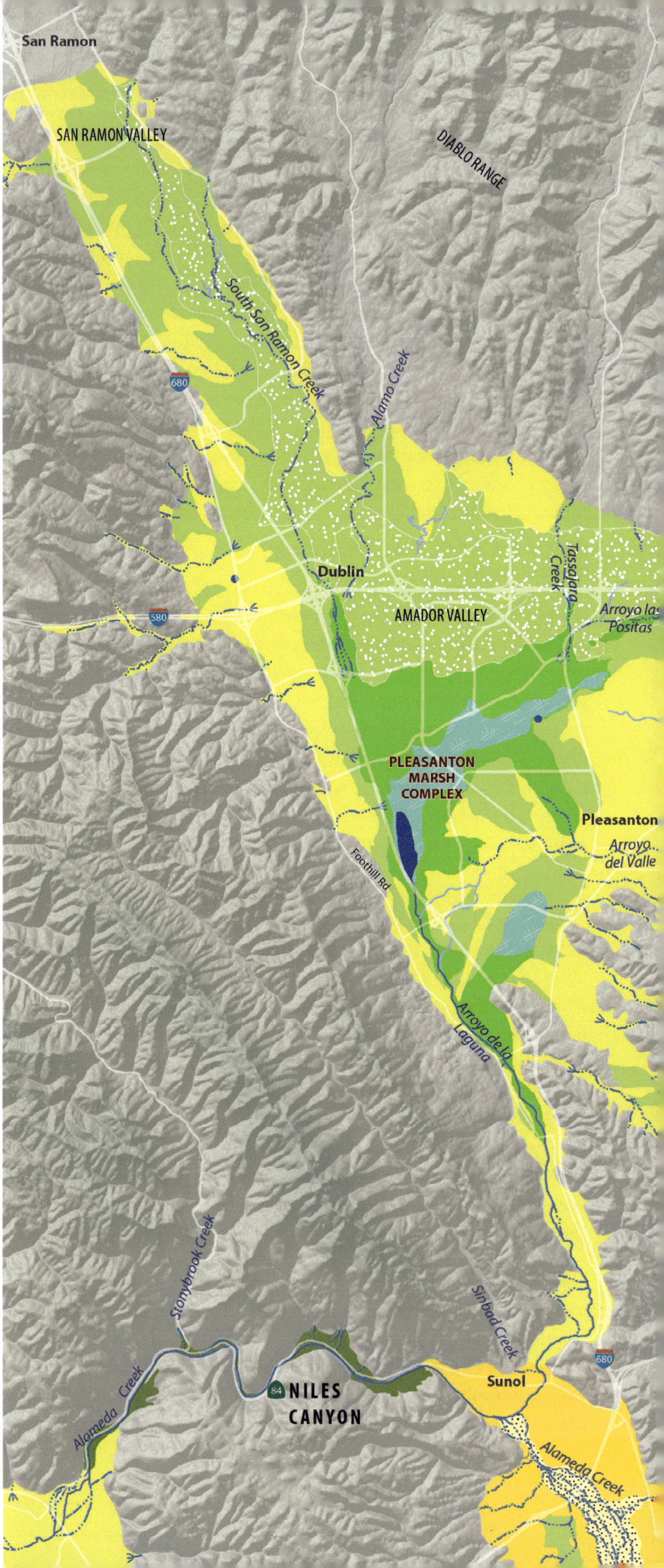

Figure 4.2a. Livermore-Amador Valley ca. 1800. Patterns across Livermore-Amador Valley varied from north to south and from east to west. The two wetland complexes—Pleasanton marsh complex in the west and Springtown alkali sink in the east—were prominent historical features.

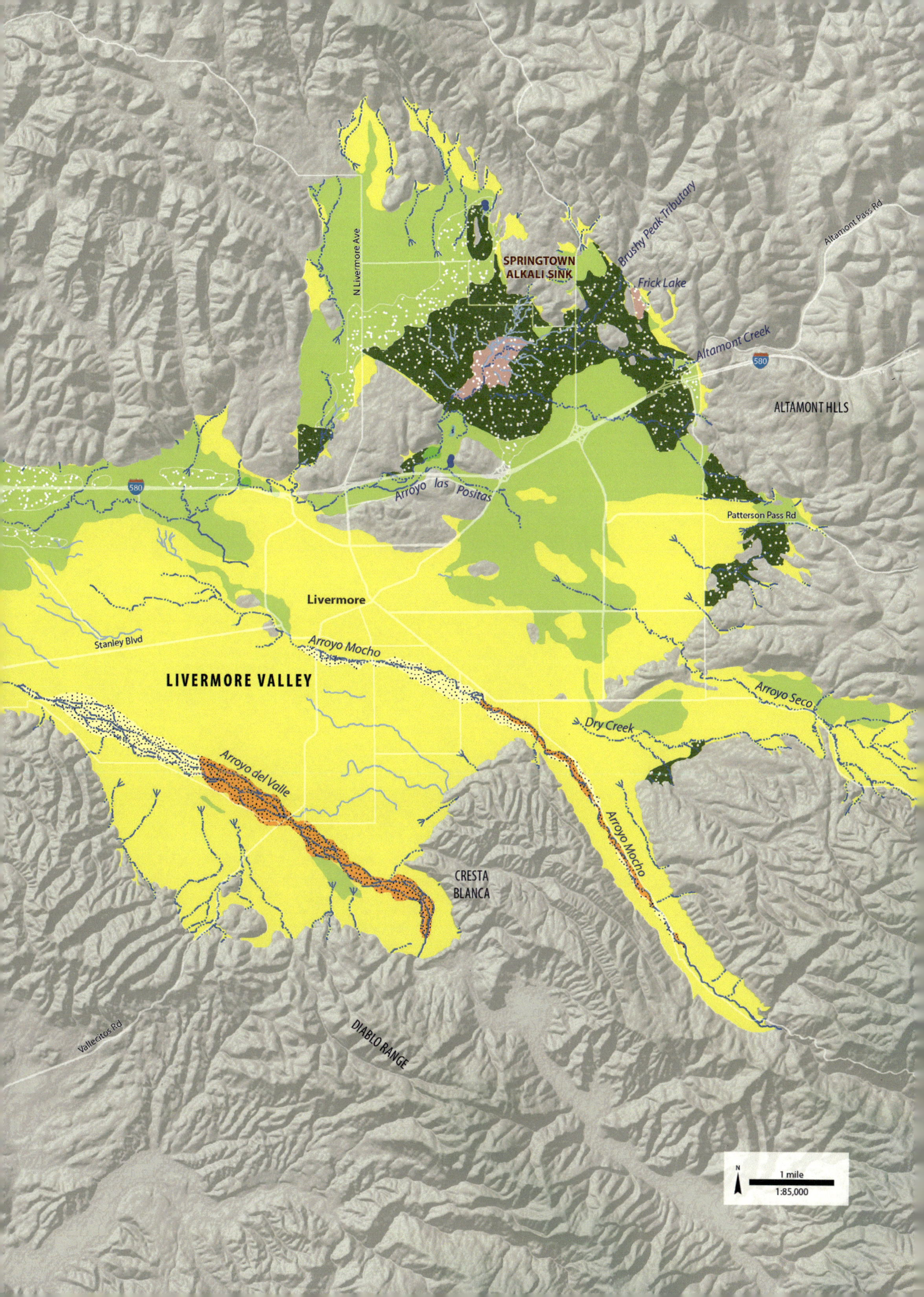

N Livermore Ave
SPRINGTOWN
ALKALI SINK
Brushy Peak Tributary
Frick Lake
Altamont Pass Rd
Altamont Creek
580
ALTAMONT HLLS
580
Arroyo las Positas
Patterson Pass Rd
Livermore
Stanley Blvd
Arroyo Mocho
LIVERMORE VALLEY
Arroyo Seco
Dry Creek
Arroyo del Valle
Arroyo Mocho
CRESTA
BLANCA
Vallecitos Rd
DIABLO RANGE
N
1 mile
1:85,000

Figure 4.2b. Livermore-Amador Valley 2009. The red outline represents the study area. Blue lines represent BAARI stream mapping (SFEI 2011) within the study area. (USDA 2009)

SPRINGTOWN ALKALI SINK
Brushy Peak Tributary
Frick Lake
Altamont Creek
580
Altamont Pass Rd
ALTAMONT HLLS
N. Livermore Ave
Arroyo las Positas
Patterson Pass Rd
Livermore
Stanley Blvd
Arroyo Mocho
LIVERMORE VALLEY
Arroyo Seco
Dry Creek
Arroyo del Valle
Arroyo Mocho
CRESTA BLANCA
SYCAMORE GROVE PARK
LAKE DEL VALLE
DIABLO RANGE
Vallecitos Rd
1 mile
1:85,000

Figure 4.3. Gravel beds in Livermore Valley, early 1900s. These coarse deposits contributed to the rapid drainage in southeastern Livermore Valley and were frequently noted by early visitors. These gravels were transported from the headwaters of Mocho and Del Valle. (courtesy The Bancroft Library, UC Berkeley)

interrupting the flow of groundwater and creating springs. (For more complete treatment see Vickery 1925, Fisher et al. 1966, Ford and Hills 1974, Carpenter et al. 1984, Unruh and Sawyer 1997, Sawyer 1999, Figuers and Ehman 2004.)

Topography in turn influenced habitat distribution. Bedrock exposures protrude above the valley floor at two significant points in the study area. These hills interrupt the flow of ground- and surface-water, creating poorly-drained areas, which facilitated the development of wetlands. East of Livermore, the Springtown alkali sink was located behind the knolls of the Springtown anticlines, formed by late Quaternary uplift (Sawyer 1999). The basin west of Pleasanton, confined by the East Bay hills and

BOX 4.1. FORMATION OF THE LIVERMORE-AMADOR VALLEY

The Livermore-Amador Valley is an east-west down-dropped depression between the Diablo Range and the East Bay hills (Howard 1979, Carpenter et al. 1984, Sloan 2006). Because it was created by geological processes rather than carved by streams, the valley is oversized for its fluvial inputs, providing streams space to spread and sink. The valley is filled with sediments deposited from the Pliocene through the Holocene which consist of a series of layers of silt and clay alternating with gravels (Williams 1912, Fisher et al. 1966, Sowers 2003, Sloan 2006). These layers record the dynamic geologic and geomorphic processes over time.

Underlying what is now Livermore-Amador Valley are Sierran deposits carried from the volcanically active mountain ranges in the late Pliocene (Ferriz 2001, Sloan 2006). Livermore-Amador Valley formed when the surrounding hills began to lift between 6 and 2.5 million years ago (Ferriz 2001, Sloan 2006). The drainages created by this continued uplift contributed sediment, known as the Livermore Gravels, which resulted in deposits up to 4,000 feet thick in places (Howard 1979, Sloan 2006). Over the millennia, watercourses shifted between flowing north through San Ramon Valley and flowing south through Sunol (Williams 1912, Ferriz 2001). At times the San Ramon Valley route was blocked, allowing sediment to settle in the low points of the valley and resulting in the deposition of layers of silt and clay (Williams 1912, Ferriz 2001, Sloan 2006).

The shape of the alluvial Livermore-Amador Valley floor has changed over time in response to tectonic activity and paleo-fluvial processes, which altered the depositional patterns across the valley floor. Researchers suggest that Pliocene-age uplift of the Diablo Range created alluvial fan deposits along the edges of the Livermore-Amador Valley. Concurrently, Arroyo Mocho, the largest stream in the valley at that time, deposited Livermore Gravels in a northwest trending direction and finer sediment along the northern edge of the valley (Figuers and Ehman 2004). The ancestral stream took a broad turn to the south where the East Bay hills blocked its course. Although Arroyo del Valle is currently the largest stream entering the valley, it cuts perpendicularly across the Livermore Gravels, indicating that the early depositional environment was controlled primarily by Arroyo Mocho (see fig. 9.2c; Figuers and Ehman 2004). In more recent times, Arroyo del Valle has dominated deposition across the valley, spreading coarse gravels along the length of its fan (Williams 1912, Fisher et al. 1966:17).

alluvial fans formed in the Pliocene, developed into the Pleasanton marsh complex (fig. 4.4). The Pleasanton marsh complex occurred at the lowest point in Livermore-Amador Valley, and as a result received more ground and surface water inputs and was more consistently saturated than the Springtown sink.

At a local scale, microtopography influenced habitat distribution. The presence of small mounds and depressions altered the flow of water and saturation level in the soils. Areas of slightly lower elevation within wetlands developed into ponds; patches of higher elevation supported willow thickets rather than marsh.

The dry climate of Livermore-Amador Valley also shaped vegetation patterns. On average, the Livermore-Amador Valley experiences higher temperatures than the Niles Cone, with similar rainfall, resulting in higher evaporation rates and contributing to the existence of alkali habitats.

Figure 4.4. Livermore-Amador Valley Feb 20, 1955. This image, taken in the aftermath of the 1955 flood, clearly shows the courses of Arroyo del Valle (at left) and Arroyo Mocho (center) through the valley. Towards the top (west) of the photograph, darker areas reveal traces of the Pleasanton marsh complex, even though the marsh itself has been completely drained and cleared by this time. (P-226, courtesy Pleasanton's Museum on Main)

People have also shaped the ecology of Livermore-Amador Valley. At the beginning of the historical period, the Chochenyo Ohlone managed the Livermore-Amador Valley landscape (Treutlein and Fages 1972, Crespí and Bolton 1927). As Europeans moved in, residents of Mission San José used the valley to pasture cattle; grazing persisted as the dominant land use for decades. The mid-1800s saw waves of early immigrants settling to raise grain after returning from the Gold Rush. Although a few orchards and row crops were introduced in the late 1800s, grain culture continued across most of the valley as late as the 1940s, a trajectory that differs from that seen in neighboring valleys such as the Santa Clara Valley, or even the Niles Cone—which converted to orchard culture and row crops by the early 1900s (USDA 1940, Grossinger et al. 2008). Today Livermore-Amador Valley is home to over 200,000 people, and growing rapidly.

Combined, these factors of geology, soils, topography, climate, and land use controlled the distribution of habitats (box 4.2). In this section, we discuss each of the major historical habitat types in Livermore-Amador Valley: oaks and herbaceous cover; wetlands; and creeks and riparian habitats (table 4.1, fig. 4.5).

Table 4.1. Historical land cover in Livermore-Amador Valley. The table below summarizes our findings for land cover patterns in Livermore-Amador Valley circa 1800. Numbers were rounded to the nearest 50 acres.

Figure 4.5. Simplified chart of historical land cover in Livermore-Amador Valley. The pie chart presents a simplified summary of the data in the table. Wetland types have been grouped into seasonal and perennial, and riparian habitat types are not shown.

Habitat Type	Acres
Grassland	***25,900***
Seasonal Alkali Wetland	***8,250***
Alkali Meadow	5,150
Alkali Sink Scrub-Vernal Pool Complex	2,900
Alkali Playa Complex	200
Seasonal Freshwater Wetland	***11,350***
Wet Meadow	11,350
Perennial Freshwater Wetland	***650***
Perennial Pond	50
Valley Freshwater Marsh	600
Riparian Vegetation and Willow Thicket	***3,250***
Sparsely Vegetated Braided Channel	650
Sycamore Alluvial Woodland	600
Willow Thicket	2,000

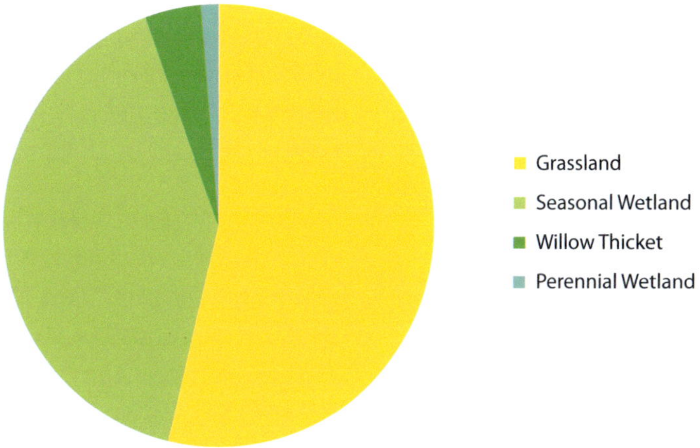

BOX 4.2. ASSESSING EARLY MODIFICATIONS TO LIVERMORE-AMADOR VALLEY

By the time of the earliest detailed mapping in the 1870s, Livermore-Amador Valley had already experienced some impacts of Euro-American settlement. Although wholesale drainage of the Pleasanton marsh complex was not yet initiated by this time, many previously discontinuous channels had already been connected. Cattle from Mission San José and the early Spanish ranchos grazed in the area beginning around 1800, likely contributing to increased erosion. Settlers had begun clearing trees for firewood and diverting water for irrigation and drainage.

The Pleasanton marsh complex is an example of a feature that could conceivably have developed due to high rates of anthropogenic erosion and the resulting sedimentation of a former open water lake. However, the historical record contains strong evidence that this feature was not created as a result of Euro-American modifications. Even the earliest historical records contain evidence of a large freshwater marsh, rather than a lake. In 1772 Spanish explorer Fages, one of the first Europeans to enter the valley, recorded the presence of "many tulares and lakes" in Amador Valley (Fages 1772 in Treutlein and Fages 1972:353). This description matches mid-19th century descriptions (see Pleasanton marsh complex p. 87, see box 4.5) and early maps. Comparing different types of data from multiple sources and timeframes helps in assessing the persistence and variability of features through the historical record.

Oaks and grassland

This is one of the prettiest places in the State ... The beautiful carpet of green, dotted by oak trees and pretty country houses, and sprinkled with fragrance exhaling flowers of variegated hues.

—*SAN FRANCISCO CHRONICLE* 1870, DESCRIBING LIVERMORE-AMADOR VALLEY

A first impression of Livermore Valley in 1800 was of vast grasslands. Native grasses, wildflowers, and other low herbaceous cover grew across the well-drained gravelly land of the southern valley, while herbaceous wet and alkali meadows covered the moister northern half of the valley (fig. 4.6). Early accounts contain glowing descriptions of the native wildflower displays across the valley floor. In May 1860 a traveler described a valley "covered with a deep green carpet of grass and clover, here and there hidden by the abundance of flowers" (*Daily Alta California* 1860a), while in 1878 there were "beautiful variegated wild pansies, the lupin[e] and California poppy" (Williams et al. 1878:257).

In addition to the colorful wildflower displays, there were a mix of grasses and forbs (Minnich 2008). A traveler passing through described the "fine grassy camping plain" near Livermore's house (Lyman 1848). As late as 1860, a visitor described six distinct species of clover in a handful of flowers and grasses (*Daily Alta California* 1860b). Rhizomatous ryegrasses were likely a dominant component of these communities, particularly in clay loam soils (see Holstein 2000, Holstein 2001, Minnich 2008).

The most popular of the indigenous flowers is the escholtzia, *or California poppy, and during the months of April, May, and June the uncultivated fields and hills are covered with this beautiful flower, often remaining in bloom until July and August.*

—COQUHOUN 1893, DESCRIBING LIVERMORE-AMADOR VALLEY

These rich and varied grasslands contained only sparse tree cover. The absence of trees and scrub may have been shaped in part by Ohlone burning. Travelers making their way to the mines complained about a lack of firewood (e.g., Ryan 1850). This was more extreme towards eastern Livermore Valley; travel author Bayard Taylor (1850) describes the transition from the oak savanna of Sunol Valley to the dry alkaline eastern valley edge:

> The first twenty miles of our journey passed through one of the most beautiful regions in the world. The broad oval valleys, shaded by magnificent oaks and enclosed by the lofty mountains of the Coast Range, open beyond each other like a suite of palace chambers, each charming more than the last ...
>
> We passed from these into hot, scorched plains, separated by low ranges of hills, on one of which is situated Livermore's Ranche.

Oaks also grew on the hillsides surrounding the valley, as they do today, but non-riparian oaks on the valley floor were sparse. Descending through San Ramon Valley into Livermore-Amador Valley in 1772, Spanish explorer Fages reported that "the entire valley was of very good soil and very grassy, and with some white-oaks [valley oaks] and live-oaks on the hills on both sides of it," although he did not describe oaks on the valley floor (Treutlein and Fages 1972). Most of the dense tree cover visible in early photography occurred along the major creeks, especially Arroyo Mocho and Arroyo del Valle (fig. 4.7).

The lack of oaks and other trees may have been partially due to the soils. In the northern half of the valley, seasonal wetlands underlain by a claypan created soil saturation levels too high for oaks. In the southern valley the

coarse, gravelly soils may not have retained enough moisture for oaks (Davis et al. 2000; see p. 95 for discussion of seasonal wetlands). The plain was described in the late 1800s as "very gravelly" (Likins 1874), containing "much thin, gravelly land, unfit for tillage and of little value for pasture" (*Daily Alta California* 1860c). Surveyor Sherman Day noted "gravelly plain. No bearing trees ... Surface level and rather hard in summer ... Light growth of pasture grasses" (Day 1853:252). Few maps show oaks, an indication that there were no groves of local significance; in nearby valleys that had substantial oak cover, high density groves were consistently mapped (see Grossinger et al. 2008, Beller et al. 2010, Stanford et al. 2011, Grossinger 2012).

Figure 4.6. Grasslands of Livermore-Amador Valley. A mix of grasses and wildflowers likely covered the floor of Livermore-Amador Valley historically. Today (clockwise from top left) annual grasses, goldfields, and clover can be found on the valley floor and hillsides. (photos by Amy Wolitzer, 2011)

The coarse soil and lack of readily available water across the southern valley restricted agriculture as well as oaks, resulting in a relatively late conversion to more intensive crops (e.g., orchards). As late as 1890 the

Figure 4.7. Corridors of trees are visible along Arroyo Mocho and Arroyo del Valle in this 1889 image, looking southwest across Livermore-Amador Valley. Additional oaks are scattered thinly in the distance, intermixed with agricultural fields. In the foreground, planted exotic trees line the streets of Livermore. (courtesy Livermore Heritage Guild)

San Francisco Chronicle reported that hay and grain were the "principal industry in the valley."

However, occasional oaks did grow across Livermore-Amador Valley. A report from the 1880s describes the valley "dotted over with oak trees" and an 1870s article attempting to promote settlement described Livermore "beautifully nestled amid sturdy oaks" (Williams et al. 1878:257; fig. 4.8, fig. 4.9). In addition to these scattered oaks within grassland, General Land Office (GLO) surveyors recorded non-riparian oak groves in three places, suggesting the presence of small stands of higher density. One GLO survey transect crossed two oak groves east of Pleasanton, to the north and south of Arroyo del Valle, with notes including "in oak grove on gravelly plain" and "line passes again across level plain among scattered oaks" (Day 1853:253). A second transect occurred at the far eastern edge of the valley, and recorded a few additional oaks as bearing trees—"timber white [valley] oak" (Dyer 1862b) and "timber scattering oak" (Dyer 1869). While these data certainly indicate oaks as a component of the Livermore-Amador Valley landscape, we find much less consistent and extensive descriptions of a valley oak dominated landscape here than in other Bay Area valleys (see Grossinger et al. 2008, Grossinger 2012, Point of Timber in Stanford et al. 2011), suggesting a more sparse pattern (box 4.3).

Wetlands

In the mid-1800s, Livermore-Amador Valley contained an estimated 19,600 acres of seasonal wetlands, with an additional 650 acres of perennial wetlands and 2,000 acres of willow thicket or swamp. These wetlands

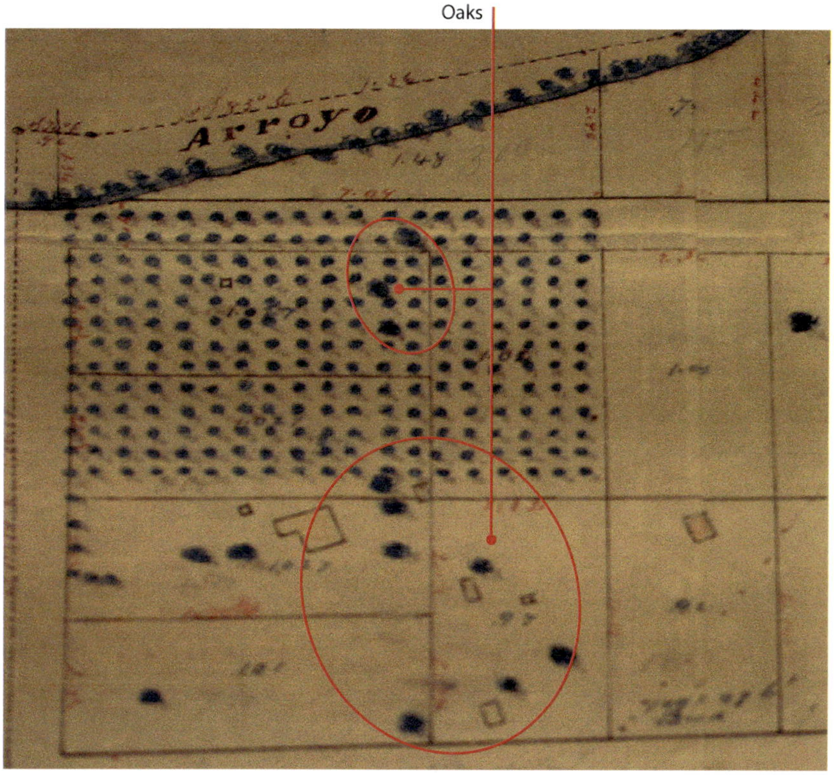

Figure 4.8. Scattered trees. This map shows trees on a property near First Street and Hopyard Road in Pleasanton. Riparian trees line Arroyo del Valle, and additional scattered trees (likely oaks) spread further south. Some oak trees were left standing within the orchard. Low density clusters of oaks did occur in Livermore-Amador Valley, but were so sparse that we classified them as grassland (a class that can contain occasional oaks). Note the irregular pattern of the oaks contrasted with the orchard. (courtesy Pleasanton's Museum on Main)

Figure 4.9. Low density oaks persisted over time. Most remnant oaks visible on historical aerial photographs occur at such low densities that they are best classified as grassland (which contains occasional oaks). This image looks east across the valley from the hills west of Pleasanton at a patch of low density oaks scattered through agricultural land and along a swale. Using additional photographs, we calculated these oaks at a density of fewer than three trees per acre. (#87511, © San Francisco Public Utilities Commission)

dominated the northern side of the valley, stretching from east to west in a nearly continuous swath.

Variation in rainfall both seasonally and from year to year created substantial spatial variation in the extent of wetlands. In the winter, much of the valley would have been saturated, while in summer the seasonal wetlands dried to hard, grassy land.

Extensive wetland complexes developed in the two large basins of the valley. The salt-influenced Springtown sink in the eastern basin supported seasonal alkali wetlands with low scrub and herbaceous vegetation. The Pleasanton marsh complex in the western basin contained many springs and supported a large area of open water, seasonal wetlands, and extensive willow thickets and freshwater marshlands. In both wetland complexes, the most saturated habitats occurred at the lowest point of the basin. We use our habitat types to capture some of the variation in species assemblages created by these complex spatial patterns and physical drivers. This section describes these

BOX 4.3. WOODCUTTING

Early accounts of Livermore-Amador Valley described a lack of trees that could be converted to lumber for construction (e.g., Halley 1876:493). Wood from gnarled and branching oaks and sycamores was more suitable for fuel. People reported streamside oaks and sycamores that were "chiefly valuable for fuel ... being generally too brittle for building or mechanical purposes" (Faulkner 1866).

Although building-quality wood was historically sparse, this clearing for fuel did result in a loss of trees, likely primarily along creeks (fig. 4.10). Historian Wood stated that "along the banks of the creeks are many good-sized trees, mainly oaks and sycamores, the wood from the latter having given much satisfaction, while generally the timber is used as fuel for home consumption and foreign export" (Wood 1883:458). By 1910 the soil survey described that in several locations "the native vegetation, except a few valley oaks ... [has] been removed," and "in former years it supported a few valley oaks" (Westover and Van Duyne 1910:53). However, these descriptions refer mostly to riparian oaks and sycamores, and no accounts indicate that there was ever substantial timber across the valley.

Figure 4.10. Woodcutting. This undated photo shows wood harvest in San Ramon Valley.

two large wetland complexes, finishing with a short discussion of the surrounding wet and alkali meadows.

Springtown alkali sink

The Springtown alkali sink was composed of 3,500 acres of alkali wetlands east of Livermore. The sink contained a variety of alkali wetland types, with the highest alkali concentrations and most saturated habitats in the middle (alkali playa, alkali sink scrub) and the less saturated habitats with lower alkali concentrations along the edge (vernal pool complex, alkali meadow).

Habitat types in the sink were intermixed. Many small mounds and depressions covered the surface, creating small-scale variation in alkalinity and inundation frequency, and resulting in complex variations in habitat type. The mounds and depressions of the vernal pool landscape were responsible for much of this complexity, along with swales created by small drainages (fig. 4.11). This variability resulted in small patches of intermixed habitat types within the larger gradient from the wet, alkali center of the basin to the edge. Each of the alkali habitat types we show on our map represents a combination of intermixed habitat types, or complexes.

The Springtown alkali sink formed in a small basin that was divided from the rest of Livermore Valley by small hills and a fault (Springtown anticline and fault; Ferriz 2001, Sawyer and Unruh 2004, Unruh and Sundermann 2006). Water and salts flowing downstream were constricted between the small hills. This limited passage created a backup behind the hills so that water spread, forming wetlands. As with other wetland areas, an underlying hardpan or compact clay layer was found across much of the area's soils. As a result, they were slow to drain (Westover and Van Duyne 1911).

A defining characteristic of the wetland complex was its high concentration of alkali salts. Alkali levels in this region were greater than 0.2% (in the first six feet) over an almost 3,000 acre area, and they ranged up to over 1% in some places (Westover and Van Duyne 1911; fig. 4.12). Alkali salts originated in marine sedimentary rocks with relatively high salt concentrations to the north and east of Livermore (Westover and Van Duyne 1911, Carpenter et al. 1984, Edwards and Thayer 2008, Mikesell et al. 2010). These marine sediments eroded and flowed downstream through Altamont Creek and other small drainages. As these creeks overflowed and spread during high flows, they spread minerals across the land of the basin. Over time, high evaporation rates and low rainfall resulted in highly concentrated salts (Coats et al. 1988, US DOE 2004). The salts percolated down through the groundwater—one account described a well that was bored to 640 feet, only to produce water that "shot up through the well and rose to 40 feet above ground level ... the water was strongly alkaline, and killed all vegetation in the vicinity" (Williams 1912:48).

The limestone formation is probably quite extensive in this region, as the water and the soil in the northeasterly section of the valley are quite alkali, the water from some wells being unfit for domestic use, for stock, or for irrigation.

—WILLIAMS 1912:47

SPRINGTOWN SINK HABITATS The resulting alkali-influenced habitat mosaic included several habitat types that were historically common in parts of California but are becoming increasingly rare (Coats et al. 1993, Holland

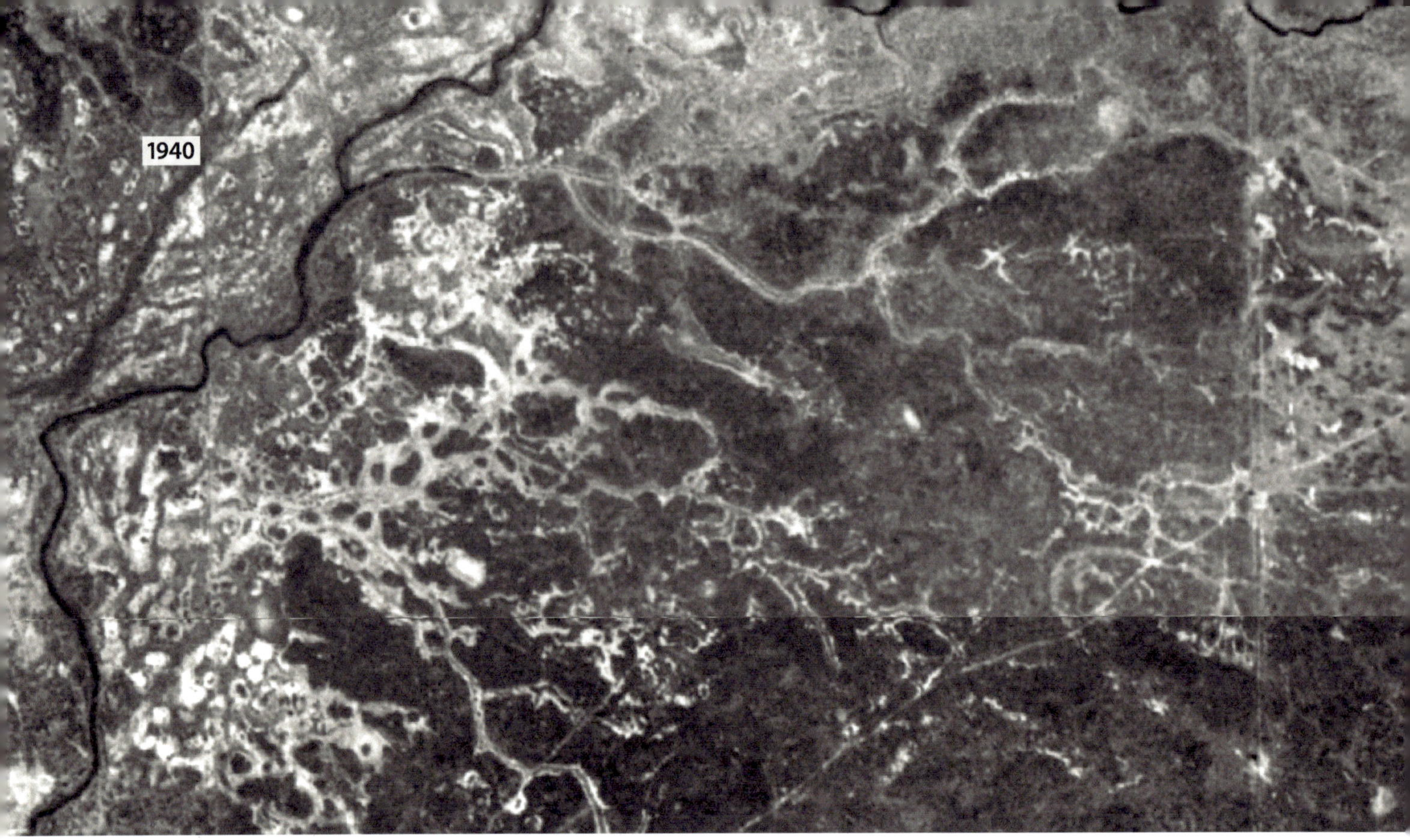

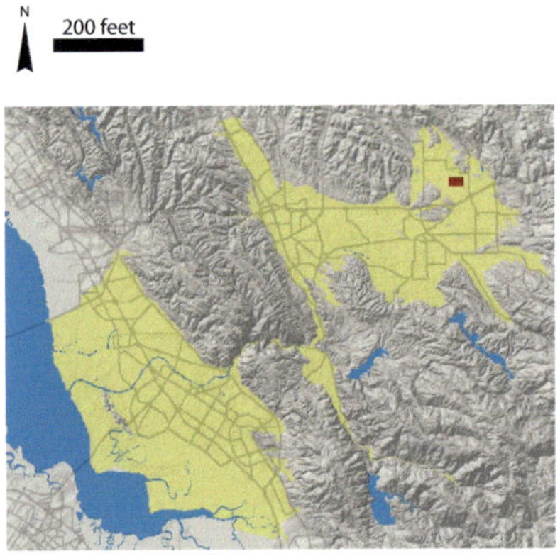

2009, ICF International 2010a). These habitat types were arranged from the lowest, most alkaline center, to the fresher, less saturated edge (Center for Conservation Biology 1992). At the center of the basin, high alkali concentrations created alkali playas and scalds (bare, unvegetated areas due to the extreme mineral content of the soil) mixed with patches of alkali meadow. Surrounding this was alkali sink scrub-vernal pool complex, which consisted of three intermixed habitat types: alkali meadow, alkali sink scrub, and vernal pool complex. A fringe of wet meadow occurred towards the edge of the basin. Remnants of each of these habitats are still present today. The following paragraphs describe each of these habitat types, working from the center of the basin out.

Figure 4.11. Potential vernal pool swales (above) with evident salt deposits are visible to the southeast of Springtown in this aerial. These swales are a common component of vernal pool complexes. (USDA 1940)

Figure 4.12. Juxtaposition of native and exotic vegetation. High salt concentrations limited agriculture and slowed the process of development in Springtown. Remnants of the historical alkali complex persist today, including a number of native species found in swaths of relatively undeveloped land. Palm trees in the background contrast with the scrubby native vegetation. (photo by Bronwen Stanford)

At the center of the Springtown sink was an area of extreme salt accumulation, with concentrations of alkali salts over 1% (Westover and Van Duyne 1911). Altamont Creek flowed through the center of the area of highest concentration and contributed additional salts through overflows. The high concentration of alkali salts resulted in large scalds and seasonally flooded bare patches along drainages, which contrast with the regular patterns of the surrounding vernal pool complex. Patches of alkali sink scrub, alkali meadow, and even vernal pool complexes surrounded alkali playas. Botanist Joseph Burtt Davy described the center of the sink in 1898:

> Bad alkali sink, nearly devoid of vegetation; margin characterized by *Frankenia*, *Hordeum* & *Bromus mollis* [*B. hordeaceus*] patches of *Frankenia* and *Bromus* & patches of *Frankenia* and *Hordeum*; the *Hordeum* & *Bromus* not growing together.
>
> Towards the center, *Frankenia suadea [salina]* & *Allenrolfea* occur; the latter not luxuriant.
>
> Salt grass & *Centromadia* local, on slopes. They also occur on the flat, but not on the lowest wettest places.
>
> *Allenrolfea* [iodine bush] follows the line of the stream bed. (Burtt Davy 1898)

Allenrolfea occidentalis *[iodine bush] scrub w/alkaline barrens, scattered claypan vernal pools & intermittent drainages. W/* frankenia, salicornia, lasthenia, hordeum geniculatum. *Rare taxa incl* cordylanthus mollis hispidus *&* C. Palmatus.

—JOKERST 1986

The complex of habitat types surrounding the center of the sink included alkali sink scrub, vernal pool complex, and alkali meadow. A number of now-rare, alkali-tolerant species were recorded by botanists working in the area in the late 1800s, including San Joaquin spearscale (*Atriplex joaquinana*; 1888), alkali milk-vetch (*Astragalus tener* var. *tener*; 1892), and caperfruited tropidocarpum (*Tropidocarpum capparideum*; 1897) (CNDDB 2010). Other potential focal species for the EACCS that have been found in this alkali complex more recently include brittlescale (*Atriplex depressa*), heartscale (*Atriplex cordulata*), and hispid bird's-beak (*Chloropyron molle* ssp. *hispidum*) (CNDDB 2010). Other more widely occurring species recently observed in Springtown include downingia (*Downingia* spp.), California waterwort (*Elatine californica*), and flowering quillwort (*Triglochin scilloides*) (Ertter 1997).

The efficiency of flooding, with the establishment of underground tile drainage, in the reclamation of alkali lands has been demonstrated, but the dense, impervious nature of the subsoil and the lack of an adequate water supply makes the reclamation of alkali soils in this area a difficult problem.

—WESTOVER AND VAN DUYNE 1910:62

Alkali sink scrub (or valley sink scrub) was present through much of this area. It once covered 260,000 acres through the San Joaquin Valley, of which less than a fifth remains (Coats et al. 1988, Center for Conservation Biology 1992, ICF International 2010a). Alkali sink scrub frequently develops on the edges of poorly drained basins in highly alkaline soils, with shrub cover of iodine bush (*Allenrolfea occidentalis*) and seep weed (*Suaeda* spp.) and an herbaceous understory (Holland 1986, Coats et al. 1988, Ornduff et al. 2003). Joseph Burtt Davy (1898) noted iodine bush, a key indicator for alkali sink scrub, describing that "*Allenrolfea* follows the line of the stream bed" in the alkali sink east of Livermore. Almost a century later, botanist Jokert noted "*Allenrolfea occidentalis* scrub w/alkaline barrens, scattered claypan vernal pools & intermittent drainages" (CNDDB 2010). Alkali sink scrub typically occurs along streams or the edges of scalds, and was probably more dominant towards the center of the complex. Palmate-bracted bird's beak (*Cordylanthus palmatus*), an endangered species found in the Springtown alkali sink, is associated with alkali sink scrub (fig. 4.13, fig. 4.14). We mapped sink scrub in a complex with vernal pools across much of the Springtown area.

Scattered within alkali sink scrub and alkali meadow habitats were broad areas of vernal pools, some of which persist today (Keeler-Wolf et al. 1998, Holland 2009, SFEI 2011). The vernal pools in the Springtown complex represent the most intact remnants of a large alkali wetland complex extending from Byron Hot Springs to the Mount Hamilton Range (Keeler-Wolf et al. 1998). Holland (2009) reported a 27% decline in vernal pool area in the Springtown complex between 1986 and 2005. However, the historical extent was much greater according to soil and species records. The historical and contemporary soil surveys repeatedly described undulating vernal pool topography in this area (described as "hog-wallow topography"; Westover and Van Duyne 1911, Welch et al. 1966). Species records also note a number of vernal pool species through the area, including vernal pool fairy shrimp (*Branchinecta lynchi*) and goldfields (*Lasthenia* spp.) (fig. 4.15, fig 4.16; CNDDB 2010).

Seasonally wet, salt-affected herbaceous grasslands and forblands—which we refer to as alkali meadow—were interspersed with alkali sink scrub and vernal pools. These alkali meadows covered the higher mounds of the vernal pool complexes and likely intermixed with wet meadow and grassland in less alkaline regions of the sink, particularly at the periphery of the sink. Alkali meadow species also may have formed the understory to alkali sink scrub (Coats et al. 1993). The early soil report noted the presence of salt grass (*Distichilis spicata*), an alkali meadow species, across many of the alkali-affected soil types (Westover and Van Duyne 1911). Additional prominent species would have included forb and wildflower species such as tidy-tips (*Layia* spp.).

Figure 4.13. Rare species of the alkali sink. Alkali sink scrub supports the two endangered plant species currently found in Livermore Valley. These are palmate-bracted bird's beak (*Cordylanthus palmatus*) shown at left, and Livermore tarplant (*Deinandra bacigalupii*) shown at right. (left photo by Shannon Lindquist 2009; right by Bronwen Stanford 2011)

Figure 4.14. Alkali sink scrub. Remnant sink scrub vegetation can still be seen in the Springtown sink in eastern Livermore. Much of the undeveloped area still functions as a seasonal wetland. (photo by Bronwen Stanford, April 2, 2011)

Figure 4.15. Vernal pools and goldfields. The distinctive rings of vegetation that form around vernal pools can still be seen in this alkali area south of Frick Lake in eastern Livermore Valley. (photo by Bronwen Stanford, April 2, 2011)

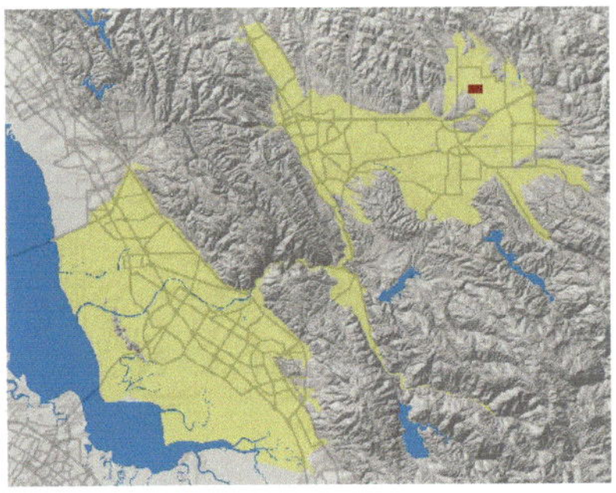

Figure 4.16. Vernal pool persistence. Vernal pools are clearly visible in 1940s aerials of eastern Livermore Valley, and some remain even in 2009 imagery. The intersection of Hartford Avenue and Lorraine Street is visible. (USDA 1940; USDA 2005)

N 500 feet

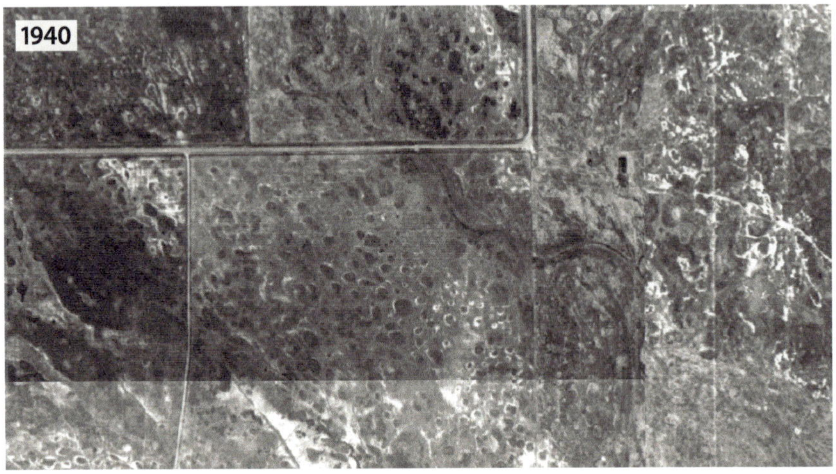

In addition to these rings of seasonal wetland habitats, a few distinct wetland features added complexity to the landscape. Located at the eastern edge of the basin, Frick Lake was a seasonal alkali lake or playa, drying in summer and showing up on maps as early as 1857 (fig. 4.17; e.g., Higley 1857a, Whitney 1873, Allardt 1874, Thompson and West 1878). The lake had no outlet historically, so alkali salts from the nearby hills would have concentrated in this lake over time. This habitat is present in relatively unchanged form today (Kohlmann et al. 2008).

At the bottom of the Springtown sink (near the confluence of Altamont Creek and Las Positas), a series of freshwater wetland features developed, including a willow thicket, springs, and a small 0.8 acre perennial pond surrounded by a three acre marsh (fig. 4.18). The transition towards fresher conditions was gradual. Upstream of the confluence with Las Positas, a series of sources recorded salty conditions. Altamont Creek was described in the summer of 1853 with "brackish water in pools" (Day 1853:281). At the same place, Dyer in March of 1869 described the creek as "muddy and strongly impregnated with alkali" (Dyer 1869), and a circa 1840 land grant map shows *agua salada* (salty water). Further downstream, a few patches of alkali wetland persisted, but water from Las Positas springs was historically fresh enough to use for

Figure 4.17. Frick Lake. Photo (A), looking southeast, shows Frick Lake in the spring, after the wet season. Photo (B) is taken from the same spot later the same year, when the lake had become completely dry. Photo (C) shows a view looking northwest from the opposite side of the lake, showing iodine bush (*Allenrolfea occidentalis*) and goldfields (*Lasthenia* spp.). (A and C: April 2011, B: August 2011, by Lance Storm)

irrigation (Wood 1883:461). Following the creek downstream, GLO surveyor Sherman Day recorded a perennial pond, a willow thicket, and "water in springs and lagunas" (1853:281); these features are no longer present.

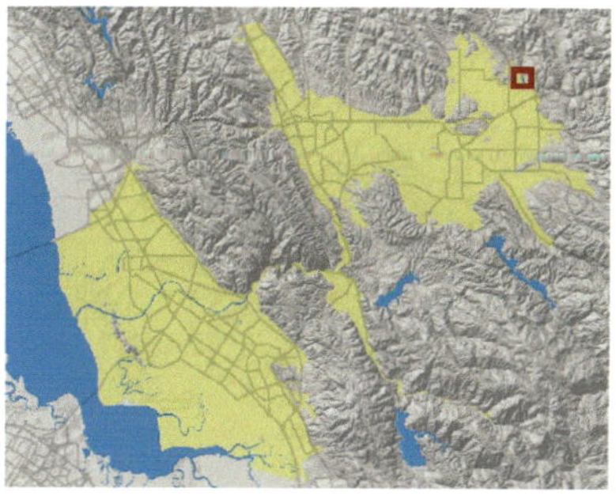

Pleasanton marsh complex

The Pleasanton marsh complex spread across 2,600 acres at the western edge of Livermore-Amador Valley (or 10,000 acres including the surrounding seasonal wetlands). This wetland extended from Interstate 580 on the north to the intersection of Sunol Boulevard with Interstate 680 on the south, and from Foothill Boulevard east beyond Santa Rita Road, covering much of modern-day Pleasanton (fig. 4.19). It was one of the largest freshwater, non-tidal wetlands in the Bay Area, over twice as large as other large freshwater wetland complexes documented by historical ecology studies, including Willow Glen (790 acres) and Laguna Seca (1,000 acres) in San Jose and the Yountville wetlands (880 acres) in Napa Valley (Grossinger et al. 2006, Grossinger et al. 2008, Grossinger 2012).

The Pleasanton marsh complex was composed of a variety of intermixed wetland types, which we mapped as three broad classes: perennial open water, valley freshwater marsh, and willow thicket or swamp, all surrounded

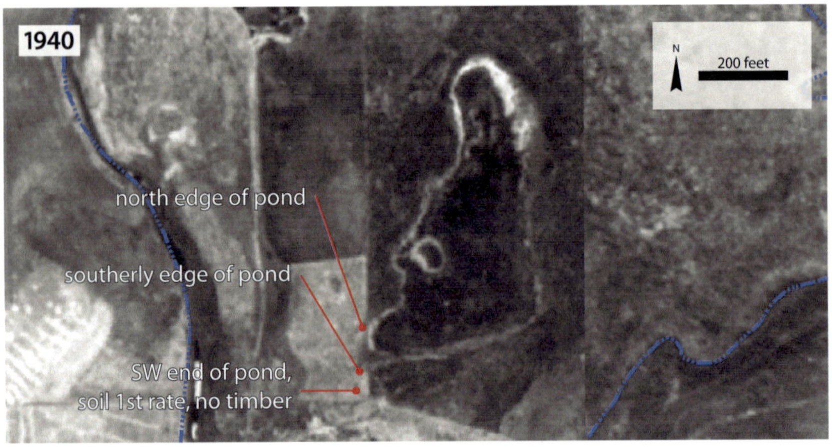

Figure 4.18. Historical pond. In March 1869, surveyor Dyer recorded a pond near the confluence of Las Positas and Altamont Creek. The quotes from the survey, overlaid on the aerial, indicate a larger pond than that visible in the 1940 aerial. We mapped this feature as open water surrounded by valley freshwater marsh, using these data and a number of historical maps, including detailed USGS mapping. This pond, today covered with a development, was likely one of several permanent, year-round features within the largely seasonal Springtown alkali sink. (USDA 1940)

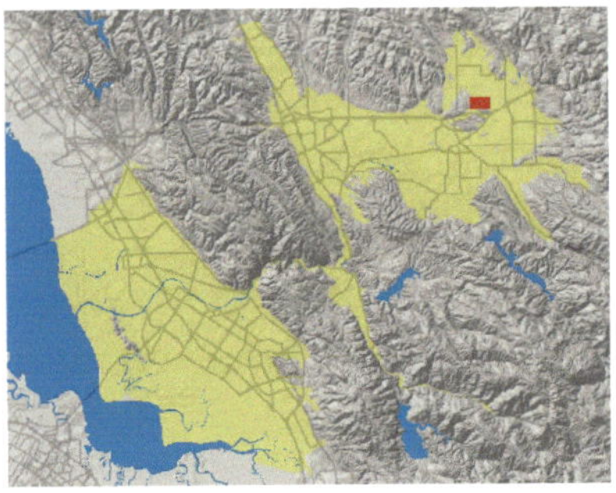

by a fringe of wet meadow and alkali meadow (fig. 4.20, box 4.4). However, as with the alkali habitats, the boundaries between wetland types were inevitably more complex than shown by our mapping. Patches of tule marsh and small ponds with pondweeds (*Potamogeton* spp.) formed within the willows, and small willow groves grew in slightly drier patches within valley freshwater marsh. Early visitors recorded this pattern: surveyor Howe recorded "A pond in the edge of the tule marsh 12 chains E" (Howe 1851), while explorer Fages described "many tulares and lakes" in 1772 (Treutlein and Fages 1972:353). The open water portions grew and shrank based on rainfall and groundwater levels.

The location and character of the Pleasanton marsh complex were controlled by the geology and hydrology of the valley. Water percolating through the gravel beds of Arroyo Mocho and Arroyo del Valle became trapped beneath the clay cap, described as "30 feet thick" at the western edge of the pinched valley, which created artesian conditions (Lee 1916b:2382). The valley floor sloped to intersect groundwater, producing ground saturation through the dry season (Williams 1912, Lee 1916b). These conditions combined to result in artesian springs: "The water exists in a Tule Lake, partly subterranean, five hundred feet above tide level, surrounded by hundreds of natural wells, which are full to the brim in the driest seasons. During ordinary wet seasons, these wells overflow and inundate a large surface" (Scott 1871).

The Pleasanton hop fields are located in an artesian belt, and there are several flowing wells on the ranch. From some of these the water flows to a height of four feet above the surface of the ground. A large portion of the property had to be surface drained before it was available for cultivation, but the sub-irrigation is such that even during the dry season there is an abundant crop.

—MCGOWN 1902

Stream deposits and bedrock shaped the marsh. To the east, natural levee deposits of Arroyo del Valle restricted the extent of the marsh (fig. 4.22; similar to wetlands on lower Llagas Creek in Santa Clara Valley, see Grossinger et al. 2008:100). To the west and south, the marsh was constrained by the Calaveras Fault and abrupt upland slope of the edge of the valley. To the flatter north, the marsh gradually graded into seasonal wetlands, fed by water and fine sediment from the many intermittent streams draining from the northern edge of the valley.

The three habitat types and surrounding seasonal wetlands were arranged along a gradient from the lowest, wettest center to the slightly higher, less

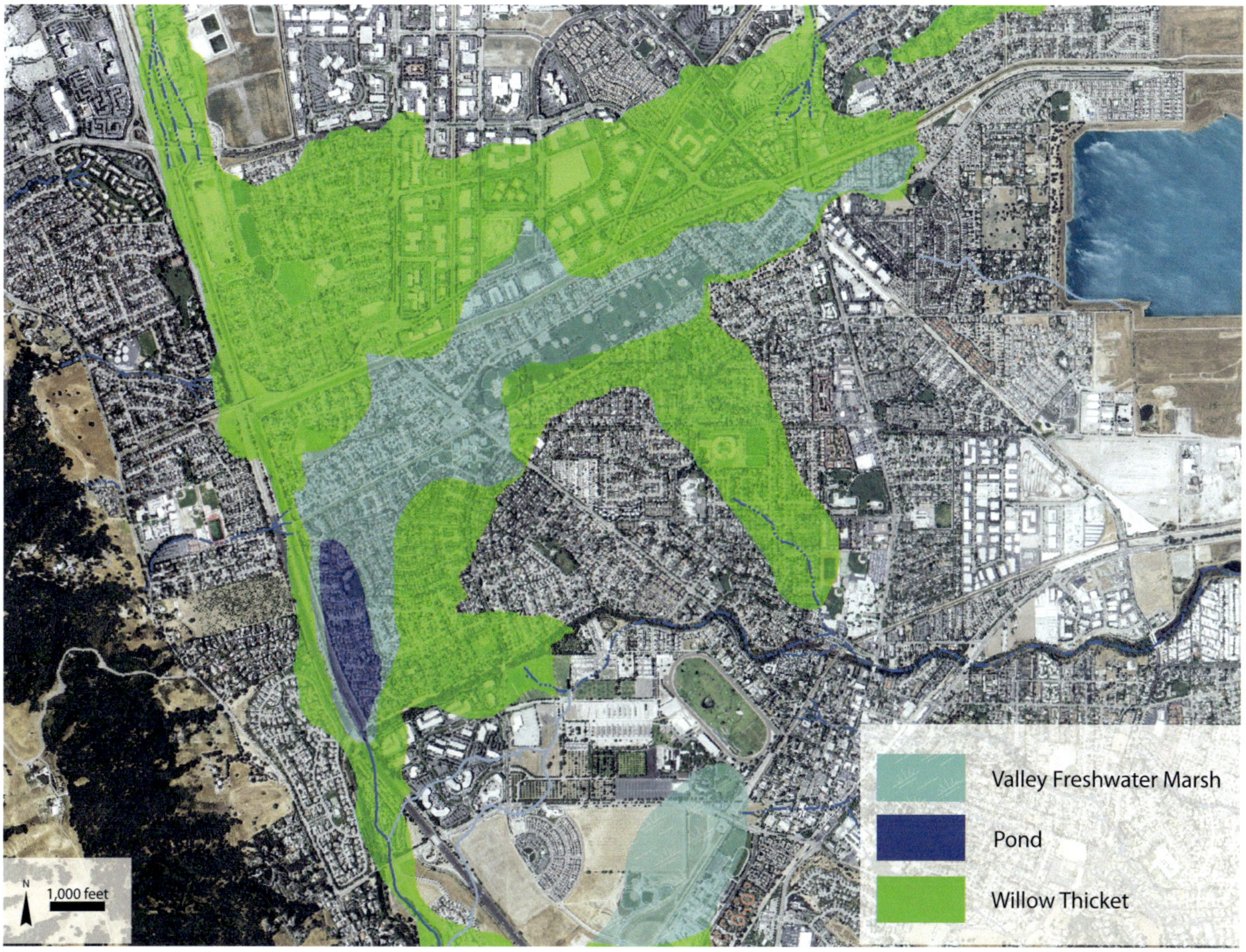

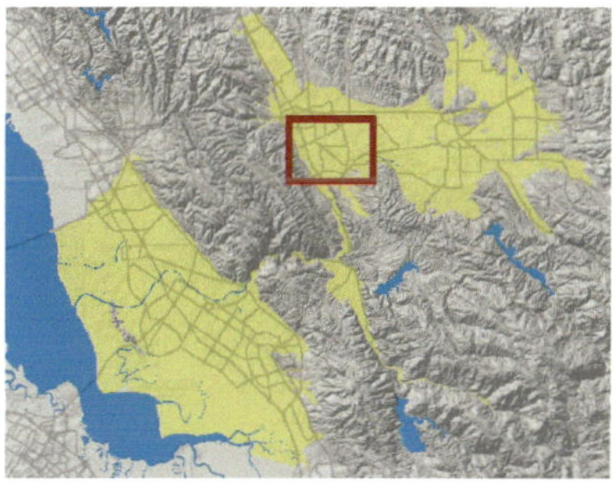

Figure 4.19. The historical extent of the Pleasanton marsh complex included much of the area now occupied by Pleasanton. Note the large playing field aligned with the historical valley freshwater marsh. (USDA 2009)

Figure 4.20. Allardt [1880]1907. Details from this 1907 map are shown below, including depictions of tules, willows, and a lagoon. This map "showing the original condition and extent of swamps and tules" was created after much of the original marsh had been destroyed. It was produced to document the historical marsh complex, based on surveys by Surveyor George Allardt in 1880, and provides the most detailed depiction that we were able to find. We refer to this map as Allardt [1880]1907 to acknowledge its history. (courtesy The Bancroft Library, UC Berkeley)

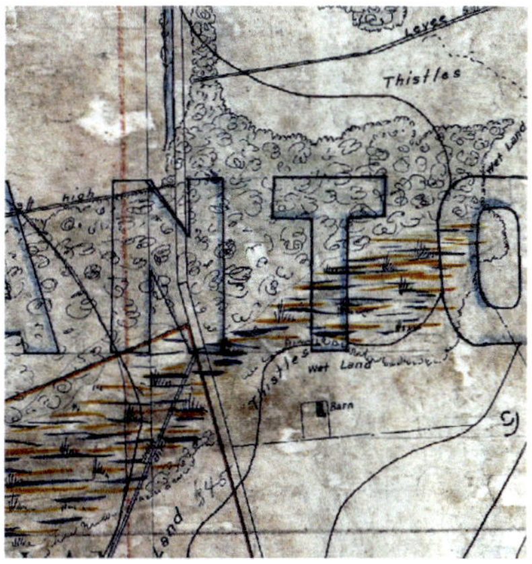

BOX 4.4. THE MANY NAMES OF THE PLEASANTON MARSH COMPLEX

The large wetland complex at the western edge of Livermore-Amador Valley had no standardized name through the historical period. Textual accounts indicate that rather than being seen as a single feature, people saw the complex as a series of distinct wetland features. Early land grant testimony refers to the *Tular* and the *Laguna*, or the *cienega* (marsh) and tule swamp (Alviso 1861, Amador 1861, Wilson 1861). Early maps often showed the complex as a wetland feature, but often labeled each portion separately, indicating its structural diversity (fig. 4.21). The earliest map with a clearly unified name is Allardt's 1880 map, which applies the name "Lake Pleasanton" to the entire feature, though it still labels each wetland type individually. Table 4.2 shows some of the names applied to the marsh complex, or portions of the complex.

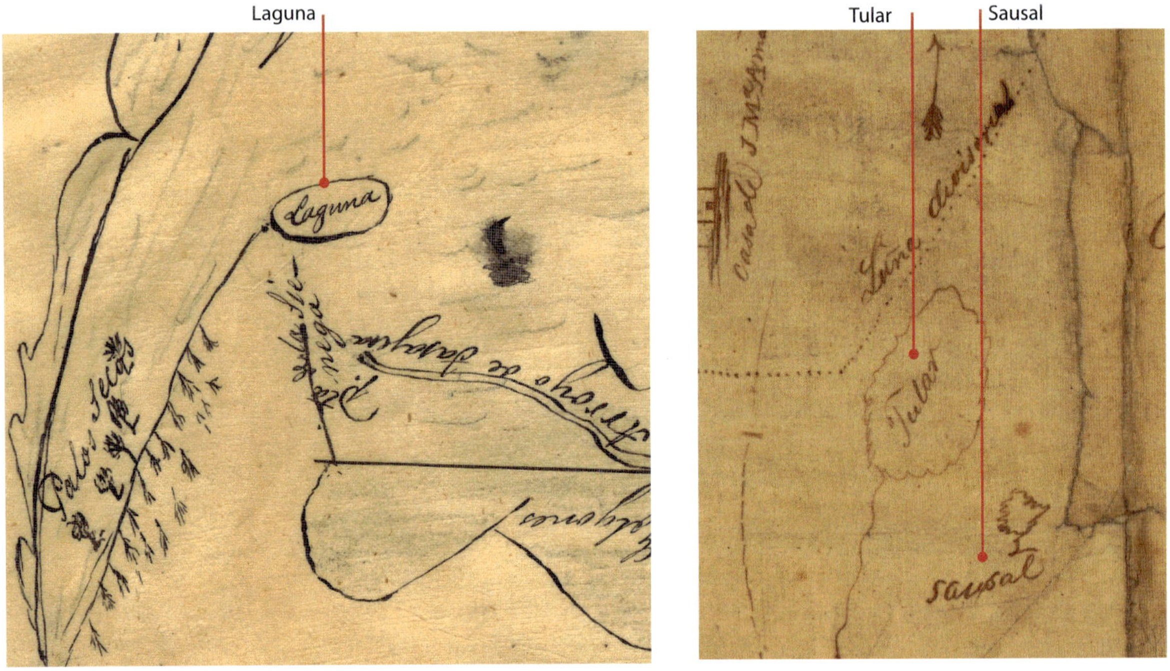

Figure 4.21. Diseño maps from the 1840s depict the marsh complex in a variety of ways, from labeling the entire feature as "laguna" to labeling one portion "tular" and the other "sausal." These two maps were made within a decade of each other, so the differences indicate likely shifts in emphasis rather than actual change. (USDC ca. 1840a,b, courtesy The Bancroft Library, UC Berkeley)

Table 4.2. Names of the Pleasanton marsh complex.

Maps	Source	Year
"Lagoon"/"Willow thicket"	La Croze 1860	1860
"The Tulare"	Dyer 1862c	1862
"Lagoon"/"Willows"	Whitney 1873	1873
"Tulare Lake or The Lagoon"/"Willows"	Allardt 1874	1874
"Swamp, Tule and Willow"	Cash 1875	1875
"Willow Marsh"	Thompson & West 1878	1878
"Lake Pleasanton" (includes "tule swamp," "lagoon," and "willows")	Allardt [1880]1907	1880 / 1907
Text		
"Tule Lake"	Scott 1871	1871
"Bolsa or swamp"	Von Geldern 1912	1912
"Lake Pleasanton"	Von Geldern 1912	1912
"a small lake called Tulare, the Lagoon, or Bolsa"	Gutmann 1919	1919

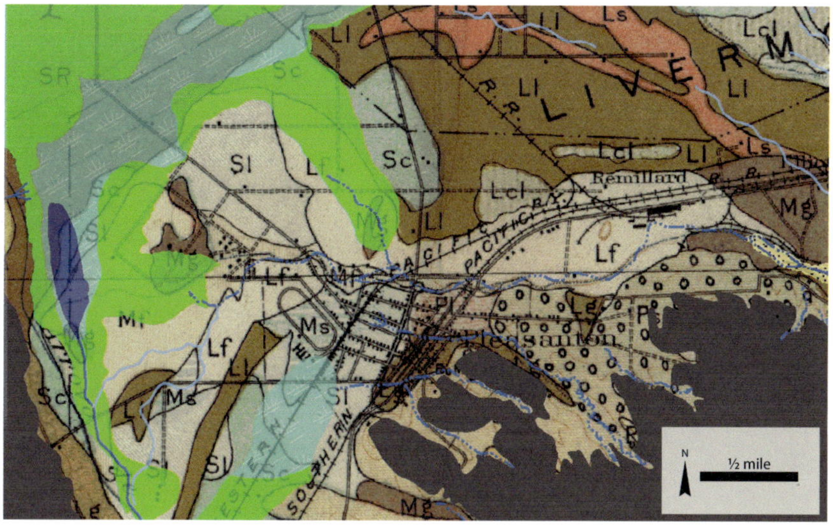

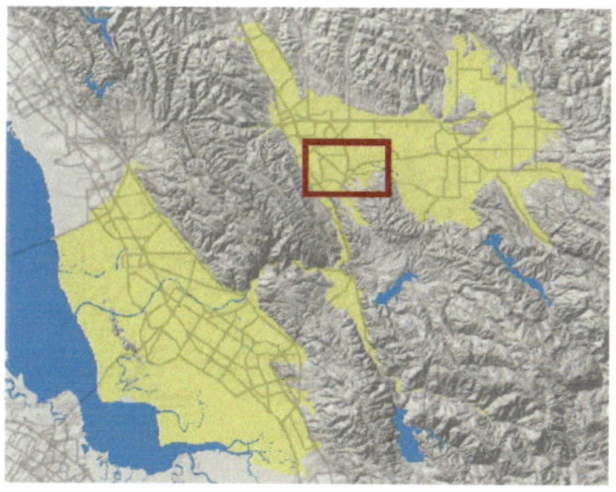

Figure 4.22. The natural levees of Arroyo del Valle. Extensions of coarse soil deposited over time by Arroyo del Valle constrain the southern extent of the marsh complex. These soils are shown in a tan color (Lf and Mf) and trace fingers of natural levees that built higher ground above the marsh. (Westover and Van Duyne 1910)

Valley Freshwater Marsh
Perennial Pond
Willow Thicket

saturated perimeter. In the center, a perennial pond occupied the lowest portion of the valley and connected with Arroyo de la Laguna to drain water to the south. Surrounding this pond was a swath of valley freshwater marsh. Further removed, the willow thickets around the border of the perennial wetland occupied areas inundated for shorter periods, grading into wet meadow and alkali meadow, which were temporarily flooded and saturated for only portions of the year.

This large wetland mosaic provided important habitat for a number of native species. As one of the few large perennial water resources in the area, the marsh complex would have supported native fishes such as thicktail chub (*Gila crassicauda*) and Sacramento sucker (*Catostomus occidentalis*), songbirds such as the yellow-breasted cuckoo (*Coccyzus americanus*), and even grizzly bears (*Ursus arctos horribilis*; Fages 1772 in Treutlein and Fages 1972). The feature was referred to as a "*cienega* [marsh] and tule swamp," with "water all the year round" that was good for snipe hunting (Coffee 1862:1030-1031, La Croze 1862b:1019-1023).

No historical sources explicitly describe the relative sizes of the different wetland habitats. Our depiction is an approximation based on available sources (in particular the highly detailed Allardt [1880]1907; see fig. 4.20), and is supported by land grant testimony describing the landscape during early Spanish settlement. Describing conditions in 1811, José María Amador states that "The Tular" (including "trees [presumably willows] and tulares") was about a league (2.6 miles) from north to south, and a mile wide (Amador 1861:232). This earliest spatial description matches the detailed mapping from Allardt's 1880 survey. However, there are some suggestions that the pond and marsh might have historically been much larger in relation to the willow thicket (see depiction by La Croze 1862a, Dyer 1862c), and that the entire marsh complex also might have extended much further north into the alkali meadow zone (fig. 4.23).

PLEASANTON MARSH COMPLEX HABITATS Each of the three primary habitat types—pond, marsh, and willow thicket—was well documented in the

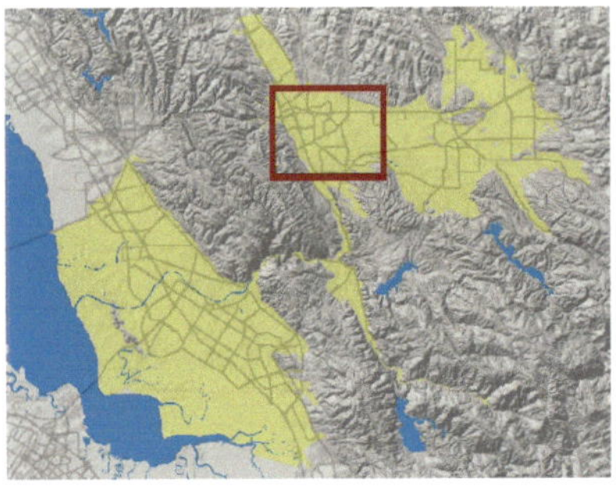

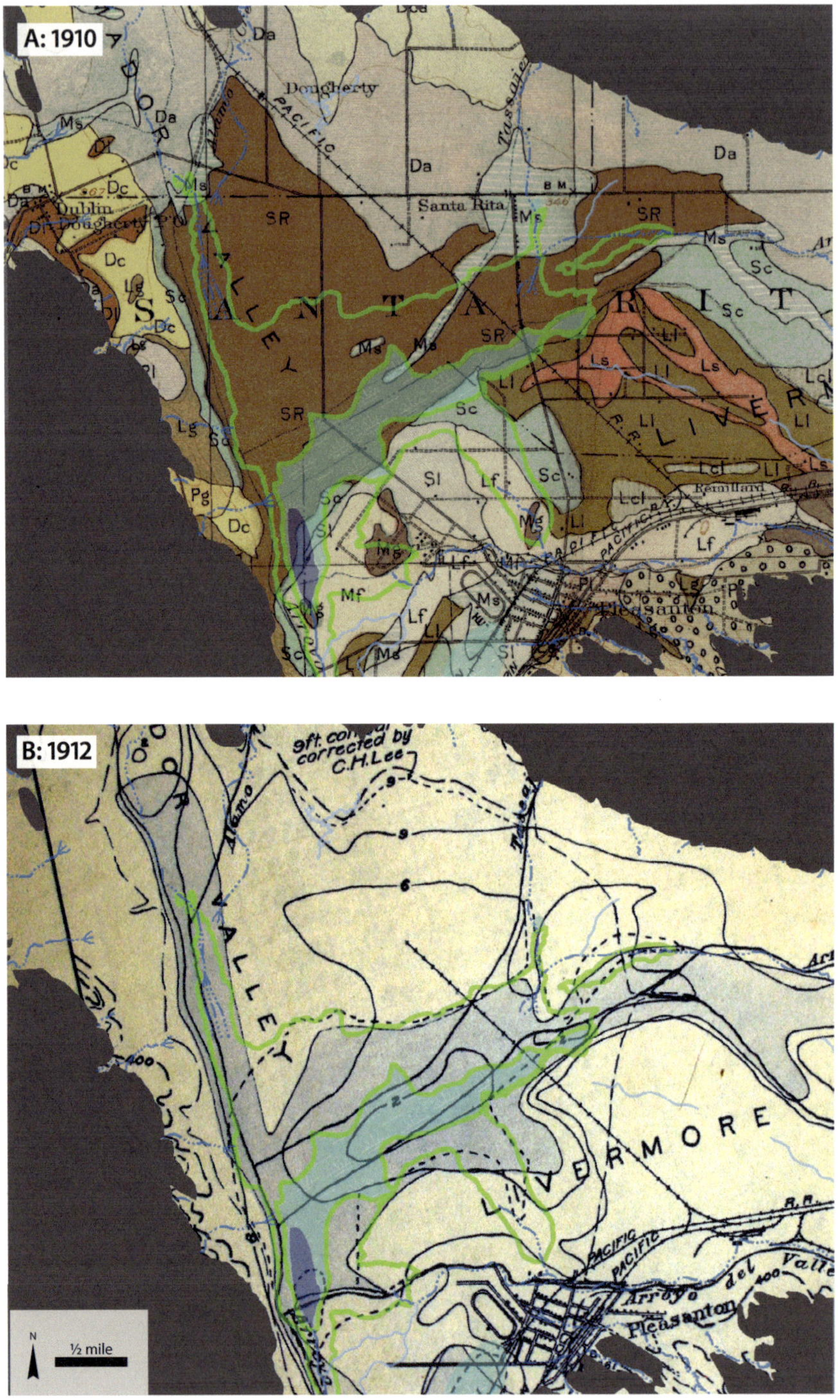

Figure 4.23. Early historical ecology research. Two early sources suggest that the perennial areas of the marsh complex may have extended beyond what we show. The 1910 soil survey map (top) shows Santa Rita clay adobe (SR) under the marsh complex and extending far north. According to the soil survey, this soil type "probably represents what was at one time a shallow lake, which gradually filled until it became a tule swamp" (Westover and Van Duyne 1911:42). A few years later, in 1912, Spring Valley Water Company mapped the historical extent of the marsh to determine zones of evaporation, in order to maintain groundwater levels below the depth of evaporation. Groundwater contours on this map indicate that the zone of relatively high groundwater extended north of the marsh complex even as late as 1912, when the marsh had been almost entirely cleared. The images at right show these two maps overlaid with the outline of the historical Pleasanton marsh complex (outline in green). (A: Westover and Van Duyne 1910; B: from SVWC 1912b, PWC 009, SFPUC Archives)

historical record. The perennial pond in the southern portion of the Pleasanton marsh complex, at the head of Arroyo de la Laguna, was distinguished from the surrounding freshwater marsh in a number of early maps. These maps clearly show a lagoon—indicating open water—within the surrounding marsh (e.g., LaCroze 1860, Whitney 1873, Allardt 1874, Allardt [1880]1907). Early narrative accounts record "a lake or bulrush pond"(Edwards 1932), "a lagoon filled with tule" which extended "a considerable distance from its borders" (California Legislature Assembly 1854:54), "a permanent laguna of water" (La Croze 1860) and "a small lake called Tulare, the Lagoon, or Bolsa, into which the Alamo, Tassajara, and Positas Creeks used to flow" (Gutmann 1919:25).

While this pond was permanent, the extent of open water varied seasonally and interannually. In wet seasons, it likely extended across the marsh and into the wet and alkali meadows to the north, contracting in late summer (fig. 4.24). In 1912, a series of reports estimated the extent of the recently drained Laguna, describing that historically, "With the advancing season the lake became smaller, until it contracted in area to its minimum outlines of a perpetual swamp, which, in this condition, covered a space of about 2 square miles" (Von Geldern 1912:18). In this regard, the wetland complex could be considered a vegetated intermittent lake with a small perennial pond at its center.

A large body of freshwater marsh surrounded this perennial pond, dominated by tules and extending north and east from the Laguna (fig. 4.25). GLO surveyors in the area described surveying through the "tulare," "tulares," or "tule swamp" (Howe 1851, Dyer 1862b, Healy 1863). A map based on an 1880 survey of the marsh depicted a "tule swamp abounding with copious living springs" (Allardt [1880]1907; see fig. 4.20). Early descriptions document the presence of tules as well; in 1772, Fages described passing "many tulares and lakes, and close to them many bear diggings. The shores of the lake were much overgrown" (Treutlein and Fages 1972:353), and land case testimony accounts repeatedly refer to a large "tular" and the "Point of the Tular" (see Amador 1855, Crockett and Chittenden 1861, Amador 1861). While the dominant vegetation was likely tule (*Schoenoplectus* spp.), the marsh would have included a suite of other freshwater emergent species, including rushes (*Juncus* spp.) and sedges (*Carex* spp.).

A large willow thicket surrounded the open water and freshwater marsh, covering the slightly less saturated ground on the margins of the complex.

At the western extremity of the Livermore Valley, is a tract variously known as the Bolsa, or Lagoon, some fifteen hundred acres in extent, swampy in character, and covered with a thick growth of willows, which, during the winter season is generally under water, and to some extent preserving this character in summer.

—WOOD 1883:454

Figure 4.24. Seasonal variation in open water area. The images below show the approximate flooded area of the ca. 1800 Pleasanton marsh complex (from left to right) immediately following rain events in winter, in spring, and in late summer. In summer the only large open water area would have been the perennial pond at the center, while for periods in winter all of the surrounding alkali and wet meadows would flood. This dramatic annual variation in size (which would also vary by rainfall year) may help explain some of the differences in

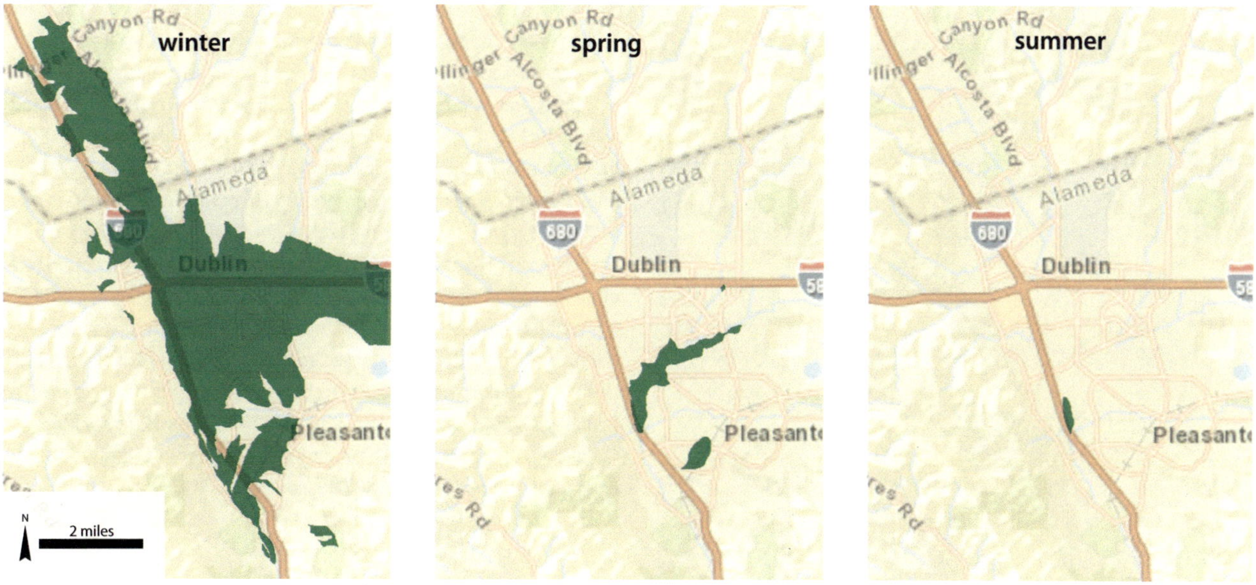

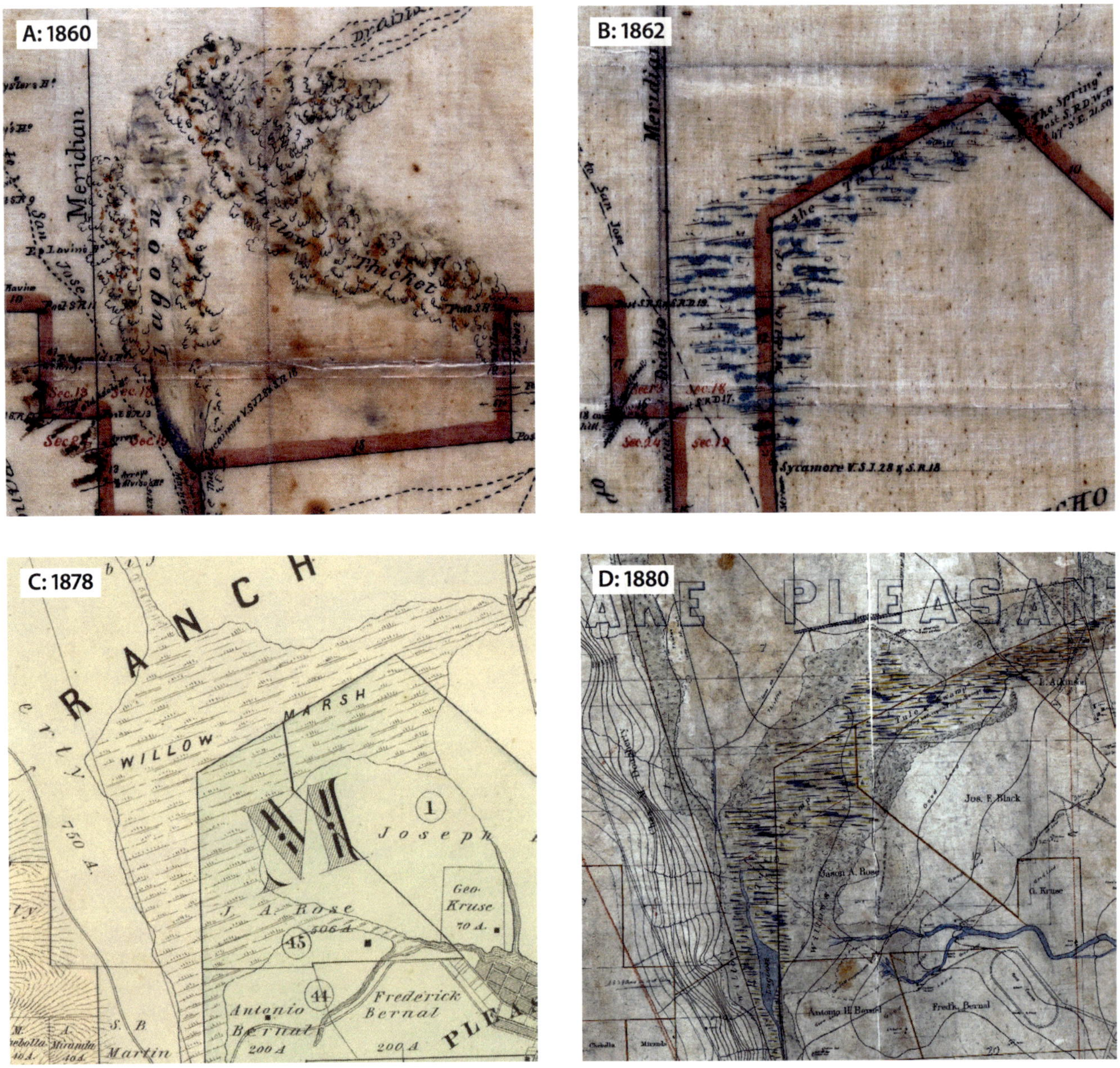

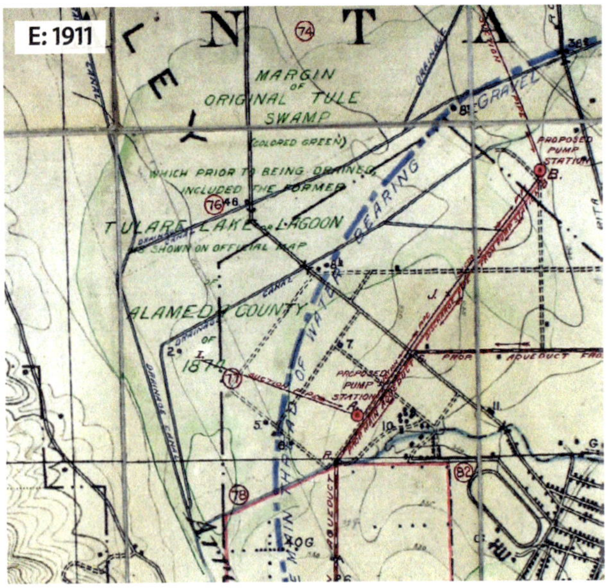

Figure 4.25. Habitats of the Pleasanton marsh complex. Each of these maps provides a different perspective on the marsh complex, emphasizing a different aspect of its ecology or history: the lagoon and willow thicket (A), freshwater marshes (B), willows (C), rings of discrete habitat types (D), and finally canalization and drainage (E). Maps A-D represent the same relatively unmodified feature, but each focuses on a different aspect, resulting in widely different depictions. By using a combination of sources such as these we can assemble a robust picture of the historical marsh complex. The red line in B represents the land grant boundary that ran through the center of the complex, and is visible in C, D, and E as well. (A: La Croze 1860, courtesy The Bancroft Library, UC Berkeley; B: Dyer 1862c, courtesy The Bancroft Library, UC Berkeley; C: Thompson and West 1878, courtesy David Rumsey Map Collection; D: Allardt [1880]1907, courtesy The Bancroft Library, UC Berkeley; E: Schussler 1911, MB 068, courtesy SFPUC Archives)

GLO surveyors noted entering and leaving a willow thicket, and described passing through "a thick willow swamp—the western border to quite an extensive tule swamp on the left" (Howe 1851). Even the early *diseño* maps make a distinction between regions covered with tules and with willows (see fig. 4.21), and some detailed later maps distinguish rings of vegetation (e.g., Allardt [1880]1907), or clumps of marsh and willows (e.g., La Croze 1860, Duerr 1872b, Cash 1875). Land grant case testimony includes descriptions of "trees and tulares," indicating the presence of willows (Amador 1861:232). A similar willow-tule mosaic characterized other major historical Bay Area wetlands (e.g., Laguna Seca and Willow Glen near San Jose, and Rutherford near Yountville in Napa, see Grossinger et al. 2008, Grossinger 2012).

Springs also created smaller pockets of valley freshwater marsh within areas dominated by willows. One well-recorded example of this was the *cieneguita* or "little marsh," which surrounded a historical spring at the eastern extent of the valley freshwater marsh. This *cieneguita* was described repeatedly in the land grant testimony: "a spring of water that rises ... and scatters over the surface of the land" (Amador 1861:230). These springs were later developed as water supplies, along with additional artesian wells that were dug in the marsh (fig. 4.26, box 4.5).

Wet and alkali meadows

Along with the two major wetland complexes described above, an additional 15,600 acres of wet meadow and alkali meadow provided seasonal wetland habitat across Livermore-Amador Valley. These wetlands formed an almost continuous band over 13 miles long along the northern edge of the valley, with additional smaller patches scattered further south.

These seasonal wetlands (wet and alkali meadows) historically supported a rich variety of species. They were flooded for days to months during the winter, drying into a hard clay surface in the summer and early fall. Salt grass was likely present across much of the alkali meadow (Westover and Van Duyne 1911), along with a series of other vernal pool and wetland species, including goldfields (*Lasthenia* spp.), tidytips (*Layia* spp.), clover (*Trifolium* spp.), downingia (*Downingia* spp.), and popcornflower (*Plagiobothrys* spp.) (Jepson 1891, Bioletti 1892, Greene 1895, Ertter 1997, Edwards and Thayer 2008). Even as recently as the 1990s and 2000s, a variety of species that were listed as the focal species for the EACCS were recorded within the alkali meadow zone north of Pleasanton, including red-legged frog (*Rana aurora draytonii*), tiger salamander (*Ambystoma californiense*), Congdon's tarplant (*Centromadia parryi* ssp. *congdonii*), hairless popcornflower (*Plagiobothrys glaber*), saline clover (*Trifolium depauperatum* var. *hydrophilum*), and San Joaquin spearscale (*Atriplex joaquinana*) (ICF International 2010a, CNDDB 2010). On wet meadows, wetland grasses and graminoids would have likely been important as well, potentially including species such as semaphoregrass (*Pleuropogon* spp.), spikerush

Figure 4.26. "Artesian well overflowing" near Pleasanton, ca. 1900. High groundwater and artesian pressure supported free-flowing wells such as this one. (image 82212, © San Francisco Public Utilities Commission)

BOX 4.5. CONVERSION OF THE MARSH

Conversion of the Pleasanton marsh complex to arable land occurred in two phases. In the first phase, which occurred in the first half of the 1800s, cattle grazed around the margin of the marsh and some residents began clearing and draining the land at the edges of their properties. In the second phase, the later 1800s through 1910, people actively worked to clear and drain the entire marsh, both to make the land accessible for agriculture, and to access and manage water stored in aquifers below the marsh.

Many early sources document the two phases. In the 1860s, the Santa Rita land grant case attempted to establish the extent of the marsh when the first settlers arrived. Landowners described how the marsh had changed. Alviso described that by 1861 the marsh was "smaller" and "the tular … is not so large, because the cattle eat it up" (Alviso 1861:853, 861). Another early landowner reported that near Arroyo de la Laguna there was "wood and a kind of Cienega" but "now it is all destroyed and dried up" (Sibrian 1861:713). José Amador described the reduced size of a portion of the marsh by 1861 as follows:

> I knew that Tular from 1811; it extends northerly [from Arroyo de la Laguna] for about a league [3 miles], a great deal of the plain, which is now clear was formerly Tular; it was covered with trees and tulares and nobody passed through it … This was in old times. The Tular has been lessened since by the travelling of the cattle. (Amador 1861:231-232)

By the 1880s, maps began to show a marsh that was reduced in size. The clearest and most detailed depiction of the entire feature comes from a survey in 1880 (Allardt [1880]1907), which does not include two extensions of marsh towards the east shown on earlier maps (e.g., La Croze 1860, Whitney 1873, Allardt 1874, Thompson and West 1878) and described in GLO survey notes of the area (Howe 1851: "thick willow swamp"; La Croze 1860: "enter willow thicket"). This map also shows some early drainage canals (fig. 4.27).

Rapid conversion of the remaining marsh occurred between the 1880s and early 1900s. Spring Valley Water Company and

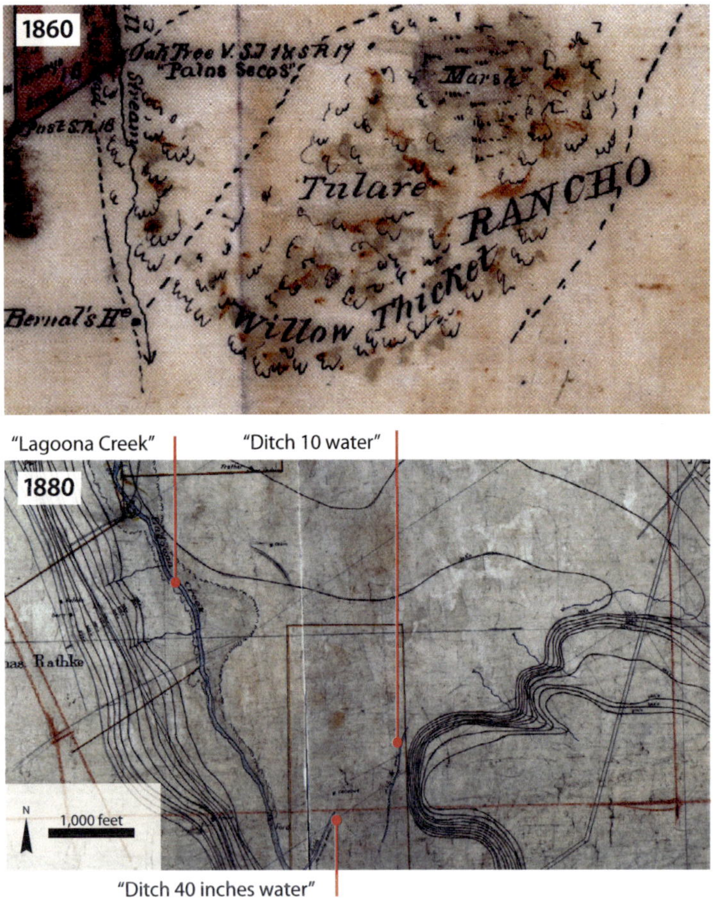

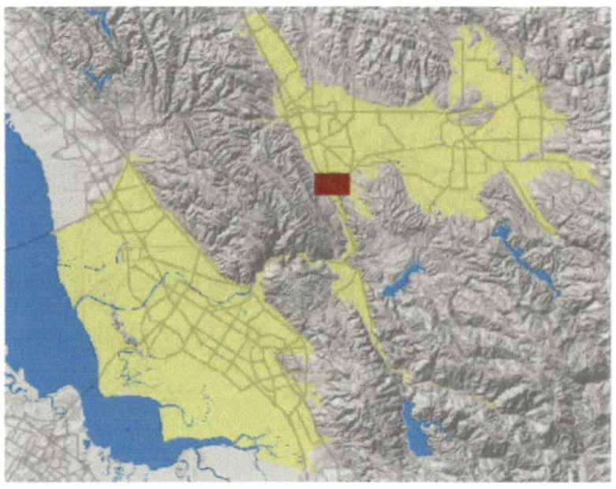

Figure 4.27. Conversion of the marsh. An 1860 map (top) shows a large patch of "tulare" and a willow thicket south of the main body of the marsh. By 1880, (bottom) this area had been cleared, and the labels "Ditch 40 inches water" and "Ditch 10 water" indicate that active drainage was taking place. (La Croze 1860 and Allardt [1880]1907, both courtesy The Bancroft Library, UC Berkeley)

the city of San Francisco commissioned many reports in the early 1900s to establish how best to store water under Livermore-Amador Valley, including investigations of depth to groundwater and research on evaporation rates (e.g., Williams 1912, SVWC 1912b, etc.). Ditches carried water from the artesian springs into Arroyo de la Laguna and then to San Francisco (for more details, see water use discussion p. 60). By 1907 only remnants of willows along canals remained (Tibbetts 1907b). The 1911 soil report noted that the underlying soil type once "supported a growth of tules and a few willows" although "at present it is all under cultivation" (Westover and Van Duyne 1911:39). By 1912 "the original tule swamp … [had] almost entirely disappeared" (Lee 1912, fig. 4.28).

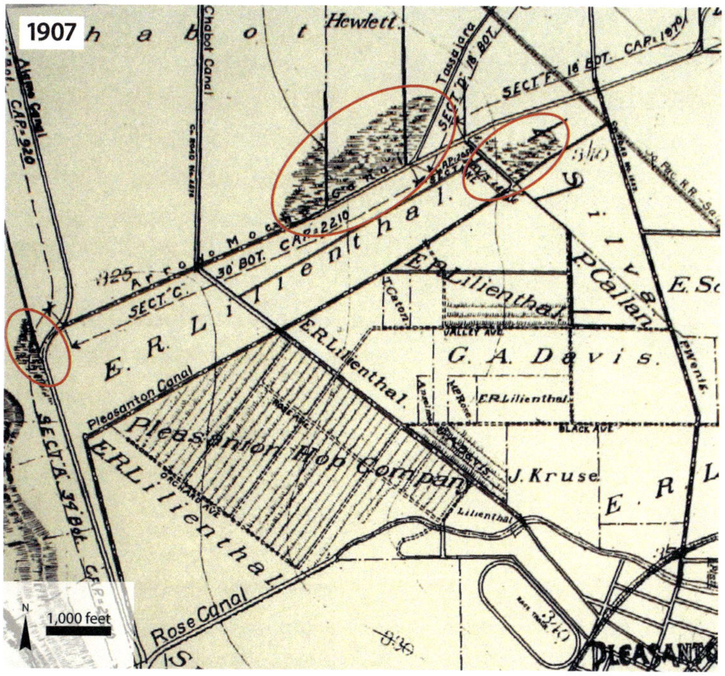

Figure 4.28. "Proposed location of the Pleasanton Canal Systems." This 1907 map shows proposed ditches that cut across the marsh complex. A few patches of remnant willow thicket remained at the time, circled in red here. (Tibbetts 1907b, courtesy Earth Sciences and Map Library, UC Berkeley)

Despite this rapid conversion, the wetland did not entirely disappear (fig. 4.29). The drainage canals initially "would easily silt up and overflow with tule and weeds and the property would be flooded nearly every year" (Gutmann 1919). Even after the area was mostly reclaimed, water remained:

> The recent drainage canals which have been put over this lower section of the valley have tended to relieve the swampy condition to a considerable extent, but still the surface of the soil, which is dark in color, is extremely moist, and during a portion of the year is more than saturated as that water is standing on the surface. (Haviland and Tibbetts 1912:40)

Figure 4.29. "Pond. Near the eastern end of the C line of wells, Pleasanton. June 1, 1916." This pond (likely artificial) is surrounded by marsh vegetation and shows an example of how natural marsh remnants may have looked in the early 1900s. (AD-866, © San Francisco Public Utilities Commission Archives)

In 1916, water was still present "4 or 5 feet below the level of the ground" providing natural "sub-irrigation" that allowed farmers to grow crops such as alfalfa without irrigation (Mortimer 1915: 256). Even today, the soil types and high groundwater are a reminder of the presence of the historical marsh.

We had water on about 1175 acres, and there was only about five or ten acres right around the house where it did not overflow.

—RASMUSSEN 1915:407, FARMER NEAR PLEASANTON

(*Eleocharis* spp.), *Triglochin* spp., and *Juncus* spp. Some of the wildflower descriptions on p. 76 also likely describe the vegetation on these seasonal wetlands.

We mapped seasonal wetlands based on descriptions in the historical soil surveys, modified for topography and supported by a range of additional sources. The soil surveys described soils that were "sticky" where "water stands for days during the rainy season" (Westover and Van Duyne 1911; for detail on mapping methods, see Chapter 2).

These seasonal wetlands were recorded by a number of early sources, perhaps because of their massive extent. Surveyors traveling across the valley in the 1850s during the summer and fall described crossing land that would be inundated in the wet season. About a mile north of the main body of the Pleasanton marsh complex, surveyor Richard Howe (1851) described "the middle of a prairie—a beautiful flat. This valley seems to spread out towards the east, and has the appearance of overflowing in the wet weather, to the depth of from 2 to 4 feet deep." Half a mile further south, he described the effects of spreading streams: "land here is rich, but overflows in wet weather from the Mountain streams that run into it, and spread over it." Two years later, surveyor Sherman Day (1853) traveled east across the northern edge of the valley in July, recording the location and characteristics of streams and wetlands. As he crossed Alamo Creek, he noted, "Line crosses level plain, wet in winter ... The bottoms of this creek are annually overflowed." Just east of Tassajara Creek, he noted "rich meadow land, apparently wet in winter, but baked and cracked in summer" (Day 1853, July).

Even during the initial stages of drainage and groundwater decline, these seasonally saturated zones persisted. The map of Allardt's 1880 survey shows willows and tules, but it also shows a border labeled "wet land," and patches of thistles, which were likely growing in seasonal wetlands that were not valuable for farming. As late as 1912, a survey to establish the zone of evaporation noted "the fact of no floods in the past winter, shows that the soil is kept moist from ground water alone" (SVWC 1912b:488; see fig. 4.23). Testimony from 1915 recorded "extreme moisture" that interfered with farming, and resulted in some years with "about three or four feet of water on top" of the land (Rasmussen 1915:407).

Any attempt to irrigate these soils is almost certain to increase the concentration [of alkali] in the surface foot to such an extent as to prohibit crop growth.

—WESTOVER AND VAN DUYNE 1910:62

A subset of the extensive seasonal wetlands contained alkali salts at concentrations greater than 0.2% in the first six feet (Westover and Van Duyne 1911); we map these areas as alkali meadow. A salt content of 0.2% or greater was the threshold that early soil surveys "recognized as having sufficient concentration of salt to be injurious to crops" (Carpenter and Cosby 1939:68); several soils were described with "so much alkali that it is fit only for grazing" (Westover and Van Duyne 1911; fig. 4.30). East of Livermore in an alkali area, botanist Burtt Davy described "no trees ... hills barren and bare" (Burtt Davy 1898). More recently, longtime resident Paul Banke recalled that in alkali-influenced areas northeast of Livermore

Figure 4.30. Extent of alkali in Livermore-Amador Valley. This map shows alkali concentrations in soils across Livermore-Amador Valley as recorded in the 1911 soil survey report. The high intensity alkali area in the eastern portion of the valley underlies Springtown, and the western patch of alkali soils corresponds to the Pleasanton marsh complex and surrounding seasonal wetlands. The marsh complex was mostly drained by this time, which may have led to a zone of elevated alkali levels. (data Westover and Van Duyne 1911)

Alkali Concentration

200 to 400 parts per 100,000

400 to 600 parts per 100,000

600 to 1,000 parts per 100,000

over 1,000 parts per 100,000

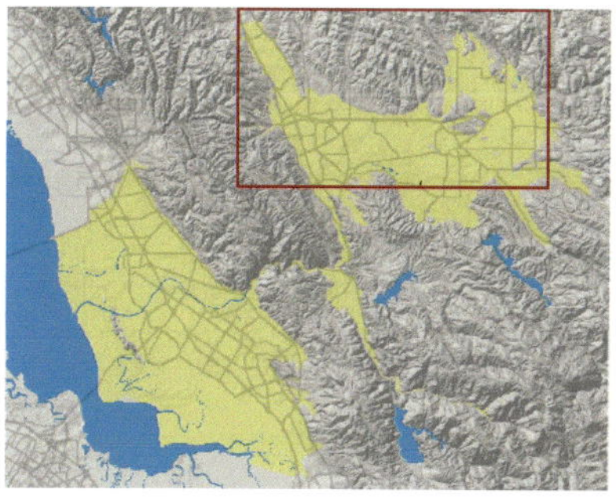

growing trees was "near impossible. Really tough, because of soil" (Banke pers. comm.).

Salts in the alkali meadows of Amador Valley likely built up through a combination of alkaline runoff from surrounding hills and high evaporation rates. The area immediately surrounding the historical marsh was described as "damp, more or less alkali soil" and a "circumscribing zone of saltgrass and alkali land"(SVWC 1912b:485-6). Testimony by a landowner in 1915 described "alkali, more than anything else that prevents a good crop" near the Pleasanton marsh complex (Rasmussen 1915:407). An appraiser went on to state that "some areas there won't grow crops under any conditions, because there is alkali in the soil" (Callaghan 1915).

Alkali levels also increased in western Livermore-Amador Valley as the marsh was cleared. Areas that had previously supported valley freshwater marsh and willow thicket were particularly affected. Even before the conversion and drainage of the marsh, areas on the fringes of the perennial wetland would have had high evaporation rates, leading to higher levels of salts (SVWC 1912b). As the marsh was cleared, and irrigation introduced in some places, evaporation rates increased. In 1910, the areas of highest alkali concentrations in the western half of the valley were concentrated to the north of the Pleasanton marsh complex, which had recently undergone intensive drainage and conversion to agriculture (Westover and Van Duyne 1911).

Streams and riparian habitats

The network of creeks that fed and drained the wetlands varied in form across the valley. Most streams flowed through discontinuous channels, although in many cases they maintained some connectivity to the marsh complex or other streams through subsurface flow. The creeks provided groundwater recharge and transported water and sediment to support the wetlands of the valley. This section provides a general discussion of streams draining into the Livermore-Amador Valley and then focuses on the larger streams of the northern sub-watershed—Arroyo del Valle, Arroyo Mocho, Arroyo las Positas, Tassajara Creek, and Arroyo de la Laguna.

The streams feeding into the valley from the north and the south flow over different geologic types in their watersheds, resulting in differing geomorphic forms (Leopold et al. 1964, Graymer et al. 1996; fig. 4.31). The southern streams, Arroyo del Valle and Arroyo Mocho, drain large watersheds composed of erosive and tectonically active uplands, and historically deposited large amounts of coarse alluvium onto their fans in the southern portion of Livermore Valley, developing into broad, braided systems (Williams 1912, Carpenter et al. 1984). The northern tributaries

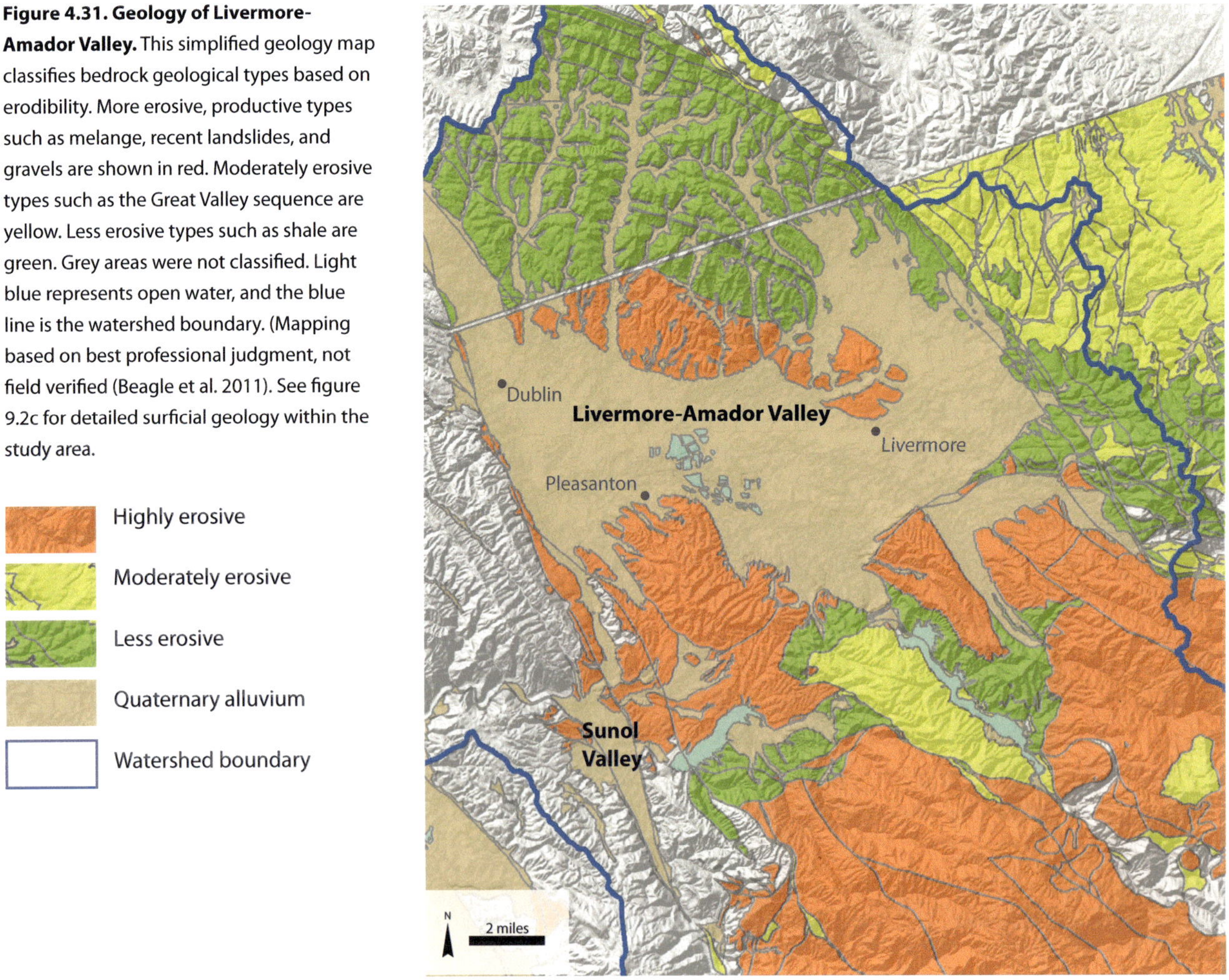

Figure 4.31. Geology of Livermore-Amador Valley. This simplified geology map classifies bedrock geological types based on erodibility. More erosive, productive types such as melange, recent landslides, and gravels are shown in red. Moderately erosive types such as the Great Valley sequence are yellow. Less erosive types such as shale are green. Grey areas were not classified. Light blue represents open water, and the blue line is the watershed boundary. (Mapping based on best professional judgment, not field verified (Beagle et al. 2011). See figure 9.2c for detailed surficial geology within the study area.

drain smaller watersheds, composed largely of Cretaceous and Tertiary rocks, which are more fine-grained (Fisher et al. 1966). The fine-grained, less erosive geology resulted in much smaller, less active fans and less rapid percolation into groundwater (Williams 1912).

These differing types of geology and morphology led to distinct patterns of flooding, vegetation, and active channel widths (see also box 1.1). Arroyo Mocho and Arroyo del Valle, in the southern valley, developed broad, braided channels that supported sycamore alluvial woodland and were prone to flooding across the surrounding valleys. They shifted in course over thousands of years, building up broad gravelly alluvial fans that allowed for rapid percolation to groundwater. By contrast, the smaller northern watersheds draining towards Arroyo las Positas (e.g., Cayetano, Altamont, Cottonwood, Collier, Tassajara) followed narrow single-thread channels that spread and lost definition across the valley. These streams were supported by small watersheds and typically did not carry enough flow to build defined channels across the valley floor. The clay substrate underlying these northern tributaries prevented most percolation through to the underlying aquifers and held flood waters at the surface, helping support the seasonal wetlands. Riparian vegetation was mostly herbaceous, with occasional oak and sycamore trees or clumps of willows. Arroyo de la Laguna, the one waterway draining away from the valley, transported water from all of these drainages towards Alameda Creek. (For a summary of stream and reach characteristics, see table 4.4.)

In addition to these larger systems, a series of smaller drainages flowed into Livermore-Amador Valley from the hills (fig. 4.32). These largely ephemeral streams carried water only during high flows, and many of their channels lost definition as they reached the valley floor, maintaining a surface connection to Arroyo las Positas only in wet years. Several drainages spread into wetlands and then re-formed into defined channels, and a series of swales drained towards Arroyo las Positas.

Figure 4.32. Traces of historical streams. Many sinuous small tributaries drained towards Arroyo del Valle from the south, in most cases sinking into the coarse soil before reaching Del Valle. Today those that remain have been straightened or replaced with development. At the Ruby Hill Country Club, the golf course follows the course of the historical creek. This image shows the historical stream network overlaid on contemporary imagery. (USDA 2009)

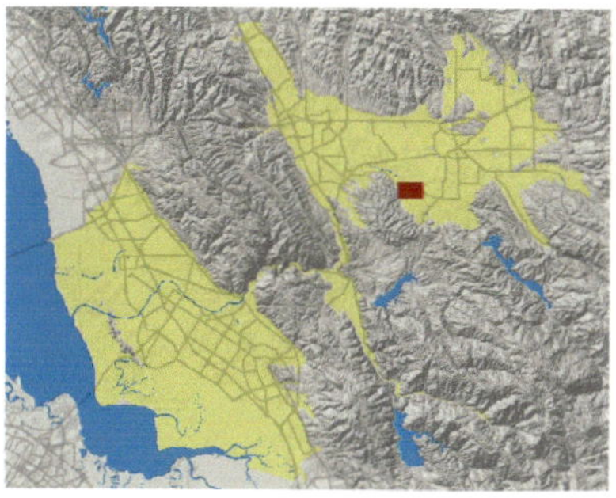

Summer water was rare in creeks across the valley. In the hills many reaches were wet year-round, but upon entering the valley, water had dispersed or disappeared into underlying gravels (Day 1853, Williams 1912, Freeman 1912). GLO surveyors repeatedly described crossing "dry creek" and "dry bed" (fig. 4.33, table 4.3) and even Arroyos Mocho and del Valle sank rapidly in dry seasons (Freeman 1912). However, pools and springs did provide limited water resources (box 4.6). Summer pools were noted in four locations across Livermore-Amador Valley: directly below the perennial portions of Arroyo del Valle and Arroyo Mocho, near Las Positas springs along Altamont Creek, and on Collier Creek, within mapped wet meadow (Day 1853, Bunshah 1910, Fuller 1912). Dry season flow was also present along upper Del Valle and Mocho, extending into the study area

Figure 4.33. GLO flow information for Livermore-Amador Valley. (below) Almost all of the GLO survey data recording flow patterns in Livermore-Amador Valley were recorded by Sherman Day in the summer of 1853, a relatively wet year (150% of average precipitation). His survey followed the four sides of the valley, with an additional transect through the middle, providing reasonably complete coverage. Although the majority of the streams were described as dry, he noted summer flow and pools in a few scattered locations. The following map and table 4.3 summarize flow information for Livermore-Amador Valley.

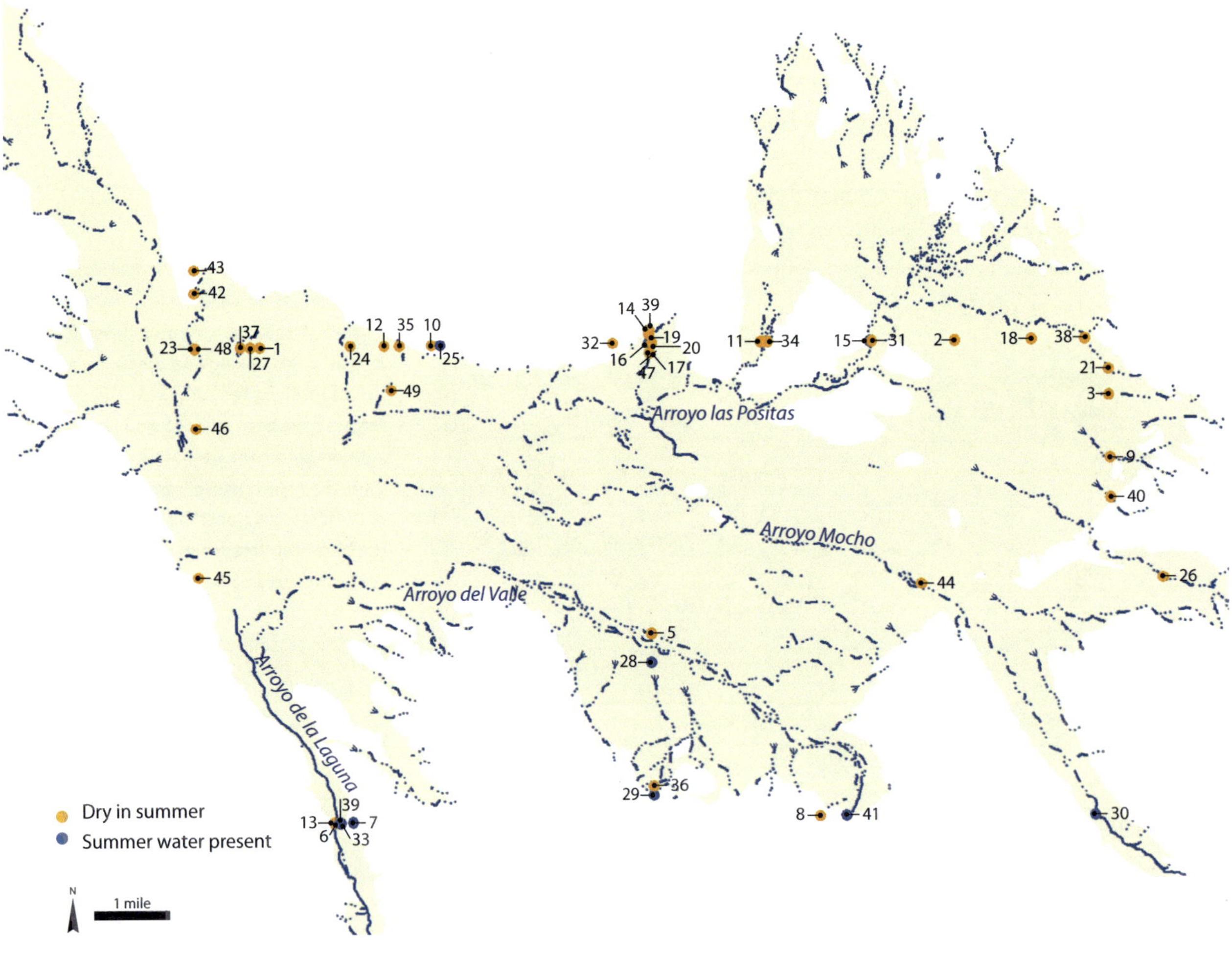

Table 4.3. Summary of GLO flow information for Livermore-Amador Valley.

#	Excerpt	Surveyor	Date	Year
1	leave small dry run	Sherman Day	9-Jul	1853
2	level plain, soil 1st, 2d rate, no timber, no surface water in summer	Sherman Day	28-Jul	1853
3	dry arroyo (4.6 ft) wide	Sherman Day	4-Aug	1853
4	[dry run]	Sherman Day	2-Jul	1853
5	RB dry creek Arroyo Vaya, lined with sycamores	Sherman Day	5-Jul	1853
6	main branch (23.1 ft)wide, copious stream, enough to turn mill, sycamore, thicket in creek valley	Sherman Day	13-Jul	1853
7	abundance of timber and water	Sherman Day	13-Jul	1853
8	dry run	Sherman Day	8-Aug	1853
9	small dry arroyo (49.5 ft) N, opposite shallow level part where channel spreads out, below arroyo cut deep	Sherman Day	4-Aug	1853
10	deep gully (16.5 ft) wide, little water to S, empties into laguna, opposite end of laguna	Sherman Day	9-Jul	1853
11	Left bank dry run (16.5 ft) wide	Sherman Day	28-Jul	1853
12	small dry gully	Sherman Day	9-Jul	1853
13	dry bayou (26.4 ft) wide	Sherman Day	13-Jul	1853
14	[dry run]	Sherman Day	2-Jul	1853
15	creek (46.2 ft) wide, brackish water in pools	Sherman Day	28-Jul	1853
16	white oak in bed of creek at foot of bluff, pools and springs, some hills alkali, no timber	Sherman Day	2-Jul	1853
17	sandy bottom of dry creek	Sherman Day	5-Jul	1853
18	open plain, soil 1st, 2nd rate, no surface water in summer	Sherman Day	28-Jul	1853
19	dry run (13.2 ft) wide	Sherman Day	2-Jul	1853
20	[dry run]	Sherman Day	2-Jul	1853
21	dry arroyo (6.6 ft) wide	Sherman Day	4-Aug	1853
23	small sandy bend, bed of creek, bottoms of creek annually overflowed	Sherman Day	9-Jul	1853
24	small dry gulch	Sherman Day	9-Jul	1853
25	deep gully (33 ft) wide, running water	Sherman Day	9-Jul	1853
26	dry creek (19.8 ft) wide	E. H. Dyer	16-Aug	1862
27	dry deeply cut channel (16.5 ft) wide, sandy bottom	Sherman Day	9-Jul	1853
28	no water in summer, 2 milies above in pools, live oaks and sycs nr cr and on plain	Sherman Day	5-Jul	1853
29	dry run to NE, water in spring on run	Sherman Day	5-Jul	1853
30	Arroyo Mocho (66 ft) wide, copious stream, rocky gravelly bed, lower down pools, then dry	Sherman Day	9-Aug	1853
31	creek (6.6 ft), muddy and strongly impregnated with alkali	E. H. Dyer	19-Mar	1869
32	dry gulley	Sherman Day	11-Jul	1853
33	oak, deep bayou (18.5 ft) wide	Sherman Day	13-Jul	1853
34	cross same run	Sherman Day	28-Jul	1853
35	dry gully (9.9 ft) wide	Sherman Day	9-Jul	1853
36	dry branch	Sherman Day	5-Jul	1853
37	cross small dry run 4 times	Sherman Day	9-Jul	1853
38	dry arroyo (9.9 ft) wide	Sherman Day	28-Jul	1853
39	small bayou with water	Sherman Day	13-Jul	1853
40	dry run (5.3 ft) wide, soil 1st, 2d rate, no timber, no water on surface late in summer	Sherman Day	4-Aug	1853
41	RB creek, water, dam of sandstone rocks, gravelly bed, sycs, alders, willows	Sherman Day	8-Aug	1853
42	dry creek (99 ft) [E]	Richard Howe	11-Oct	1851
43	small dry creek (5.3 ft) E, rich soil, wild oats	Richard Howe	11-Oct	1851
44	dry bed of Arroyo Mocho	E. H. Dyer	14-Nov	1871
45	dry brook running into swamp	Richard Howe	15-Oct	1851
46	overflows wet weather, clover, prickley reeds, small pond, reeds, tuleys, willows	Richard Howe	15-Oct	1851
47	middle of dry brook	W.H. Carlton	1-Feb	1879
48	dry creek (49.5 ft) wide	Richard Howe	15-Oct	1851
49	dry arroyo (13.2 ft) wide	John La Croze	2-Nov	1860

(Day 1853). Surveyor Sherman Day described this pattern along Arroyo Mocho in August: "copious stream running here over a rocky and a gravelly bed—lower down it is in pools, and then entirely dry" (Day 1853). The combination of low flows and pools can support thermal stratification, allowing the bottom of these pools to maintain low water temperatures well into the summer (Hanson Environmental 2002; see fig. 8.10).

Riparian cover along creeks in Livermore-Amador Valley ranged from broad sycamore alluvial woodland to occasional valley oaks and grassland to iodine bush and alkali meadow. Larger streams—such as Arroyo las Positas, Arroyo Mocho, and Arroyo del Valle—were bordered by oaks and sycamores along much of their length, with notable stands of sycamore alluvial woodland along Mocho and Del Valle (fig. 4.34). A few glowing 1880s accounts describe substantial riparian forest along creeks in the valley: "a belt of timber a half mile in width...along the principal creeks. The varieties found here are mainly oak and sycamore. The trees are of good size, few being less than one, and many exceeding four feet in diameter"

Figure 4.34. Riparian trees along Mocho, Del Valle, and Positas. In the largely treeless plain, the riparian trees along the major creeks were an important feature. In this land case map, note the clearly depicted trees along each creek. (USDC ca. 1840e, courtesy The Bancroft Library, UC Berkeley)

BOX 4.6. ARTESIAN SPRINGS

There were three sets of historically significant springs in Livermore-Amador Valley: Las Positas northeast of present-day Livermore near the confluence of Las Positas and Altamont Creek; the *cieneguita* or "little marsh" on the northeastern edge of the Pleasanton Sports & Recreation Park; and Alamilla Springs, still a local feature along San Ramon Road near the intersection of Dublin Boulevard at the southern end of San Ramon Valley (fig. 4.35). Additional springs certainly existed; however, these three were culturally significant features.

All three springs were valued by early settlers and became important local landmarks. Robert Livermore settled near Las Positas and used the springs to water his orchards (Wood 1883). Ownership of the *cieneguita* was a major point of discussion in the distribution of land grants in Livermore-Amador Valley:

> Was anything said ... in respect to the use of the waters of the Cieneguita?
>
> At first Dolores Pacheco wanted all the water and the Messrs Bernal and Pico insisted that it should belong to them; The Governor had told them that he would not grant the land at all, until they agreed among themselves, and then they did agree that the line should run through the middle of the water and Tular. (Amador 1861:229)

Alamilla Springs is depicted on a number of early *diseños* as an *Ojo de Agua* (USDC 1840a,f), which translates as an "eye of water" or natural well, and was used to define the boundaries of the San Ramon land grant (USDC 1840h). Landowner José María Amador stated that "the augmentation [of area] that I asked was to the south of the Alamillos," the spring beside which he built his second house in 1832 (1861:189, Stokle 1968:122).

Springs also occurred near the distributary of Arroyo Mocho, likely due to the presence of the Mocho Fault. A spring on the south side of Oak Knoll was depicted in a map from 1874 (Allardt) and described in a local newspaper: "Water, as if from a large spring, rises from the ground a short distance this side of Oak Knoll& [sic] forming quite a stream, flows into the Arroyo Mocho" (*Echo* 1890).

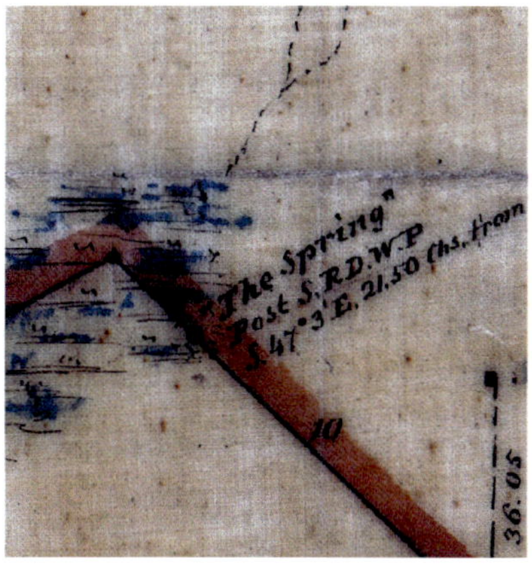

Figure 4.35. Artesian water of Livermore-Amador Valley. Above, "Pleasanton artesian well," possibly part of the historical Pleasanton marsh complex. At left, a *diseño* and confirmation map show springs as prominent features. Top left: Alamilla springs in Dublin labeled "ojo de agua." Bottom left: "The Spring" in the Pleasanton marsh complex. (above: P-2314, courtesy Pleasanton's Museum on Main; left: USDC ca. 1840f and Dyer 1862c, courtesy The Bancroft Library, UC Berkeley)

Along the wooded creeks, in the valley, one might fancy they were riding through a private English park. Immense oaks and sycamores tower overhead, their wide spreading branches, thickly covered with dark green foliage, forming a natural canopy, under which one may ride for miles.

—FAULKNER 1866, DESCRIBING LIVERMORE-AMADOR VALLEY

(Faulkner 1866). Most accounts, however, were more moderate in their descriptions. Riparian forest half a mile wide would likely have been only along the broad braided channel of Arroyo del Valle, if it occurred at all. More commonly, riparian cover consisted of irregularly spaced oak trees and low herbaceous cover.

The following section provides detail on historical form and function of the large creeks tributary to Livermore-Amador Valley. For each of these creeks, we describe the distinct reaches, as well as characteristics such as dry season flow and riparian vegetation.

Arroyo del Valle

Arroyo del Valle shifted from a fairly narrow, single-thread system to a broad, braided channel and then back to a single channel before bifurcating into multiple distributary channels feeding the Pleasanton marsh complex. The transition from braided to single-thread channel reflected the shift in transport capacity and sediment supply in the channel and the corresponding change in soil type from gravelly soils to the clay-based soils of the marsh (Holmes and Nelson 1917).

A large body of evidence documents the shifts in form along Arroyo del Valle. Directly downstream of the reservoir, near the white cliff faces named *Cresta Blanca* ("white crest"), the creek was sinuous and bordered by large riparian oaks and sycamores. There was a nearby vineyard and a local picnicking spot along Arroyo del Valle (both also named Cresta Blanca), and many early photographs document people picnicking along the creek (fig. 4.36). Riparian growth through this reach was fairly dense; GLO surveyor Day described "many sycamores, alders and willows growing in the wide gravelly bed of the creek" (Day 1853:289). Nearby, he found "many sycamores, alders, poplars and willows, and water in pools" (Day 1853).

Going down stream from the Cresta Blanca winery the creek forks into several channels, varying in width from about 40 feet to 100 feet. These channels are separated by higher banks of gravel, standing 4 or 5 feet above the beds of the channels.

—WILLIAMS 1912:296

Del Valle began to split into multiple channels shortly after entering the valley, approximately where the Veteran's Hospital is located today. Historical maps show Arroyo del Valle broadening to develop a braided pattern, with clearly depicted islands between the multiple channels of the creek (Boardman 1870, Duerr 1872a, Allardt 1874, Gibbes 1878, Thompson and West 1878, USGS 1906). These islands could be quite wide: one survey recorded an "island 50 lks [33 ft] wide" (Castro 1876). In the 1940 aerial imagery of this area, multiple scoured channels can be seen. This braided form corresponded with a coarse gravelly substrate and large sediment load; through much of Livermore Valley the strip of soil underlying the creek was characterized in the historical soil survey by "numerous abandoned channels," an underlying "bed of coarse gravel many feet in thickness," and in the contemporary soil survey as "porous sandy soil," or "riverwash" (Westover and Van Duyne 1910:35, Welch et al. 1966; fig. 4.37). Water sank through these gravels, so that much of the flow of the creek continued subsurface (fig. 4.38). This rocky, braided system was described prior to the construction of the dam by resident Paul Banke:

Figure 4.36. "Robertson summer camp in Arroyo del Valle." Many people enjoyed picnicking along the shaded reaches of Del Valle just before it entered Livermore Valley. George Patterson remembered that "It was nice bottom country, with a beautiful creek with sycamore trees, and in the spring we used to go up and have great picnics up there." (Patterson 1987). (image dated "about 1898," 392-138 courtesy Livermore Heritage Guild)

> As you went up the arroyo, the whole landscape particularly the trees, began to change. And you got to parts where there would be deep cuts or washes ... And then the arroyo would be 300-400 feet across and there would be a part that would still have water, and parts that would not have water if it was the summer, and then in the winter you might have a hard time getting down in there because it would run pretty good at times.
>
> It was one of those arroyos that was more like a delta, or various channels, than an arroyo. It wasn't one big channel. There were big rocks, big boulders, and there would be islands that might have a bunch of trees, and sometimes dense brush and poison oak, and there would be this little island that would be its own little microcosm.

Through this reach, Del Valle shifted from a perennial to an intermittent stream. At the edge of the valley, a mile downstream of the reservoir, Sherman Day noted "a fine stream of water, running over a dam of sandstone rocks" in August 1853 (Day 1853:289). Another mile and a half downstream, in Sycamore Grove Park, he described water "in pools." As water continued to sink through the gravels, he found "no water in summer" at Isabel Avenue, another two miles further downstream (1853). The pools were part of the gradual transition as water sank further below the surface.

In the braided reach of the creek, the riparian corridor may have been up to 1,500 feet wide. In some places, even wider outer relic floodplain terraces are still visible in the LiDAR survey and historical aerials, extending the potential corridor width up to 3,000 feet (fig. 4.39). In some places, this wide, braided reach supported sycamore alluvial woodland in the channel and possibly valley oaks (*Quercus lobata*) further away from the active channel on terraces (fig. 4.40; Duerr 1872a, USDA 1939, Keeler-Wolf et al. 1996:65). Other species, such as sedges (*Carex* spp.) and annual grasses,

Figure 4.37. Arroyo del Valle in 1897 and 2011. The top image shows Cresta Blanca, just downstream of the current reservoir. Note the dry gravelly stream bed with scrub cover. Large riparian trees grow away from the active channels. The bottom image shows Sycamore Grove Park in 2011, when water was relatively high. (top: image courtesy Livermore Heritage Guild; bottom: by Amy Wolitzer 2011)

The water is confined to a very narrow channel in most places ... here is a little island, and there is a little island, and the stream runs on each side and so on; you cannot make any reductions for the land that is used by the stream. In some places the islands in the Arroyo Valle are quite large.

—SCHWEEN 1915:244, DESCRIBING ARROYO DEL VALLE

would have flourished seasonally. Remnants of the sycamore alluvial woodland are still present (Olson 1991).

The sycamore alluvial woodland depicted on the habitat map includes areas of varying densities, from mostly grassland or unvegetated gravels to true sycamore alluvial woodland. Rather than capturing individual stands, we mapped zones where sycamore alluvial woodland was the dominant habitat type. As a result, this area cannot be compared directly with contemporary sycamore alluvial woodland mapping (see Keeler-Wolf et al. 1996).

In contrast to the braided reach, the portion of Del Valle in the vicinity of Pleasanton was a single-thread meandering channel (Boardman 1870, Allardt 1874, Thompson and West 1878, USGS 1906). Historically, this

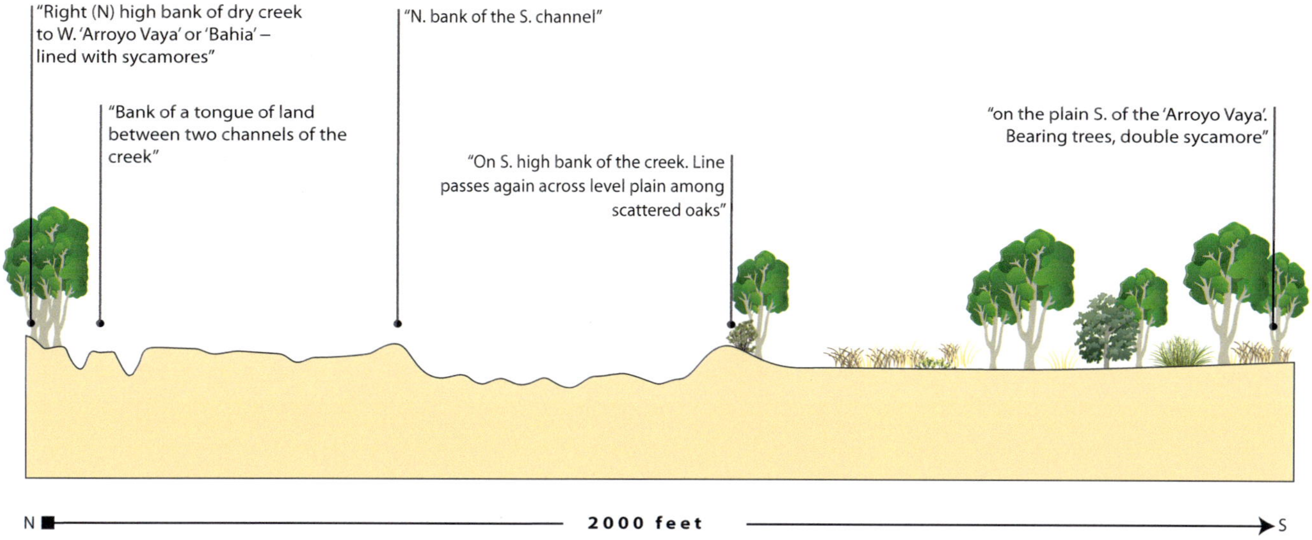

Figure 4.38. General Land Office survey transect by Sherman Day, crossing Arroyo del Valle from north to south on July 5, 1853. In his summary of the line he surveyed, Day noted, "Soil 2nd and 3d rate, might be much improved by cultivation. No water in summer near the line. On the same creek two miles above water stands in pools. Surface level, fine oaks, live oaks and sycamores near the creek and on the plain."

lower reach began in what is now Shadow Cliffs Regional Recreational Area, where the dominant substrate shifted from gravel to clay (mapped as fine-grained Livermore silty fine sandy loam; Westover and Van Duyne 1910). By this point, the stream had dropped its load of coarse gravels on its fan and lost most surface flow. Cyril Williams described the transition in 1912 from a "gravel channel several hundred feet wide" to a narrowing channel with "outcrops of yellow clay…in the bed and sides of the channel." Then, "clayey exposures" increased "until, at point almost opposite the brick factory, gravel disappears entirely, the channel of the creek consisting of almost a pure clay with some loam and other sediment, and the banks being almost vertical and consisting of a yellowish clay" (Williams 1912:302). Williams noted that the brickyard and gravel company—industries based on these distinct substrates—were located on either side of the point of transition from gravel to clay in the bed of Del Valle, evidence that this was a well-recognized (and utilized) transition. The consistent placement of this transition point over time suggests that geologic and geomorphic factors were controlling this transition. The sediment types sorted naturally so that larger, heavier gravels were deposited as the stream lost power upstream, and fine sediments washed closer to the marsh.

Through this reach, sycamores still dominated the riparian cover: "The town [Pleasanton] was formerly called *Alisal*, or the Sycamores, on account of the numerous large trees of this species that lined the bed of the Arroyo Valle in its course to the Laguna close by" (Halley 1876; see also Platt 1910 in Davis 1976; fig. 4.42). Flow in this reach may have resurfaced in places, but was largely intermittent and subsurface. In areas of high groundwater and perennial flow, the channel was flanked with willow groves (e.g., red willow (*Salix laevigata*), arroyo willow (*S. lasiolepis*), etc.) and other aquatic plant species (La Croze 1860). The gravel beds continued underground, carrying water to the Pleasanton marsh complex (Williams 1912, Fisher et al. 1966).

Figure 4.39. Abandoned terraces. Distinct terraces are visible beyond the active, vegetated corridor in this imagery from 1940 (A) and are still visible in 2007 detailed elevation mapping (B). These terraces extend beyond our mapped banks and may represent a much more extensive historical spread of the creek, or, more likely, earlier epochs of river floodplain. Our fairly conservative mapping captures only the more clearly defined central zone. (C: USDA 2005; A: LiDAR 2007; B: USDA 1940)

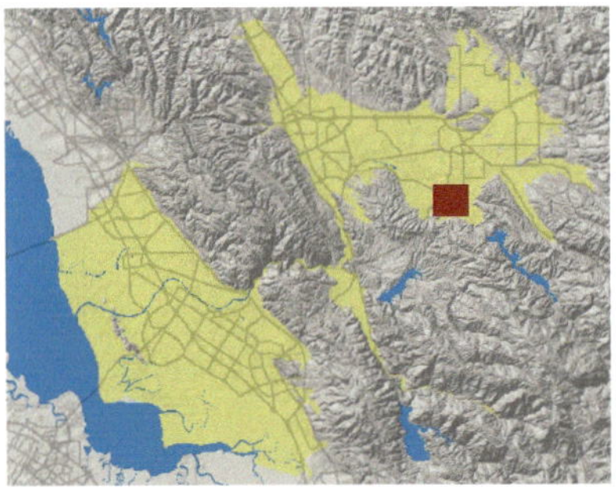

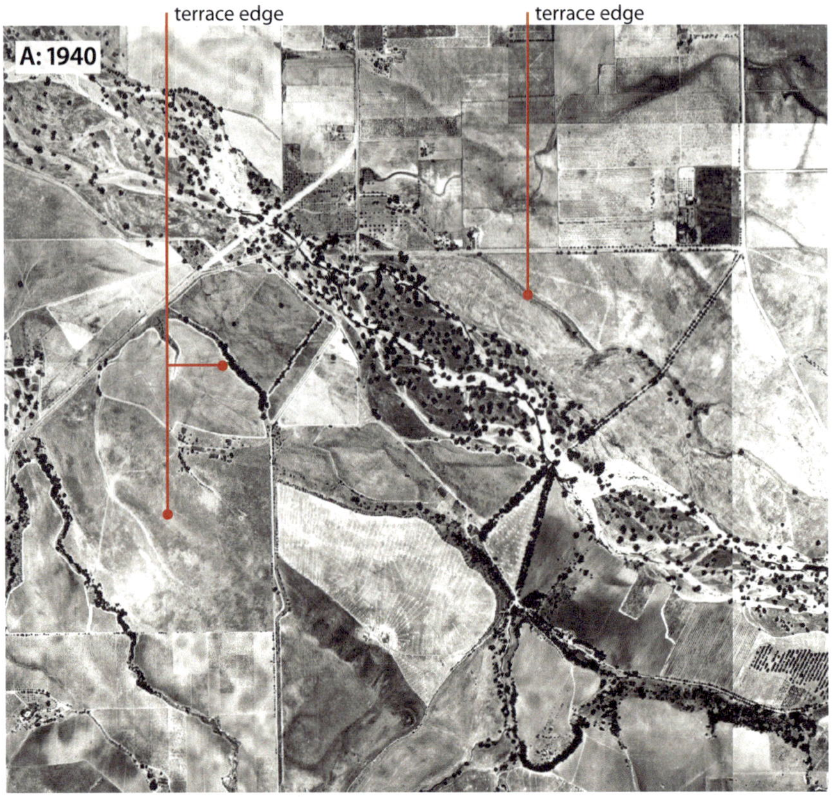

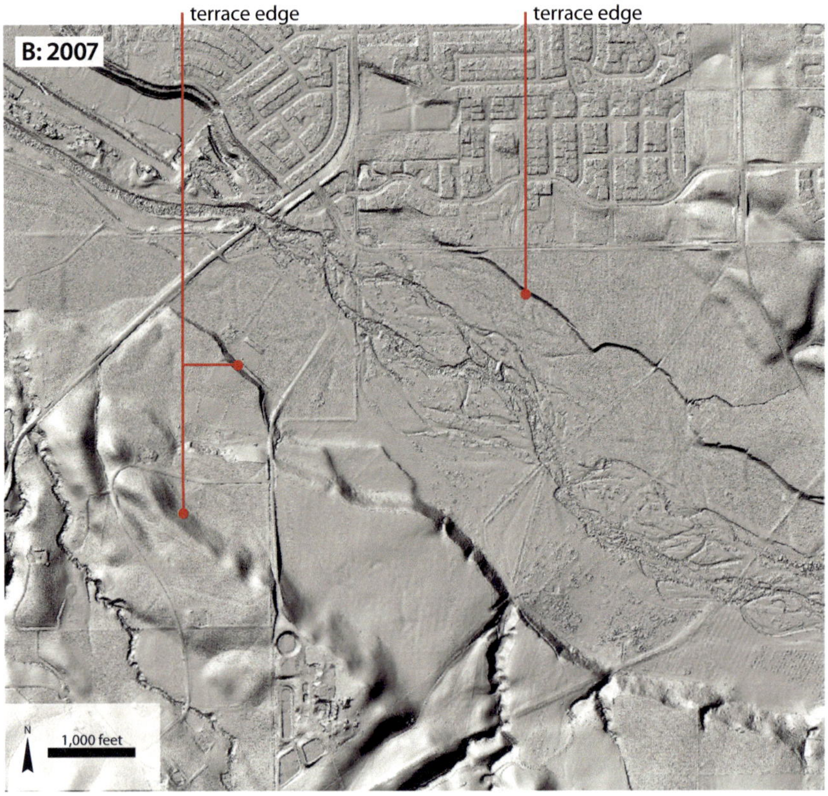

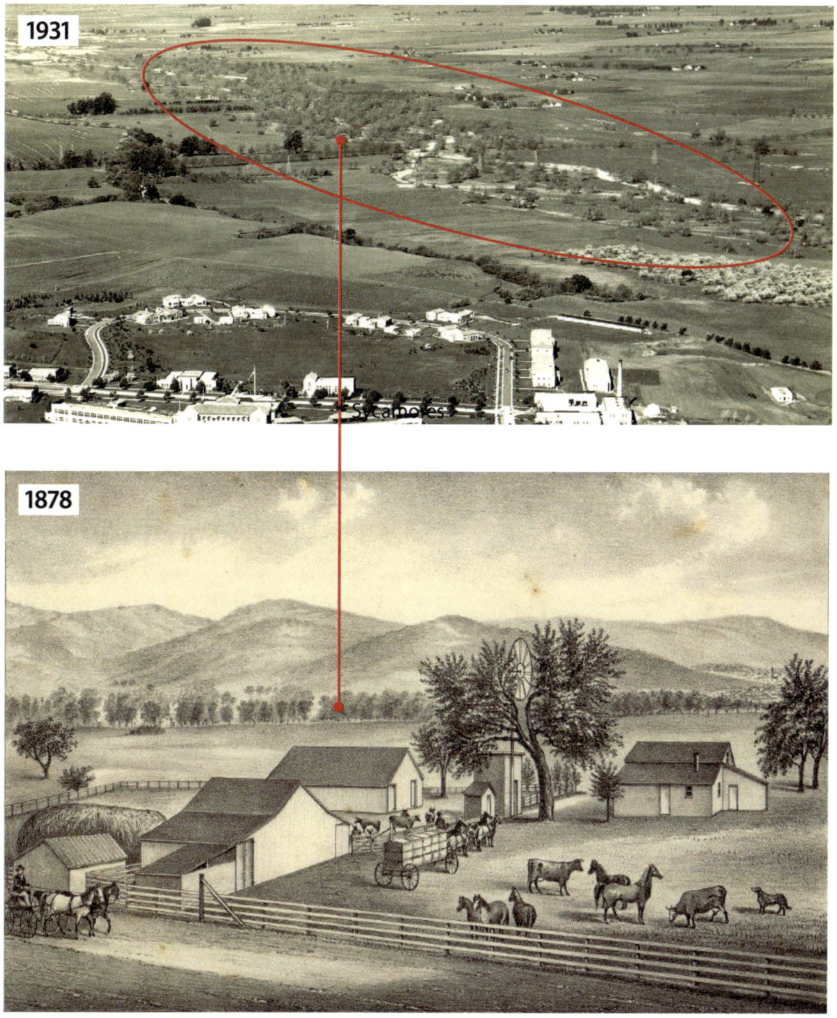

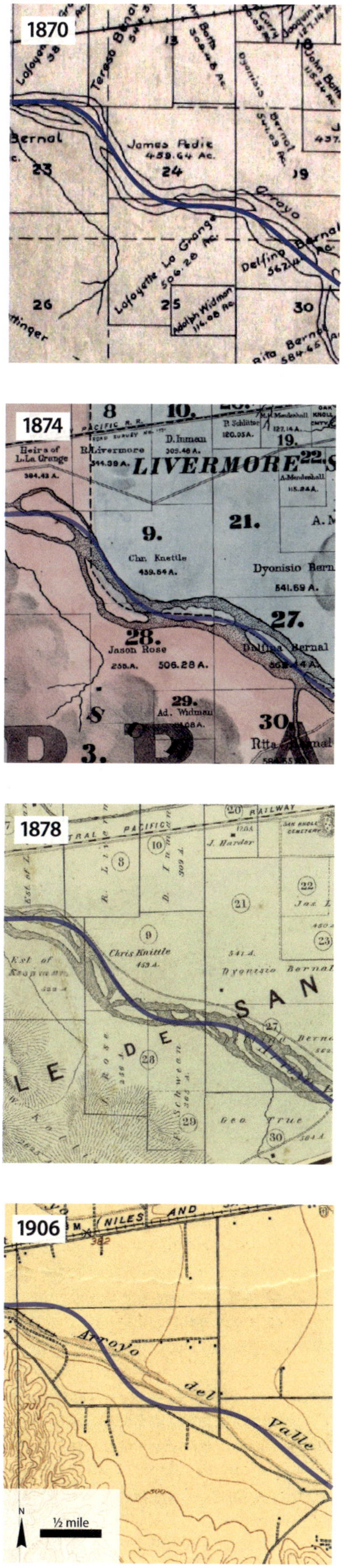

Figure 4.40. Two images of sycamores along Arroyo del Valle. (above) Sycamores along Arroyo del Valle stand out in the otherwise relatively treeless Livermore Valley. In the top image, dated February 12, 1931, dotted sycamores can be seen marking the course of the creek leading north away from the Veteran's Hospital. The bottom image shows the property of a landowner living along the creek, near the intersection of Isabel Ave and Stanley Blvd. Sycamores along Del Valle can clearly be seen in the distance. (top: courtesy California Historical Society, CHS2012.934.tif; bottom: Thompson and West 1878, courtesy David Rumsey Map Collection)

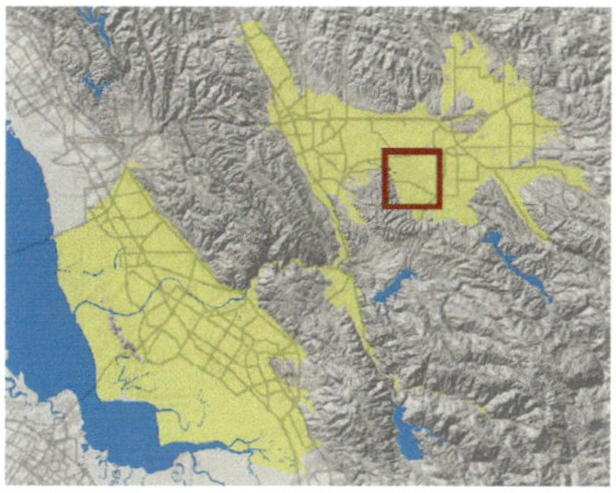

Figure 4.41. Early gravel mining likely caused a shift in the course of Arroyo del Valle (right). Maps from the 1870s show a large S-bend that is no longer shown in the similarly detailed 1906 USGS quad. (top to bottom: Boardman 1870, courtesy Museum of Local History, Fremont, CA; Allardt 1874, courtesy The Bancroft Library, UC Berkeley; Thompson and West 1878, courtesy David Rumsey Map Collection; USGS 1906)

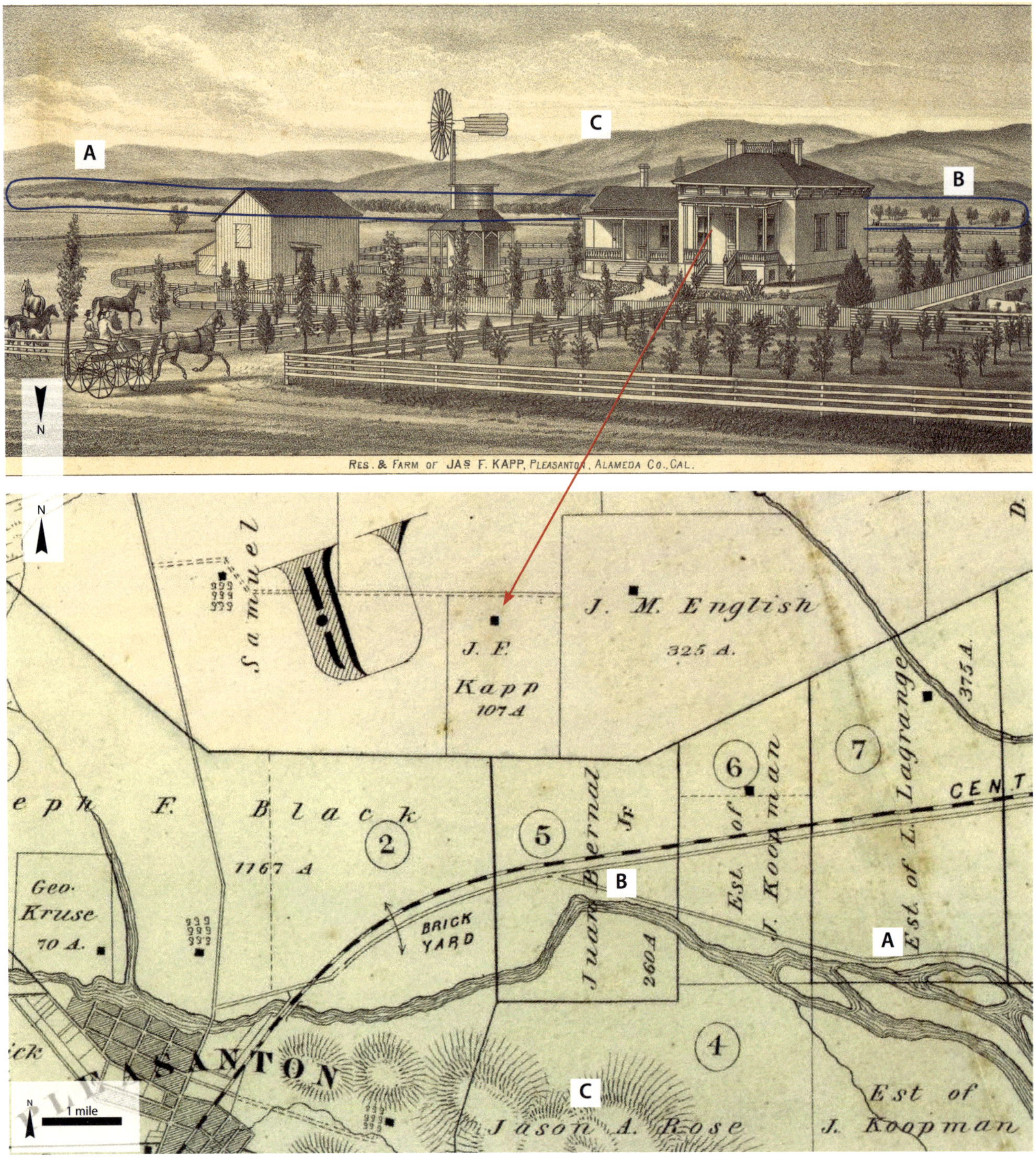

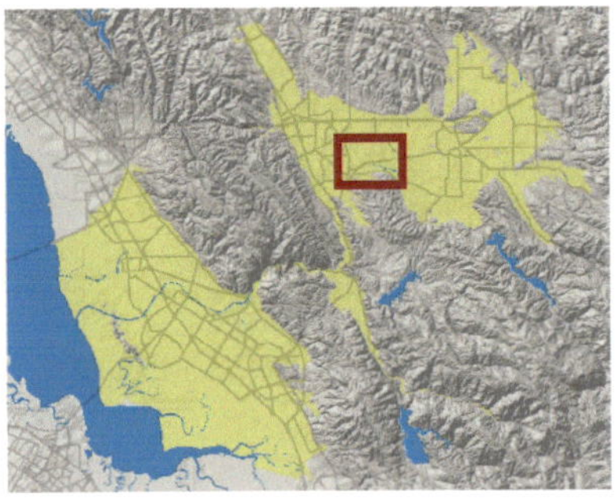

Figure 4.42. Riparian transition along Del Valle, 1878. The paired images above show Arroyo del Valle at the point where riparian cover shifts from sycamore alluvial woodland to more sparse cover. Using the map at bottom we were able to place the lithograph, which looks south towards the creek. The line of riparian trees in the background represents the sycamore alluvial woodland along Arroyo del Valle (A). When the creek becomes visible again on the right side of the house, further downstream (B), riparian cover has shifted to much more sparse cover as the creek has re-formed to a single thread. The hills in the background (C) help orient the image. (Thompson and West 1878, courtesy David Rumsey Map Collection)

As Arroyo del Valle neared the Pleasanton marsh complex, it spread into broad distributaries leading north and west towards the marsh (see Boardman 1866, Boardman 1870, Allardt 1874, Cash 1875, Thompson and West 1878, Allardt [1880]1907). These distributaries marked the downstream extent of Del Valle's natural levees and confined the southern portion of the Pleasanton marsh complex to a narrow strip along the western hills of Livermore-Amador Valley. Finer sediments continued to wash along the course of Arroyo del Valle, finally reaching the Pleasanton marsh complex during high flow events as sheet flow.

After entering the marsh, Del Valle quickly lost channel definition, although the exact endpoint varied with the year (in dry years Del Valle did not maintain surface flow to the marsh at all—see Williams 1912:33). In addition to the large distributary channels shown on maps, the creek may have had many smaller distributary channels. There is no indication that Del Valle connected directly with Arroyo de la Laguna as it does today; in fact many maps show how this connection was created over time as the marsh was ditched and drained, with a substantial connection only in place by 1910 (e.g., SVWC 1910; fig. 4.43).

... where the walks and drives along the sycamore and willow lined banks of the winding Arroyo Mocho would render life ten thousand times worth the living.

—SAN FRANCISCO CALL 1865

Arroyo Mocho

Arroyo Mocho historically maintained a similar pattern as Del Valle, and shifted from a narrow single-thread channel in its semi-confined box canyon to a broad stream corridor with occasional braided reaches across

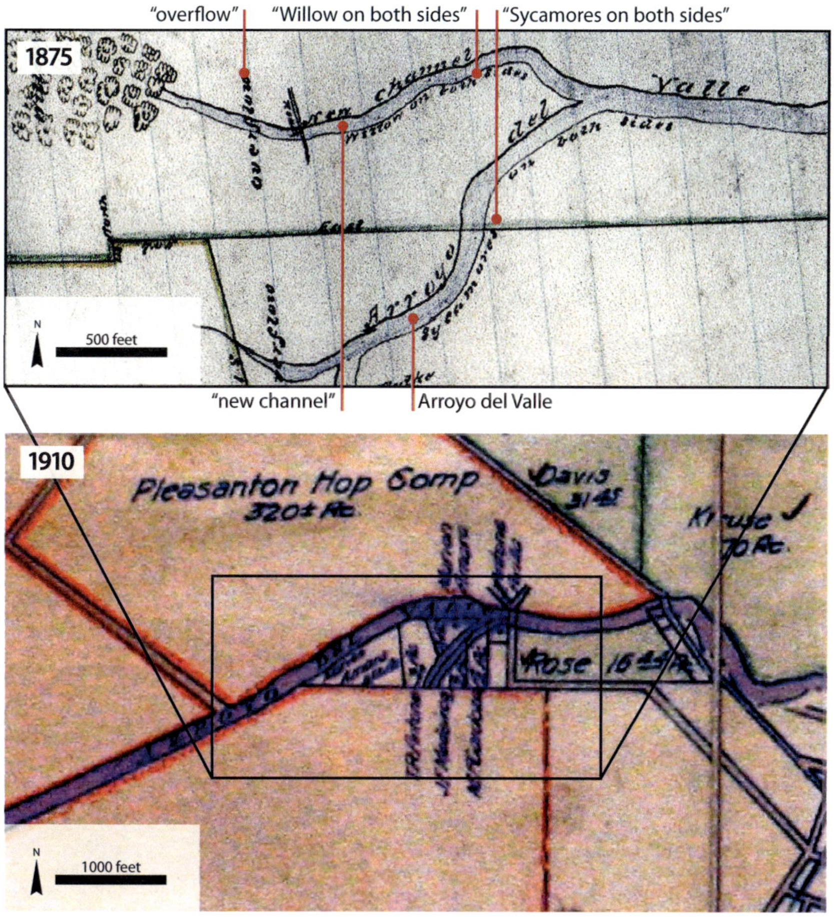

Figure 4.43. The primary channel conveying flow from Del Valle to the marsh likely shifted over time, and even shifted during the historical period. This 1875 map labels new and old beds of Del Valle. By 1910 this had shifted again, as the flow of Del Valle was directed into the Rose Canal. The top image is zoomed to the boxed area in the lower image. (Cash 1875, courtesy Alameda County Department of Public Works; SVWC 1910, courtesy Earth Sciences and Map Library, UC Berkeley)

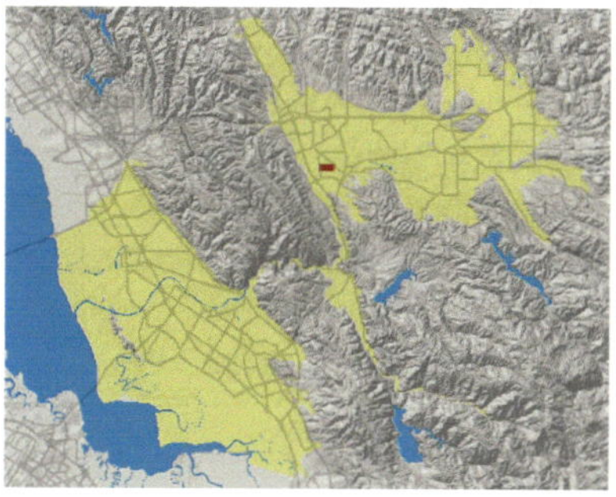

its fan, ending in a series of distributaries. However, Mocho was supplied by a smaller watershed (60 square miles to Del Valle's 165 square miles) and was described as "narrower and more shifting" than Arroyo del Valle (Fuller 1912:41). These shifts are still visible as terraces in the 2007 LiDAR survey as the creek meanders across a long narrow valley before entering Livermore Valley.

[Mocho's] upper channel is shallow, overflows at very high flood and is liable to shift at every overflow.

—FULLER 1912:55

Over time, Arroyo Mocho shifted back and forth between low banks (Fuller 1912:41) that were in some places flanked by benches as much as 25 feet high (Carlton 1874). One detailed GLO survey transect provided detailed information on the width of the historical banks. Three surveyors crossed Arroyo Mocho at the same point between March 1863 and April 1874. The most detailed survey was made by William Carlton, crossing the creek twice (once going north, and once going south) in April 1874. He described banks 200 feet wide and a "main channel" of Arroyo Mocho. Six hundred feet south of the southern bank, he also described ascending a bench 25 feet high. In the 1940 aerial photographs, this bench is still visible, and traces can still be followed in 2009 NAIP imagery and 2007 LiDAR surveys.

The historical record also indicates that at least by 1900, Arroyo Mocho was undersized for high flows and frequently flooded across the valley. It was an actively aggrading stream. The creek historically occupied a much broader zone than it does today (fig. 4.44). Cyril Williams described the undersized creek in 1912: "the channel of the creek is wide and shallow, and in times of ordinary or heavy floods these banks overflow, upon occasion even through the town of Livermore" (Williams 1912:305). As late as the 1950s, a resident described a "wide floodplain" near Livermore, where the creek spread to create ponds "out on the floodplain" (Fletcher pers. comm.).

Similar to Arroyo del Valle, Arroyo Mocho maintained perennial flow as it entered the valley, which then slowly sank subsurface through the gravels. Towards the top of the alluvial valley, surveyor Sherman Day described a "copious stream running here over a rocky and a gravelly bed" in August (1853:293). Further downstream he described pools, and then a stream that was "entirely dry." Pools were also described in a local newspaper, drying

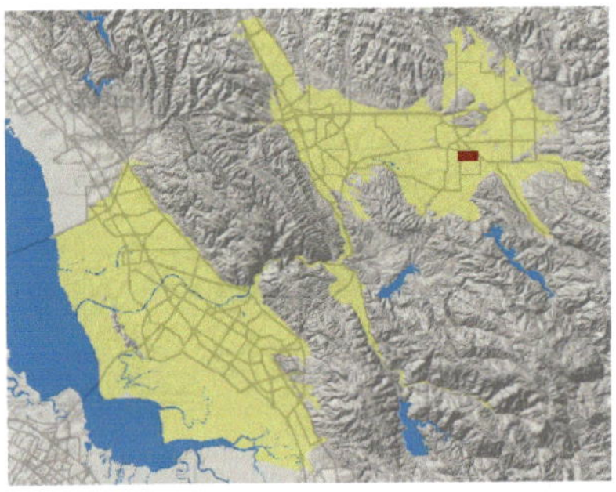

Figure 4.44. Overlay of historical banks on present-day Livermore. Over time, the town of Livermore has encroached on Arroyo Mocho. The creek historically occupied a much broader area than it does today, and appears to have occupied swaths of land that are now covered with agriculture and residential neighborhoods. The blue line shows the historical stream course, and the brown indicates the historical location of the outer banks. (USDA 2005)

"under severe hot spell" in June 1910 (Bunshah 1910). A resident described the contrast between summer and winter in the 1950s as follows: "from June to November ... there was no water, it was very hot and dry but during the spring time, the creek would start flowing down and it would become a series of ponds, and it would fill up with moss, and there would be lots of toads and then these side ponds would have the tiger salamanders" (Fletcher pers. comm.).

Arroyo Mocho supported a riparian corridor that varied in width and density, but most likely averaged 800 feet wide, substantially narrower than Del Valle. This zone of creek influence included sycamore alluvial woodland, oaks, and bars and islands, and the creek was flanked by active benches and abandoned terraces. Surveyor Day observed white (valley) oaks in 1853 "on a level clay, flat bench, being the 3d bottom of the 'Arroyo Mocho', coming out of the mountains" (Day 1853). Scalebroom scrub (*Lepidospartum squamatum*) was also recorded along Arroyo Mocho (Sharsmith 1945). Scalebroom occurs primarily in the Great Basin and deserts and is associated with intermittently flooded alluvial deposits, such as sycamore alluvial woodland and mule fat scrub (*Baccharis salicifolia*) (Sharsmith 1945, Magney 1992, Sawyer et al. 2009).

As Arroyo Mocho approached the historical town of Livermore (near the current intersection of Holmes Street and Murietta Boulevard), it became more braided. Multiple channels can be seen in aerial photographs from 1940, and in times of flood the creek scoured beyond these channels to banks on either side, depicted in early maps as a substantially broader reach (Allardt 1874, Thompson and West 1878, USDA 1940). The creek periodically overflowed even these broader banks into the town of Livermore (Williams 1912:305). Gravel mining could exacerbate erosion—"the consequent washing away of the banks and bed of the creek ... made the crossing quite dangerous" according to the *Echo* (1894b; fig. 4.45). The dynamic and erosive nature of the stream and its proximity to Livermore may explain the more intensive ditching of Arroyo Mocho through this stretch compared with further upstream.

The arroyos Mocho and Valle were both, at times, during the late storm, unfordable, causing a deal of inconvenience. When will the powers that be give us needed bridges—instead of everlastingly hauling gravel?

—*ECHO* 1894C, COMPLAINING ABOUT GRAVEL REMOVAL FROM THE STREAM

Figure 4.45. "Hauling gravel from Mocho Creek bed." Arroyo Mocho transported large volumes of sediment, including cobbles and gravels, which early residents of the valley excavated for road surfaces long before commercial gravel mining began in the area. (N252, courtesy Livermore Heritage Guild)

Near Livermore, sycamore alluvial woodland disappeared and tree cover was likely more sparse due to the intermittent water supply, but large riparian trees continued. Near Livermore a local historian recorded that "the banks [of Arroyo Mocho] ... abound with oak and sycamore trees of great size" (Wood 1883). GLO surveyor Sherman Day crossed Arroyo Mocho just over half a mile west of Oak Knoll (fig. 4.46), and described a creek 60 feet wide, with "a wide gravel bottom ... the creek is lined with sycamores along its margin" (Day 1853). The zone of riparian influence may have been as wide as 600 to 1,300 feet across the stream, as measured from signatures in the 1940s aerial photos and terrace remnants in the 2007 LiDAR survey.

West of Oak Knoll Cemetery, downstream of the Western Pacific Railroad crossing, Arroyo Mocho crossed a thrust fault (the Livermore Fault), and historically began to lose its defined channel (Day 1853, Halley 1876). Although different sources propose different reasons for the name Mocho (translated as "cut-off"), surveyor Sherman Day described that it was so named "because it terminates about 2 miles W. of Livermore's, by spreading itself out on the plain" (Day 1853). In times of flood, surface flow continued as sheet flow or through poorly defined and discontinuous channels, but much of the flow had already sunk into the coarse gravels of the valley. Even later, when Mocho flowed through a ditch, it did not maintain surface flow through the ditch: "In ordinary or critical years the Arroyo Mocho Creek sinks into the Livermore Valley gravels, and seldom in such years reaches the Laguna Creek as surface flow" (Williams 1912:571; see also Gutmann 1919:6). This pattern continues today.

The Mocho flows a short distance northwest of the town of Livermore, near which it forks into small rivulets and disappears.

—HALLEY 1876:494

Figure 4.46. Arroyo Mocho at Oak Knoll. Southwest of the intersection of Murrieta and E Stanley boulevards is a small hill that was an important local landmark in an otherwise flat plain, located just west of the historical town of Livermore. This hill became the site of Oak Knoll Cemetery, which has existed at least since the 1870s (Thompson and West 1878). Arroyo Mocho curved north around Oak Knoll, where it followed a "vertical bank of mixed gravel and clay" (Williams 1912:305). The coarse gravels of the stream bed can be clearly seen in this circa 1912 photo. (photo from Williams 1912, courtesy Laurel Collins)

ca. 1912

Although lower Arroyo Mocho was dry and gravelly, the presence of the Livermore Fault created a series of springs (Tibbetts 1907a, Unruh and Sawyer 1997). Continuing south along current Kitty Hawk Road, surveyor Sherman Day (1853) crossed an "open gravelly plain" and described "water in a swamp slough" to the east. This "swamp slough" was likely a small perennial wetland. By July, Arroyo Mocho would have completely dried up, so a wet slough must have been fed by a local spring. This wetland occurred near a small distributary channel shown on the earliest detailed maps overflowing the path of the railroad (Unknown 1850, Boardman 1870, Allardt 1874). The presence of the perennial wetland and spring may have exacerbated need for drainage, and made this a more desirable location to direct the flow of Arroyo Mocho when it was ditched.

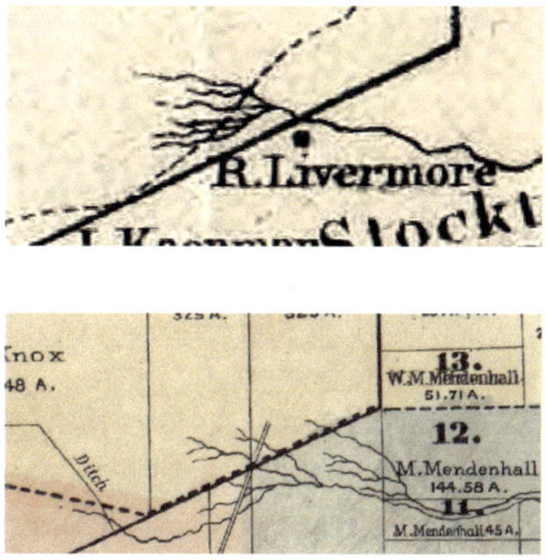

Figure 4.47. Mocho distributaries. Early maps consistently show Mocho's distributaries with many small branching channels, although the exact location varies. (top: Whitney 1873, courtesy David Rumsey Map Collection; bottom: Allardt 1874, courtesy The Bancroft Library, UC Berkeley)

The exact position at which Arroyo Mocho spread into distributary channels varied over time, but historical maps and GLO survey notes converge on a general location for this transition (e.g., Unknown 1850, Higley 1857a, La Croze 1860, Dyer 1862a, Healy 1863, Boardman 1870, Whitney 1873, Allardt 1874; figure 4.47). Early textual accounts describe a distributary occurring two miles east of the Pleasanton marsh complex (Day 1853, Halley 1876). Subsurface flow from Mocho seeped towards the Pleasanton marsh complex and Arroyo las Positas, and was seen as an important groundwater source (Williams 1912). Even today, this creek is used to recharge groundwater.

The lack of a defined channel in the lower part of Arroyo Mocho and the unpredictability and varying nature of sheet flow patterns likely motivated early efforts to ditch and redirect the creek. Historically the creek took a northerly meandering course before forming a series of distributaries (Unknown 1850, Higley 1857a, La Croze 1860, Boardman 1870, Whitney 1873, King 1880). However, by 1889 maps show Arroyo Mocho flowing further south, ditched to run parallel with the railroad before swinging north to connect with Arroyo Las Positas (Nusbuamer 1889a; fig. 4.48).

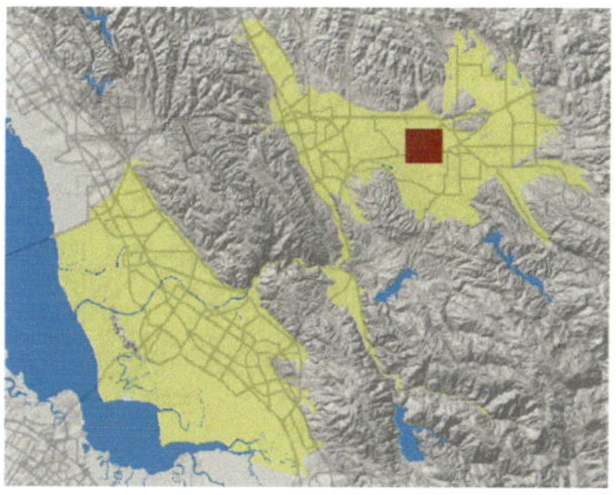

Figure 4.48. The distributary channels of Arroyo Mocho are still visible in 1940 aerials (C). Historical maps indicate two major branches, which are shown as depressions by USGS in 1906 (A), and are underlain with the same soil type (B), an indication that sediment from the creek was deposited by both branches. These traces help establish the likely historical courses of the creek (A: USGS 1906; B: Westover and Van Duyne 1910; C: USDA 1940)

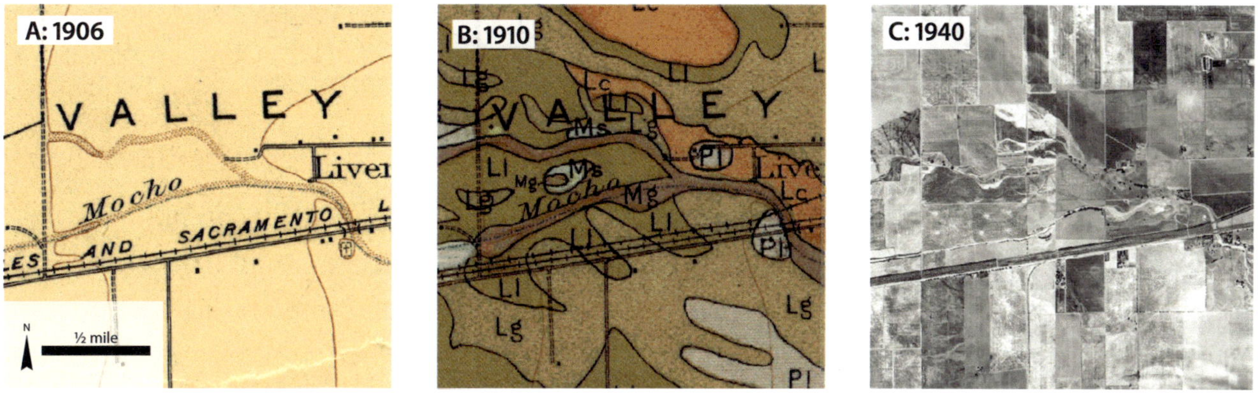

This may have occurred as early as 1878 (Thompson and West 1878). Riparian vegetation in this reach of the creek would have been minimal, consisting mainly of hydrophytic grasses as the stream was integrated into the larger matrix of oak savanna and grassland before reaching the marsh.

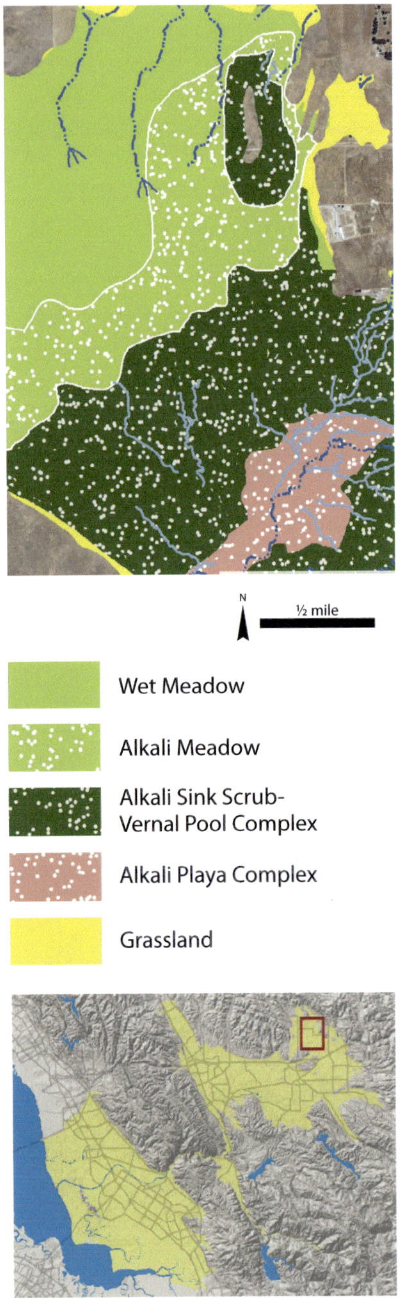

Figure 4.49. Discontinuous channels draining into Springtown. Creeks flowed south from the Altamont Hills to form distributaries in the seasonal wetlands of Springtown. Some of these channels re-formed as swales in light blue in the center of the alkali sink.

Arroyo las Positas and eastern Livermore Valley

The seasonal wetlands of eastern Livermore Valley were fed by a number of small creeks flowing through discontinuous channels. Upper Arroyo las Positas (now channelized to flow through eastern Livermore Valley) was historically one of many creeks to spread across the valley, and received its name only where a more defined channel formed at the confluence with Altamont Creek (Higley 1857a, Whitney 1873, Allardt 1874, Nusbaumer 1889a, Westover and Van Duyne 1910). The flow from these many discontinuous channels drained through surface swales and subsurface flow to reach the Springtown alkali sink and Las Positas springs (Williams 1912, Coats et al. 1988; fig. 4.49). Coats et al. (1988) describes three distinct hydrologic zones surrounding the alkali sink: upland flows through defined channels; a recharge zone through loams deposited at the foot up the hills; and the swales and wetlands of the alkali sink. (See box 1.1 for a conceptual overview of stream patterns and hydrologic transitions.)

Several of the small drainages flowing from the hills followed a similar pattern. These creeks had defined channels in the upper portion of the valley and re-formed into a channel at the bottom of the valley, but for a portion in the middle all flow sank below the surface (fig. 4.50).

These small, discontinuous streams helped create the alkaline conditions of eastern Livermore. Creeks draining from the Altamont uplands carried salts from the marine sediments deposited there, which then percolated into the groundwater and concentrated in the soils through evaporation (Trusk 1854 in Williams 1912, Westover and Van Duyne 1911, Coats et al. 1988). Historically, Altamont Creek maintained a continuous and much more sinuous channel through the sink than today, likely frequently overflowing its banks and depositing salts across the surrounding area (USDA 1940, Coats et al. 1988).

The major creek flowing from this area was Arroyo las Positas, which historically re-formed near Las Positas Springs, and joined Altamont Creek near their present-day confluence. (Las Positas springs were located just north of Interstate 580, east of Las Colinas Road.) Referred to as "Livermore's Creek" in early testimony, Las Positas, or "little pools," was named for the springs that fed it, maintaining the creek as an important water supply in the otherwise dry eastern plain (Bryant [1848]1985). Early travelers and Gold Rush 49ers stopped at Las Positas and Livermore's rancho, as this was their last source of water before the San Joaquin River as they headed east (Moerenhout [1849]1935).

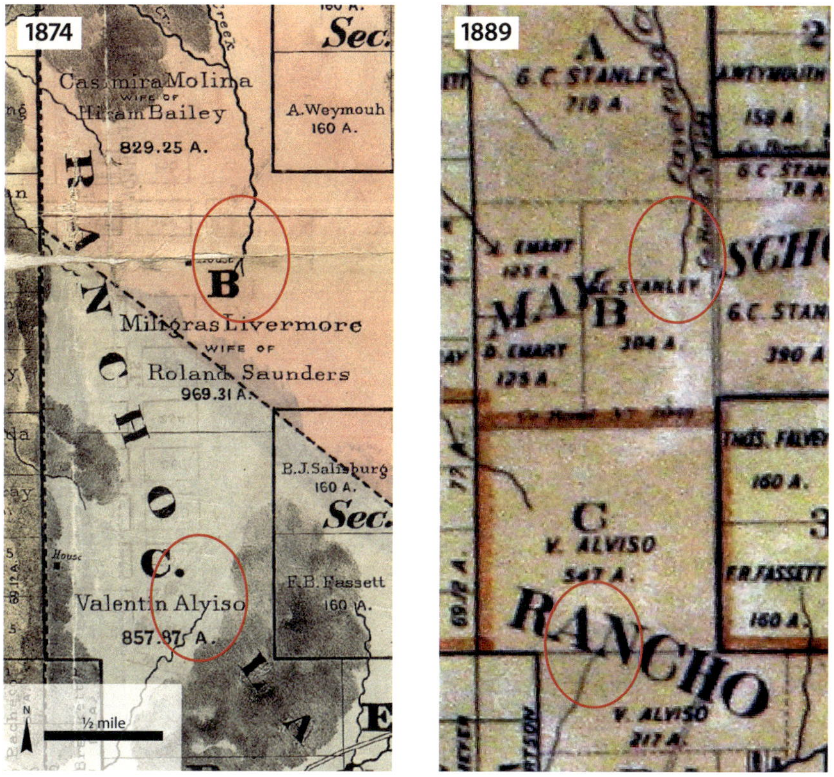

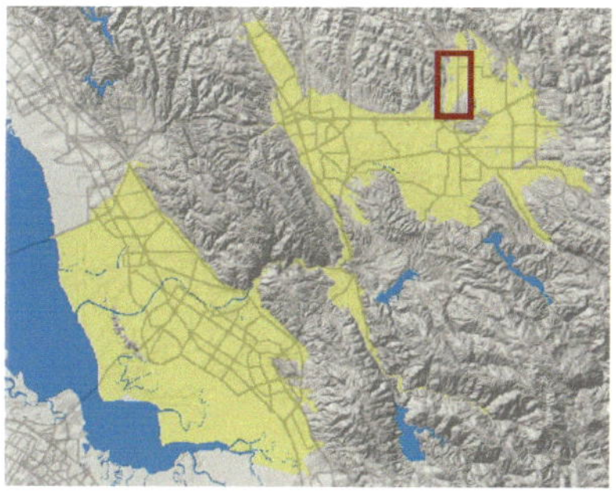

Figure 4.50. Cayetano Creek is depicted with a well-defined channel from where it exits the hills to a point over 1.5 miles downstream, just north of Hartman Road near North Livermore Ave, where it formed a distributary. Half a mile further downstream, the creek re-formed into a few defined channels. Towards the confluence with Arroyo las Positas, a series of GLO survey points document the presence of a defined channel for "San Castano" creek. This pattern of interrupted continuous channels occurred along several of the small creeks in the study area. (Allardt 1874, courtesy The Bancroft Library, UC Berkeley; Nusbaumer and Boardman 1889, courtesy Museum of Local History, Fremont)

Immediately below Las Positas springs, the creek followed a confined path between two small knolls, with a vegetated corridor 130 feet wide, likely a mix of willows and herbaceous vegetation (USDA 1940). Las Positas here was described as "a shallow stream … supplying us with excellent water" (Ryan 1850). The creek appears to have followed several distinct courses over time, and was likely ditched to provide irrigation (see multiple channels in Thompson and West 1878, USGS 1907).

Springs through this area along Las Positas were fed by high groundwater, constricted between the hills and a thick clay cap (Williams 1912:49). As the city of Livermore developed, it initially relied on water from artesian springs in this area, although water here was much more alkaline than that of the Pleasanton marsh complex (Williams 1912:227). These springs also helped support willows and a wet meadow (USDC 1840e, Allardt 1874, USGS 1907; fig. 4.51). The willow thicket may have been relatively dry—surveyors recorded that a hotel was constructed there (Day 1853).

The northeast corner [of Livermore Valley] is barren of streams, but such as flow from small springs.

—HALLEY 1876:494, DESCRIBING LIVERMORE VALLEY

Some dry-season flow may have existed along Arroyo las Positas near the Las Positas springs. Sherman Day described a "swampy water course" in July 1853 along Altamont Creek near Las Positas (1853). Surveyor Lewis recorded "running water" along a portion of the creek (1861:511).

Lower Arroyo las Positas, below the springs and Altamont Creek, maintained a continuous single thread channel to the Pleasanton marsh complex. In contrast with the coarse gravels of Arroyo Mocho and

Figure 4.51. Wetlands surrounding Arroyo las Positas. This circa 1840 land grant map shows a small willow grove (*sausalita*), springs (*pocitas*), and salty water (*agua salada*), as well as riparian trees along the creek. (USDC ca.1840e, courtesy The Bancroft Library, UC Berkeley)

Arroyo del Valle, Las Positas flowed across a bed of comparatively fine sediment, skirting the edge of the wet and alkali meadows of central Livermore-Amador Valley. This contrast was of great importance to early engineers studying the valley, because it meant that rather than providing groundwater recharge, Arroyo las Positas functioned as a water transport system (Williams 1912, Fuller 1912). Cyril Williams described this contrast:

> The feeders of this artesian basin are the Arroyo Mocho and Valle Creeks; the other streams from the watershed, viz: the Positas, Tassajara, etc., traverse a territory where adobe and tight clay predominate, within and on the side of the channel, and on the surface of the surrounding country, with the result that the streams flowing over these, is continuous to the Laguna Creek, with no measurable loss into the gravels of the valley. (Williams 1912:31)

Riparian cover along Arroyo las Positas (and the smaller tributaries) varied in width and composition as the stream flowed west. In the upper part of the watershed, the riparian vegetation was part of the larger alkali wetland. Alkali sink scrub and alkali meadow species bordered the creek, with iodine bush (*Allenrolfea occidentalis*) concentrated along the stream corridor (Burtt Davy 1898, Coats et al. 1988). As the creek flowed between the two bedrock knolls, groundwater came closer to the surface and artesian conditions developed (Williams 1912:49) supporting a wider vegetated corridor of over 100 feet (USDA 1940).

Further downstream, historical sources suggest a relatively narrow riparian corridor, widening into patches of dense vegetation near springs or reaches with higher groundwater. Between these patches, riparian vegetation likely consisted of occasional oaks and herbaceous vegetation (fig. 4.52). However, even within this relatively unvegetated reach, there were patches of more dense riparian vegetation. Directly upstream of the confluence with Collier Creek, GLO surveyors noted crossing "a swampy water course" (Day 1853) in July, and oak trees lining the creek. Aerial photos from 1940 show several

Figure 4.52. Incised channel and sparse vegetation along Arroyo las Positas. The channel had likely been modified by the time of this photograph, and the banks appear very steep, perhaps the result of incision. However, the herbaceous vegetation and occasional trees match the descriptions in the historical record, indicating that similar conditions may have occurred prior to Euro-American management. (from Williams 1912, courtesy Laurel Collins)

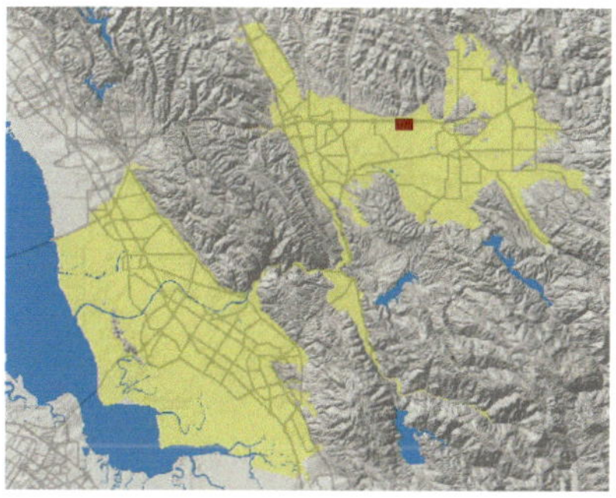

Figure 4.53. Willow patch along Arroyo las Positas. A few remnant willow groves are still visible as dark patches in 1940 aerial photography, hinting at a much wider riparian corridor historically. (USDA 1940)

remnant swaths of willows 150-250 feet wide, indicating that even within this relatively sparsely vegetated reach, there was substantial variation (fig. 4.53; USDA 1940). These may have been anomalous reaches of rising or shallow groundwater, which enabled dense willows and other riparian vegetation to survive year round. The wetness of the soil may have made these areas unattractive to farmers, explaining why they were still visible in the 1940s, at the height of agriculture in the valley (Banke pers. comm.).

Tassajara Creek

Unlike most other drainages, Tassajara Creek did not form distributaries across the valley floor. A number of early sources document the historical condition of the creek, and most depict a continuous channel to the edge of the marsh (e.g., Fallon 1861, Lewis 1861, Allardt 1874, Thompson and West 1878; fig. 4.54).

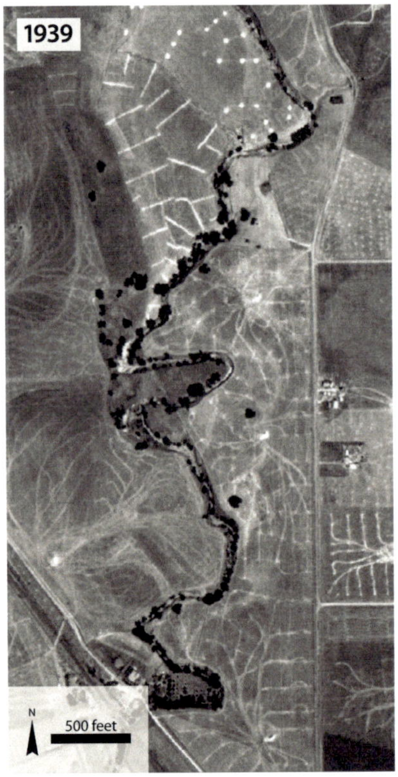

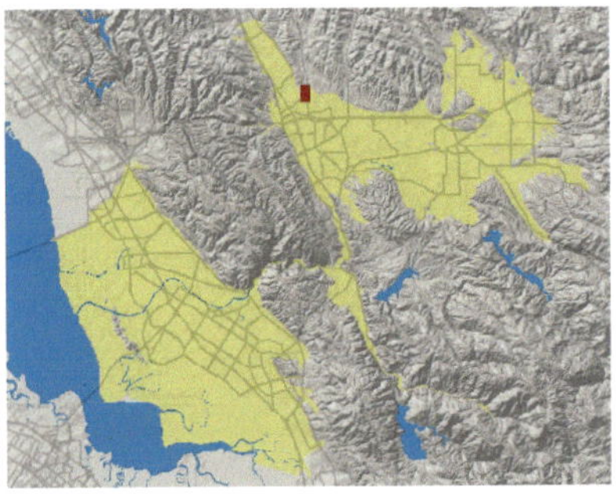

Figure 4.54. Dotted oaks along Tassajara Creek. The narrow, oak-dominated corridor visible along Tassajara Creek in 1939 may have been typical of many of the small, summer-dry tributaries in Livermore-Amador Valley. (USDA 1939)

Tassajara drains an 9500 acre area, which resulted in a sizeable stream that slowly dissipated towards the marsh. In 1853, the channel was 18 feet deep, 66 feet wide, and "lined with trees" a mile upstream of the marsh (Day 1853). Further downstream the channel became shallow, such that "a considerable amount of water must spread itself over the adjacent land" (Lewis 1861). Surveyor Lewis noted that it is "very shallow near the Tule swamp" (1861). This lower portion of the stream was ditched and dredged to control flooding, appearing ditched on maps as early as 1872 (Duerr 1872b), and a deposition from 1916 describes that even after this modification "there were a lot of willows, and they were allowed to choke the stream up, and thus turn the stream over this place, which left this deposit of fine soil on this area north of the road" (Callaghan 1915:575). The soil underlying the creek here was described as having been "laid down by the floods of the Tassajero [sic] Creek, which has been dammed at times and its waters turned over the adjacent fields" (Westover and Van Duyne 1911:36).

Limited evidence suggests that the channel may have disappeared even before reaching the Pleasanton marsh complex (box 4.7). A *diseño* map shows Tassajara spreading and ending some distance from the marsh, and a confirmation map shows a dotted line continuing south, possibly indicating that there was no clearly defined channel (see Beller et al. 2010). However, the bulk of evidence indicates that the creek continued to the marsh. GLO surveyor La Croze (1860) described crossing the Tassajara immediately before reaching the historical border of the tule marsh. One local landowner described that "there is a sanjon from the Tassajero to the tules," which may refer to an artificial ditch or a shallow channel (Higuera 1861; *Sanjón* is generally translated as "ditch," but can refer to both natural and artificial channels, and in this case likely is used to describe the broad, swale-like channel).

The creek maintained perennial flow in the hills, but appears to have become intermittent as it reached the valley floor. Land grant testimony describes the creek with mostly perennial flow until it reached the valley. After that point, in summer "there was no running water"(Lewis 1861:511).

Arroyo de la Laguna

Arroyo de la Laguna flowed south from the perennial pond in the center of the Pleasanton marsh complex to Sunol Valley, serving as the connection between the Livermore-Amador Valley watersheds and Alameda Creek. Constrained on the west side by the uplifting East Bay hills and to the east by the hills south of Pleasanton, Arroyo de la Laguna occupied a narrow valley, and maintained perennial flow through the dry season. The name of the creek ("stream of the lake") emphasizes its historical role as the outlet for the Pleasanton marsh. It functioned as the southern extension of the marsh over much of its length, with multiple channels and willows, and very gradual elevation change (only 50 feet over 4.5 miles). Maps and

BOX 4.7. ARROYO LAS POSITAS AND TASSAJARA CREEK

Las Positas and Tassajara both flowed into the Pleasanton marsh complex, eventually spreading in the valley freshwater marsh and willows. The exact nature of their confluence and position of the end of the defined channel likely changed from year to year, depending on flows, sediment, and fluctuating groundwater levels. However, it seems likely that the two creeks did form a natural confluence prior to the 1870s. Maps consistently show the creeks spreading into the marsh. Water collected to reform into a channel at Arroyo de la Laguna, over two miles away.

As early as 1860, a GLO surveyor described the confluence of the two creeks (La Croze 1860). Early maps consistently show the two creeks joining, either far inside the marsh (fig. 4.55b), or before reaching the marsh (fig. 4.55c,d; see fig. 4.25b). The two creeks were ditched and straightened as early as 1872, which suggests that the channels were not large enough or did not extend far enough into the marsh to satisfy farmers' drainage needs (fig. 4.55). Later maps show Tassajara Creek ditched (fig. 4.55a,e). Downstream, the creeks formed a delta distributary into the marsh, losing their defined channel and surface flow.

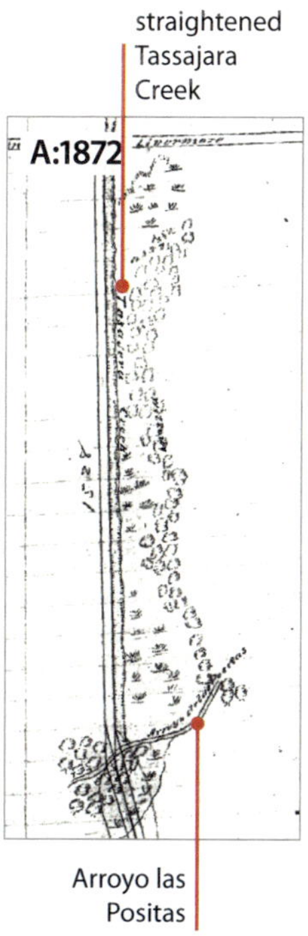

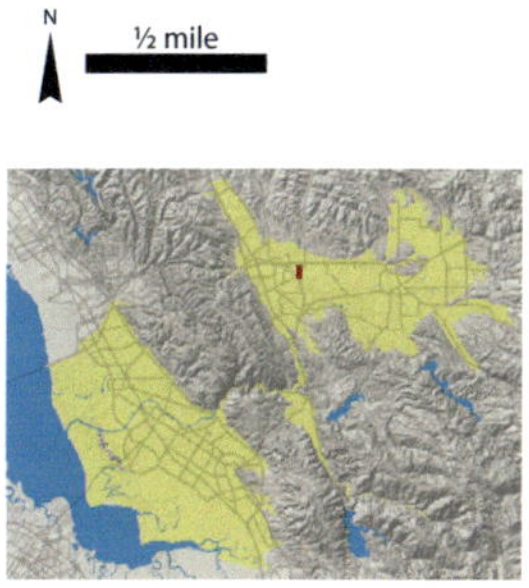

Figure 4.55. Confluence of Las Positas and Tassajara. A number of sources indicate that Las Positas and Tassajara formed a direct confluence, but this confluence is shown in a variety of locations over time, sometimes within the marsh complex, and sometimes far upstream. This graphic shows some of the maps listed in the text above. The four maps below are zoomed to the same location. (A: Duerr 1872, courtesy Alameda County Department of Public Works; B: Higley 1857a, courtesy David Rumsey Map Collection; C: La Croze 1860, courtesy The Bancroft Library, UC Berkeley; D: Whitney 1873, courtesy David Rumsey Map Collection; E: Thompson and West 1878, courtesy David Rumsey Map Collection)

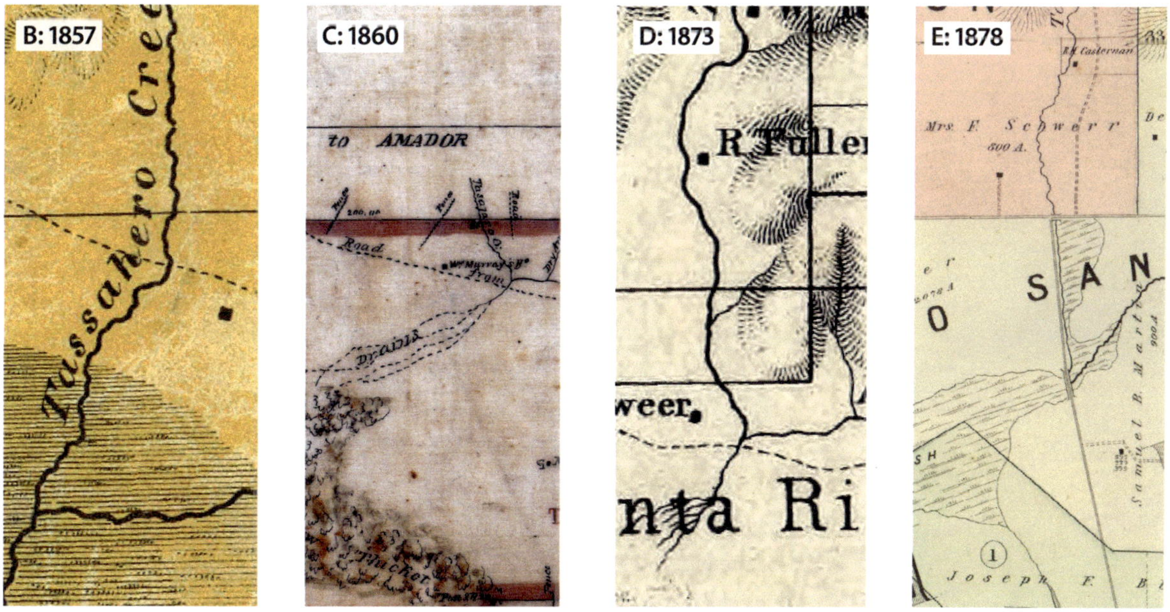

Figure 4.56. Arroyo de la Laguna, 2011. These photos were taken at a restoration project just downstream of Verona Bridge, in an area with very high incision. Note the steep banks and large bars within the channel. (photos by Bronwen Stanford)

survey notes depict a relatively broad, flat, multi-threaded system, perennial and very wet in places. As the marsh complex was drained, the creek rapidly incised and transformed from a broad and marshy system to the deep channel seen today (fig. 4.56).

The upper portion of Arroyo de la Laguna drained from the open water pond through the southern marshes and willow thicket of the Pleasanton marsh complex. GLO surveyors were explicit about this connection: Surveyor Dyer followed the "general course of a small stream" from the "southern edge of the Tular" (Dyer 1862a), and another surveyor, Charles Healy, described Arroyo de la Laguna as "a small stream which flows from a Laguna on the North" (Dyer 1863). The creek had a clay bed, likely overflow from the fine-grained sediment filling the Pleasanton marsh (Williams 1912).

Arroyo de la Laguna was historically distinct from nearby streams in Livermore-Amador Valley, with perennial flow and more lush riparian cover. Spanish explorers Crespí and Fages described Arroyo de la Laguna as "an abundant stream" with "an abundance of water and trees" (Treutlein and Fages 1972:353, Crespí and Bolton 1927:298). The willow corridor along the creek is depicted on a series of maps (Allardt 1874, Thompson and West 1878, Allardt [1880]1907, etc.), and GLO surveyors noted that the creek "overflows the valley in wet weather" (Howe 1851). Just south of the southern extent of the marsh (approximately at present-day Verona Bridge), surveyor Sherman Day described crossing a "copious stream" as well as several "bayous" in a "dense thicket [i.e., willows] … occupying the valley of the creek" (Day 1853:256). "Bayou" can be defined several ways, but it typically refers to a swale-like channel, shallow and containing water.

At least one early map shows a transition in the character of the stream just south of this point (about half a mile south of Verona Bridge; Thompson and West 1878). In 1772 Spanish explorer Fages described a change in the landscape, "the canyon becoming narrower and the land more rough than before, with small descents into little gullies, but all the hollows overgrown with groves of live-oaks and white-oaks" (Treutlein and Fages 1972:353). Even here, as late as 1900 survey maps show a multi-thread channel (Allardt 1874, Nusbaumer ca. 1900), indicating that the creek maintained a complex pattern through to Sunol Valley.

The stream maintained summer surface flow throughout its length. Sherman Day described the main stem as 23 feet wide and "a copious stream, enough to turn a mill" in July (Day 1853:25) with additional smaller side channels. Even after the destruction of the marsh, engineer Von Geldern observed in 1912 that "a constant percolative flow by the way of the canyon is continuous during the entire season, from the valley's constriction, at the head, down to Sunol" (Von Geldern 1912:8). Another report describes the modification to Arroyo de la Laguna:

> Prior to Spring Valley's well pumping operations Laguna Creek was a live stream the year around it being fed from both the Pleasanton Lagoon and artesian flow from the underground reservoir. Subsequent to 1889 all this dry season flow was diverted and used as a domestic water supply for San Francisco by Spring Valley. (Espy 1950)

Riparian vegetation along upper Arroyo de la Laguna was dense willow thicket, a continuation of the Pleasanton marsh complex. Surveyor Howe described "enter thick woodland. Brush &c. thick as hair" between Sunol Boulevard and Bernal Ave along Interstate 680 (Howe 1851). Further downstream, a narrow band of dense vegetation continued, ranging from 300 to 700 feet wide, and likely composed of a mix of valley oak, sycamore, and willow (fig. 4.57). GLO surveys recorded valley oak and sycamore along the channel (Day 1853, Healy 1863). Even as late as the mid-1900s, a local rancher described "a huge amount of vegetation in the banks ...lots of willows, cottonwoods, sycamores, oaks ... it was a sedate, seasonal, meandering stream, and today we see it as a hugely flashy, undercut stream" due to land use changes and resulting changed hydrology (Koopman pers. comm.).

Arroyo de la Laguna was heavily modified as part of the draining of the Pleasanton marsh complex. The creek was needed to transport water rapidly out of Livermore-Amador Valley, and formed one of the key pipelines for Spring Valley Water Company development of the Alameda Creek watershed. The canals constructed to drain the Pleasanton marsh complex acted as funnels, flushing water and sediment downstream which had historically been trapped in the sediment basin of the marsh, and causing significant down-cutting (see fig. 4.56). Early engineering reports indicate that the flat, slow-draining channel posed problems: Williams described that in 1880 the creek "followed a very indefinite course at a much higher elevation than at present, that the lower valley was silted up, and that the inflowing waters had no unobstructed outlet channels" (Williams 1912:218). He also reported that "since the clearing of Laguna channel the

Figure 4.57. Riparian trees along Arroyo de la Laguna. This circa 1840 map shows two types of riparian vegetation along Arroyo de la Laguna, with small, brushy trees (possibly willows) to the east, and larger *Palos secos* ("dry trees"—possibly oaks) to the west. This map suggests that the tree cover was a notable feature of Arroyo de la Laguna. (USDC ca. 1840a, courtesy The Bancroft Library, UC Berkeley)

In recent years the marsh has been drained by the construction of reclamation ditches and by the deepening and clearing of a portion of the Laguna Creek channel, which allowed the flood waters a free course. The channel, now less obstructed, is greatly cut down.

—CYRIL WILLIAMS 1912

creek has worn down its channel at a rapid rate" of three feet in ten years (Williams 1912:218). Water from the Pleasanton wells flowed down ditches to Arroyo de la Laguna, and from there was piped south through Sunol Valley to flow through the Sunol Filter Gallery (see box 5.1). Two channels are clearly visible in lower Arroyo de la Laguna in the 1906 USGS mapping, one flowing near the historical course, and one diverting water south to the filter galleries (see fig. 5.9). In addition to this diversion, the channel was enlarged and cleared, leading to rapid incision. In the 1900s, the land on either side of the channel was used for grazing (Koopman pers. comm.). Today the channel continues to incise, and is a source of primarily fine sediment for the watershed (Bigelow et al. 2008). Large in-channel bars are forming within some reaches that exacerbate erosion on the bank opposite the bar (Bigelow et al. 2008).

Stream characteristics summary

Table 4.4 summarizes the morphology, flow, and riparian vegetation characteristics for each of the fluvial reaches described above. We developed broad classes that could apply to streams throughout the study area, and focused on the large stream systems. The many small tributaries draining in to Arroyo de las Positas and the Pleasanton marsh complex from the north and the east were grouped together so that they could be characterized as a whole. These and many other small named and unnamed tributaries were not included individually here, but would have contributed to overall watershed processes.

The data presented below were synthesized from a variety of sources, including narrative records, historical aerials, historical maps, digital elevation datasets, and geology and soils data. For further discussion of riparian width classes and geomorphic processes, see Chapter 9.

Table 4.4. Historical stream reach characteristics. This table describes stream characteristics within the study area, and so is restricted to the valley floor. The upper reaches of the northern tributaries, which historically functioned as sediment sources, are not included in the table, and the lower reaches of these tributaries are listed only as transport and depositional reaches.

Creek	Reach	Watershed Area (sq. miles)[1]	Dominant Morphology[2]	Dominant Process [2] (Geomorphic)	Substrate[3]	Dry Season Flow	Riparian Corridor Width Classes[4]	Riparian Vegetation[5]
Arroyo las Positas	Valley floor	81	Single-stem, meandering	Transport, depositional	Sand, silt, clay	Perennial/ Intermittent	<200 ft (<60 m)	Sparse oak, Willow-cottonwood
Arroyo Mocho	Upper Mocho (canyon to Tesla Road)	45	Single-stem, meandering	Production, transport	Boulders, gravels, sand	Perennial/ Intermittent with pools	200-660 ft (60-200 m)	Sycamore alluvial woodland
	Middle Mocho (Tesla Road to Oak Knoll)	55	Braided	Transport, depositional	Boulders, gravels, sand	Intermittent	660-1320 ft (200-400 m)	Sycamore alluvial woodland
	Lower Mocho (Oak Knoll to distributaries)	59	Distributary channels	Depositional	Gravels, silt, clay	Intermittent	200-660 ft (60-200 m)	Sycamore, Sparse oak
Arroyo del Valle	Upper Del Valle (mouth of canyon to Shadow Cliffs)	165	Braided	Production, transport	Boulders, gravels, sand	Perennial/ Intermittent with pools	>1320 ft (>400 m)	Sycamore alluvial woodland
	Lower Del Valle (Shadow Cliffs to marsh)	168	Distributary channels	Transport, depositional	Sand, silt, clay	Intermittent	200-660 ft (60-200 m)	Sycamore, Sparse oak
Arroyo de la Laguna	Valley floor	412	Single-stem, meandering	Transport, depositional	Gravels, silt, clay	Perennial	200-660 ft (60-200 m)	Willow-cottonwood, Sycamore
Northern Tributaries	Valley floor	57	Single-stem, distributary	Transport, depositional	Cobbles, gravels, fines	Intermittent	<200 ft (<60 m)	Sparse oak, Willow-cottonwood
Eastern Tributaries	Valley floor	45	Single-stem, distributary	Transport, depositional	Sand, silt, clay	Intermittent/ Ephemeral	<200 ft (<60 m)	Alkali sink scrub, alkali meadow

[1]Watershed area was calculated at the downstream endpoint of each reach using contemporary USGS Streamstats, but is likely representative of the historical drainage area.

[2]Dominant morphology and process were determined from the mapped form of the creek, our understanding of fluvial geomorphic processes, and geology data.

[3]Substrate classes were developed from historical descriptions, soils and geology mapping, and historical photographs.

[4]Riparian corridor width measures from the outer edge of stream-associated vegetation on one side of the stream to the other, including the stream bed. These widths were based largely on the width of a representative reach in the historical aerial imagery and current geomorphic features in the 2007 LiDAR data (USDA 1939-40, LiDAR 2007).

[5]Broad riparian vegetation classes were developed from those species data that exist, and describe the inner corridor of riparian vegetation. Further from the creek, riparian vegetation would have included valley oaks and/or sycamores.

5 • UPPER ALAMEDA CREEK, SUNOL VALLEY, AND NILES CANYON

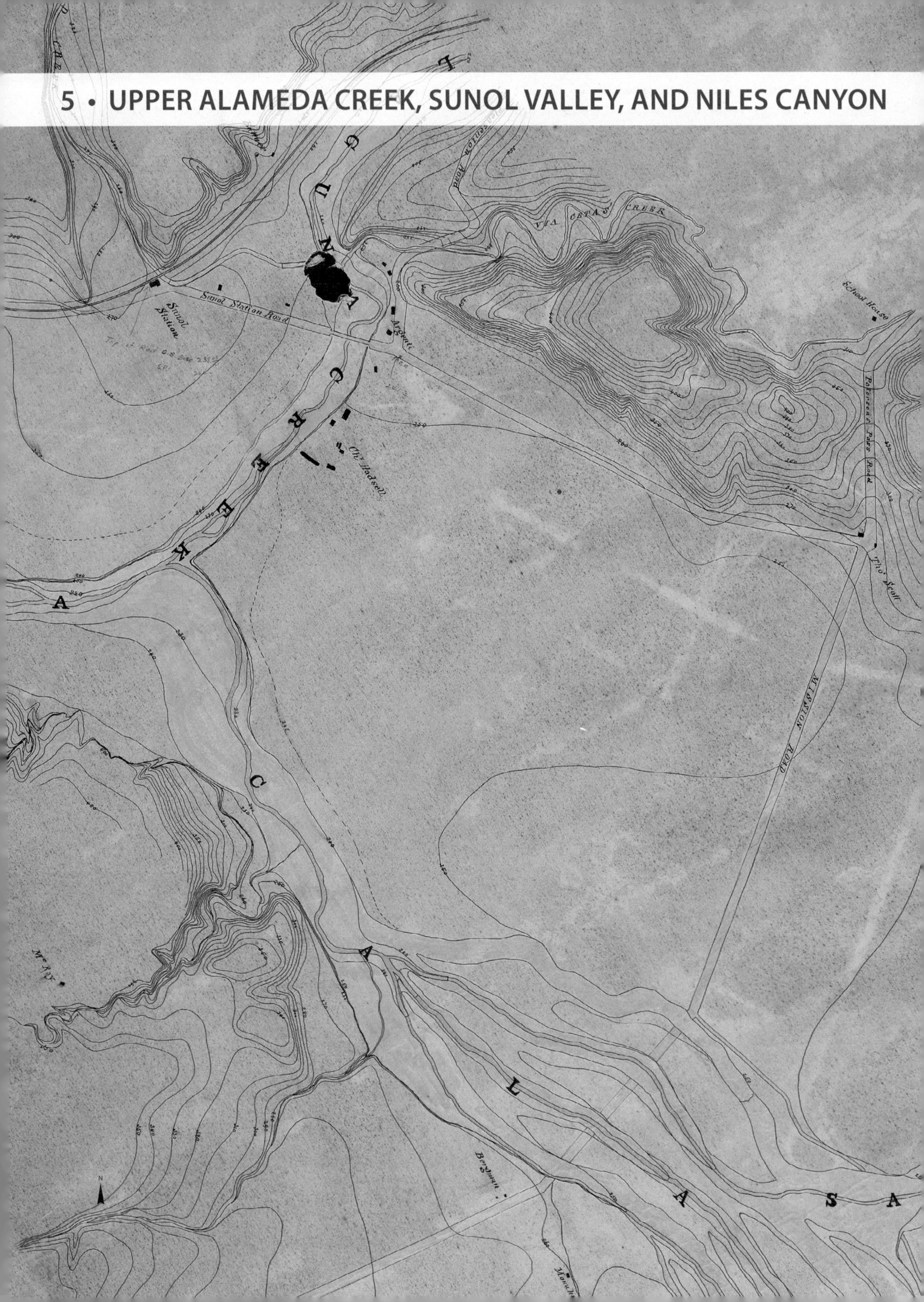

Alameda Creek flows through miles of canyons and the long and narrow Sunol Valley before reaching Fremont and the Bay (fig. 5.1). This section focuses on the almost 20 mile reach of Alameda Creek from the Alameda county line downstream to Niles (fig. 5.2). Along this course, the creek moves through three distinct geomorphic settings: the upper canyon from the Alameda County boundary to Sunol Valley, Sunol Valley, and Niles Canyon. The southern watershed (defined as south of the confluence with Arroyo de la Laguna) drains a 200 square mile area.

Sunol Valley is a narrow alluvial valley, described in 1898 as a "deep gorge in the inner Coast range" (Morse 1898). It is bounded by two faults and was formed during a period of intense tectonic activity during the mid-Pleistocene, after the initial formation of Livermore-Amador Valley (Fisher et al. 1966:23; fig. 9.2c). Calaveras Fault, a right lateral strike-slip fault which runs along the east side of Sunol Valley, underwent major activity during this time period, which resulted in down-dropping, and the subsequent formation of the Sunol Valley graben (Fisher et al. 1966:23). Sunol Valley then filled with deposits that eroded from the tectonically active East Bay hills and the Calaveras Fault (Clark 1924). Like Livermore-Amador Valley, Sunol Valley was created by geological rather than fluvial processes. The size of the valley influenced its geomorphology and the resulting habitat distribution by providing the stream with space to spread.

In contrast, Niles Canyon is a confined, narrow, alluvial valley. The creek historically lost little if any water through this reach, and flowed year-round to Niles. Alameda Creek is an antecedent stream, draining Livermore-Amador Valley before the East Bay Hills were uplifted. As the hills rose, the creek incised through the hills and carved Niles Canyon (Sloan 2006).

In more recent times, Sunol Valley and the canyons upstream and downstream have been shaped by Alameda Creek. Alameda Creek carries substantial loads of coarse sediment from the Franciscan Complex, much like arroyos Mocho and Del Valle. Flows from the creek have deposited gravels and sands across the valley, resulting in the "bed of coarse gravel many feet in thickness" (Westover and Van Duyne 1910:35) that forms the Sunol groundwater basin (Fisher et al. 1966).

Large swaths of riparian sycamore alluvial woodland characterized Alameda Creek's broad braided stream bed through Sunol Valley, while mixed riparian forest was found through the canyons both upstream and downstream. A sparse oak savanna grew across the northern Sunol Valley, where perennial flows supported a forested riparian corridor along Arroyo de la Laguna (table 5.1, fig. 5.3).

These habitats occupied reaches with distinct morphology and flow. Above Sunol Valley, bedrock surface geology kept water at the surface year round. As groundwater levels dropped below the creek bed at the head of Sunol Valley, Alameda Creek entered a losing reach with a braided channel. Near the head of Niles Canyon, Alameda Creek joined Arroyo

Cathcart canyon (now known as Sunol glen) ... is a deep gorge in the inner Coast range ... Riding through this deep canyon the scenery is very beautiful. In places the mountains almost come together, overtopping one hundreds of feet. In the bed of the ravine a stream of clear sparkling water rushes merrily along over a gravelly bottom, making its way to the greater Alameda canyon, and thence into the waters of the bay of San Francisco.

—HARRY MORSE 1898

Figure 5.1. Alameda Creek at the confluece with Arroyo de la Laguna. This image, rotated so that east is up, shows detail from the Spring Valley Water Company map. See also figure 5.7. The darker low flow channel is visible within the broader channels of Alameda Creek and Arroyo de la Laguna. (Unknown ca. 1890, © San Francisco Public Utilities Commission)

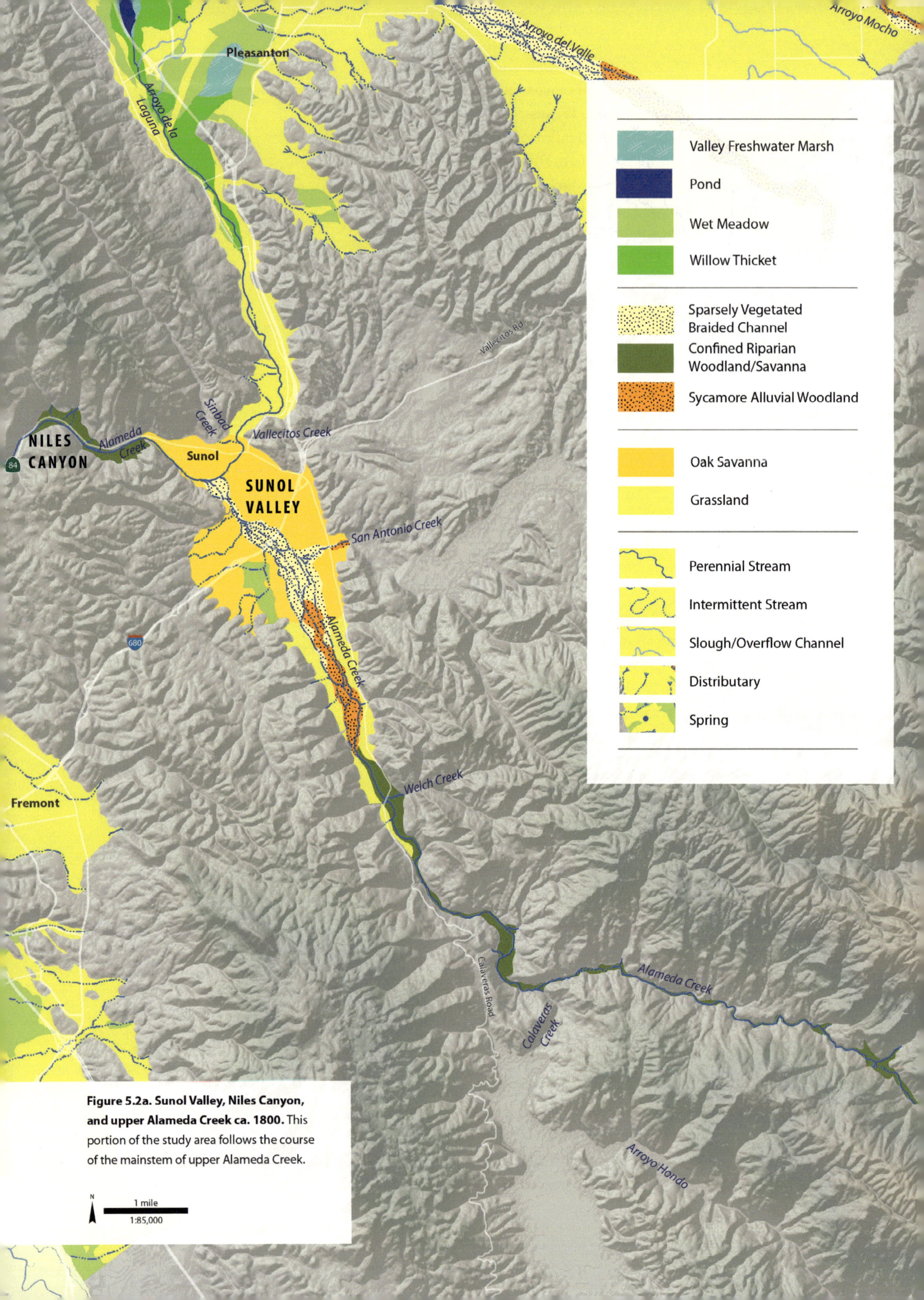

Figure 5.2a. Sunol Valley, Niles Canyon, and upper Alameda Creek ca. 1800. This portion of the study area follows the course of the mainstem of upper Alameda Creek.

Figure 5.2b. Sunol Valley, Niles Canyon, and upper Alameda Creek 2009. Red outline represents the study area. (USDA 2009)

Table 5.1. Historical land cover in upper Alameda Creek, Sunol Valley, and Niles Canyon. The table below summarizes our findings for land cover patterns in Sunol-Niles circa 1800. Acreages are rounded to the nearest 50.

Figure 5.3. Simplified chart of historical land cover in upper Sunol Valley. The pie chart presents a simplified summary of the data in the table. Riparian cover types are excluded, so this only covers non-riparian habitats of Sunol Valley.

Habitat Type	Acres
Grassland/Savanna	***1,750***
Grassland	350
Oak Savanna	1,400
Riparian Vegetation	***1,350***
Sparsely Vegetated Braided Channel	400
Confined Riparian Woodland/Savanna	750
Sycamore Alluvial Woodland	200
Seasonal freshwater wetland	***100***
Wet Meadow	100

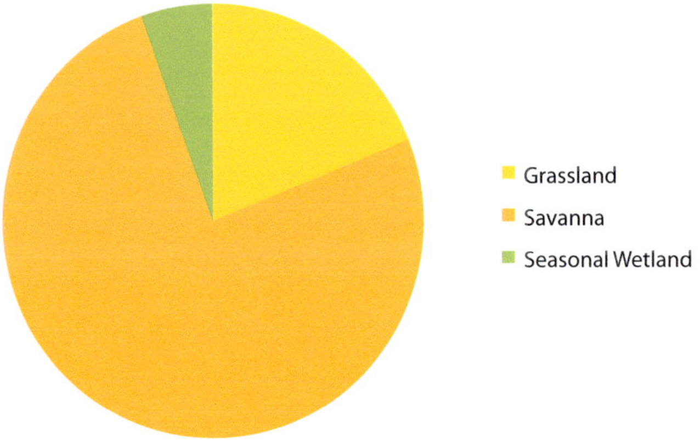

de la Laguna and flows became perennial again. Through Niles Canyon bedrock again maintained base flows and perennial flow (see box 1.1). This heterogeneous pattern of surface hydrology and corresponding stream and riparian habitat, controlled by geology, matches that seen elsewhere, both in the project area (lower Alameda Creek, Arroyo del Valle) and regionally (Coyote Creek; Grossinger et al. 2006). (For a summary of stream and reach characteristics, see table 5.2.)

Historical records for this portion of Alameda Creek are relatively sparse. This was due in part to the early acquisition of much of the valley floor by Spring Valley, which resulted in fewer property maps and descriptions by early settlers. The small size of the valley also limited the number of records.

Sunol Valley was historically an important part of San Francisco's water supply system. Through the 19th century, the pattern of grazing and limited agriculture here matched that of Livermore-Amador Valley. However, with the completion of the Sunol Dam in 1900, the Sunol Filter Galleries a few years later, and the Hetch Hetchy Aqueduct in 1934, Sunol Valley became one of the major water arteries for the Bay Area (fig. 5.4). Similarly, the construction of the railroad through Niles Canyon in 1869 made it an important regional transportation corridor.

Two large dams have altered the movement of water and sediment through this portion of Alameda Creek. The Calaveras Reservoir, constructed between 1913 and 1925, and the San Antonio Reservoir, completed in 1965, cut off 138 square miles of drainage area, trapping water and coarse sediment that would otherwise have supplied Sunol Valley. In addition, the Calaveras Reservoir captures flow from Alameda Creek through the diversion dam. Most of the coarse sediment for the lower watershed was supplied by these large systems (see fig. 9.13), while Arroyo de la Laguna supplied primarily fine sediment. The capture of flood flows and coarse sediment may have helped to stabilize the channel through Sunol Valley and reduce scour.

Figure 5.4. Modifications to the water system. A: This aqueduct was installed to transport collected water through Sunol Valley. Note the sparse tree cover (likely valley oaks) in the far distance. B: Men working on the Niles Dam at the mouth of Niles Canyon, which was used by the Spring Valley Water Company to divert water for San Francisco beginning in 1889. (A: #53520; B: #62757, both © San Francisco Public Utilities Commission)

The following section describes the major reaches of Alameda Creek and its associated riparian habitats from upstream of the diversion dam to the mouth of Niles Canyon. This section also touches briefly on the valley floor habitats of Sunol Valley.

Upper Alameda Canyon reach

Upstream of Sunol Valley, Alameda Creek flowed through a confined canyon, mostly in a single channel. At times the canyon was only as wide as the creek itself. In other places, the canyon widened to almost 1,000 feet, creating small patches of oak savanna (fig. 5.5; USDA 1939-40). Through most of this upper reach, Alameda Creek had perennial flow. This appears to have been interrupted by at least one intermittent reach where the creek flowed through two channels (USGS 1906). The canyon transported large amounts of coarse sediment and carried high flows. Riparian cover included sycamores, oaks, box elder, buckeye, sedges, and willows. Calaveras Creek and Arroyo Hondo contributed water and sediment to this upper reach.

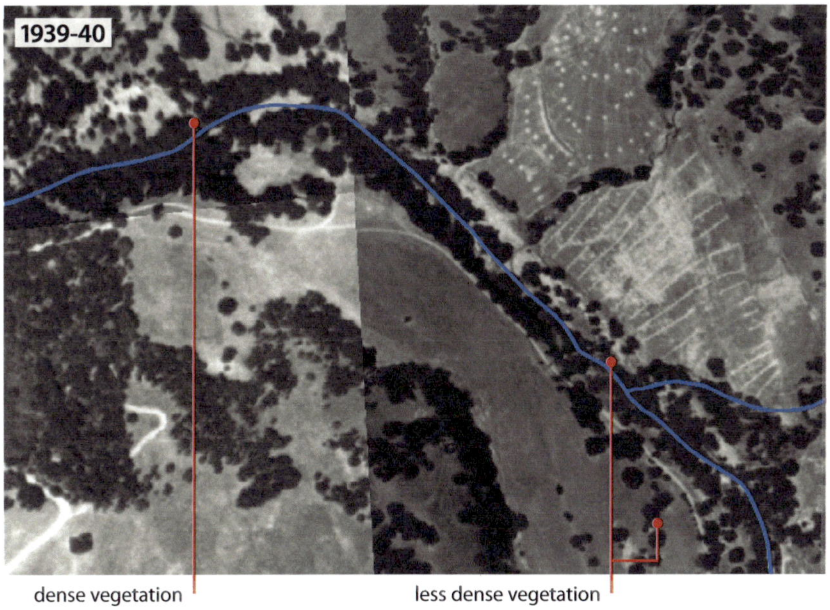

Figure 5.5. Small clearings in canyon vegetation. Swaths of sparse hillside tree cover along upper Alameda Creek were classed as "savanna." Some of these clearings were likely the result of grazing or farming, but within the wider portions of the canyon small patches of sparse oak savanna likely occurred within woodlands even in the absence of grazing. The lines are likely traces of haying. (USDA 1939-40)

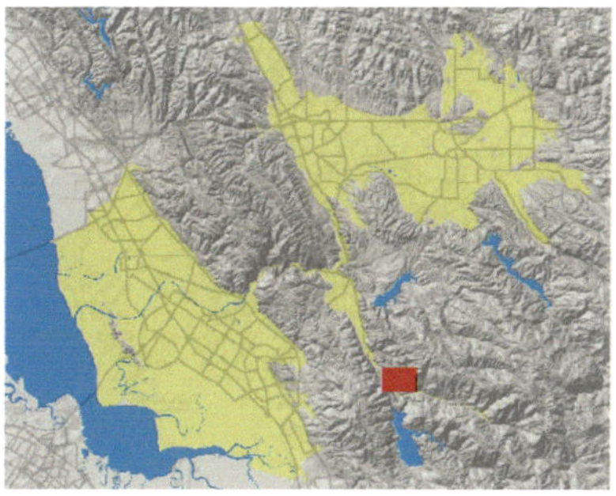

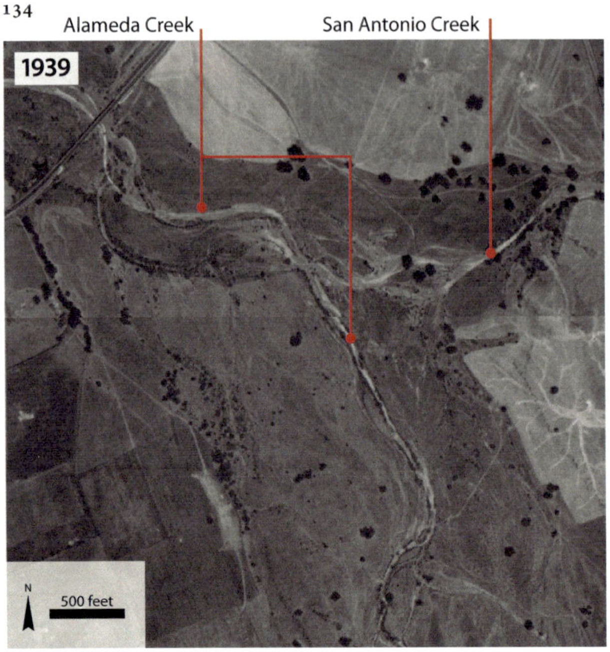

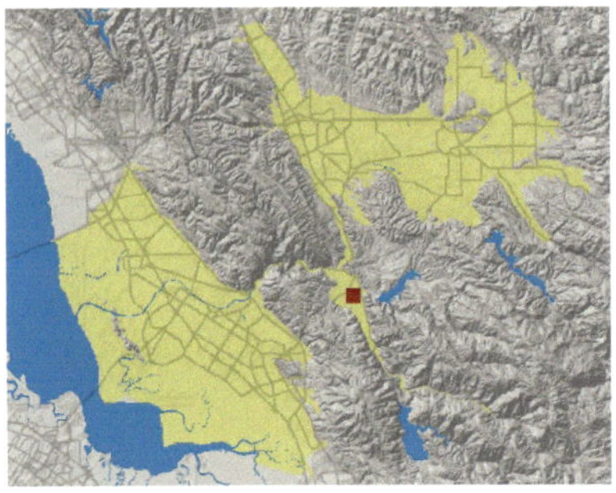

Figure 5.6. Gravel pits. These paired images show northern Sunol Valley in 1939 and 2009. The braided historical channel has been replaced through northern Sunol Valley with gravel quarries, making use of the creek's substantial gravel deposits. (USDA 1939; USDA 2009)

Observations made along the banks of the Alameda Creek in the Sunol Valley indicate that the stream is at present a cutting one, and also seems to indicate that in the past it meandered freely from one side to the other, laying down gravel beds.

—MULHOLLAND AND LIPPINCOTT 1912:13

Sunol Valley reach

Alameda Creek braided as it entered Sunol Valley, flowing through multiple channels separated by broad bars and islands (fig. 5.6). The creek deposited coarse sediment across the valley floor, and high sediment inputs entered the creek from both Calaveras and San Antonio creeks. Towards the northern end of the valley, Arroyo de la Laguna formed a confluence with Alameda Creek, contributing fine sediment from the Pleasanton marsh complex.

Historical maps consistently show a transition point in Alameda Creek from single thread to multi-thread at the head of the valley (e.g., Allardt 1874, Thompson and West 1878, Nusbaumer and Boardman 1889, USGS 1906). Some of the islands that formed between these channels appear to have been relatively persistent at the decadal scale. In particular, near the confluence of San Antonio Creek with Alameda Creek a number of maps show large and defined islands with consistent shapes across a span of forty years (e.g., Boardman 1870, Allardt 1874, King 1880, Haviland 1910). This pattern has been called "island-braided" (Beechie et al. 2006). The construction of small summer dams and road crossings may have contributed to braiding observed in the historical period.

Distinct outer banks formed around the gravelly braided channel, separating it from the surrounding valley floor. These banks broadened to a third of a mile in places (Allardt 1874, Unknown ca. 1890). Near the confluence of Alameda Creek and Arroyo de la Laguna, surveyor Healy described "the foot of a high bluff bank," possibly representing the outer bank of the creek (Healy 1863). Upstream, near the confluence of San Antonio Creek with Alameda Creek, Surveyor Howe (1851:12) described the "bank of a creek from the mountains SE. Which has changed its bed from time to time until it has cut and deposited a width of 50 rods [825 ft] or more." This width matches the visible creek bed width on the 1939 aerial (fig. 5.7; USDA 1939).

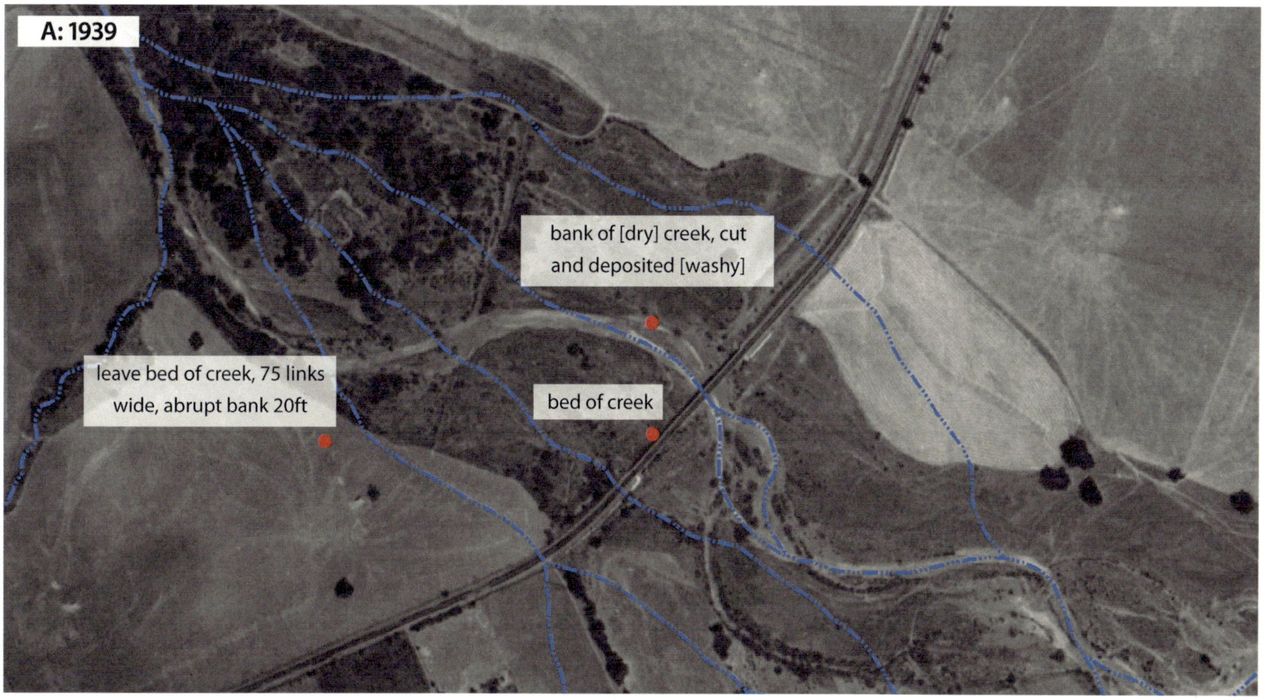

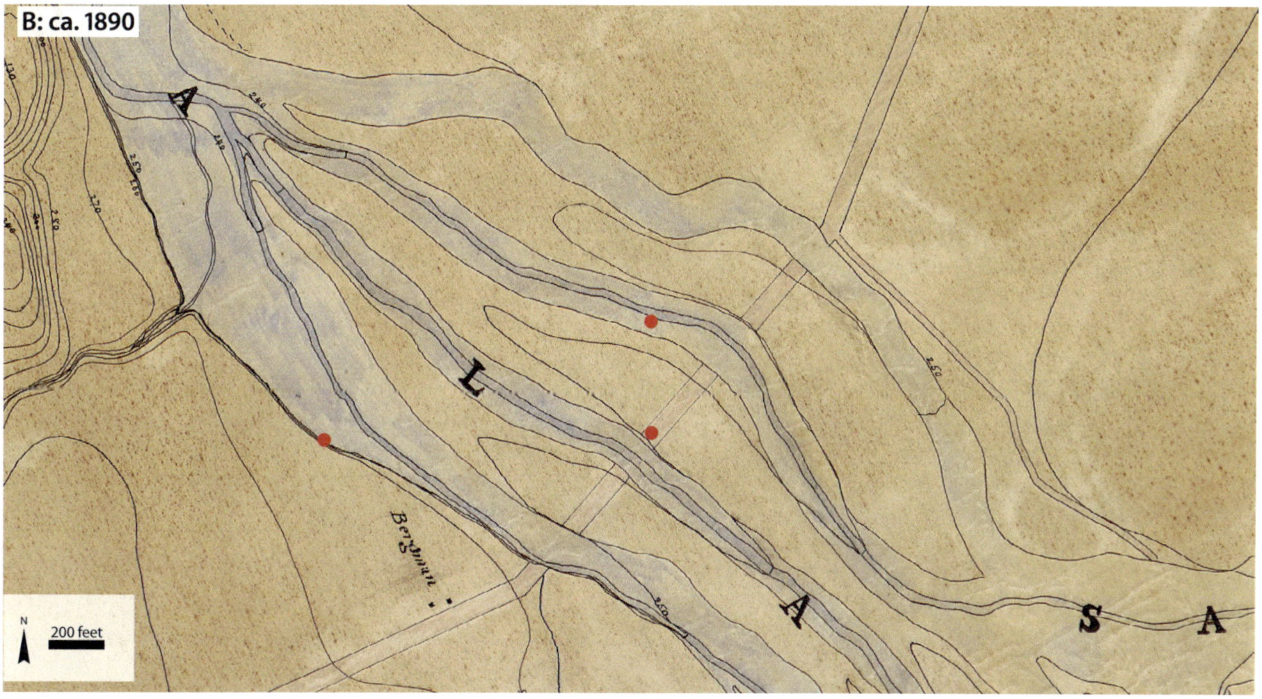

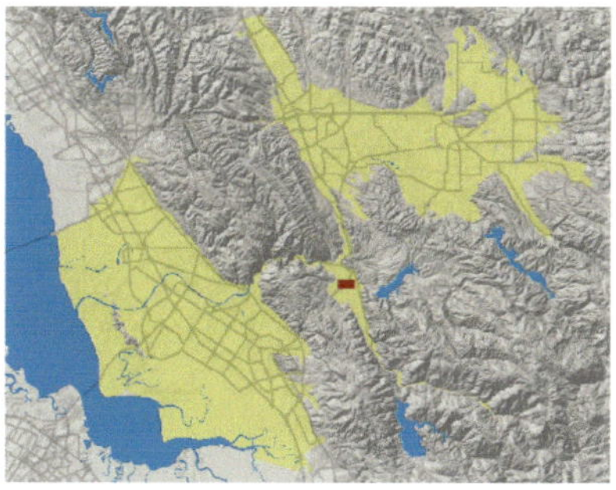

Figure 5.7. The outer banks of Alameda Creek. In 1851 GLO surveyor Richard Howe described crossing the active channel and the outer banks of Alameda Creek near present-day Interstate 680. (I-680 follows the course of the road visible in the images above.) Excerpts from his description have been placed on the 1939 aerials (A), confirming that the aerials represent historical patterns, albeit with some expected intervening changes. A late 19th century map (B) provides another view of the same reach, again with a fairly consistent channel pattern. This map is undated, but the presence of the railroad dates it between 1869 and 1906. The property ownership indicates that it was likely from the 1890s. This map may have been made to help Spring Valley Water Company design the filter galleries and dam in Sunol Valley. (A: USDA 1939; B: Unknown ca. 1890, © San Francisco Public Utilities Commission)

This valley may be likened to a large basin ... filled with gravel, partly topped with loam. The large streams of water flowing from the adjacent mountains sink into this vast gravel bed, which, while acting as a subterranean reservoir, also filters all the water prior to its entering into the Sunol aqueduct.

—SPRING VALLEY WATER COMPANY N.D.

BOX 5.1. SUNOL VALLEY FILTER GALLERIES

Layers of gravel and sand extended the length of Sunol Valley, and were used as water storage throughout the early 1900s (fig. 5.8; Williams 1912). This system was a remarkable example of the efficient re-engineering of a fluvial system. As demands for water increased in San Francisco, the Spring Valley Water Company began to use the naturally occurring gravels as filter beds to remove sediment and temporarily store water from Alameda Creek before piping it across the bay (Williams 1912). An 1899 article in the *San Francisco Chronicle* described the need for the project:

> At present the Alameda creek supply can only be used a portion of the year for the reason that in the winter season the floods wash the soil from the hillsides down into the creek and render the water unsuitable for domestic purposes... an immense natural filter will be excavated out of a large gravel bed at a point above the company's Sunol dam. (*San Francisco Chronicle* 1899b)

The first pipeline had been run across the Bay to supply water from Alameda Creek to San Francisco in 1888 (West 1937), and in 1889 the Niles Dam was built across the bottom of Niles Canyon to facilitate water capture. In 1901, the flow of Arroyo de la Laguna, including water from the Pleasanton wells, was diverted and ditched so that it flowed through the filter beds (fig. 5.9), and Sunol Dam, completed the year prior, was constructed across the top of Niles Canyon to trap both ground water and surface flow within the valley (Williams 1912, Bartell 1914, Fisher et al. 1966). An underground filter gallery was installed 10-12 feet under the surface of the filter basins (fig. 5.10). The filter gallery was built with a

(CONTINUED, PAGE 138)

Figure 5.8. Sunol gravel beds. Gravels were deposited as Alameda Creek slowed from a high energy stream in the confined upper canyon to a braided stream, spreading across Sunol Valley. These coarse alluvial gravel deposits created a natural water filter in Sunol Valley. (from SVWC ca. 1900-29, courtesy The Bancroft Library, UC Berkeley)

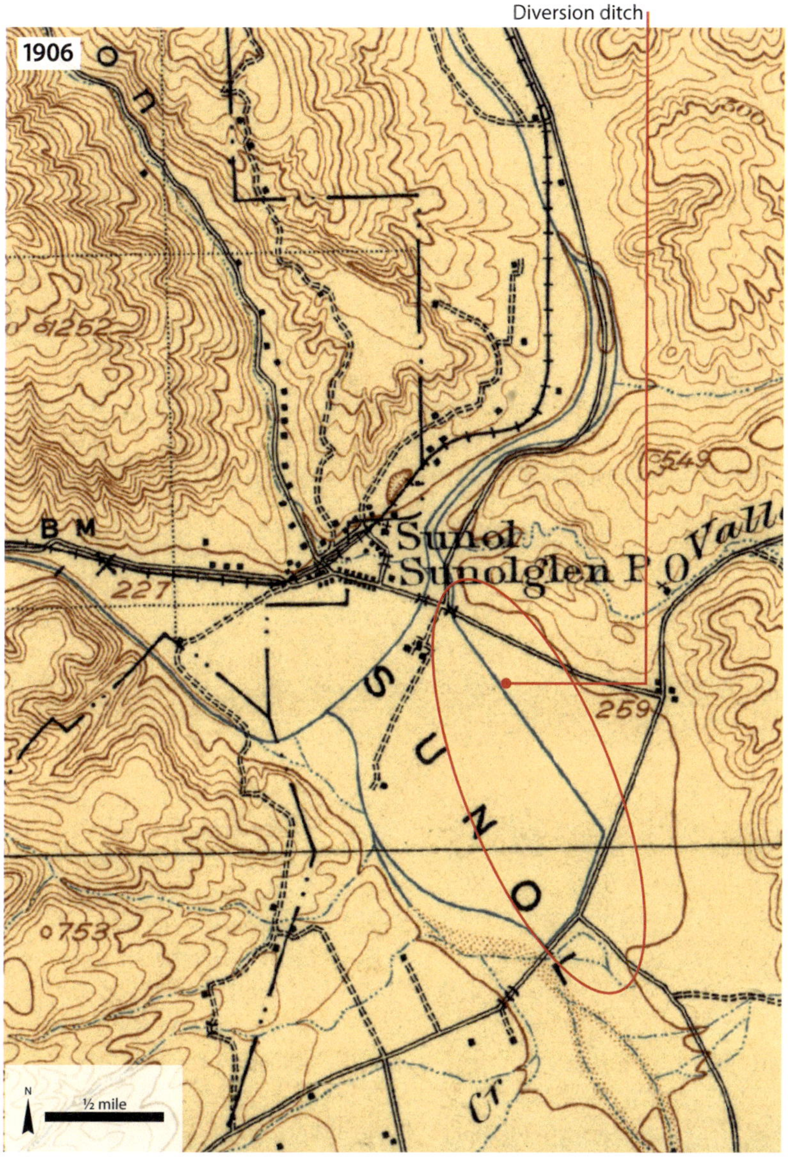

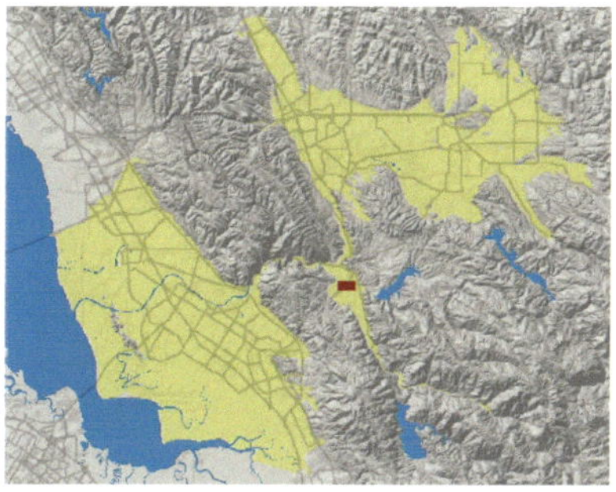

Figure 5.9. Diversion of Arroyo de la Laguna. This 1906 USGS quad sheet shows Sunol Valley and Arroyo de la Laguna after a diversion dam was built in 1901. This dam directed flows east into a ditch, which is depicted as a second blue line running parallel to the natural course of Arroyo de la Laguna. As it enters Sunol Valley, this ditch runs south to feed the water into the upper portion of the Sunol filter beds. This allowed for more complete capture of water by the filter beds. The ditch was abandoned in 1909 when a pipe was installed to send water directly to the Sunol Water Temple from the Pleasanton wells (see also Williams 1912, Lee 1916a). (USGS 1906)

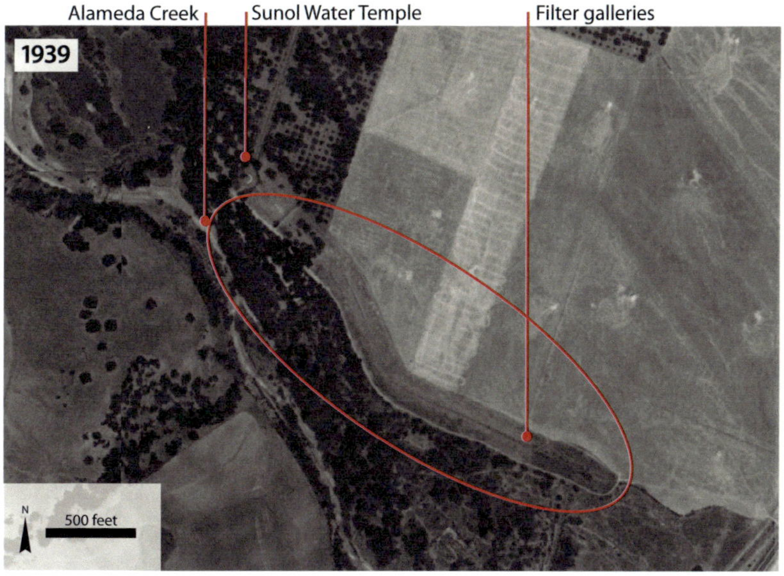

Figure 5.10. Aerial view of Sunol filter galleries. The filter galleries were constructed to collect water underground after it filtered through the gravels of the valley floor. At left, the filter galleries are visible stretching along the eastern side of Alameda Creek and ending at the Sunol Water Temple. The galleries continued north of the Water Temple as well. (USDA 1939)

Figure 5.11. Sunol filter galleries, early 1900s. Image (A) shows the top of the galleries extending across the valley floor. The filter galleries drew from the high groundwater associated with Alameda Creek. The stretch shown here measures 2,600 feet long. Image (B) shows a view from inside of the filter galleries, taken in the early 1900s. (A: from SVWC ca. 1900-29, courtesy The Bancroft Library, UC Berkeley; B: from San Francisco Water and Power 2005)

(CONTINUED, BOX 5.1, SUNOL VALLEY FILTER GALLERIES)

concrete top and walls with weep holes, and the bottom was open to the gravels to allow groundwater to percolate up and collect inside (fig. 5.11). Groundwater was reported at 8-12 feet below the surface (*San Francisco Chronicle* 1899b). The underground filter stretched from just north of Interstate 680 to the Sunol Water Temple and then further north towards the Sunol Dam, and directed the water into an aqueduct for conveyance to San Francisco (Williams 1912, Fisher et al. 1966, *San Francisco Chronicle* 1899b). Additional perforated pipe drains were installed up to 20 feet under the bed and bank of Alameda Creek; these pipes collected subsurface water and passed it into the filter gallery. The Sunol Dam was an important part of this system: "the yield of the galleries is increased by impounding surface water behind Sunol Dam downstream from the galleries" (Fisher et al. 1966). As a result of the water extraction in Sunol Valley, flows in Alameda Creek downstream of the dam and across the Niles Cone decreased, leading to litigation (e.g., Clough case in 1901, see box 6.1).

Figure 5.12. Sunol Water Temple and Sunol Valley. Constructed in 1910, the Water Temple (A) marked the place where water from Alameda Creek, Arroyo de la Laguna, and the Pleasanton wells was piped together. In the early 20th century this provided half of San Francisco's water supply. The Water Temple still stands in Sunol Valley. (B) A view of Sunol Valley from above. (A: from SVWC ca. 1900-29, courtesy The Bancroft Library, UC Berkeley; B: photo by Robin Grossinger)

By 1909 water from Pleasanton bypassed Arroyo de la Laguna and was instead piped from Pleasanton directly to the Sunol Water Temple (fig. 5.12; Lee 1916a). By 1925, the Calaveras Dam was completed and allowed for more direct capture of water. After 1934, the Hetch Hetchy pipeline carried water to San Francisco from the Sierra, and the filter galleries decreased in importance (California Department of Transportation 1998). Today the filter galleries are still in use, although to a much more limited extent due to the decline in groundwater levels through the basin. Some of the water collected in the galleries is used to water the Sunol Valley golf course, while the remainder is pumped to San Antonio Reservoir or released back to the creek (Leonardson pers. comm.).

Figure 5.13. Sunol Dam. Sunol Dam was constructed in 1900 (A; shown under construction) to trap water from Alameda Creek in the Sunol filter galleries. Previously, water had been diverted further downstream at the Niles Dam. (B) Floodwaters flowing over Sunol Dam. (A: #158383, © San Francisco Public Utilities Commission; B: from SVWC ca. 1900-29, courtesy The Bancroft Library, UC Berkeley)

The active channel only occupied a portion of the space between the outer banks. Surveyor Healy measured the creek at 100 feet wide just before its confluence with Arroyo de la Laguna (Healy 1863). Surveyor Howe described the "bed of creek" only 50 feet wide within banks almost 1,000 feet across (1851:12; see fig. 5.7a). During high flows, the creek spread and branched into additional channels within the outer banks. This was most clearly depicted by a detailed map from the late 1800s (see fig. 5.7b). This map shows the multiple channels through Sunol Valley and distinguishes between channels that carried low flows and channels or portions of channels that were engaged only during high flow events (Unknown ca. 1890). Sycamores growing along the creek likely helped stabilize these islands and supported the formation of pools.

Alameda Creek flowed through a losing reach in Sunol Valley until the confluence with Arroyo de la Laguna. In winter, flows could rage through the valley, but summer water was limited (fig. 5.13; Brewer 1974:352). Although we mapped perennial flows extending into Sunol Valley, this was based solely on USGS mapping (1906), and we were unable to find corroborating evidence. However, it is clear that surface water existed only for a portion of the year by the time the creek reached the location of the current Hetch Hetchy Aqueduct (where Calaveras Road becomes Calaveras Avenue; USGS 1906). In October of 1851 (a very dry year), General Land Office (GLO) surveyor Howe (1851:12) recorded that there was "no water running in it at this time" in the bed of the Alameda Creek near San Antonio Creek. However, pools formed within the bed of the creek and could retain cool water as late as July (*San Francisco Call* 1900).

Sycamore alluvial woodland dominated the riparian corridor through the losing reach in southern Sunol Valley (fig. 5.14). Sycamore alluvial woodland grew along alluvial terraces and islands within the banks of Alameda Creek and would likely have included live oak (*Quercus agrifolia*), valley oak (*Quercus lobata*), and California bay (*Umbellularia californica*) in addition to sycamores (*Platanus racemosa*), while the understory included plants such as poison oak (*Toxicodendron diversilobum*), mule fat (*Baccharis salicifolia*), and sedges (e.g., *Carex* spp.) (USDA 1939, Keeler-Wolf et al. 1996). GLO surveyor Tracy (1852b:2) noted entering "the head of a narrow valley watered by a creek bordered by oak and sycamore," and the 1910 soil survey describes "many large sycamore trees" on the soil type that spans southern Sunol Valley (Westover and Van Duyne 1911:35). In 1772 Spanish explorer Crespí described heading north across San Antonio Creek and seeing Alameda Creek "full of trees" (Crespí and Bolton 1927:298).

If the country was as nature made it, if the railroads had not gone through and villages built, a large dam might have been built at Sunol and the Sunol Valley itself might have been a reservoir ... it would have been a very advantageous development. Apparently the time when that could have been done has gone.

—ALLEN HAZEN 1916:2444

In contrast to the southern portion, the northern end of the valley did not historically support sycamore alluvial woodland. Alameda Creek became narrower towards northern Sunol Valley and the riparian corridor narrowed as well, likely including a mix of grassland, scrub, and valley oaks (Callaghan 1915:628). Sycamore alluvial woodland thinned downstream to occasional sycamores (Howe 1851:12, USDA 1939). In addition to stabilizing islands, the sycamores contributed large woody debris, which fell in the stream and helped create more complex channel morphology.

At the northernmost edge of the valley, flows from Alameda Creek and perennial Arroyo de la Laguna converged towards Niles Canyon and combined to make Alameda Creek perennial again. Riparian cover was much more dense here along both Arroyo de la Laguna and Alameda Creek. It was "always cool and delightful" beneath the "shade of the big trees" (fig. 5.15; *San Francisco Chronicle* 1892). Trees recorded through this area include valley oaks, live oaks, box elders, cottonwoods, and sycamores (Unknown 1913).

Oak savanna and grassland

A sparse oak savanna covered northern Sunol Valley, consisting of oak trees dotted at irregular intervals across grassland. From the earliest written record of the valley, people commented on these scattered oaks. In 1772, Crespí described coming to "good land forested with trees" (Crespí and Bolton 1927). Nearly a century later, GLO surveyor Howe described "a beautiful flat studded with scattering oak timber" (Howe 1851:12), and 1860s newspaper accounts described "much oak and sycamore timber" and "fine oak trees that ornament it like a park" (*Daily Alta California* 1860c and 1863). Botanist William Brewer in 1860 described "a lovely little valley—the Suñol Valley—a little plain of perhaps 1500 or 2000 acres, studded with scattered oaks, large and of exceeding beauty" (Brewer 1974:351-352). These accounts contrast with descriptions of Livermore-Amador Valley and the Niles Cone, both of which were described without significant oak cover (see p. 76 and p. 159). Fine organic sediment carried from the Pleasanton

Suñol (Soon-yole) valley, nearly circular in shape, and three miles in diameter, a fertile vale, still in a state of nature, with a luxuriant growth of grass, fine oak trees that ornament it like a park, and a surrounding of picturesque mountains on all sides.

—*DAILY ALTA CALIFORNIA* 1860C

Figure 5.14. Trees along Alameda Creek through lower Sunol Valley. This early 20th century photograph (A) shows scattered oaks along barely visible Alameda Creek. Further up the creek, remnant sycamores persist today (B). (A: from SVWC n.d., courtesy San Francisco History Center, San Francisco Public Library; B: photo by Alison Whipple 2011)

Figure 5.15. Riparian cover along Arroyo de la Laguna, 1890 and 1912. These images show the vegetation along Arroyo de la Laguna as it enters (A) and flows through Sunol Valley (B). Note the dense riparian corridor (B), which contrasts with the open sycamore alluvial woodland found further south in Sunol Valley along Alameda Creek. (A: from Williams 1912, courtesy Laurel Collins; B: P-59, courtesy Pleasanton's Museum on Main)

marsh complex may have helped support oak savanna across the northern Sunol Valley. As with other regions of the study area, the sparse tree cover was likely due in part to careful management and burning by the Ohlone.

The oaks appear to have been largely cleared by the late 1800s. As early as 1880, accounts described northern Sunol Valley as "originally dotted over with oaks" (Porter 1884), suggesting that significant clearing had already occurred. The 1911 soil survey also stated that "formerly there were quite a number of valley oaks" in the northern portion of the valley (Westover and Van Duyne 1911). Very few oaks remained in 1939, when the first full-coverage aerial photographs were taken (USDA 1939).

Sunol Valley oaks likely grew at low densities. The descriptions of "scattered oaks" that look like they had been planted for a park (see above quotes) match descriptions of savannas in Santa Clara Valley (Grossinger et al. 2008, Beller et al. 2010, Whipple et al. 2011). However, historical sources do not provide sufficient information to allow us to estimate densities beyond this broad classification. Only one description noted a higher density grove of oaks (Howe 1851:12), and this was along the edge of a hill, possibly referring to upland vegetation.

Beneath the scattered oak trees, grasses and forbs formed the matrix of Sunol Valley vegetation (Treutlein and Fages 1972). The gravelly soils of the southern portion of the valley in particular supported few trees, and were covered primarily with grassland (*Daily Alta California* 1860c, Crespí and Bolton 1927:298, Brewer 1974). This grassland likely consisted of a mix of species, including a variety of clovers, perennial bunchgrasses, and forbs. Sunol Valley also contained wet meadow areas supporting more water-tolerant species, likely including *Elymus* and *Juncus* spp. Historical sources provide few details on the exact species composition of the grasslands and meadows of Sunol Valley.

Niles Canyon reach

Passing slowly up Alameda Cañon, with his eyes open and appreciative, one may secure a gallery of mental pictures, which in richness of color, variety and beauty, will rival any collection of landscape paintings in the world ... When we remember that this rare bit of rural beauty and wildness is almost on the threshold of a populous city, our appreciation is quickened and heightened.

—BARTLETT 1883

Below Sunol Valley, canyon walls and shallow bedrock again constrained Alameda Creek, which maintained perennial flow through Niles Canyon (USGS 1906). Just upstream of Sunol Dam a 1915 report describes Niles Canyon as follows: "a steep rock bank on the west side, a deposit of fine gravel on the bottom and a low bank of sandy clay on the east side" (Unknown 1915). Through some reaches the canyon widened, and settlers were able to practice agriculture.

The creek bed was historically filled with coarse sand, cobbles, and gravels, and developed "many fine pools" (*Morning Call* 1892), as well as bars and riffles (fig. 5.16; see also *San Francisco Chronicle* 1885). In 1795 Lieutenant Hermenegildo Sal walked upstream from Niles to Stonybrook Canyon and described a stream that was "exceedingly rocky ... the stream has half a yard [1.5 ft] of water, and about four [12 ft] wide. In various parts it makes pools.

Here it widens and more water is collected" (Sal 1795). In places the creek broadened and split to form islands and large bars (Haviland 1913). One well documented island measuring 300 feet across occurred just upstream of the Stonybrook Creek confluence, and was depicted on maps spanning 40 years (fig. 5.17; Allardt 1874, Thompson and West 1878, Nusbaumer and Boardman 1899, USGS 1906, Dillman 1911a).

Mixed riparian forest with brush and many trees bordered Alameda Creek through Niles Canyon. Accounts from the late 1800s describe the "willow-and-sycamore-lined stream," banks "overhung with alders and buckeye trees" and campsites where "shade trees abound" (Brotherhood of Locomotive Engineers 1875:547, *San Francisco Chronicle* 1885, *Morning Call* 1892; fig. 5.18). The canyon was a favorite picnicking spot, resulting in a large collection of photographs and descriptions.

Riparian trees here and throughout the project area grew in expected patterns such that the most hydrophilic species (such as alders and willows), grew near the creek, and the more drought tolerant species (such as oaks) spread towards the uplands (fig. 5.19). Milicent Shinn, a psychologist who spent her childhood in Niles, described finding trees "arranged with

Figure 5.16. Boulders and pools through Niles Canyon, ca. 1908. Pools through Niles Canyon served as swimming holes and picnicking spots. Note the large boulders. Although dense riparian vegetation surrounds the creek in this image, there is no closed canopy and the creek is relatively open to the sky. (courtesy California Historical Society, CHS2012.930)

Alameda's willow-and-sycamore-lined stream is fair to see at this time of the year. Its pools and "reaches"—wide, blue expanses of tree-shaded water, parted each from each by rocky "riffles" (another rural phrase)—are loved by artists and full of comfort for tired city dwellers.

—*SAN FRANCISCO CHRONICLE* 1885

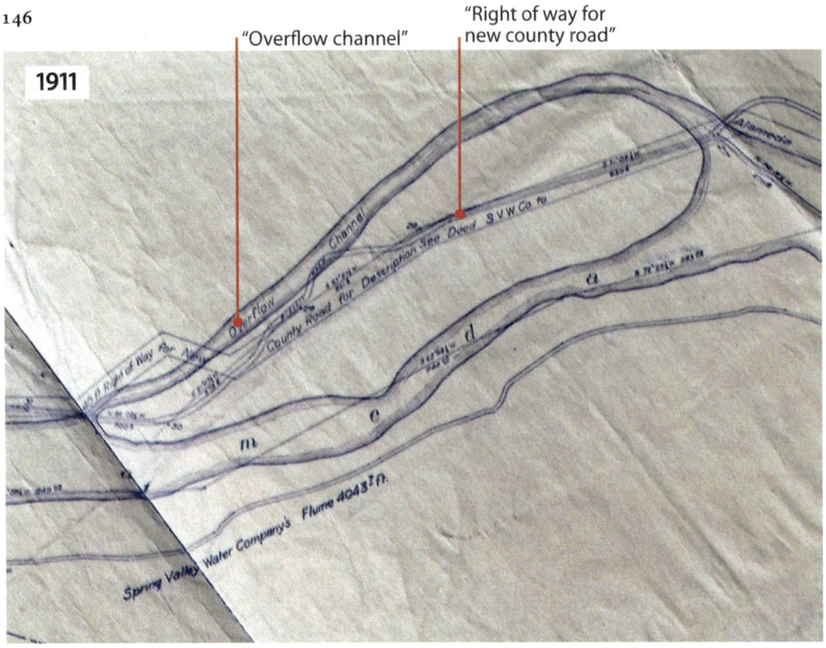

Figure 5.17. Island in Niles Canyon. The northern branch of the creek splits off the main channel, and is labeled "overflow channel" in this 1911 map, which also shows a planned county road crossing the creek using the island at this point, indicating that the island was dry and permanent enough to support a road (Dillman 1911a). Eventually a dam was built to force the creek into the southern channel (Anian 1993:11). A smaller but similarly persistent "bar" occurred two miles further upstream. (Dillman 1911, courtesy Museum of Local History, Fremont)

Figure 5.18. Upper Niles Canyon. This image, ca. 1900, shows the suspension bridge constructed 200 feet upstream of Sunol Dam at the upper end of Niles Canyon. Large riparian trees shade the channel, but do not completely cover it. The men appear to be crossing the creek on a narrow bar. ("Suspension Bridge" P-2191, courtesy Pleasanton's Museum on Main)

Figure 5.19. Footbridge across Alameda Creek. Large trees developed along the creek through Niles Canyon. Note the relatively open canopy and large, well-developed trees. (Alameda Creek and foot bridge, Fisher Scrapbook Collection, courtesy Museum of Local History, Fremont)

much precision" as follows: alders "cling resolutely to the edge of the summer channel, pushing away other trees" while the sycamore, slightly more removed, "reaches its best stature and breadth in good soil, away from a perennial stream but where the ground is well wet in winter." Oaks, meanwhile, are found "further back than the sycamores, scattered over the hillsides themselves" (Shinn 1883).

While a variety of trees grew along the banks of Alameda Creek, the sycamore was particularly notable. In 1795, Sal described the lower portion of Niles Canyon as "thickly covered with sycamores, some willow, madrone, laurels" (Sal 1795). Almost 100 years later, describing this same place, the *San Francisco Chronicle* (1885) proclaimed "if the coast range has finer specimens of the 'spotted sycamores,' we know not where to find them" and another account described how Alameda Creek "emerges between ranks of alder and sycamore" (Burns 1888). Sycamores were also noted by Jepson as the "most abundant tree" along Alameda Creek, and were the one bearing tree used in GLO surveys in Niles Canyon (Tracy 1857, Jepson 1905). However, many other species were also noted, including black cottonwood (*Populus balsamifera* spp. *trichocarpa*) and seep willow (*Baccharis salicifolia*) (Jepson 1909). One popular picnic grove at the mouth of Stonybrook Canyon was notable in the 1870s because "the grove was of oaks, not of sycamore" (*San Francisco Chronicle* 1885).

Summary

Table 5.2, following, summarizes the morphology, flow, and riparian characteristics for each of the reaches of Alameda Creek described above. The many tributaries to Alameda Creek, including Calaveras, San Antonio, Stonybrook, and Sinbad creeks, are not included here because they barely extend into the study area, although they contributed to overall watershed processes.

The data presented below were synthesized from a variety of sources, including narrative records, historical aerials, historical maps, digital elevation datasets, and geology and soils data (both modern and historical). For further discussion of riparian width classes and geomorphic processes, see Chapter 9.

Table 5.2. Historical stream reach characteristics. This table describes stream characteristics within the study area, which is restricted to the valley floor and Alameda Creek mainstem.

Creek	Reach	Watershed Area (sq. miles)[1]	Dominant Morphology[2]	Dominant Process [2] (Geomorphic)	Substrate[3]	Dry Season Flow	Riparian Corridor Width Classes[4]	Riparian Vegetation[5]
Alameda Creek (upper canyon)	upper Alameda canyon	148	Single-stem, confined	Transport	Boulders, gravels, sand	Perennial	200-660 ft (60-200 m)	Mixed riparian forest
Alameda Creek (Sunol Valley)	southern Sunol Valley semi-confined *(upper valley north to 680)*	193	Braided	Depositional	Boulders, gravels, sand	Intermittent	>1320 ft (400 m)	Sycamore alluvial woodland
	mid-Sunol Valley *(680 to confluence with ADLL)*	208	Braided	Depositional	Gravels, sand	Intermittent	660-1320 ft (200-400 m)	Sparse oaks, sparse sycamores, grassland
	northern Sunol Valley *(confluence of ADLL to Niles Canyon)*	619	Braided	Depositional	Gravels, silt, clay	Perennial	200-660 ft (60-200 m)	Mixed riparian forest
Alameda Creek (Niles Canyon)	Niles Canyon	640	Single-stem, confined	Transport	Boulders, gravels, sand	Perennial	<200 ft (<60m), 200-660 ft (60-200 m)	Mixed riparian forest

[1] Watershed area was calculated at the downstream endpoint of each reach using contemporary USGS Streamstats, but is likely representative of the historical drainage area.

[2] Dominant morphology and process were determined from the mapped form of the creek, our understanding of fluvial geomorphic processes, and geology data.

[3] Substrate classes were developed from historical descriptions, soils and geology mapping, and historical photographs.

[4] Riparian corridor width measures from the outer edge of stream-associated vegetation on one side of the stream to the other, including the stream bed. These widths were based largely on the width of a representative reach in the historical aerial imagery and current geomorphic features in the LiDAR data (USDA 1939-40, LiDAR 2007).

[5] Broad riparian vegetation classes were developed from those species data that exist, and describe the inner corridor of riparian vegetation. Further from the creek, riparian vegetation would have included valley oaks and/or sycamores.

RANCHO
DE LA
Alameda
Creek
Alameda Creek
Henry Currah
Union City Flour Mill
Benson's Ware house
ALVARADO
E. Farley
Bridge
Brick house
Road
Beginning
Post C.R.1
Low flat Land
Slough
Salmon
Slough
LOS
MARSH
T. 4 S.
R. 2 W.
Augustin Alviso
Road and Fence
House
Sanjon de los Alisos
Post C.R.4
Post
Slough
63.15
LANDS
OF
THE
T. 5 S.
R. 2 W.
rritos
in Alviso
ERAL
Sur.

After exiting Niles Canyon, Alameda Creek flowed across its alluvial fan, through grasslands and seasonal wetlands, towards San Francisco Bay. A one to three mile wide strip of tidal marshland formed the border between the seasonal wetlands of the lower fan and the open water of the Bay (fig. 6.1). This chapter covers this lower reach of Alameda Creek and the surrounding habitats of the Niles Cone. It also discusses areas just outside of the cone that are included in this study, extending both north and south (fig. 6.2).

Alameda Creek flowed from Niles Canyon to the Bay across the Niles Cone, the large alluvial fan built by the creek. Alluvial fans or cones are common features of arid and semi-arid climates with infrequent, heavy precipitation. They often result from abrupt changes in the gradient of a stream, which cause the stream to lose power and spread sediment and water across the valley floor (Blair and McPherson 1994). The fan shape is built as the stream deposits successive loads of sediment in patterns radiating from the canyon mouth (see fig. 9.2c). Deposition can happen in large events as giant pulses of sediment, or yearly as constant rates of sediment discharge deposit on the fan. In this watershed, coarse sediment was disproportionately sourced from the southern watershed, and fines from the northern watershed.

The Niles Cone was built by Alameda Creek over the 2.5 million years since the East Bay hills formed (Ferriz 2001). One 1912 map described this feature as the "Delta of Alameda Creek" to describe the many channels branching off the creek (Dockweiler 1912a). Over the millennia, boulders, gravels, sands, and silts pulsed through Sunol Valley and Niles Canyon, and were deposited on the short, steep gradient between the top of the Niles Cone and the Bay (Clark 1924:16). As the channel filled with sediment over time, it lost channel capacity until the creek was forced to abandon its channel and make a new course towards the Bay (Blair and McPherson 1994, Knighton 1998, Lajoie 2003). Under the fan surface are layers of coarse and fine sediment, likely deposited during alternating periods of cool, wet climate (depositing more coarse sediment) and warmer, drier periods (Koltermann and Gorelick 1992). The most recent course of Alameda Creek (prior to the course observed in the 19th century) was along Sanjón de los Alisos, which was likely abandoned around the 1600s. This recent course was still was marked with shellmounds and sycamores in the 1800s (Whitney 1873, Shinn [1889]1991:100, Country Club of Washington Township [1904]1950:85, Sowers 1999).

The habitats of the Niles Cone reflected the sorting action of the fan as well as the accompanying shifts in water and sediment supply. A number of overflow channels and abandoned former channels crossed the plain, and Alameda Creek frequently overflowed across its fan, depositing sediment. The creek washed fine sediment toward the tidal marsh, creating seasonal wetlands. Towards the edge of the tidal marsh, the Coyote Hills trapped sediment and water, contributing to the formation of wetlands (Koltermann and Gorelick 1992).

As the voyager penetrates the valley, ascending the little Alameda, the whole perspective is in the highest sense beautiful. A broad, level carpet, of bright green, is fringed in the misty distance with a high-raised bordering of shining yellow. The vessels on the winding stream, and the few small trees that wave on its shores, resemble embossed figures on a magnificent ground-work; and the growing crops that are scattered over the whole, in various colors and stages of vegetation, give to the picture the charm of almost endless variety. It seems impossible that even nature, with all her skill, could have painted a more delightful or instructive scene.

—CAPRON 1854:175

Figure 6.1. Alameda Creek and the tidal marshes along Niles Cone. (opposite page) This 1860 map emphasizes the connection between the creek and the tidal marshland. Los Cerritos are the Coyote Hills, and the slough to the north of these is the current mouth of the Alameda Creek Flood Control Channel. Note the historical mouth of Alameda Creek, over two miles to the north. (Lewis 1860, courtesy The Bancroft Library, UC Berkeley)

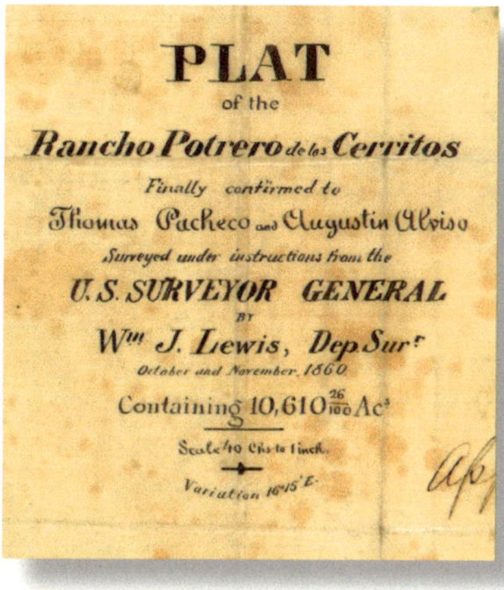

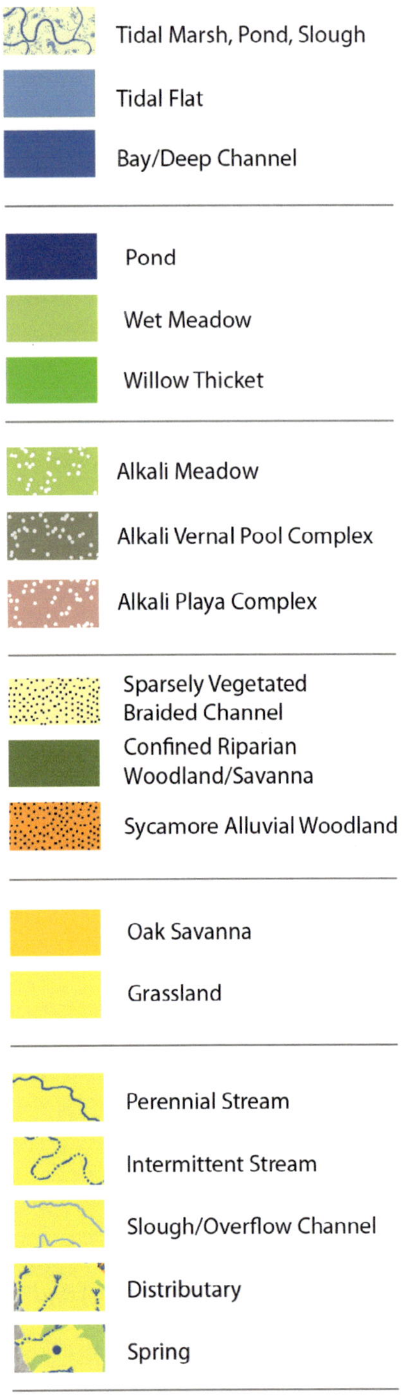

Figure 6.2a. Niles Cone ca. 1800. The study area includes Niles Cone and extends beyond it both to the north and south. Alameda Creek traced a sinuous course through the grasslands and seasonal wetlands of the Niles Cone, and finally flowed to the Bay through the extensive fringing tidal marshlands. Historical placenames have been used on this map—for contemporary placenames see fig. 6.2b.

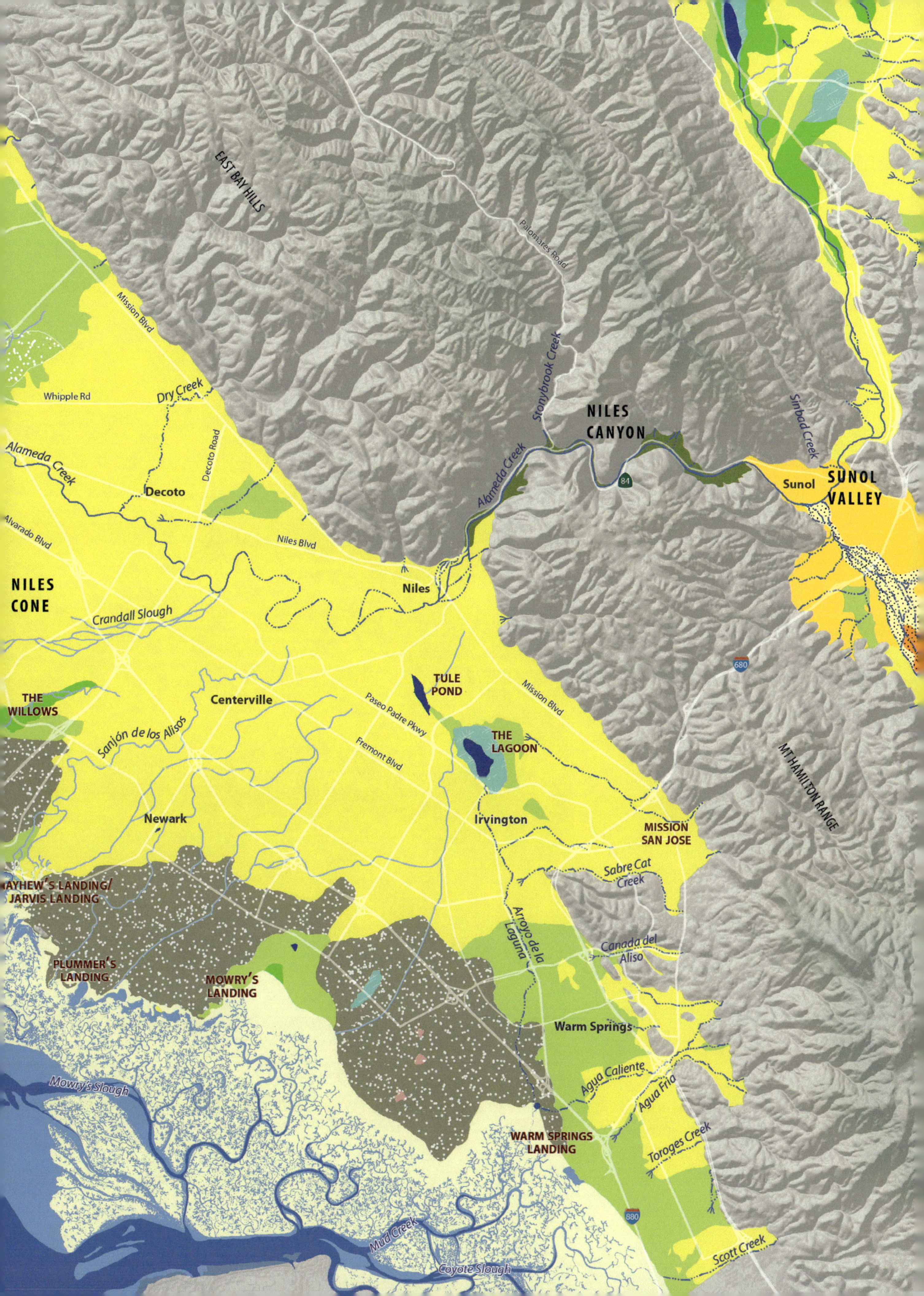

EAST BAY HILLS
Palomares Road
Mission Blvd
Stonybrook Creek
NILES
CANYON
Sinbad Creek
Whipple Rd
Dry Creek
Decoto Road
Alameda Creek
Alameda Creek
84
Sunol
SUNOL
VALLEY
Decoto
Alvarado Blvd
Niles Blvd
NILES
CONE
Niles
Crandall Slough
680
TULE
POND
Mission Blvd
THE
WILLOWS
Centerville
Paseo Padre Pkwy
THE
LAGOON
Sanjón de los Alisos
Fremont Blvd
MT HAMILTON RANGE
Newark
Irvington
MISSION
SAN JOSE
Sabre Cat
Creek
AYHEW'S LANDING/
JARVIS LANDING
Arroyo de la
Laguna
Canada del
Aliso
PLUMMER'S
LANDING
MOWRY'S
LANDING
Warm Springs
Agua Caliente
Agua Fria
Mowry's Slough
WARM SPRINGS
LANDING
Toroges Creek
880
Mud Creek
Scott Creek
Coyote Slough

Figure 6.2b. Niles Cone, 2009. BAARI stream mapping (SFEI 2011) for the study area is shown in blue. Tidal channels are not shown. The red line indicates the study area boundary. (USDA 2009)

EAST BAY HILLS
Palomares Road
Mission Blvd
Dry Creek
Decoto
Decoto Road
Niles Blvd
Stonybrook Creek
NILES CANYON
84
Alameda Creek
Sinbad Creek
Sunol
SUNOL VALLEY
NILES CONE
Niles
Crandall Slough
THE WILLOWS
Centerville
TULE POND
Mission Blvd
Paseo Padre Pkwy
Fremont Blvd
THE LAGOON
680
MT HAMILTON RANGE
Sanjon de los Alisos
Newark
Irvington
MISSION SAN JOSE
Sabre Cat Creek
JARVIS LANDING
Arroyo de la Laguna
PLUMMER'S LANDING
MOWRY'S LANDING
Canada del Aliso
Warm Springs
Caliente
Agua
Agua Fria
Mowry's Slough
WARM SPRINGS LANDING
Toroges Creek
880
Mud Creek
Coyote Slough
Scott Creek

The gravel bed I am speaking of, forms almost a semi-circle, something like a big fan, with the apex near the old Vallejo Mill site. The radius of the fan reaching and including Centerville. The radius, the distance would be, I suppose, four miles.

—KELLOGG 1901

The Hayward Fault also affected habitat distribution by controlling the movement of water across the Niles Cone. The fault extends across the eastern edge of the cone and interrupts subsurface water movement, forming a groundwater barrier. This effect is so pronounced that a 20 foot difference in groundwater levels across the fault was observed (Clark 1915, Pierce 1948, Morris et al. 1960). In the first half of the 20th century, Clark described that

> the fault plane acts as an underground dam and prevents the passage of the ground water, except in small amounts ... Where the fault cuts through the large gravel deposits near the apex of the Niles cone it is probably less effective as a dam than where the proportion of fine sediments is greater, and some of the ground water derived from Alameda Creek may therefore cross the fault through underground channels. (Clark 1915:150)

The Niles Cone was more rapidly developed and modified than other portions of the study area. At the time of European contact, the cone was the site of a number of Ohlone villages (Font and Bolton 1933:356), including the Alson and Tuibun (Holmes 1997). In 1797 it became the site of the 14th mission, Mission San José. Soil fertility and access to San Francisco markets (via landings at the Bay margin) helped the Niles Cone

Figure 6.3. Alameda Creek at Niles, 1878. This lithograph shows Alameda Creek as it flows out of Niles Canyon (looking upstream). This property adjoined the creek just downstream of the braided reach at the Niles Canyon mouth. Clearing had narrowed the riparian corridor by this time, and some trees may have been omitted for artistic reasons. (Thompson and West 1878, courtesy David Rumsey Map Collection)

become an important agricultural center (fig. 6.3). Agriculture on the cone shifted from cattle and grain in the 1850s to orchards and row crops (such as asparagus and tomatoes) by 1900, a transition that occurred here much earlier than in Livermore-Amador Valley (Morse and Westdahl 1897). Alameda Creek itself has been exploited as a water resource on the cone for over 150 years (fig. 6.4). By the mid-20th century, a combination of diversions from Alameda Creek and groundwater pumping resulted in saltwater intrusion to the Niles Cone aquifers (Morris et al. 1960; for more information see Chapter 3).

Today, the Niles Cone is densely urbanized. Major cities include Newark and Union City, but the plain is dominated by the city of Fremont, historically five distinct cities (Niles, Warm Springs, Centerville, Irvington/Washington Corners, and Mission San José) that incorporated in 1956.

The first half of this chapter describes each of the major historical habitat types of the Niles Cone, including grasslands, non-tidal wetlands, and the tidal wetlands that fringed the cone (table 6.1, fig. 6.5). The second half covers the historical morphology, flow, and riparian characteristics of Alameda Creek and other drainages.

Table 6.1. Historical land cover in the Niles Cone region. Acreages are rounded.

Habitat Type	Acres
Grassland	***27,000***
Seasonal Alkali Wetland	***7,520***
Alkali Meadow	1,050
Alkali Playa Complex	20
Alkali Vernal Pool Complex	6,450
Seasonal Freshwater Wetland	***7,150***
Wet Meadow	7,150
Riparian Forest and Willows	***440***
Willow Thicket	400
Confined Woodland/Savanna	40
Perennial Freshwater Wetland	***275***
Pond	75
Valley Freshwater Marsh	200
Tidal Wetland	***25,750***
Bay Associated Tidal Flat	1,400
Marsh Associated Tidal Flat	2,400
Panne/ Salina	1,150
Shallow Channel	550
Vegetated Tidal Marsh	20,250

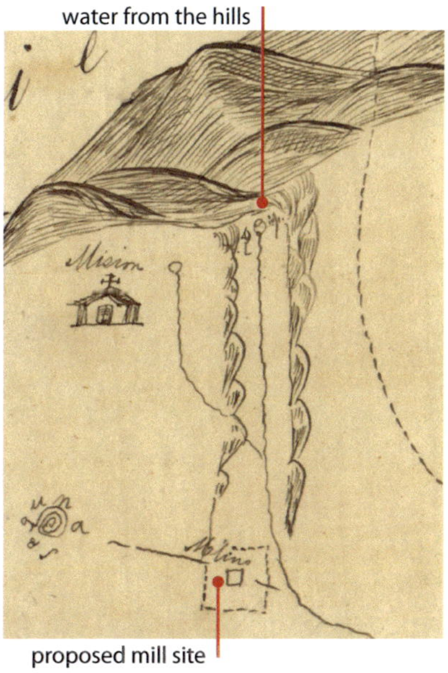

Figure 6.4. Early water use. This 1840s land case map proposes the construction of a mill powered by water from Arroyo de la Laguna and streams in the hills. The claimant requested permission to make use of "the three sources situated in the Sierra towards the SE ... uniting them by means of a dam" (Vallejo 1840). (USDC ca.1840g, courtesy The Bancroft Library, UC Berkeley)

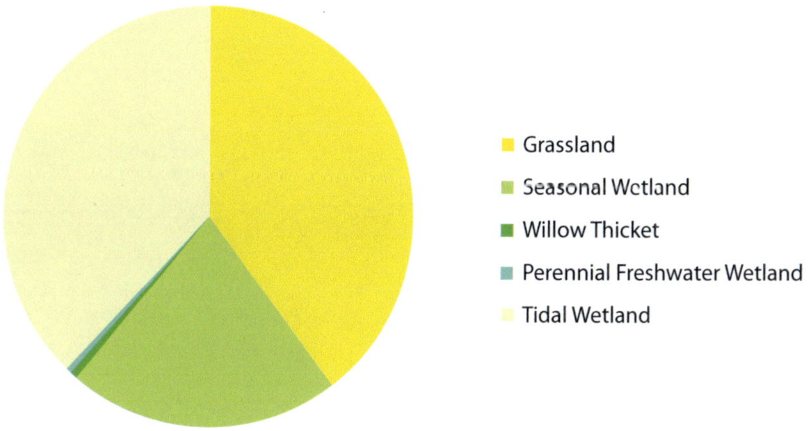

Figure 6.5. Historical land cover in the Niles Cone region. This simplified pie chart summarizes our findings for land cover patterns on the Fremont Plain circa 1800. Non-tidal wetlands have been grouped into seasonal and perennial wetlands. All tidal wetland types were combined. Riparian types are not shown.

The alluvium of this township is broader in extent and of greater depth than anywhere else about the bay. It is, in general, a rich black loam, which is found to rest upon a substratum of gravel and sand at a depth of from six to thirty feet, which latter depth is found at Niles. From the presence of this layer of gravel a natural system of subsoil drainage is maintained of such excellence that the soil may be turned up from any depth and yet be found always moist without being saturated.

—COQUHOUN 1893:58

Niles Cone habitats

Both wetland and grassland habitats developed on the Niles Cone. Grassland grew across the eastern portion of the cone. Seasonal wetlands extended along the western edge of the Niles Cone, bordering the tidal marsh (fig. 6.6). A willow thicket was found in the artesian zone within the seasonal wetlands. Perennial freshwater ponds and marsh occurred along the Hayward Fault south of Niles. This section describes each of these habitat types.

Grasslands

The majority of the Niles Cone was covered with low herbaceous vegetation. Historical records describe grasslands and wildflowers interrupted by the trees of Alameda Creek's riparian corridor. Crossing the plain in spring, explorers in 1776 described the plain as "green with grass and … thickly covered with various wild flowers" (Anza and Bolton 1930) and "very level country, green and flower-covered all the way to the estuary" (Font and Bolton 1933). An 1885 newspaper article recalled past wildflower displays near Niles: "the growth of wild flowers was

Figure 6.6. Historical Niles Cone. This image from an 1840s land case map shows the variety of habitats found on the Niles Cone—riparian vegetation along Alameda and Dry creeks, tidal marsh (lower left), and the hills. (USDC ca.1840c, courtesy The Bancroft Library, UC Berkeley)

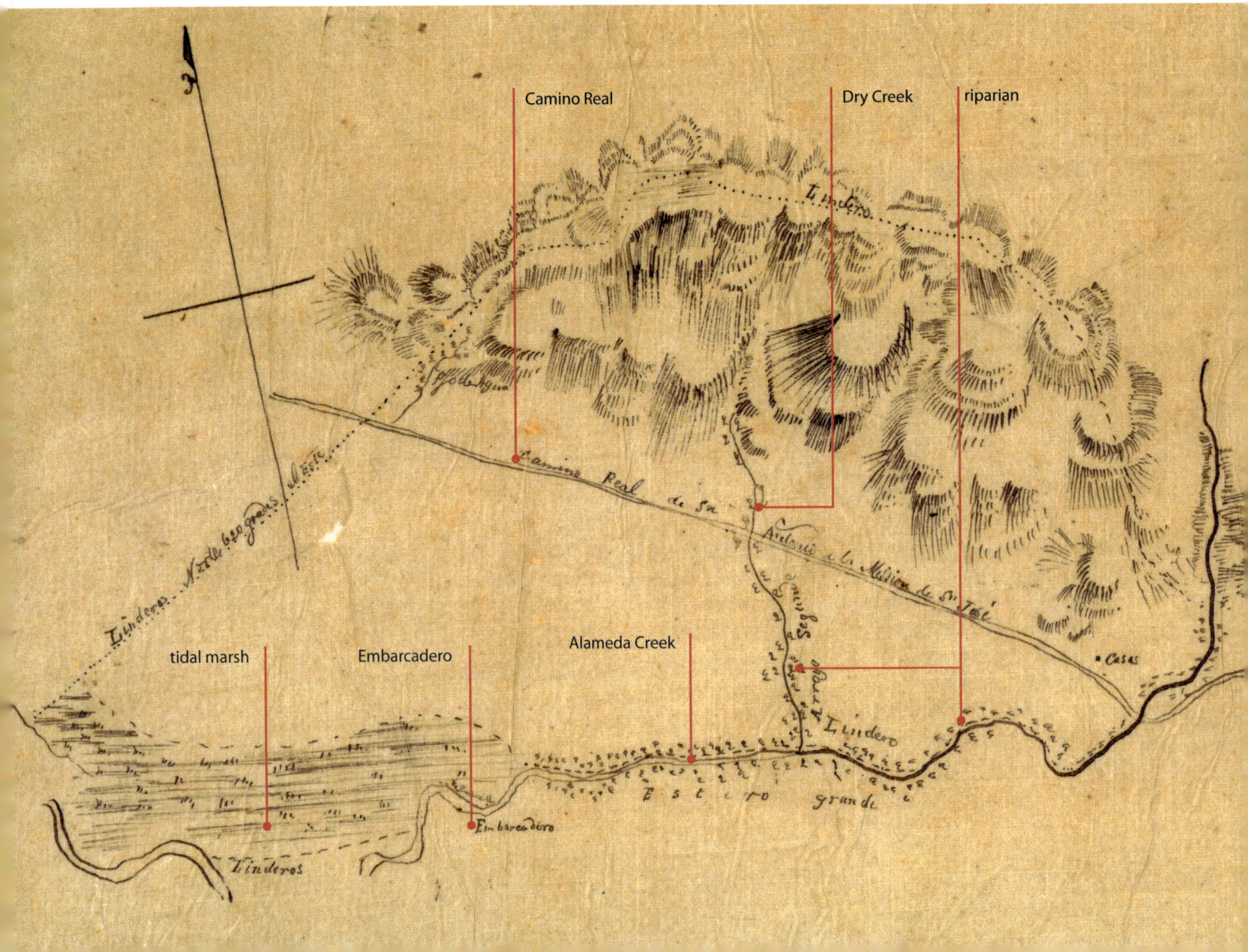

infinitely superior to anything now known in the region" (*San Francisco Chronicle* 1885). These descriptions of abundant wildflowers support emerging research, showing that broad lowland regions of California may have been dominated by forbs such as wildflowers rather than perennial bunchgrasses as previously theorized (see Minnich 2008, Martinez 2010). Research also suggests that rhizomatous ryegrasses may have been dominant on relatively flat clay and loam soils without hardpans or claypans, which occurred across much of the Niles Cone (Holmes and Nelson 1917, Holstein 2001). The grasses and wildflowers described in these accounts may have occurred in either grasslands or wet meadows, and many may have historically grown in both settings.

All the land is level, black, and very well covered with good grass, mallows, and other herbs.

—CRESPÍ 1772, IN CRESPÍ AND BOLTON 1927:286

Unfortunately, historical species composition is not well documented (see Hamilton 1997, Holstein 2000, Holstein 2001, Minnich 2008). Grassland species composition was likely similar to that found in Livermore-Amador Valley, and would have likely included rhizomatous ryegrass species (e.g., *Elymus* spp.), as well as purple needlegrass (*Stipa pulchra*) and clover (*Trifolium*) and poppy (*Eschscholzia*) species.

As in Livermore-Amador Valley, early visitors commented repeatedly on the lack of trees. Dantí was careful to specify that both large trees (for lumber) and smaller trees (for firewood) were lacking: "the timber in this place is scarce, as is also the firewood" (Dantí 1795 in Cook 1957; lumber referred to large, sturdy trees useful for construction, such as redwoods, while trees not suited for building, such as willows and oaks, were potential firewood). Other explorers reinforced this description: Crespí described in 1772 how "the land is all good, although it has not a single tree or any firewood, except what is in the beds of the arroyos" (Crespí and Bolton 1927), and Font similarly reported "no other timber or firewood than that afforded by the trees in the arroyos" (1776 in Font and Bolton 1933). Much later, in 1854, Elisha Capron described the lack of timber as a benefit for agriculture: "The husbandman has no labor to perform preparatory to fitting the land and putting in the seed. No dense forests of heavy timber are to be felled, —nature having cleared the soil" (Capron 1854:175). Although we do find records of a riparian tree corridor along Alameda Creek (see p. 200), this was the exception on the otherwise treeless plain.

The fields about Centerville were wonderfully blue, a deep blue, with the flower of Phacelia ciliata.

—W.L. JEPSON APRIL 5, 1920

The picture contrasts sharply with nearby Santa Clara Valley (bordering the Bay to the south), where early paintings, maps, surveys, and descriptions consistently documented the presence of valley oaks (Grossinger et al. 2008, Beller et al. 2010, Whipple et al. 2011). The Niles Cone appears to have lacked even the clusters of low density oaks found in Livermore-Amador Valley—early images show very few or no trees (fig 6.7). Almost no remnant oaks were visible in the 1939 aerial photography (USDA 1939;).

Seasonal wetlands

Wet meadow, alkali meadow, and vernal pool complex formed the seasonal wetlands on the western fringe of the Niles Cone. All three seasonal

Figure 6.7. Grasslands and agriculture on the Niles Cone. These paired images show different views of agriculture on Niles Cone. The lithograph (A) looks east from Newark back towards the hills. The only trees visible are obviously planted, with the exception of the line of trees in the far distance, which likely represent riparian forest along Alameda Creek. The 1910s photograph (B) looks west from Niles Canyon and shows the same pattern of orderly orchards with few native trees. (A: Thompson and West 1878, courtesy David Rumsey Map Collection; B: photo #325710, © San Francisco Public Utilities Commission)

wetland types occurred on the fine sediment deposited at the toe of the fan bordering the tidal marsh. This fine sediment, along with high groundwater and occasional flooding by Alameda Creek, created a zone of saturated soils and seasonal wetland vegetation.

These were all alluvial flood plains from the Niles Canyon, and they formed seasonally like they did from the Tigris and Euphrates rivers or the Egyptian delta. They flooded every winter, and you would go out after the flood waters had gone away and plant crops and do wonderfully.

—BUCK 1986

Though three seasonal wetland types were interspersed, we mapped them as three distinct zones based on each area's dominant characteristic: wet meadow covered areas with very limited alkali influence, alkali meadow occurred where there were substantial alkali effects on vegetation, and alkali sink scrub with vernal pool complex was mapped in alkali areas that also historically supported vernal pools and sink scrub vegetation.

Wet meadow covered 7,150 acres between Hayward and Coyote Creek, almost all of it along the lower margin of the Niles Cone. Although most wet meadows occurred along the edge of the tidal marsh, additional wet meadows extended east along the northern and southern edges of the Niles Cone in low-lying areas between the fan of Alameda Creek and the fans of San Lorenzo and Coyote creeks (to the north and south, respectively). The underlying soils in these areas were described as "exceedingly sticky when wet" so that "water stands over large areas for days during the rainy season" and "in places the surface is covered with water for short periods" (Holmes and Nelson 1917). Wet meadows supported some of the same species as grasslands (including *Lasthenia* spp.), but would have included a greater proportion of more hydrophilic species as well, including *Juncus* spp. The extent and duration of saturation for the wet meadows would have varied with rainfall and flooding patterns.

In addition to the persistently moist wet meadows, a broad, more well drained zone around Alameda Creek would have been inundated for days at a time during periods of high flow (fig. 6.8), creating ephemeral surface waters and occupying sloughs leading to the tidal marsh (see overflow p. 203):

> In the rainy season Alameda Creek ... overflows the country in the vicinity of Alvarado ... and on both sides of the filled in county road northward to the swamp, thence finding its way to the sloughs westward of the railroad. Before the Survey ended in July, this branch was dried up and the adjacent land was being prepared for a crop of sugar beets. (Westdahl 1896c)

The alkali-influenced habitat types (alkali meadow and vernal pool complex) together covered 7,500 acres, intergrading with the tidal marsh along much of their length. As we were unable to map individual pools, alkali areas with evidence of vernal pools are shown as alkali sink scrub

It would flood in the winter months, but then in the summer it became grazing pasture.

—BORGHI 1987B, DESCRIBING FARMLAND ON THE NILES CONE

Figure 6.8. Temporary flooding. This image from January 26, 1914 shows farmland flooded by overflow from Alameda Creek near the Patterson ranch. (AD-141, © San Francisco Public Utilities Commission)

with vernal pool complex, and areas without documented vernal pools are alkali meadow (see methods p. 36). All of the alkali-affected areas south of Coyote Hills contained vernal pool complexes, although those further north near Alameda Creek did not.

Here, as in Livermore Valley, vernal pools occurred only on alkali-influenced soils. These seasonally flooded pools occurred in patterns of small mounds and depressions (fig. 6.9). The early soil surveyors described this pattern near Newark where they noted a "hog-wallow surface" (Holmes and Nelson 1917). The Niles Cone was one of only three prominent areas along the Bay that historically contained extensive vernal pools (Goals Project 1999). Even today the vernal pool microtopography is visible in undeveloped remnants, although only 170 acres of vernal pool complex remain (SFEI 2011; fig. 6.10).

Figure 6.9. Remnant vernal pools at Don Edwards Refuge. Vernal pools are distinguished by the rings of flowers that grow at the pool margins, as can be seen in these photographs of present day vernal pools at Warm Springs near Newark. Exclosure experiments at this site have found that grazing can help control invasive species. (photos April 16, 2011, top by Bronwen Stanford, bottom by Alison Whipple)

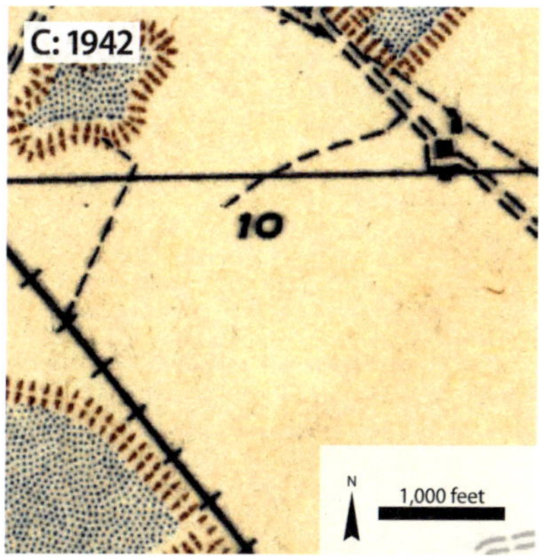

Figure 6.10. Agriculture and vernal pools. Although early agriculture on the Niles Cone destroyed many vernal pools by 1939, the textured pattern of small circles surrounded by lighter colored areas that is indicative of vernal pools can be seen in both (A) and (B) above. USGS mapping in 1942 (C) does not show the pools, but it does show some ponds, which may have functioned as vernal pools. The pond at the top left can be seen in (A) as well; we mapped it as a playa. Levees around the duck ponds may have helped to keep these areas wet into the later 20th century. (A: USDA 1939; B: USDA 2009; C: USGS 1942)

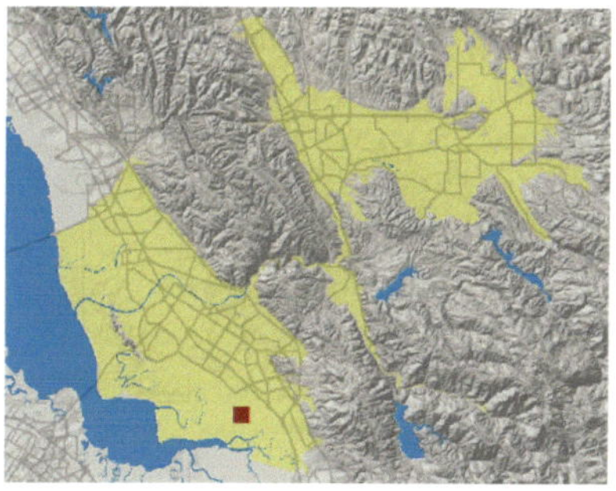

Botanical records for this zone of alkali meadow and vernal pool complex date back to the 1890s and include many alkali-affiliated species and vernal pool species, including alkali milkvetch (*Astagalus tener* var. *tener)*, hairless popcornflower (*Plagiobothrys glaber*), Congdon's tarplant (*Centromadia parryi* ssp. *congdonii*), Contra Costa goldfields (*Lasthenia conjugens*), tidy tips (*Layia* spp.), and Downingia (*Downingia pulchella)* (CNDDB 2010). In 1895, botanist Joseph Burtt Davy wrote a particularly detailed description of this area, while speculating how it might have already changed (contemporary species names are in brackets):

> Newark, Alameda Co. Plants collected beside the railroad track between Newark and the Drawbridges Alameda Co, May 6, 1895. This is a level marshy country, bordering the marshes with a good deal of alkaline soil about, apparently. This stretch of about 8 miles is the richest in flowers of the whole 52 miles from Alameda [illegible] to San José and shows how gorgeous the whole plain bordering the marshes probably was before the introduction of foreign weeds and the grazing of cattle and horses. On either side of the track outside the fences, but few wild plants are left, and in places within the fence mustards, oats, and other foreigners are becoming established in patches, bidding fair to monopolize the whole before long. The general impression in color is a mass of yellow owing to the abundance of *Lasthenia* and *Blepharipappus* [*Layia*], though in places this gives way to the masses of green and white of *Trifolium fucatium*. In places the yellow is dotted with the white heads of *Navarretia cotulaefolia* ... *Bolelia* [*Downingia*] *pulchella* (Lindl) Greene. Dried out pools ... *Lasthenia conjugens*. (Burtt Davy 1895)

Figure 6.11. Vernal pool wildflowers at Don Edwards Refuge. Clockwise from top right: popcornflower, yellowray goldfields, owl's clover, downingia, and alkali milk vetch. These tiny flowers were photographed around vernal pools at Warm Springs, within the Don Edwards Refuge. (photos 2011, Bronwen Stanford)

Many of these species can still be found today in vernal pool remnants (see fig. 6.9). Alkali milkvetch (*Astagalus tener* var. *tener*), described by Jepson (1911) as growing in "alkaline fields" and associated with alkaline vernal pools, was recently rediscovered near Newark (in Baye et al. 2000). Other vernal pool species recently recorded in this area include vernal pool tadpole shrimp (*Lepidurus packardi*) and prostrate vernal pool navarretia (*Navarretia prostrata*), which were recorded in remnant wetlands in 1996 and 2001 respectively (fig. 6.11; CNDDB 2010).

Artesian zone and the Willows

At the margin between the seasonal wetlands and the tidal marsh was an artesian zone, kept under pressure by groundwater inputs from Alameda Creek trapped below the clay soils on the lower alluvial fan (fig. 6.12). Groundwater seeped across the surface from natural springs and flowed from artificial artesian wells (Patterson 1977). Springs were especially concentrated in an area known as the Willows just east of the Coyote Hills (see fig. 6.14), but also spread north towards Alvarado: "The flowing artesian wells of Alvarado are one of the features of our county. The town is

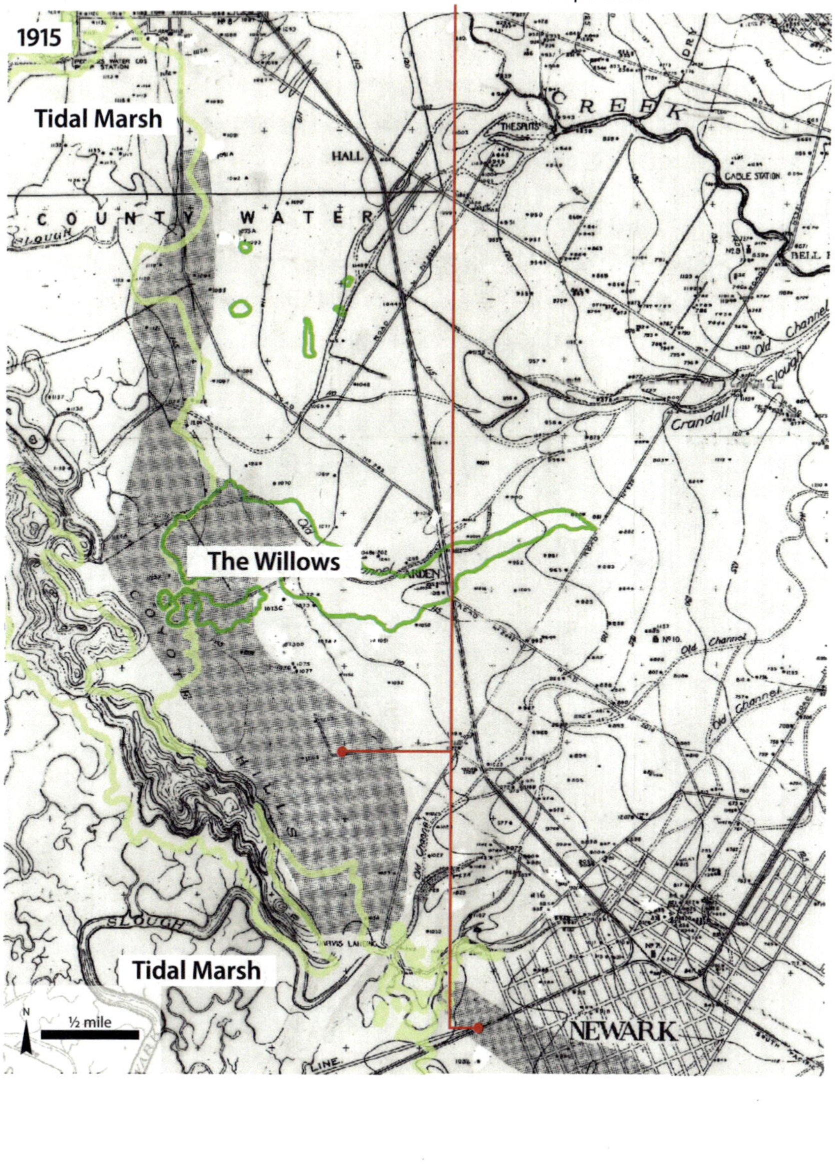

Figure 6.12. Zone of evaporation. This 1915 map shows a broad "zone of evaporation" represented by the shaded area along the tidal marsh. This zone occurred along the eastern edge of the Coyote Hills, clearly depicted on the map. The tidal marsh boundary is indicated by the line in light green, and the Willows are indicated by the dark green outline. Note that the evaporation zone includes areas straddling the tidal boundary and would have included both tidal marsh and seasonal wetlands. (Stoner et al. 1915, courtesy ACWD)

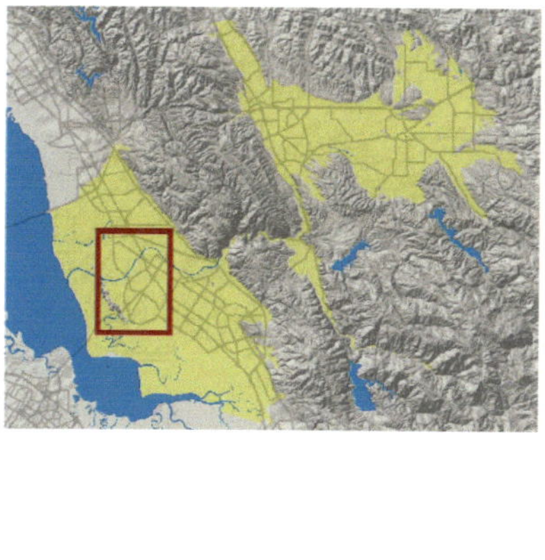

supplied with them and so are all of the prominent farmers in the adjacent country" (*Oakland Tribune* 1875). The artesian zone extended beneath the tidal marsh as well; a surveyor stated that "fresh water is obtained almost anywhere on the salt-marsh by sinking artesian wells to a depth of 80 to 200 feet" (Westdahl 1896c).

The flowing well of Captain Barron is one of the most remarkable in the state. It discharges a volume of water at least ten inches square and equaling about 100 gallons a minute. The Captain having no more than domestic use for it, the water flows up and makes a lake of four or five acres in extent, whence it flows into the bay.

—*THE OAKLAND TRIBUNE* 9/16/1875

High groundwater and artesian pressure were important resources for farmers along the Niles Cone. Fields were "subirrigated," meaning that crops absorbed water underground from high groundwater levels rather than from surface water inputs (Patterson 1977). Early settlers dug artesian wells to access freshwater aquifers (fig. 6.13; Westdahl 1896c:55, Dockweiler 1912b:57). Water companies began to take advantage of this water source as early as 1860, eventually shipping water north to Oakland and west to San Francisco (Figuers 1998). By the early 1900s, due to a combination of increased use of groundwater for irrigation, diversions and dams in Alameda Creek, and groundwater pumped to Oakland and San Francisco, the water level in the Niles Cone had been drastically reduced. (See more in land use history, Chapter 3.)

Settlers also used artesian well water to wash sediment onto the marsh and assist reclamation efforts. One resident described the rapid transformation: "Already, in a single season, that portion of land which he has completely enclosed and submerged in fresh water, shows a fine growth of flags, grass and willows, and will be excellent meadow land in another season, even without cultivation" (Browne 1872).

THE WILLOWS The artesian zone supported at least one very notable local feature. The Willows was a 400 acre willow thicket east of Coyote Hills and just south of the present-day Flood Control Channel (fig. 6.14). Swales threaded through the Willows (Westdahl 1896a, USDA 1939). The

Figure 6.13. One particularly notable spring near Alvarado was used to create an artificial lake (see above quote). This well was dug to a depth of 185 feet, and was claimed in a county history to "probably [flow] more water than any well in the State, the stream having a body of about eight inches square" (Halley 1876:487). (image courtesy California Historical Society, CHS2012.933)

report accompanying the 1896 USCS T-sheet described the Willows as a "wooded area ... a thicket made up of willow, poplar [likely cottonwood], alder and other deciduous trees ... there are numerous springs within this area" (Westdahl 1896c). Another early source described patches within the Willows containing "many fine sycamore and oak trees, some of them rivaling in size and beauty any that are to be found in the valley" (Country Club of Washington Township [1904]1950:100). These larger trees likely grew on remnant natural levees created along a previous course of Alameda Creek when it flowed close to the Coyote Hills.

Within the Willows there were "numerous springs ... conveyed in ditches to irrigate the adjoining fields ... new springs frequently break out and fill the ditches with water" (Westdahl 1896d). The high level of saturation through this area created a "slowly flowing marshy area year round" with "very heavy yields both in grain and vegetables" (Patterson 1977).

Ornithological records reflect the diversity of habitats available in and around this small wetland. On August 10, 1919, pioneer California biologist Joseph Grinnell spent an hour recording bird species in the Willows, which he described as "a large tract of dense willow, alder and sycamore, with big live oaks adjacent" (Grinnell 1919). He recorded 18 species, including species associated with oak habitats (Hutton's vireo *(Vireo huttoni)*, oak titmouse (*Baeolophus inornatus*)), brushy and marshy habitats with dense cover (song sparrow *(Melospiza melodia)*, Bewick's wren *(Thryomanes bewickii)*, spotted towhee *(Pipilo maculatus)*, Wilson's warbler (*Cardellina pusilla*)), and open and mixed habitats (red-shafted flicker (*Colaptes auratus cafer*)) (Bousman 2007, American Ornithologists' Union 2011).

In addition to its significance as a hotspot for local biodiversity, the Willows was an important local cultural landmark. In 1795 Hermengildo Sal noted that native people had built rancherias in "a large willow grove in a mire" (Mayfield 1978). In the 1860s, multiple witnesses used this feature to define the borders of the Potrero de los Cerritos land grant (e.g., Edmundson 1858, Dyer 1861b, Munyon 1861, Lewis 1862). In the early 1900s, Chinese laborers worked this soil, which proved to be highly demanding because while "extremely rich and productive ... if it was too wet, it was sticky; if it was too dry, it broke into large clods" (Patterson 1977). People farming nearby land diverted water from the Willows for irrigation (Country Club of Washington Township [1904]1950).

By the late 1800s, the Willows began to disappear as it was cleared and groundwater declined. The 1896 T-sheet report stated that "within the wooded area ... many patches have been cleared and drained" (Westdahl 1896d), and distinct remaining clumps are shown in other historical maps map (see fig. 6.14). As groundwater was drawn down in the later 1800s, the springs began to dry up, further reducing the willow thicket area. A 1904 history described this process:

> This willow swamp, covering over one hundred acres, was fed by fresh water springs that kept the ground wet and supported a dense

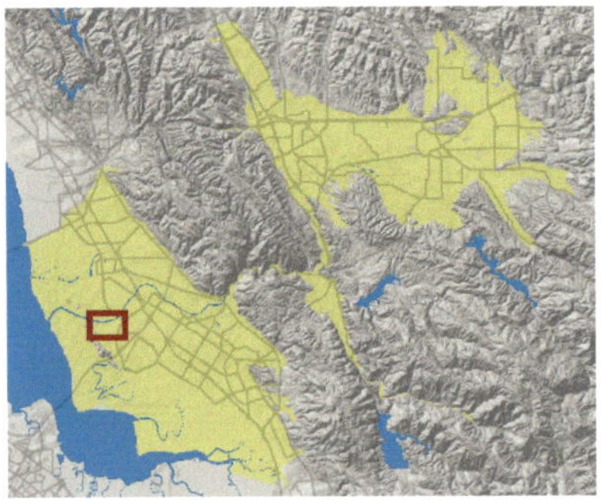

Figure 6.14. The Willows. This 1878 lithograph shows the Willows (A), with a few additional isolated clumps of willow scattered to the north. The feature was captured consistently in maps of the area, and remains wet today. The green outline in D shows the extent on 2009 imagery. (A, C: Thompson and West 1878, courtesy David Rumsey Map Collection; B: Allardt 1874, courtesy The Bancroft Library, UC Berkeley; D: USDA 2009)

N 1,000 feet

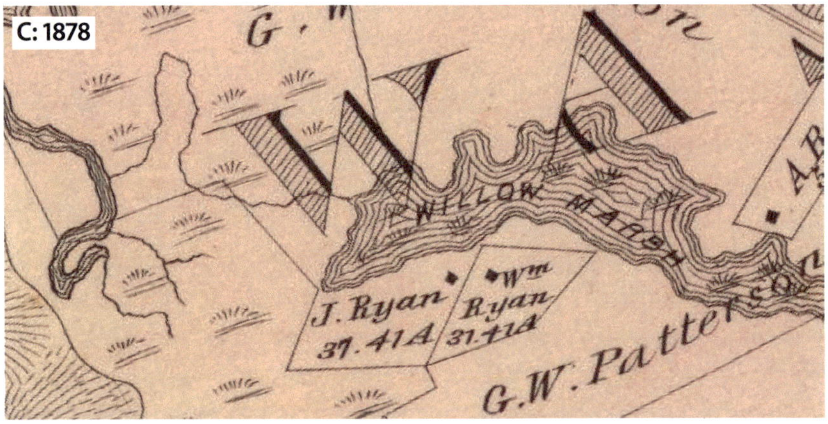

> undergrowth. Since the establishment of pumping plants and reservoirs for supplying the bay cities with water, these springs have disappeared, the trees are dying, and the swamp will before many years become only a memory. (Country Club of Washington Township [1904]1950:100)

Further south, along the tidal marsh boundary south of the Coyote Hills, there were historically several additional small willow groves and freshwater wetlands. These features were poorly documented, particularly in comparison with the Willows, but would have provided additional habitat (fig. 6.15; Tracy 1852, Morse and Westdahl 1896). Local historian Charles Shinn provided the clearest description of these features, describing "clumps of willows" near Mowry's landing, and "tules and cattails … on all sides of the island of hard ground where the wharves were" (Shinn [1889]1991:47). As early settlers developed routes to the landings and settled on the edge of the tidal marsh, they likely cleared these willows, and often planted eucalyptus in their stead (Shinn [1889]1991). Only a few traces remained by 1939 (USDA 1939).

Orchards and cultivated fields now extend across many an old tule swamp east of the old landing.

—SHINN [1889]1991:53

Figure 6.15. Mapped willow remnants at Mowry's Landing. The patch of willows shown by Tracy (B) match almost exactly with the "spring area" mapped by Stoner et al. (1915) and fall within a patch of wet meadow, surrounded by alkali wetlands. This patch of willows was likely supported by overflow from the springs. (Tracy 1852, courtesy The Bancroft Library, UC Berkeley)

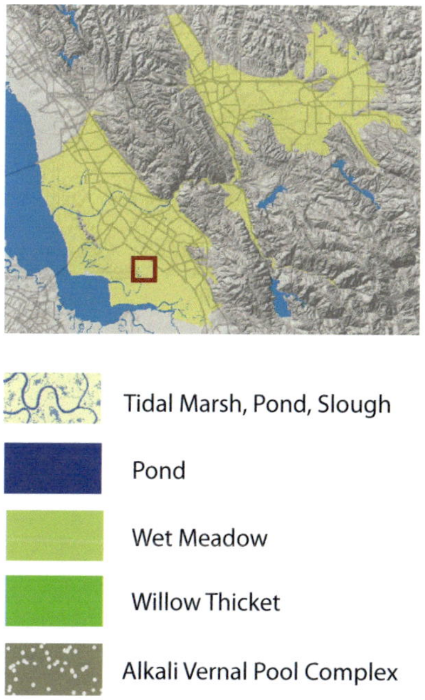

Tule Pond and the Lagoon

Movement along the Hayward Fault resulted in two additional perennial wetland features. Because of their unique origin, these features were located close to the hills, away from all other local wetlands. These were Tule Pond, a sag pond that persists today, and the Lagoon, a large freshwater marsh with a pond at the center that has been converted to Lake Elizabeth (fig. 6.16). Mission Creek drained towards the Lagoon and a band of seasonal wetland connected Tule Pond to the Lagoon (Holmes and Nelson 1917).

These features are documented repeatedly in the historical record. Spanish explorer Fages recorded in 1770 passing "a very good fresh lagoon whose margin was adorned with rushes, cat-tails, and extensive meadows, where there was an abundance of geese, of which we succeeded in killing seven" (Fages and Bolton 1911:151). One intriguing account by explorer Font in 1776 describes a "somewhat salty lagoon" near the hills, presumably the Lagoon (Font and Bolton 1933:356). The description of salt water conflicts

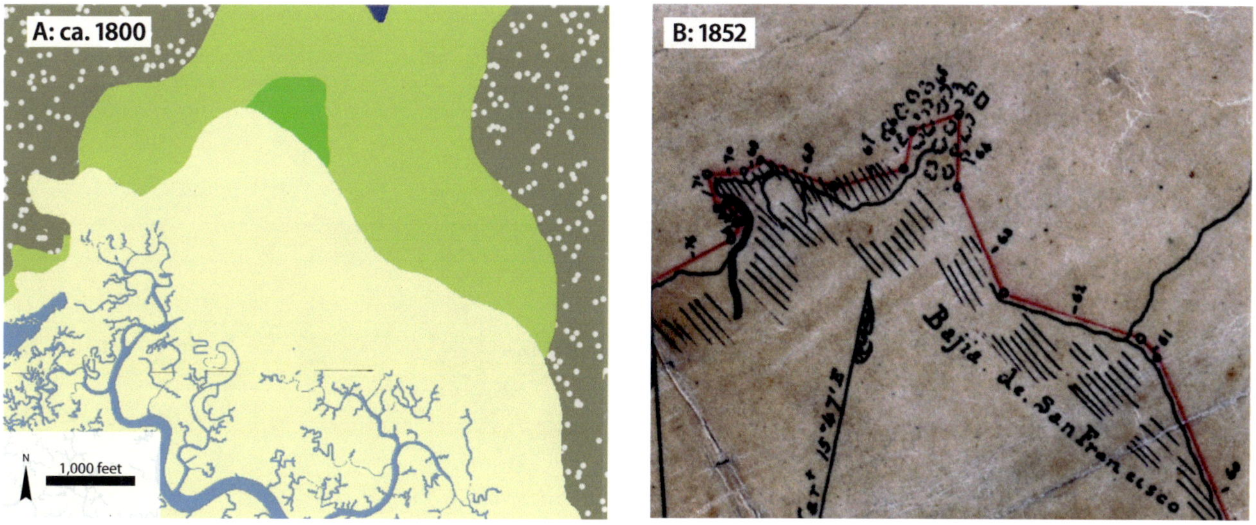

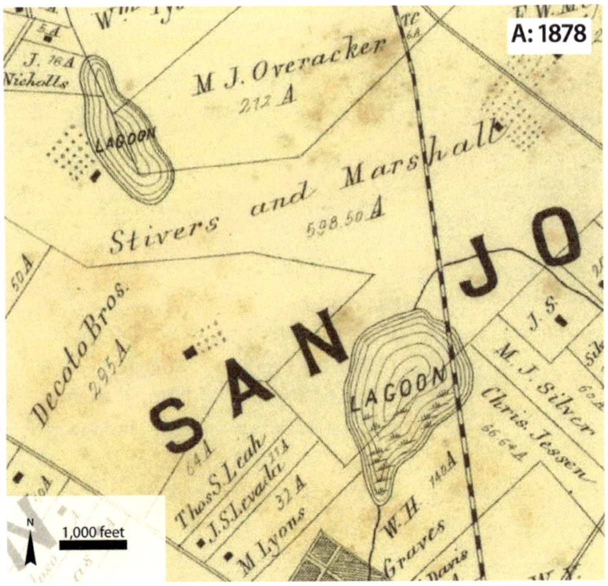

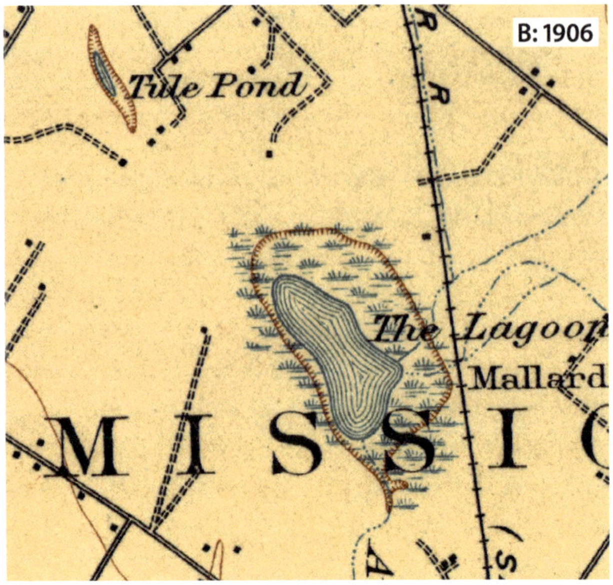

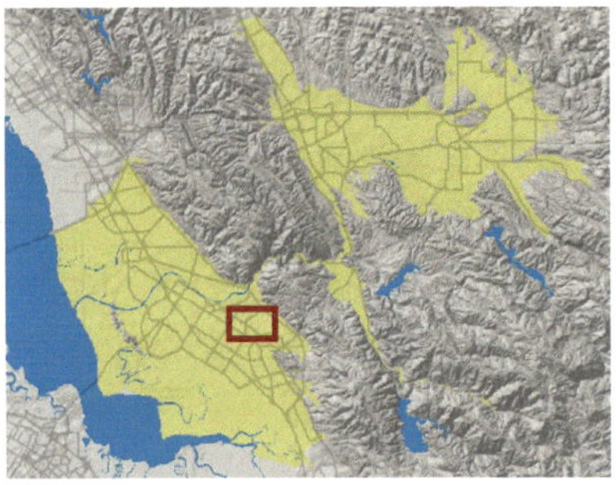

Figure 6.16. Tule Pond and the Lagoon Early maps identify Tule Pond and the Lagoon (A, B), which were popular for duck hunting in the 19th century (D). The Lagoon has been replaced with Lake Elizabeth today, while the Tule Pond (C) is actively managed to provide natural habitat. (A: Thompson and West 1878, courtesy David Rumsey Map Collection; B: USGS 1906; C: January 28, 2010 by Bronwen Stanford; D: "Among the Tules" by Thomas Hill 1894)

with other accounts, such as the account by Fages, directly above, but may have referred to oligohaline conditions—slightly saline but nearly freshwater.

Many different names were applied to these features over time. Labels included "large lake pond" (Unknown 1865) and "Clear Lake" for the northern feature (Lewis 1850b), while the southern feature was "Horner's swamp" (Shinn [1889]1991:62) and "tule pond" (Allardt 1874). "Lagoon" was applied to both (Thompson and West 1878). However, the most common names appear to have been "Tule Pond" or "Tyson Lagoon" for the northern feature, and "the Lagoon" or, today, "Stivers Lagoon" or "Lake Elizabeth" for the larger, southern feature (Holmes and Nelson 1914, Stoner et al. 1915, Clark 1924, Metsker n.d.). Like the Willows, these wetlands provided concentrated areas of unique habitat, particularly important here for waterfowl and likely for native fishes (see Chapter 8).

Tidal wetlands

Extending along the western edge of the Niles Cone, the band of tidal marshland along the Bay margin was historically from one to three miles wide, covering 25,800 acres, or over 20% of the total study area. The tidal marshlands were composed of a mix of habitat types, including shallow tidal channel networks, pannes, salinas, and deeper subtidal channels (fig. 6.17). About 2,400 acres of marsh-associated tidal flats—unvegetated mud, shell, or sand surfaces—bordered the marshlands and extended into tidal channels. Tidal flat also extended into the bay, beyond the tidal marsh border. On the landward side, 70% of the tidal marsh bordered terrestrial wetlands, forming a marsh-wetland ecotone.

Tidal marshlands provided important habitat for a broad range of native species. As early as 1913, biologists visited the tidal marsh to observe the rich community of bird species (table 6.2). Several of these species,

N

500 feet

Figure 6.17. Salinas and alkali habitats are observable at the edge of this marsh remnant. In this 1939 image, the sinuous tidal channels are visible as dark lines cutting through the surrounding marshland. At the edges of the tidal marsh, white salt scalds are visible. These form part of the alkali meadow-salt marsh ecotone. (USDA 1939)

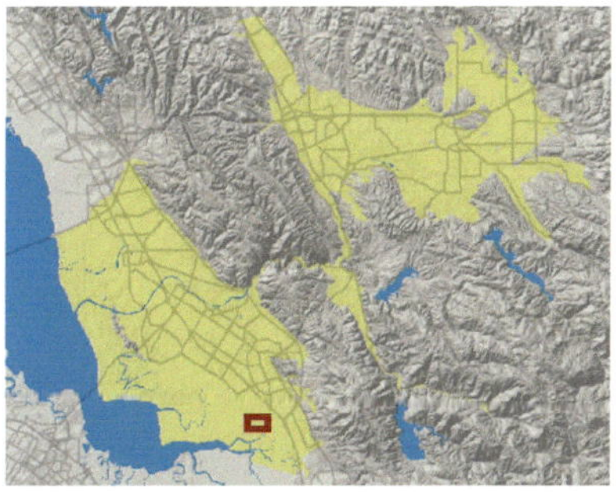

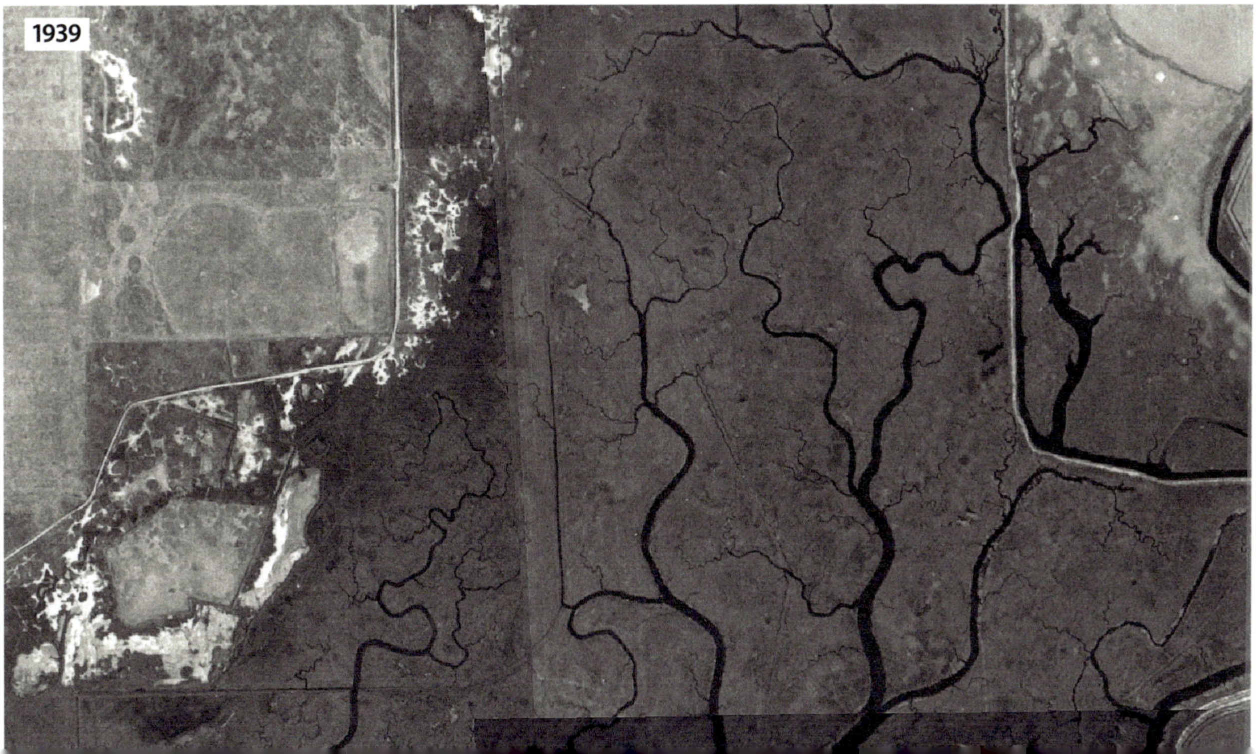

Table 6.2. Birds observed near Alvarado, 1913. This list represents the species named by Grinnell and Bryant over a few hours of observation at duck ponds near Alvarado. Bryant visited the site on May 22, 1913, and Grinnell visited July 28, 1913. Many individuals of a particular species were often noted; each species is listed only once here. Brackets indicate modern names where different from the name used in 1913. Question mark indicates that writing was illegible.

Species	Scientific name	Source	Native? (Y/N)
Barn swallow	*Hirundo rustica*	Grinnell 1913	Y
Brown black bird		Bryant 1913	
Bryant marsh sparrow		Grinnell 1913	
Butcherbird	*Cracticus* spp.	Grinnell 1913	N
Cinnamon teal	*Anas cyanoptera*	Grinnell 1913	Y
Clapper rail	*Rallus longirostris*	Bryant & Grinnell 1913	Y
Cliff swallow	*Petrochelidon pyrrhonota*	Grinnell 1913	Y
Dowitcher	*Limnodromus* spp.	Grinnell 1913	Y
English sparrow	*Passer domesticus*	Bryant 1913	N
Greater Yellowlegs	*Tringa melanoleuca*	Grinnell 1913	Y
Herring (gull)		Grinnell 1913	
Hudsonian Curlew	*Numenius phaeopus hudsonicus*	Grinnell 1913	Y
Killdeer	*Charadrius vociferus*	Grinnell 1913	Y
Least or Semi-plamated Sandpiper	*Caldris* spp.	Grinnell 1913	Y
Least Sandpiper	*Calidris minutilla*	Grinnell 1913	Y
Linnet	*Carpodacus mexicanus / Carduelis cannabina*	Bryant 1913	Y
Long-billed Dowitcher	*Limnodromus scolopaceus*	Grinnell 1913	Y
Marsh hawk [Northern harrier]	*Circus cyaneus*	Grinnell 1913	Y
Mudhen	*Fulica americana??*	Grinnell 1913	Y
Red-backed sandpiper		Grinnell 1913	Y
Red-winged blackbird	*Agelaius phoeniceus*	Bryant 1913	Y
Ring-billed gulls	*Larus delawarensis*	Grinnell 1913	Y
Salt-marsh [Alameda] song sparrow	*Melospiza melodia pusillula*	Grinnell 1913	Y
Shirker		Bryant 1913	
Soaot[?] Sandpiper		Grinnell 1913	
Song sparrow	*Melospiza melodia*	Bryant 1913	Y
Tule [Marsh] wren	*Cistothorus palustris*	Bryant & Grinnell 1913	Y

including the Alameda song sparrow (*Melospiza melodia pusillula*) and California clapper rail (*Rallus longirostris obsoletus*), are strictly dependent on tidal marshes and native tidal marsh plants for nesting habitat, while others, such as the red-winged blackbird (*Agelaius phoeniceus*) or harrier ("marsh hawk"; *Circus cyaneus*), can use a variety of wetland habitats (Grinnell and Wythe 1927, Bousman 2007, Shuford and Gardali 2008). Salt marsh-dependent mammal species recently found in this area include salt marsh harvest mouse (*Reithrodontomys raviventris*) and salt marsh wandering shrew (*Sorex vagrans halicoetes*) (1998 and 1985; CNDDB 2010). (For a detailed treatment of fish, see Chapter 8.) The marsh and channels provided a wave energy buffer between the water of the Bay and the low-lying wetlands and grasslands of the shore. Tidal marshlands also played important roles in water and sediment processes, including trapping of fine sediment and temporary storage of floodwaters, ecosystem services of recognized value today (Goals Project 1999, Steere and Schaefer 2001).

The historical tidal marshlands of south San Francisco Bay have been described previously by the Goals Project (1999) and Collins and Grossinger (2004). Here we describe the particular features documented along the shores of the Alameda Creek watershed and how they have changed during the historical period. We also present some evidence of wildlife use of the tidelands.

Tidal marsh

Tidal marsh was composed of at least three distinct zones with different plant assemblages. These were salt marsh (including high elevation marsh); brackish and freshwater marsh at creek mouths, springs, and in artesian zones; and the terrestrial-estuarine ecotone (fig. 6.18; Grossinger 1995).

The marsh fringing the Niles Cone was heavily influenced by the salty waters of the Bay. Grinnell described the marshland as a "pickleweed" and "salicornia-bay marsh" (1913, 1919); indicating the salt marsh associations, (Grossinger et al. 2006). In 1901, Jepson recorded "salt marshes near Alvarado," and "subsaline fields" near Newark (in Baye et al. 2000). The salt marshes supported a variety of salt-associated marsh species that are rare or no longer present today (e.g., *Symphyotrichum subulatum, Chloropyron maritimum, Castilleja ambigua, Puccinellia nutkaenis;* see also Baye et al. 2000).

A wide marsh of from 1 to 2 miles in width intervenes between the shore and the waters of the Bay, but is threaded by small tide creeks, which allow the approach at high tide of vessels of small burden.

—TRACY 1852B

Joseph Grinnell visited the marsh at Alvarado in 1913 with Harold Bryant, recording the bird life he encountered there. At this time, the baylands were largely diked but still included substantial areas of undiked tidal marshlands as well as artificial freshwater duck ponds. He described the bird life in great detail (see also table 6.2):

> Several small [illegible] of ducks fly past in the distance; also a cloud, fully 100, of Red-Backed Sandpipers, wheeling and undulating en masse over the marsh. Bryant just flushed a Cinnamon Teal from bunch of tules at other end of pond; also a lone Mudhen. The family of Cinnamon Teal are now up on dike waddling along in the salt grass, plainly outlined against the sky. Another flock of five Cinnamon Teal have just dropped into pond … Many salt-marsh song sparrows and Bryant marsh sparrows sit on salicornia flats; also tracks of Clapper Rail in plenty on mud of sloughlets. (Grinnell 1913)

Tidal marshlands are younger than most features of the natural landscape, although they much predate Euro-American modifications to the land. In the southern San Francisco Bay, tidal marshlands likely developed over the past 2,000 years (Atwater et al. 1979, Goals Project 1999). Descriptions of extensive tidal marsh along the Niles Cone date to the earliest records.

Alameda Creek
Coyote Hill Slough
Coyote Hills
Union City
Alvarado
Alameda Creek
T.4S. R.3W.
12
13
15
16
17
18
19
20
21
22
27
28
32
33
34
4
3
Swamp & O. Land
Surveyed for
H. P. Jones
Sept. 29, 1859
Margin of Dry Land
Peruvian I.
Hunters I.
S. & Overflowed
Land Survey
No. 22
S. & O. L.
Long's Landing
Bay of San Francisco
Rancho de los Ceritos

Marshes along Alameda Creek's shore do not appear to be the result of increased sedimentation following European contact, as has been recently suggested for some coastal systems (Kirwan et al. 2011, Mudd 2011). Explorers repeatedly commented on the presence of a large tidal marsh: Crespí recorded that "the lakes and little inlets surrounding the big estuary were many and very hard going because very miry" (1769 in Stanger and Brown 1969), while a few years later Anza noted that "the lowlands generally are flooded by the tides" (1776 in Anza and Bolton 1930). Early maps show a wide swath of marsh along the Bay (e.g., De Mofras 1844), and as early as the 1850s, maps began to accurately depict the extent of the tidal marshes (Lewis ca. 1850, Kerr 1857a,b, Rodgers and Kerr 1857). Depositions for the Potrero de los Cerritos land grant case described the three to four square miles of tidal marsh along the western side of Coyote Hills, which matches the roughly four square miles of diked baylands present today (USDA 2009). These marshes were "generally at the elevation of ordinary high tides, but are overflowed twice a month at the new and full moon" —during the spring tides (Lewis 1862). Another account describes less frequent tidal overflow, stating that "all of the marsh-lands shown upon [the T-sheet] are covered with salt water at the periods of extreme high waters in June and December" (Morse and Westdahl 1896-7). Surveyor James Stratton described the tidal marsh and Coyote Hills in 1862 as follows:

> It is of the usual character of the salt marsh land surrounding the Bay of San Francisco, and is generally but two or three inches above ordinary high tide and is only overflowed during Spring tides ... [The northern Coyote Hills] are isolated hills surrounded by the salt marsh above described, they are sometimes called islands, but not properly so, because they are only surrounded by water at the Spring tides. (Stratton 1862)

Despite the persistence of the tidal marsh through the historical period, we find some evidence that land use changes did affect the marsh, so that it expanded slightly during the latter portion of the 19th century (fig. 6.19; Westdahl 1896a). The 1896 T-sheets show the marsh extending about 1,000 feet (one-tenth of its overall width) further west than those from 1857, particularly along the stretch between Coyote Hills Slough and Dumbarton Point. Surveyor Westdahl (1896c) attributed this shift to debris from planted oyster beds and modifications to the Bay causing changes in currents. In 1889 historian Shinn also described this effect: "The old shore line of the bay, many years ago, was nearly a third of a mile east of where it now is along in the Dumbarton region. The marsh lands have slowly and steadily grown by the wash of rains and storms. The lowlands are filling up" (Shinn [1889]1991:145). The T-sheet for this region distinguishes between the new and prehistoric marsh, which had distinct characteristics:

> The new marsh formed since conditions favorable to its growth began is entirely different in character from the old and the line of demarcation between them is very distinct. The old marsh is smooth, spongy, full of salt ponds, and covered with short marsh grass with bushy growths along the sloughs. The new marsh is harder, un---- [illegible], full of lumps and holes, has no ponds, and is covered with long grass more like the tules of Suisun Bay and lighter in color than the old growth. (Westdahl 1896d)

Figure 6.18. Tidal marsh and major channels along the edge of the Niles Cone. (opposite page) This 1861 survey was made by surveyor E. H. Dyer, showing tidal marsh and major channels. He noted that the marsh boundary line "was not all made from actual survey. Part of it was, and the rest from observation, and I think it is substantially correct" (Dyer 1861b:150). (USDC 1861, courtesy The Bancroft Library, UC Berkeley)

Figure 6.19. Persistence of tidal marsh. In many places, tidal marsh was quickly filled. This 1878 lithograph shows patches of remnant marshland persisting near Roberts Landing, just north of the study area. (Thompson and West 1878, courtesy David Rumsey Map Collection)

Marsh expansion during the late 19th century in Alameda County is consistent with expansion in other parts of the region as a result of upstream sediment-producing processes and loss of tidal prism (e.g., Atwater et al. 1979, Dedrick and Chu 1993, Collins and Grossinger 2004, Gedan et al. 2009). Today, following some erosion, the marsh-bayland boundary is located generally about halfway between the 1850 and 1900 positions.

Tidal channels

Sinuous tidal channels were a dominant feature of the tidal marshlands, carrying water into and out of the marsh with each tidal cycle. Channel density was highest near major sloughs, but mapped tidal channels threaded through almost all of the tidal marsh (fig. 6.20). This high density of channels historically created navigational problems:

> The many little channels intersecting this flat land make it an absolute labyrinth, and as we were not acquainted with the terrain, we were mistaken at many times, and had to turn back, often missing the main by turning into a side channel. (Langsdorff 1806 in Clarke 1952:155)

At least ten channel networks branched directly off the mainstem of Alameda Creek. These distributary channels branched into a total of 172 miles of tidal channel, providing an extensive amount of habitat, refuge, and sediment storage directly connected to the creek (Collins and Grossinger 2004).

A series of major sloughs extended through the tidal marsh at fairly regular intervals, generally from west to east. From north to south, the most notable of these were Union City Slough (the mouth of Alameda Creek, also known as Mount Eden Slough), Coyote Hills Slough, Beard Slough (also known as Newark Slough), and Mowry Slough (fig. 6.21). Major landings were associated with each of these sloughs, as they provided the easiest route between the Bay and the Niles Cone, which otherwise was essentially blocked by the broad marshes and mud flats. All of these large sloughs persist in channelized form today. The sloughs (particularly Coyote Hills Slough) extended far inland towards the wetlands that bordered the tidal marsh, providing access for estuarine fishes to the range of habitats along the saline-brackish-fresh gradient.

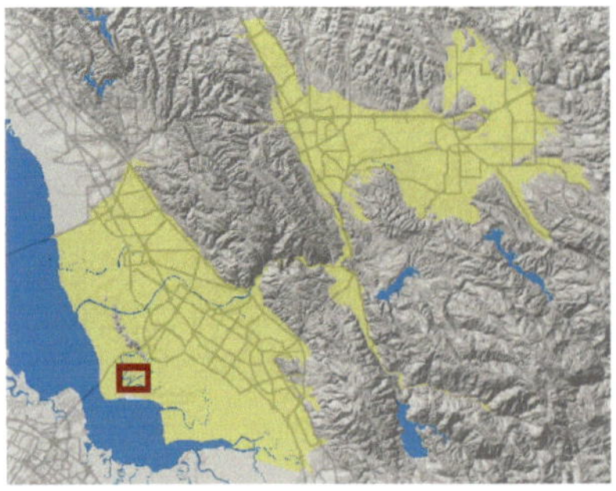

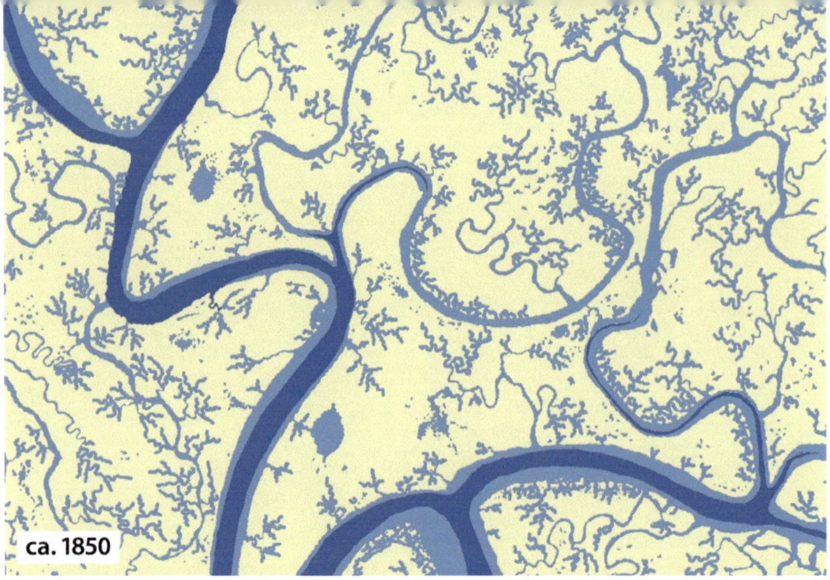

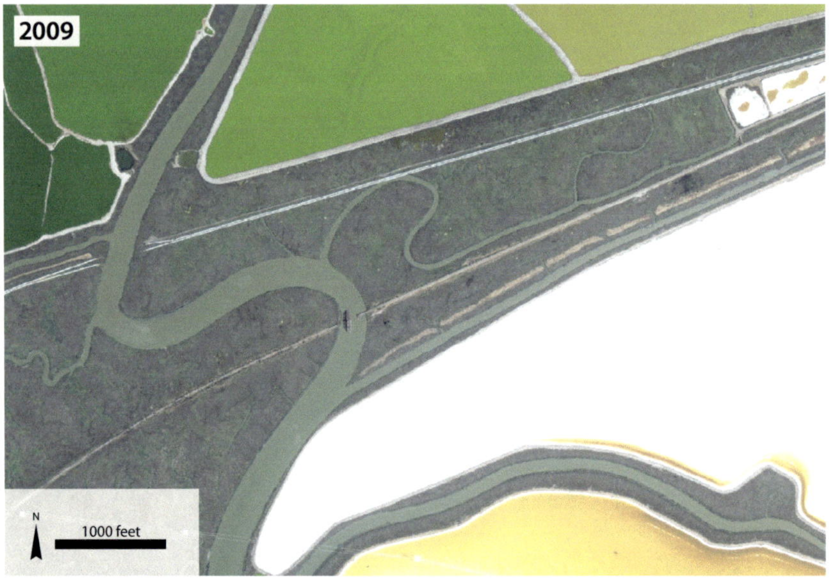

Figure 6.20. Detailed T-sheet mapping allowed us to map tidal channels in much more detail than we were able to achieve for other habitat types. Compare the many tidal channels visible this portion of marsh in the mid-1800s with the same area today. (USDA 2009)

Marsh pannes and salinas

Scattered through the tidal marsh in between these hundreds of miles of tidal channels were hundreds of open water salinas and pannes (fig. 6.22). These shallow features included the large, elongate features at the backshore edge of the tidal marsh (known as salinas or salt ponds) and the smaller bodies scattered through the marsh (marsh pannes; Goals Project 1999, Baye et al. 2000). Marsh pannes tend to form on the marsh plain away from tidal channels, typically at slightly higher elevations (Collins and Grossinger 2004). Pannes and salinas may be largely unvegetated, or may include some vascular plants or algaes, particularly those with more freshwater influence (Baye et al. 2000).

Pannes and salinas ranged in area from under 20 square feet (less than one-thousandth of an acre) to over 63 acres. However, most were small, with an average size of only a tenth of an acre. Mapped pannes and salinas covered 1,150 acres within the study area, making up four percent of the tidal marsh area. However, early T-sheet surveys for this area did not extend inland far enough to capture the large salinas along some portions of the marsh, particularly in the south, so we were not able to map salinas in these areas.

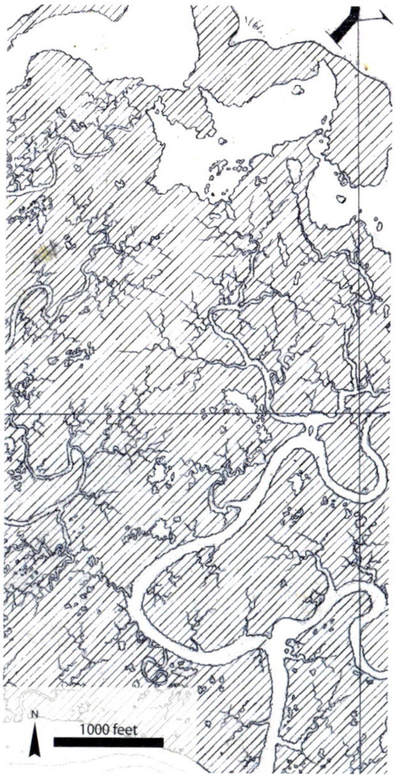

Figure 6.22. Pannes and salinas. (above) Bodies of open water form in areas that are not drained by tidal channels, either at the backshore edge of the marsh or in areas of slightly higher elevation. In the map above, pannes are visible as large, irregularly shaped unshaded areas. (Kerr 1857a, courtesy NOAA)

Figure 6.21. Coyote Hills panoramas (below) show views of tidal marsh around the Coyote Hills in August and September 1916. The top image looks north across Coyote Hills Slough, now the mouth of Alameda Creek. In 1916, the caption states that "most of the flood waters of Alameda Creek course to the Bay through this slough." The bottom image looks southwest at the southern end of the Coyote Hills and across the tidal marsh, portions of which have already been leveed. In addition to the channels, these images show the textured tidal marsh and some small pannes, or ponds, within the marsh. (AD-1005-1009 and AD-944-948, © San Francisco Public Utilities Commission)

[Coyote Hills Slough] extends up a long distance to the Eastward forming in its upper part a sausal *or willow swamp. No other between that and Alameda Creek extends much if any distance into the higher land.*

—LEWIS 1862

Some salinas may have had species assemblages that overlapped with those found in vernal pools (Baye et al. 2000). In 1901 Jepson described downingia (*Downingia pulchella)*, typically a vernal pool species, as "abundant and of rank growth in salt marshes near Alvarado" (Jepson 1901 in Baye et al. 2000). This area was historically surrounded by pannes, salinas, and alkali meadow, but has no evidence of vernal pools.

Tidal marsh ecotone

The edges of tidal marsh habitats were transitional zones that often intergraded with surrounding habitats (Collins and Grossinger 2004). Along the Niles Cone over 60% of the total tidal marsh edge was bordered by wetland habitats (rather than grassland); one-quarter was bordered by the steep uplands of the Coyote Hills (fig. 6.23). Wetland habitats bordering the tidal marsh included wet and alkali meadow, vernal pool complex, valley freshwater marsh, and willow thicket. These gradually sloping wetland habitats historically allowed the tidal marsh room to grow inland and upland as sea level rose (fig. 6.24; Grossinger 1995, Baye et al. 2000, Collins and Grossinger 2004).

In many places the gently sloping transition from tidal marsh to wet meadow and other wetlands was not clearly defined (Conerly 1931). One surveyor referred to this zone as "the debatable area immediately adjoining the Salt-marsh, which is sometimes covered at high tides" (Morse and Westdahl 1896-7:3), and much of the Potrero de los Cerritos land grant case testimony attempted to establish an appropriate border to the salt marsh. South of Coyote Hills, near Warm Springs, the hummocky topography of vernal pools can be seen intergrading with the tidal marshland (fig. 6.25).

The upper marsh edge and transition to adjoining grasslands and wetlands historically provided some of the most important marsh habitat (Goals Project 1999:A-16). The mix of habitat types helped support a rich

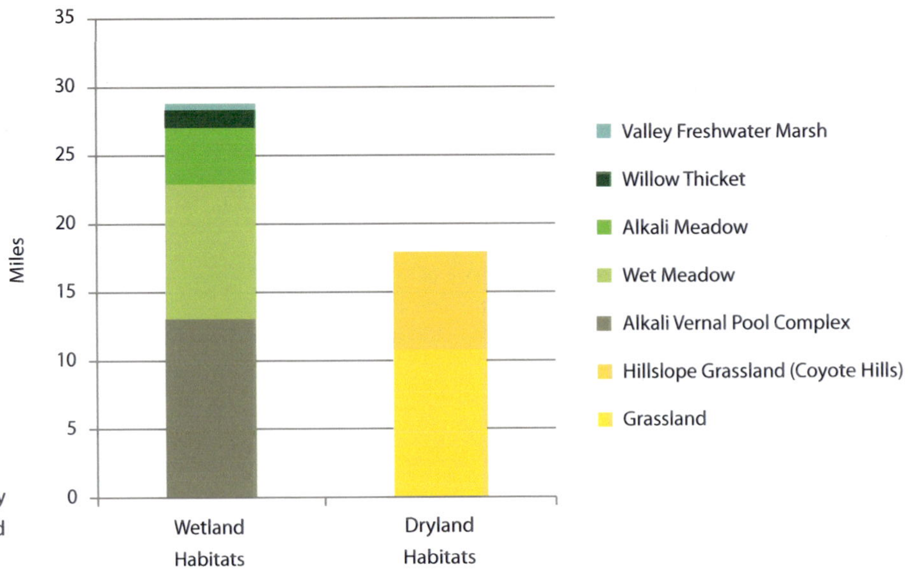

Figure 6.23. Habitats bordering tidal marsh, by miles of tidal marsh length. Sixty percent of the upland bordering the tidal marsh is one of five wetland habitats, mostly seasonal wetlands. These wetlands provided a gradual transition to the tidal marshland.

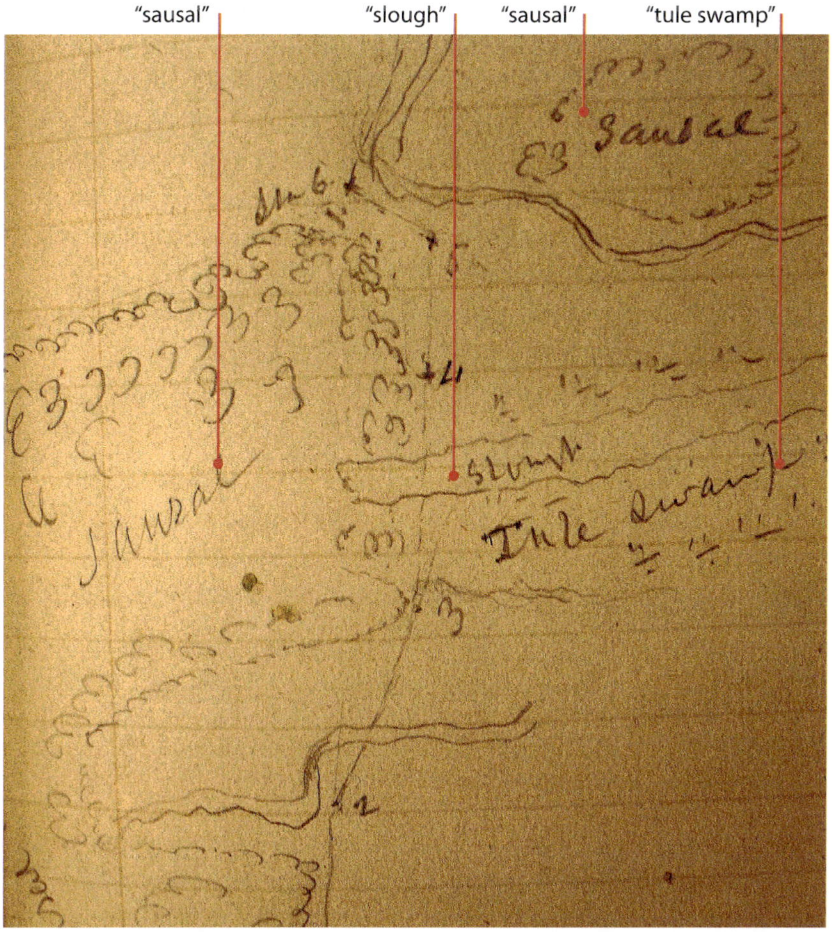

Figure 6.24. Mix of wetland habitats along the edge of a willow thicket. This detail of a map from 1852 shows the mix of distinct habitat types found along the edge of the Willows (labeled "sausal"). Habitats visible in this image include sloughs, tule swamp, and willows. (Lewis 1852, courtesy Office of the Santa Clara County Surveyor)

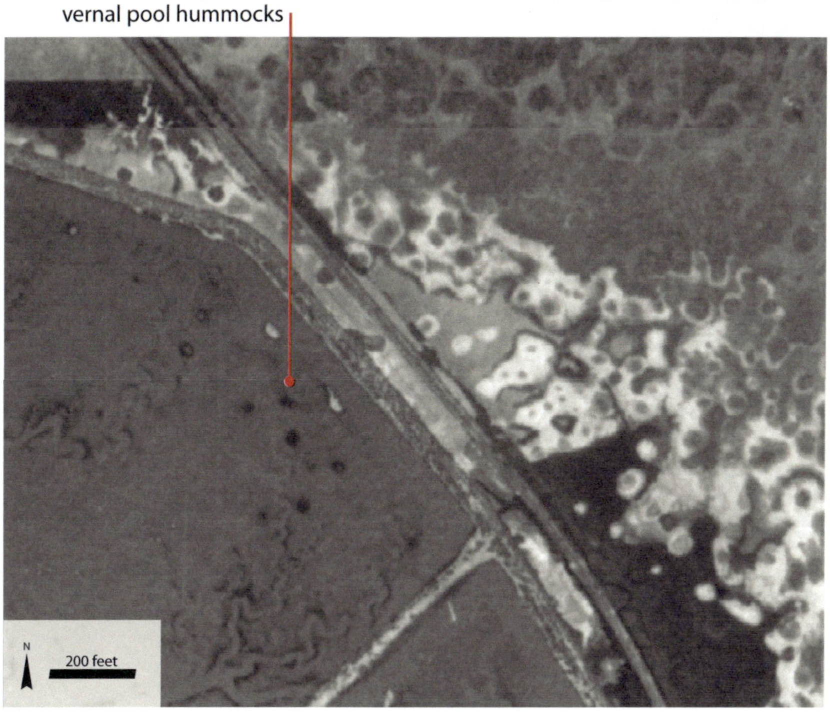

Figure 6.25. Tidal marsh-vernal pool ecotone. As tidal marshes extended inland over time, they covered areas that had formerly supported vernal pools. The tops of hummocks between ancient pools can still be seen in the salt scalds and even protruding through a salt pond in this imagery. (USDA 1939)

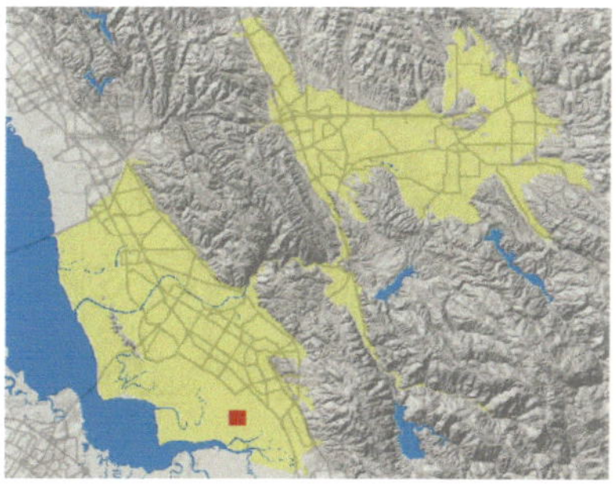

community of bird species (see table 6.2) as well as diverse plants, many of which are no longer present or are very rare (such as alkali pepperweed (*Lepidium dictyotum*), forked pepperweeed (*L. oxycarpum*) and hairless popcornflower (*Plagiobothrys glaber*); Greene 1894 and Jepson 1911 in Baye et al. 2000). The springs and seeps that dotted the upland edge of the marsh created small zones of freshwater influence. The fresh/brackish gradient likely expanded greatly at the mouth of Alameda Creek and along the artesian zone stretching south from the creek, and may be partly responsible for the large pannes in this area (Grossinger 1995).

The most interesting feature of this sheet is the change wrought by man on the otherwise valueless marsh bordering the bay and sloughs.

—WESTDAHL 1896D, DESCRIBING THE ACCOMPANYING T-SHEET

Transformation of tidal marsh habitats

Tidal marshes were generally seen as "valueless" (Westdahl 1896b) and were converted and filled by a variety of processes over the 19th and 20th centuries (fig. 6.26, fig. 6.27). Farmers used the marshes for grazing and built dikes to prevent water from flowing across the marsh (fig. 6.28):

> A larger area of former saltmarsh has been dyked and is now used for grazing cattle. The old sloughs within this area still remain but will soon disappear from the constant tramping of cattle in the dry season when the fresh water accumulated in them during winter has drained off or evaporated. (Westdahl 1896d)

Farmers worked to raise land above the reach of tides by filling marshes with sediment. Residents described digging ditches in the early 1900s to direct water towards the land to be reclaimed so that deposited sediment "made that land down there … it reclaimed a lot of this marshland here" (McKeown 1975a). They described marshes "filled up over six feet" (McKeown 1975b) with "sediment from these floods that were diverted" (Patterson 1977). The process of "warping," as this technique is known in other parts of the world,

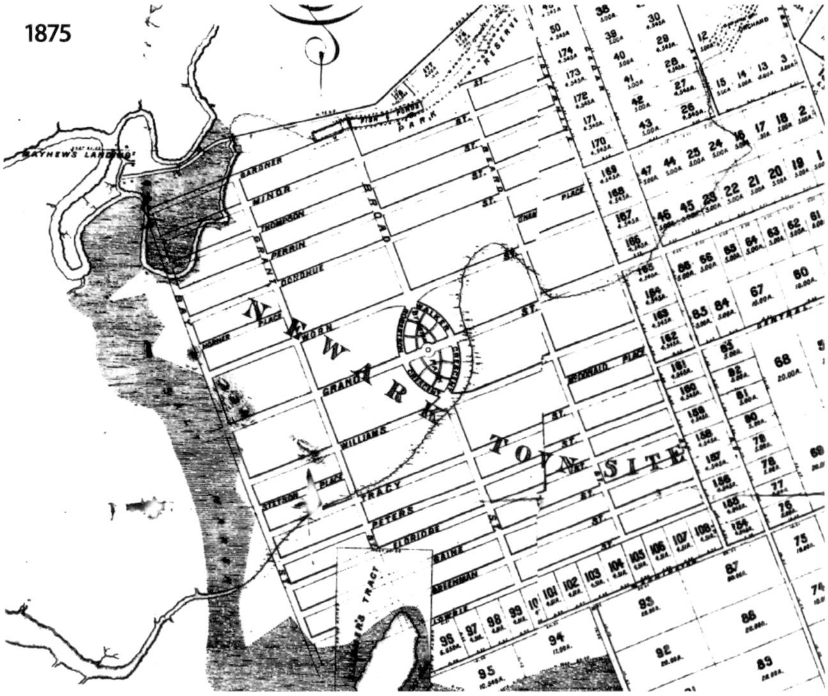

Figure 6.26. Lots were sold along the edge of the tidal marsh, as shown on this 1875 map of Newark. The dark shaded area represents the tidal marsh. The map states "this magnificent property … will be offered … at auction to the general public." (Thompson 1875, Fisher Scrapbook Collection, Newark/Alvarado, Museum of Local History, Fremont)

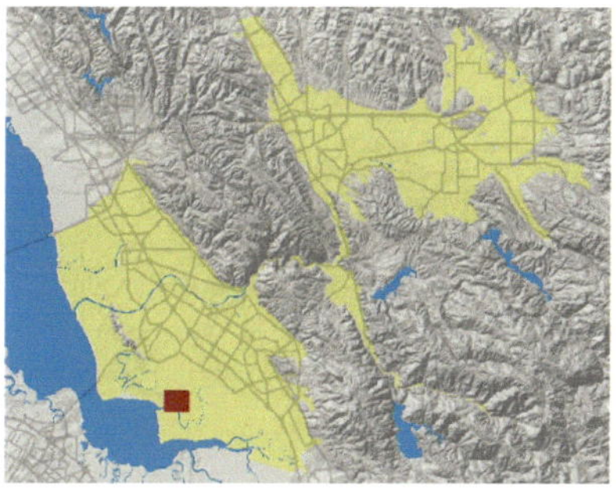

Figure 6.27. Remnant tidal channels in 1939. The sinuous tidal channels are still visible in this image from 1939, even though levees are already in place. The channels match those shown by mid-19th century T-sheets. (USDA 1939)

used natural processes to raise marsh elevation and has been previously documented elsewhere in the South Bay (Collins and Grossinger 2004). The technique may be a useful precedent for contemporary efforts to direct sediment deposits to the marsh surface to enable vertical growth in response to sea level rise (Collins and Grossinger 2004).

About one-third of the width of the valley is all that seems in grain and orchard. Two-thirds is a new Holland, waiting for its dyke-builders to shut out the sea tides.

—SHINN [1889]1991:145

Two members of the Patterson family (landowners near the Coyote Hills) recalled how overflows were used to fill in the marsh:

> By putting a levee around this area, we hoped to check the flow of water and thus allow the sediment to settle and build up the soil. We had noticed this was happening naturally, and we were trying to speed up the process. (Patterson 1955)

> In this manner, the entire lower salt marsh of the ranch has been reclaimed to a point where, at the present time, we are going to be able to farm intensively practically the entire old salt marsh. This reclamation has come about naturally, but it has taken at least fifty years to accomplish. (Patterson 1977)

Large areas of tidal marsh were converted to salt ponds as early as the 1850s, and were the site of the earliest extensive dikes constructed on the San Francisco Bay (fig. 6.29). Some of the earliest salt ponds were created from "slightly improved natural salt ponds" so that "dykes followed the windings of the sloughs to avoid the expense of damming these natural arteries" (Westdahl 1896a). These salinas and pannes had functioned as natural salt ponds for both the Ohlone and for early residents of the Mission. However, by the 1890s levees were "built in straight lines across both large and small sloughs" (Westdahl 1896a) to create salt ponds across broad areas that had formerly been tidal marshland (fig. 6.30).

Figure 6.28. Marsh conversion in 1878. These two lithographs show Eden Landing (top) and Barron's Landing at Alvarado (bottom) just to the south. Tidal marshes were filled and used as grazing land (note the cattle in the lower image, grazing former marshland). Buildings were also constructed along the edges of tidal channels for shipping. However, conversion was not yet complete at the time that these illustrations were made. Wet areas are still visible in the background of these images, and the land shown in the foreground still supports tufted, possibly marshy, vegetation. (Thompson and West 1878, courtesy David Rumsey Map collection)

Early residents also modified the marsh by drilling artesian wells and pumping water to the surface with windmills, creating zones of freshwater. Conversion to freshwater ponds and salt evaporation ponds also destroyed important habitat for a number of species, including the salt marsh harvest mouse (*Reithrodontomys raviventris*), Alameda song sparrow (*Melospiza melodia pusillula*), and the California clapper rail (*Rallus longirostris obsoletus*)(Bousman 2007). Duck ponds were built within the tidal marsh, "filled with fresh water pumped by windmills ... The larger ponds ... separated by banks contained a good growth of tules" (Bryant 1913). This freshwater had the effect of extending the fresh-brackish ecotone, and allowed the colonization of freshwater plants. Grinnell (1913) described them as "fresh water-fed tule ponds." A variety of birds were observed around these duck ponds near Alvarado:

> July 25, 1913 ... Tule wrens, many of them, young of the year and song sparrows were found in numbers among the tules. The song sparrows appeared to feed more largely on the low clumps of grass like tules in the open water but always took refuge in the thick tules. (Bryant 1913)

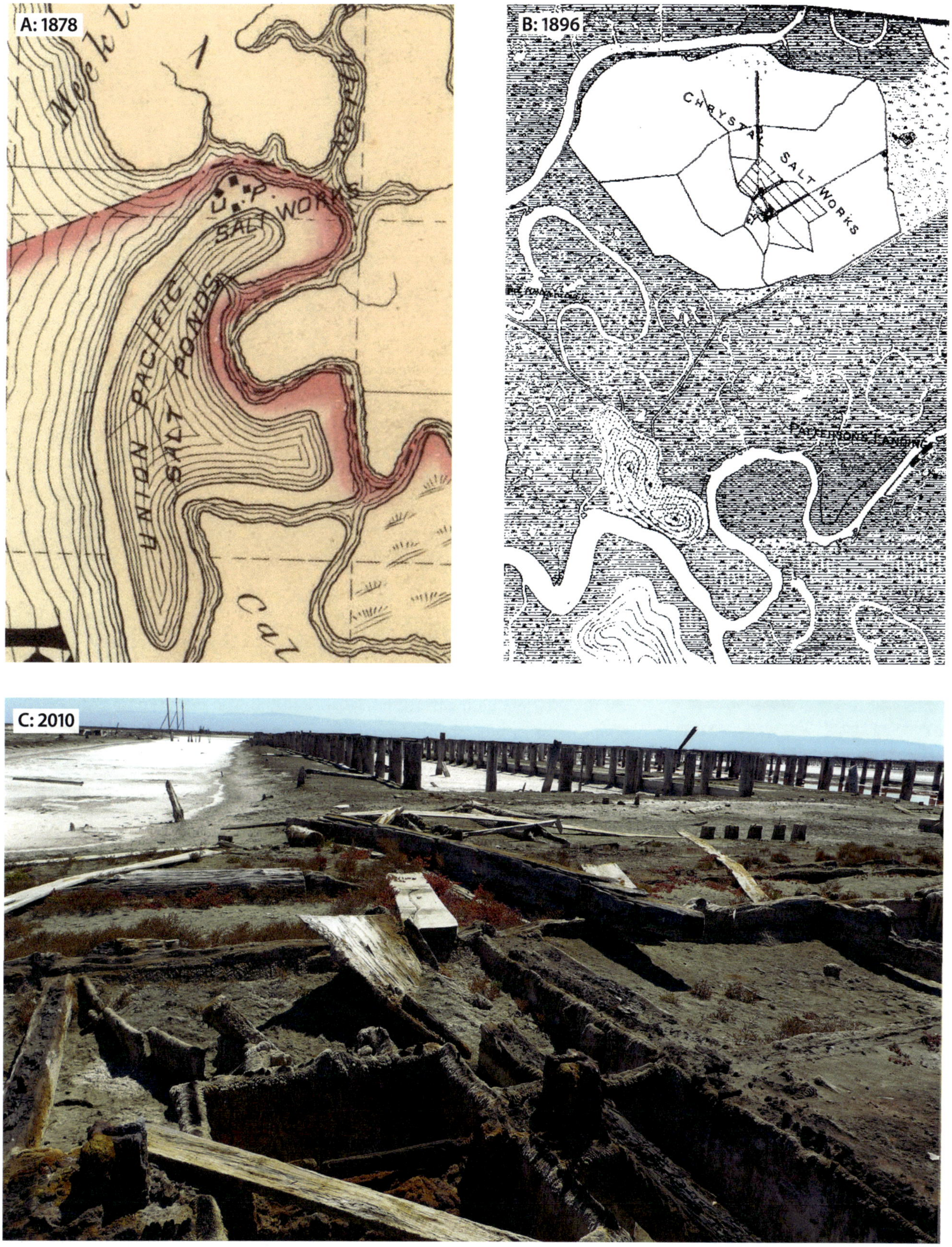

Figure 6.29. Salt ponds. These map fragments show two large salt ponds constructed in the tidal marsh: Union Pacific Salt Works near the mouth of Alameda Creek (B), and Chrystal Salt Works (A) near the Flood Control Channel. Above, the Eden Landing salt ponds today (C). These ponds are the site of a major restoration project. (A: Westdahl 1896b, courtesy NOAA; B: Thompson and West 1878, courtesy David Rumsey Map Collection; C: photo by Ruth Askevold, August 26, 2010)

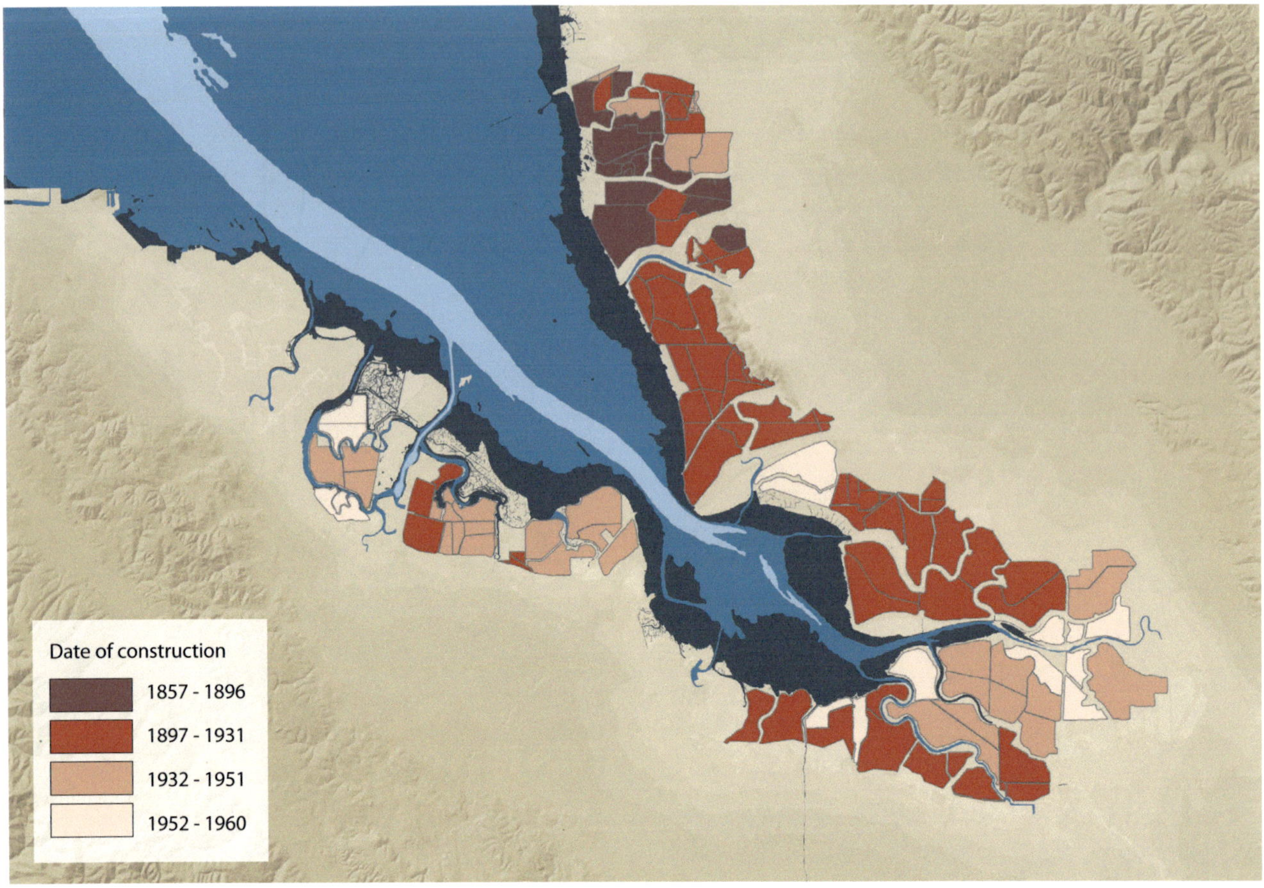

Figure 6.30. Historical development of artificial salt ponds. This figure shows the gradual construction of salt ponds in the south San Francisco Bay through the 1800s and 1900s. Large areas of tidal marshland had been converted to salt ponds by 1900.

The tracks of the California clapper rail were discovered and soon one of the dogs flushed one within two feet of where I was standing. The bird flew up about 7 or 8 feet in the air and then sailed off to the bank of a slough some 50 yds away. Here it immediately disappeared. During the day two others were flushed in the same way. They apparently frequent the banks of the sloughs and small ditches for their tracks were to be seen in the soft mud in these places.

—BRYANT 1913

As the tidal marsh transformed, tidal channels also began to fill with silt. The change in tidal channel form was due to a combination of upstream modifications resulting in more sediment (see Chapter 9) and the creation of dikes reducing tidal prism and natural scour in the remaining channels (Dedrick and Chu 1993, Collins and Grossinger 2004). The 1896 T-sheet report described these changes:

> Changes have been effected by the building of dykes, digging canals and short cuts for easier navigation which have since become the main slough, and by damming many sloughs altogether ... Alameda Creek is gradually filling up, especially the southern branch, from the large amount of sediment annually carried by freshets from the agricultural lands bordering its course. While thirty years ago bay-craft of thirty or forty tons could reach Union City vessels of only fifteen to twenty tons now find it difficult to reach the saltworks one and a half miles nearer the bay, and the bank of the creek there has to be dug out to enable them to turn at high water. (Westdahl 1896c)

By 1900, silt build-up in tidal channels limited access to landings that had been used through the 19th century. The T-sheet report noted "Beard Slough is rapidly filling up and is no longer used for navigation" (Westdahl 1896d). Donald Patterson recalled that Patterson Landing (near Coyote Hills Slough) was at one point 300 feet wide, but that by the 1960s it was "almost completely obliterated by the progressive silting, to a point now where it's not five feet wide and you have to search carefully through the grass to find it" (Patterson 1977).

Today only fragments of vegetated tidal marshland exist, totaling only 4,000 acres. Even these remnants appear very different than they did historically: channel density is much lower, largely due to much reduced tidal action (Collins and Grossinger 2004). However, projects are underway in an attempt to restore large areas of tidal marsh that are no longer used as active salt ponds (fig. 6.31).

Mr. Beard's artesian well affords a striking test of the utility of fresh water in the reclamation of these lands. Already, in a single season, that portion of land which he has completely enclosed and submerged in fresh water, shows a fine growth of flags, grass and willows, and will be excellent meadow land in another season, even without cultivation.

—BROWNE 1872:15-16, DESCRIBING CONVERSION OF TIDAL MARSHLAND

Alameda Creek and other drainages

After exiting Niles Canyon, Alameda Creek followed a sinuous course across the Niles Cone to the tidelands. This lower portion of Alameda Creek had a much more complex flow pattern and morphology historically than it does today. In addition to deep pools and a higher sinuosity, the creek showed distinct transitions in form and flow patterns across the Niles Cone. The creek shifted from broad, braided, and gravelly to narrow and meandering, and from perennial to intermittent flow and back to perennial again (fig. 6.32).

Two interconnected physical gradients caused these changes in form and flow. The first was depth to groundwater. The upper end of the cone rapidly lost water to the gravels and aquifers far below, resulting in intermittent surface flows. As the stream bed descended the alluvial fan towards sea level, however, it intercepted groundwater, changing from a losing to a gaining reach (see box 1.1).

The second major shift was in substrate. Sediment deposits sorted across the Niles Cone in a pattern typical of alluvial fans as the stream spread and lost power (Blair and McPherson 1994). The stream deposited coarser, heavier sediment towards the top of the fan, and finer materials towards the tidal marsh.

These patterns are similar to other creeks in the region (Grossinger et al. 2006; see box 1.1). However, Alameda Creek has a relatively short fan, resulting in much more rapid transitions than those seen elsewhere. Clark

Figure 6.31. Marshland restoration project at Eden Landing. (photo by Bronwen Stanford, August 26, 2010)

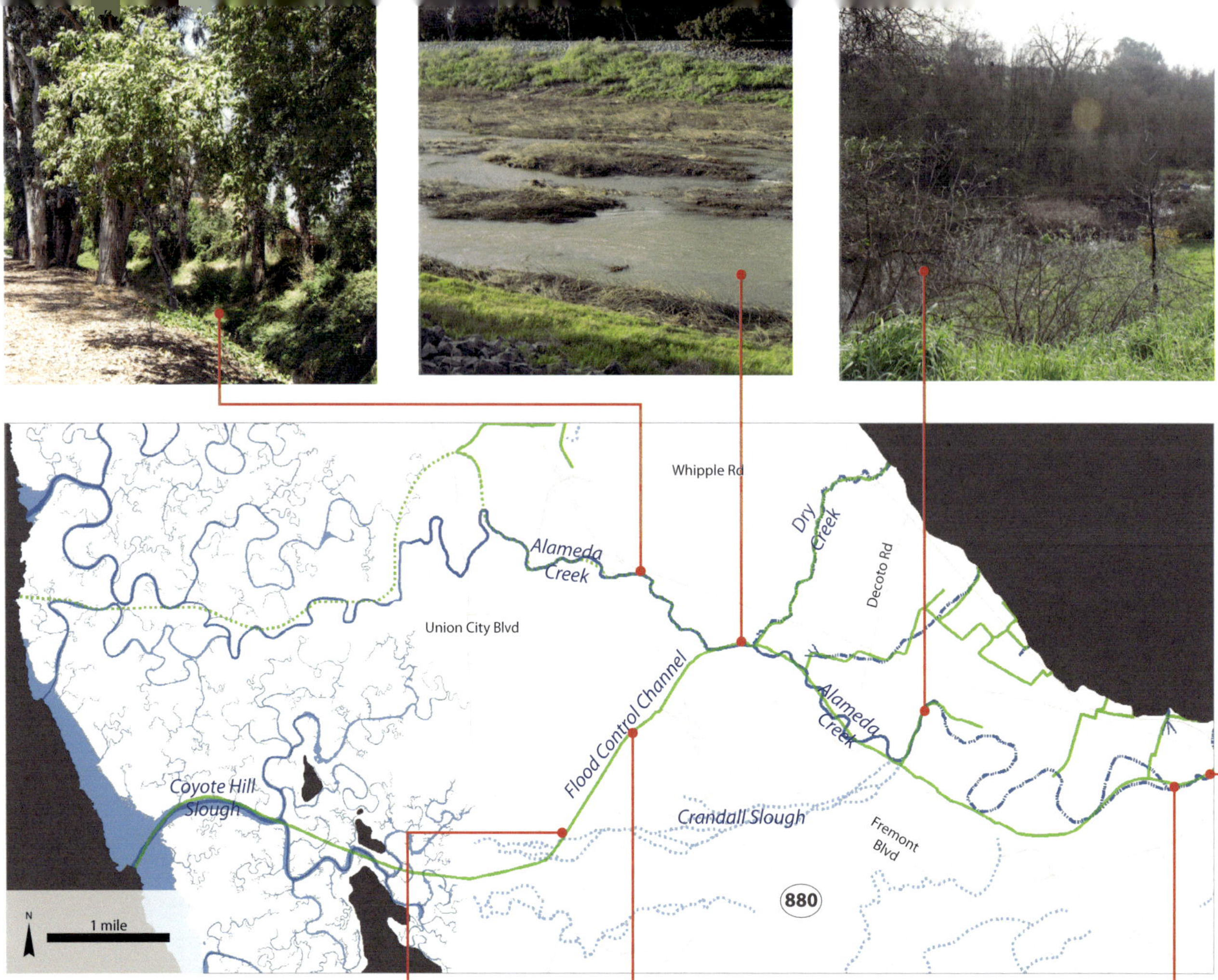

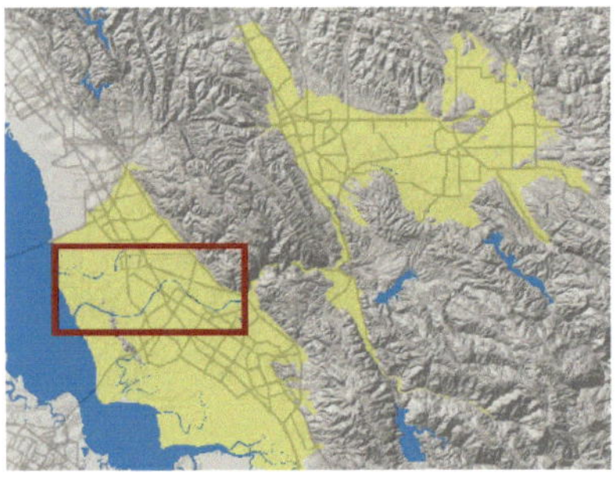

Figure 6.32. Old Alameda Creek and the Flood Control Channel. The present-day course of Alameda Creek follows the general course (although not the specific alignment) of the historical route until just upstream of Highway 880. At this point, all of the flow of Alameda Creek follows the Flood Control Channel south, while the old bed of Alameda Creek continues northwest through William Cann Neighborhood Park all the way to the Bay. Throughout this report, "Alameda Creek" refers to the historical route, while the present-day route is called the Flood Control Channel. (photos by Bronwen Stanford and Sarah Pearce, 2011)

(1924) explained that this was due to the relatively limited sediment that reached the Cone from the upstream valleys (Livermore and Sunol), both of which historically acted as sediment sinks.

These shifts in flow patterns, morphology, and substrate created four distinct reaches in this lower portion of Alameda Creek (fig. 6.33). The exact location of each of these transitions would have varied from year to year depending on rainfall.

The first reach extended from the mouth of the canyon down to just upstream of the present-day Bay Area Rapid Transit (BART) weir. This reach exhibited perennial flow within a broad, often braided channel, underlain by coarse gravels. It was also a losing reach. As Alameda Creek flowed from Niles Canyon it sank into the coarse gravels.

Immediately downstream was the second reach, which extended from the BART weir to a point half a mile upstream of Decoto Road (near Isherwood Road). This reach flowed through the area that today contains the quarry ponds. This reach had surface flow for only part of the year and contained coarse, gravelly substrate and pools. This reach had a high sinuosity, and slowly meandered as the slope of the fan became more gradual.

The third reach continued from approximately Arroyo Park, upstream of Decoto Road, down to the edge of the tidal marsh at Union City Boulevard. Perennial flow resumed through this reach, and fine clays and silts dominated the channel instead of coarse gravels. Multiple overflow channels carried flood water west to the tidal marsh.

The fourth reach flowed through tidal marsh out to the Bay, following a sinuous course that was highly connected to the surrounding tidal marsh through branching tidal channel networks. We refer to this reach as the estuarine reach.

These divisions are not absolute—local residents described smaller stretches of perennial flow within the intermittent reach, or portions of the reach that became dry later in the season than others. However, the historical record

Figure 6.33. Lower Alameda Creek reaches. The diagram below shows the four reaches we developed to show patterns along lower Alameda Creek, based on a number of historical sources. Each reach differed in morphology, substrate, and flow. Distinguishing features for each reach are included in this diagram.

Tidal Marsh, Pond, Slough
Tidal Flat
Bay/Deep Channel
Pond
Wet Meadow
Willow Thicket
Alkali Meadow

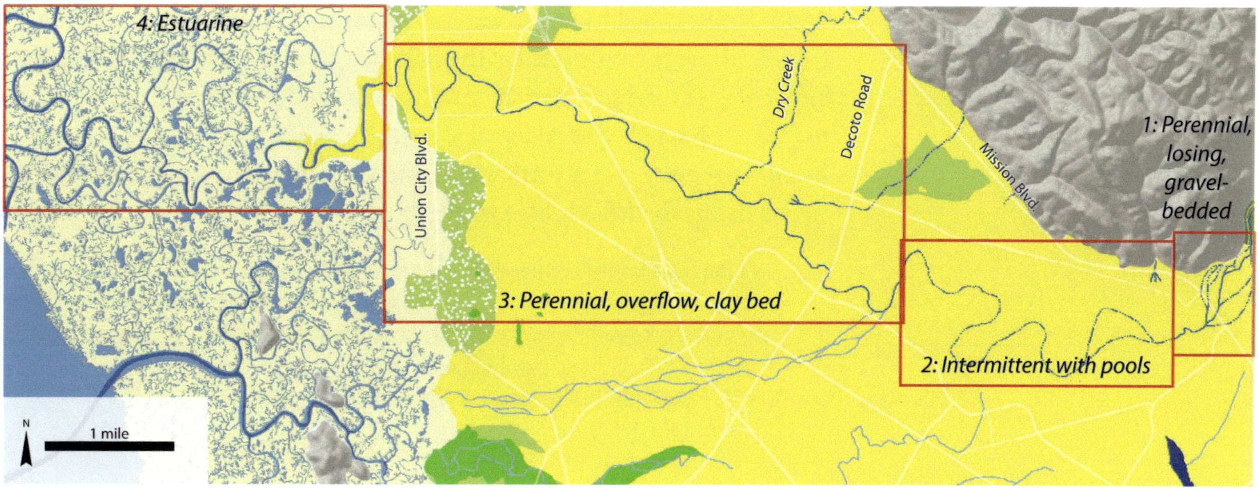

supports these broad patterns, so we organized our discussion accordingly. The following section describes the morphology and flow of each of these reaches, progressing from upstream to downstream, and then describes a few features that span multiple reaches, including pools, overflow, and riparian cover.

Morphology and flow

As Alameda Creek emerged from the bedrock confinement of Niles Canyon, it spread into multiple channels, creating a river corridor almost 2,000 feet wide (fig. 6.35; e.g., Allardt 1874, Thompson and West 1878, Boardman 1884, Nusbaumer 1885, Nusbaumer and Boardman 1889, USGS 1899a, Dillman 1911b, Haviland 1912, Ellsworth n.d.). This broad reach is discussed at length in the Clough testimony (box 6.1): SVWC engineer Schussler (1901) described "three [channels]; a little further down four. Then some smaller branches in between." Whipple, another witness, testified that "Right by the mill the creek appears to have more than one distinct channel, it settled down to about 2 or 3 channels" (Whipple 1901:6). The creek formed relatively stable islands between these channels, several of which appear to have persisted over time and are depicted with consistent form, a channel type known as "island-braided" (fig. 6.35; Beechie et al. 2001).

The main branch of Alameda Creek began to lose water to the porous alluvial gravels beginning at the canyon mouth, but typically maintained perennial flow to the present-day BART weir (Burns 1888, Hilgard 1899, Schussler 1901). The substrate here was coarse, described as a "gravel bed" by Shinn, Schussler, and others in the Clough case (fig. 6.37; see box 6.1). Whipple (1901) reported that the creek bottom was "filled up with boulders and gravel" near Niles. Engineer Schussler described how water sank through the gravels beyond the railroad (just south of present day Mission Boulevard):

> Q: At what point does the Alameda Creek empty in the gravel bed?
>
> A: It starts in not very far from the upper end of these islands, loses some of its water into the gravel beds. The main loss of water takes place below the crossing of the Southern Pacific Railroad. (Schussler 1901:3)

Residents described having to cross a flowing stream year round until the Niles Bridge was constructed at the railroad crossing in the early 1870s. The bridge was 414 feet long and 20 feet above the bed of the creek, constructed to span the broad bed of Alameda Creek (*Daily Alta California* 1872). Beard, a property owner further downstream, recalled having to ford the stream before the Niles bridge was built:

> At the Niles bridge I remember in the early days it was always flowing there. Would flow the year round right where the crossing is I have frequently crossed there. We would always have to drive through water. That was before the county bridge was built. I speak of up to 1870, 1869 –'70. In 1871 the bridge was there, you could drive across the bridge. (Beard 1901:1059)

BOX 6.1. CLOUGH CASE

Much of the detailed early information about lower Alameda Creek comes from a 1901 court case. Jane R. Clough, who owned property near Niles, sued the San Francisco-based Spring Valley Water Company (SVWC) for diverting too much water from the creek, depriving her and neighboring landowners of water for irrigation. The court transcript includes testimony by many local residents on the condition of the creek as early as the 1850s, and the sedimentation and drying of the creek over time as a result of upstream diversions. Clough eventually won the right to water for irrigation, but diversions continued. This newspaper account describes conditions in 1888 that led to the court case in 1901:

> The water is now turned into the flume and pipes of the Spring Valley Water Company from Alameda Creek at the dam in the canyon, and is flowing across the bay. Below this point the creek bed is dry, excepting a few stagnant pools. Those living on the bank of the creek near Niles are without water for themselves or their stock. The water in many of the wells has been lowered—in one or two of them disappearing entirely. (*Daily Alta California* 1888)

The court testimony provides a detailed picture of the creek during the early American period. Many of the deponents were landowners with property adjacent to the creek and had grown up along the creek. They were asked to describe the condition of the creek from their earliest memories, documenting changes as the creek responded to climate and land use modifications. Residents described many aspects of the creek, including swimming holes, creek depth, substrate patterns, dry season flow, and the effects of flooding. In their descriptions, people referred to local property boundaries, which we were able to place with the help of early property maps (fig. 6.34).

Unfortunately, only portions of the case were located. One volume resides at The Bancroft Library, but four additional volumes appear to have been lost sometime over the past century.

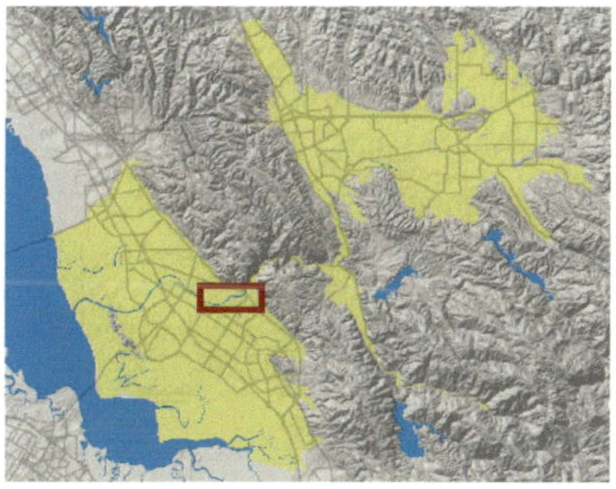

Figure 6.34. Local land ownership in 1878. The names of deponents in the land grant case can be seen on this 1878 map, including Overacker and Shinn, as well as Clough. These property lines were used as landmarks in the court case. (Thompson and West 1878, courtesy David Rumsey Map Collection)

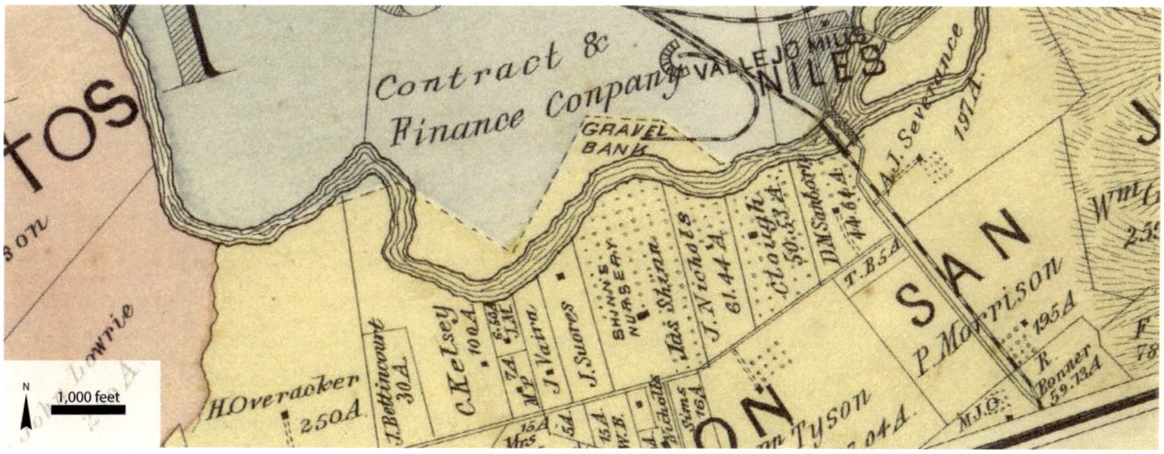

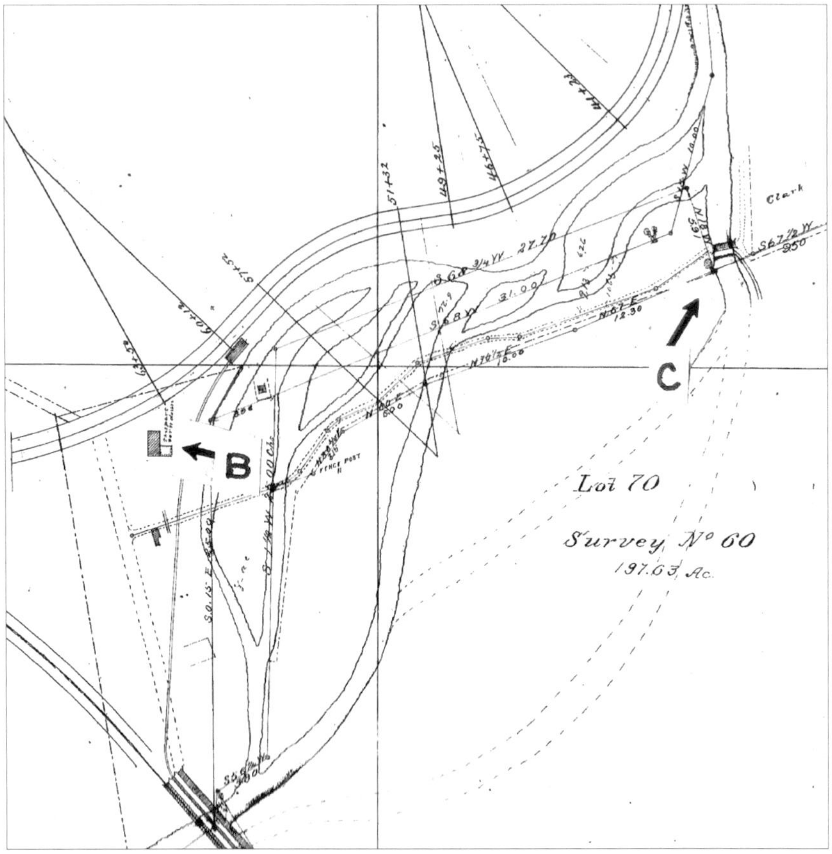

Figure 6.35. Braided channel at Niles. This early map shows the many distinct channels of Alameda Creek at the mouth of Niles Canyon. The arrow C points to a road over the eastern branch of Alameda Creek. This branch appears to be have been cut off, or proposed as a branch to be cut off, and is shown only as a dotted outline. (Survey no. 3410 H.A. Mayhew, courtesy Fisher Scrapbook Collection, Museum of Local History, Fremont)

Q. How long a distance before it [surface flow] was out of sight?

A. About a quarter of a mile. It reappeared down below Overacker's place. It would appear; disappear in other places; then appear again in other places.

—TYSON 1901:16, DESCRIBING SURFACE WATER ON ALAMEDA CREEK

Shortly after passing under the railroad, most summer flow in Alameda Creek sank below the surface. The loss of continuous summer surface flow was documented by a number of early observers over a broad period of time, suggesting that this shift was a persistent condition. Accounts were given by explorers (Dantí 1795 in Cook 1957; October), surveyors (Tracy 1853; September), engineers (Hilgard 1899, Schussler 1901), newspaper reporters (*San Francisco Call* 1903a,b), and local residents (Barry 1901, Henion 1901, Overacker 1901, Richards 1901, Shinn 1901, Tyson 1901, Whipple 1901). In this intermittent reach, the creek flowed through a sandy and gravelly bed (see fig. 6.37) with occasional short reaches of summer surface water and a few persistent deep pools.

The intermittent reach of Alameda Creek was also notable for its sinuosity, contrasting with reaches both upstream and downstream. Earlier maps show greater sinuosity than later maps, documenting the gradual straightening of the creek over time (fig. 6.38; e.g., Lewis 1860, Whitney 1873, Allardt 1874 versus later maps; e.g., Boardman 1884, USGS 1899a,b, Stoner et al. 1915). County surveyor Allardt described how the path of the creek had changed, saying "the creek has entirely changed its course, the present channel being much shorter and more direct than the channel of 1860 which is now nearly silted up and barely traceable" (Allardt 1888). Based on our historical mapping, we measured a sinuosity of almost 1.9 over the 2.2 mile distance from the beginning of the intermittent reach to Crandall Slough. By comparison, the reach directly downstream (3.2 miles from Crandall Slough to the beginning of the large meanders near

Figure 6.36. Floodwater at Niles. (above) During high flow events, water overflowed the many small channels to flow across the entire bed of Alameda Creek. This photo shows the creek "swollen" on February 21, 1914. Note the in-channel bar at the left. (AD-173, © San Francisco Public Utilities Commission)

Figure 6.37. Gravels in Alameda Creek on the upper Niles Cone. These images were taken of the channel near Niles in 1916. The top image (A) is from July 10 at the present-day BART weir; and the bottom image (B) is from May 29 at the California Nursery, near the Quarry Lakes. (A. AD-910; B: AD-843, both © San Francisco Public Utilities Commission)

Alvarado) had a sinuosity of less than 1.3. The old channel *Sanjón de los Alisos* also once connected in this reach (box 6.2).

The channel bed through this reach was broad and composed of gravels and sands, with braided channels, riffles, and one large, well documented island (see figs. 6.3, 6.38). Testimony described gravels and sands ("gravel beds," "sand flat"; Barry 1901, Brown 1901, Schussler 1901). The creek continued to flow over a broad area, and in 1871 reportedly measured 500 feet wide downstream of the present-day BART weir (Whipple 1901). Shinn (1901) reported that "the whole gravel bed opposite our place from bank to bank ... was an eighth of a mile" and contained a series of distinct channels. Whipple described the multi-thread pattern that established near Niles and continued downstream:

> Right by the mill the creek appears to have more than one distinct channel, it settled down to about 2 or 3 channels. Going downward, some places they come together again, and then spread out. The bottom of the creek is filled up with boulders and gravel ... It changed, the channel would change occasionally in high water—sometimes it would leave two channels a short distance. (Whipple 1901:747).

The creek was also deep; an early historian recounted lost travelers falling 25 feet to a dry creek bed in September of 1858 (Halley 1876:191).

Near the Decoto Road Bridge (historically Bell Ranch Bridge), Alameda Creek entered another perennial reach. This third reach extended from approximately a mile upstream of Decoto Road to the edge of the tidal marsh near Union City Boulevard. Here a clay layer replaced the coarse gravels that had filled the bed through the upper cone (Hilgard 1889, Schussler 1901), and the creek consolidated to form one channel (Whipple 1901). At approximately the same point, consistent summer flow began again in the creek, indicating that the channel was intercepting groundwater (Allardt 1888, Tyson 1901, Schussler 1901).

The transition from more coarse, gravelly substrate to clays and silts was recorded by a variety of historical sources. Hilgard (1889) described observing "in the banks a yellowish sandy clay, first overlying the gravel and finally replacing it altogether." Schussler (1901) noted that beginning just upstream of Bell Ranch Bridge the creek deposited a "clay shelf ... on top of the westerly portion of this gravel delta" so that "the visible gravel beds ... [did] not reach quite to the Bell ranch bridge."

> *Below that comes the Stevenson place where occasionally the water disappears, until it got to what is called the big bend opposite the Crandal* [sic] *Slough, and it carried water nearly all the season, steady, from there down to the Bell ranch bridge.*
>
> —BARRY 1901:850

Observers almost universally stated that perennial flows began at approximately the same point that the gravels disappeared. Dry-season surface water below the bridge was described in 1887 as "a continuous body of water, but with a very sluggish current" (Allardt 1888). Other early accounts corroborate this description, recording "summer flow in the bed of the creek beyond the edge of the clay bed" (Hilgard 1889:22) and "always water" below the bridge (Barry 1901, Beard 1901, Tyson 1901). Professor Hilgard described the gradual transition back to perennial flows:

> Water appears in pools on the general surface of the creek bed, while above it could only be reached by digging. Below Bell's Ranch bridge, the water rapidly increases and 1/8 mile below the bridge it flows a small stream between reaches of still water. (Hilgard 1889:2)

The morphology of the creek also changed below the bridge. As the channel continued downstream, it became deeper and narrower, with a current that slowed towards the tidal marsh. One account noted that flow was "sensibly

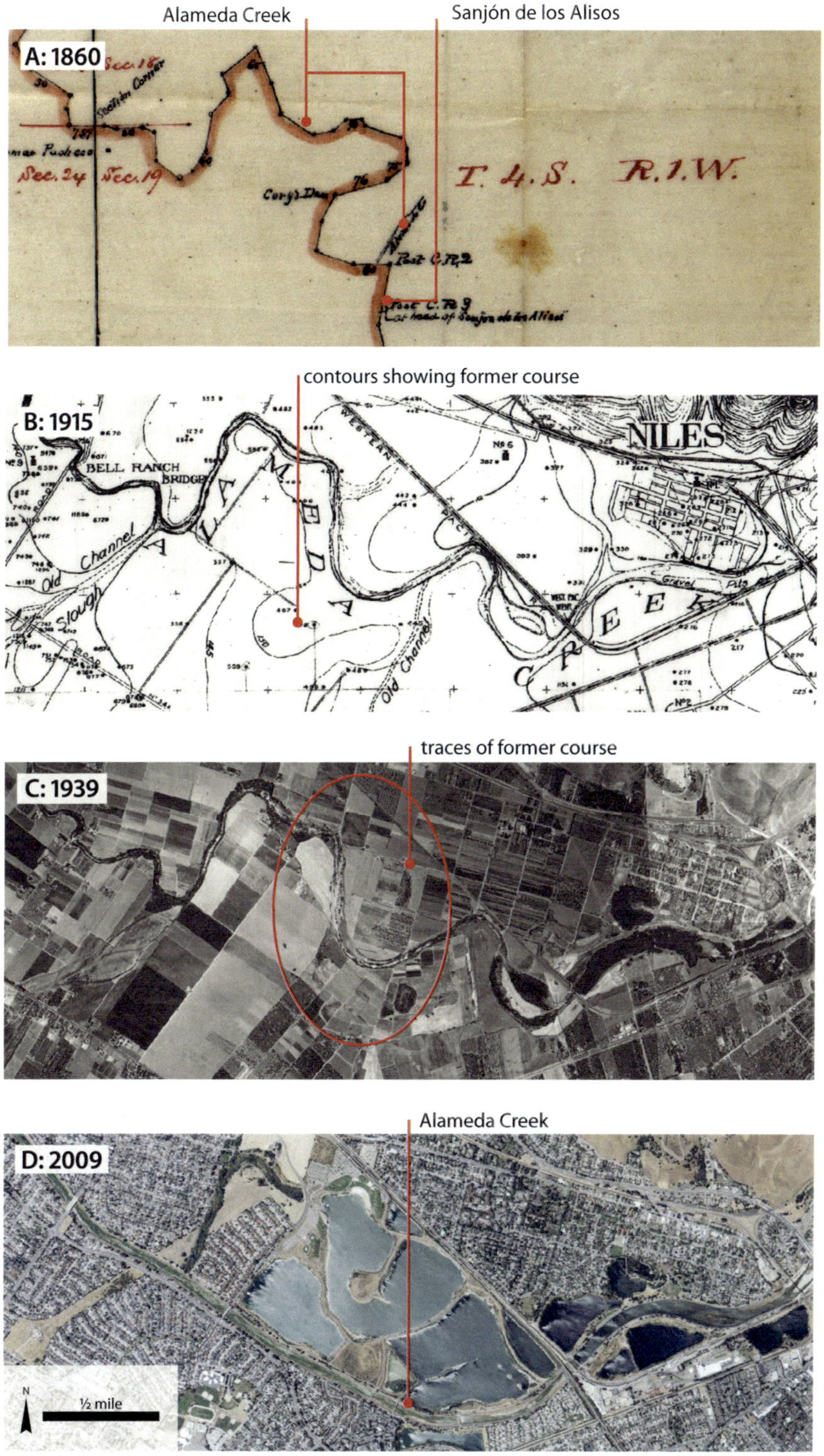

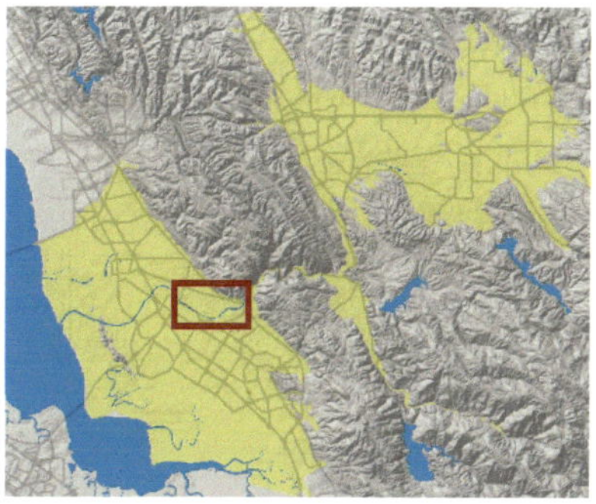

Figure 6.38. Straightening of Alameda Creek through time.

(A) The land grant boundary surveyed in 1860 followed the course of Alameda Creek, capturing the large meanders of the creek present at that time and showing the point where the creek once branched south into the former channel, Sanjón de los Alisos.

(B) Some of the larger meanders have been straightened by 1915, although they can be traced in the detailed contour lines in this map.

(C) Traces of the historical course are also faintly visible in the 1939 aerials towards the center of the image. Note the quarry ponds upstream of the circled area (the ponds appear black in the photo).

(D) Today the creek follows a straight path through this reach. Some channel fragments remain, but most of the area that was once occupied by the meanders is now covered by the Quarry Lakes.

(A: Lewis 1860, courtesy The Bancroft Library, UC Berkeley; B: Stoner et al. 1915, courtesy ACWD; C: USDA 1939; D: USDA 2009)

BOX 6.2. SANJÓN DE LOS ALISOS

Alameda Creek shifted across its fan many times over hundreds of thousands of years. Midway down the intermittent reach of historical Alameda Creek, *Sanjón de los Alisos*, or "Ditch of the Sycamores," marks the most recently abandoned course of Alameda Creek.

Sanjón, often translated "ditch," was used by early settlers to describe both natural and artificial channels (Dyer 1861b, Lewis 1862; see also Beller 2010). Surveyor Dyer (1861) described Sanjón de los Alisos as a "natural ditch," likely describing a natural channel that was shallower and less well defined than a true creek. The channel was likely named for riparian sycamores (*alisos*), some remnants of which could be seen bordering its banks as late as 1939 (USDA 1939). (Although often translated as "alder," *aliso* is best translated "sycamore" when used by early Spanish settlers in California; see *San Francisco Chronicle* 1893, Brown 2002.) A local history recorded that in 1904 sycamores still marked the channel, "though many of them have long since been cut down" (Country Club of Washington Township [1904]1950:85). Another stated that "The Indians say that it was a full stream two centuries ago ... Twice since American occupancy—in 1854 and 1863—the water from the Alameda has flowed for a few days through the old channel" (Shinn [1889]1991:100).

This channel was likely the primary course of Alameda Creek up to several hundred years ago, and served as a boundary and a marker through much of the historical period (Tracy 1852b, Lewis 1860, Shinn [1889]1991, of Washington Township [1904]1950). The 1873 Geologic Survey map shows the channel with two shell mounds along it, indicating that the channel was likely known and used by native groups in fairly recent times (Whitney 1873). However, by the historical period the direct surface connection had disappeared (e.g., Higley 1857a, Lewis 1860) or was shown on maps as an artificial channel (e.g., Whitney 1873, Allardt 1874, USGS 1899a). Limited historical records allude to an overflow connection with Alameda Creek (in 1863 and 1954; Shinn [1889]1991:100).

Despite the lack of direct surface connection, Sanjón de los Alisos appears to have carried some water in the historical period, likely fed by groundwater and runoff from the surrounding plain during high flows. Photographs taken during high flow events show water in the channel (fig. 6.39), and one of the earliest maps of the area includes a channel with trees along it, labeled *Sanjón con agua permanente* ("ditch with permanent water"), suggesting that portions of the sanjón may have had standing water in the early Spanish period (USDC ca.1840d). There may have also been a subsurface connection to Alameda Creek: in 1914 engineer Forbes noted "a draft along the old channel of Sanjón de los Alisos" underground (Forbes 1914:42).

Figure 6.39. Water in Sanjón de los Alisos. Although no direct connection remained by the time this June 1915 picture was taken, water filled the channel. This photo was taken half a mile from Mayhew's Landing. For map depictions of this channel, see fig. 6.47. (AD-308, © San Francisco Public Utilities Commission)

stagnant" near Alvarado (Hilgard 1889:5). An early resident reported that the creek was 40 feet across and 20 feet deep here in the 1870s, compared to 500 feet wide and 10 feet deep at the mouth of Niles Canyon (Whipple 1901; see incision discussion p. 211). This reach of the creek was also less sinuous than the upstream intermittent reach.

Further downstream, Alameda Creek flowed into the tidal marsh, entering a fourth distinct reach. The creek historically entered the tidal marsh just west of Union City Boulevard, over three miles east of San Francisco Bay. The creek followed a sinuous course through the marsh, traveling 6.5 miles to cover the 3.2 miles of distance. (The channel now called Old Alameda Creek has been partially straightened, and now covers that distance in only 4 miles of channel.) Numerous tidal marsh channels connected directly to the main channel of Alameda Creek, creating a dense network of aquatic habitat. Perennial flow continued through this reach with a relatively fine, silty substrate. The tidal reach of Alameda Creek provided aquatic conditions grading from fresh to brackish to saline. Miles of tidal channel also fed into the seasonal and perennial freshwater wetlands, located in the artesian zone towards the upland end of the tidal marsh.

The upper tidal reach of the creek was bordered by natural levees that prevented tidal channels from directly connecting with the creek (Kerr 1857b). This confined tidal reach continued for about half of the distance towards the Bay, over 2.5 miles of channel. The lower tidal reach, unconfined by natural levees, had direct connections to numerous small tidal channels and followed a highly sinuous course. Over a mile and a half of westward distance, the channel was directly connected to 172 miles of tidal channel along Alameda Creek (Kerr 1857b). These 172 miles occurred within 6.5 square miles of tidal marshland, a density of over 26 miles of channel per square mile of tidal marsh area (fig. 6.40). This high concentration of channel edge to marsh area would have provided valuable habitat for a number of species, including native fish and bird species (see tidal marsh p. 173). Deep water channels flowed through this reach, and dozens of marsh pannes or ponds received water at the highest tides.

After the arrival of Europeans, the tidal channel network was rapidly altered. Some tidal channels were cut off from the creek while others were directly connected to the creek. In the early historical period, Alameda Creek flowed into the San Francisco Bay through Mount Eden Slough, swinging north from the present-day Old Alameda Creek Channel (fig. 6.41; Kerr 1857b). By 1896, this course had been diked off, forcing the main flow of the channel and the major access point for navigation to the south, to exit through Union City Slough (Westdahl 1896a, USGS 1899a). Levees and ditches were created for salt harvest, and overflow from the creek was used to fill low-lying areas and reclaim them for agriculture (see overflow p. 203).

Figure 6.40. Tidal channel density. This map shows the historical tidal channel network. Channels directly connected to Alameda Creek are shown in brown and gold and measure 172 miles.

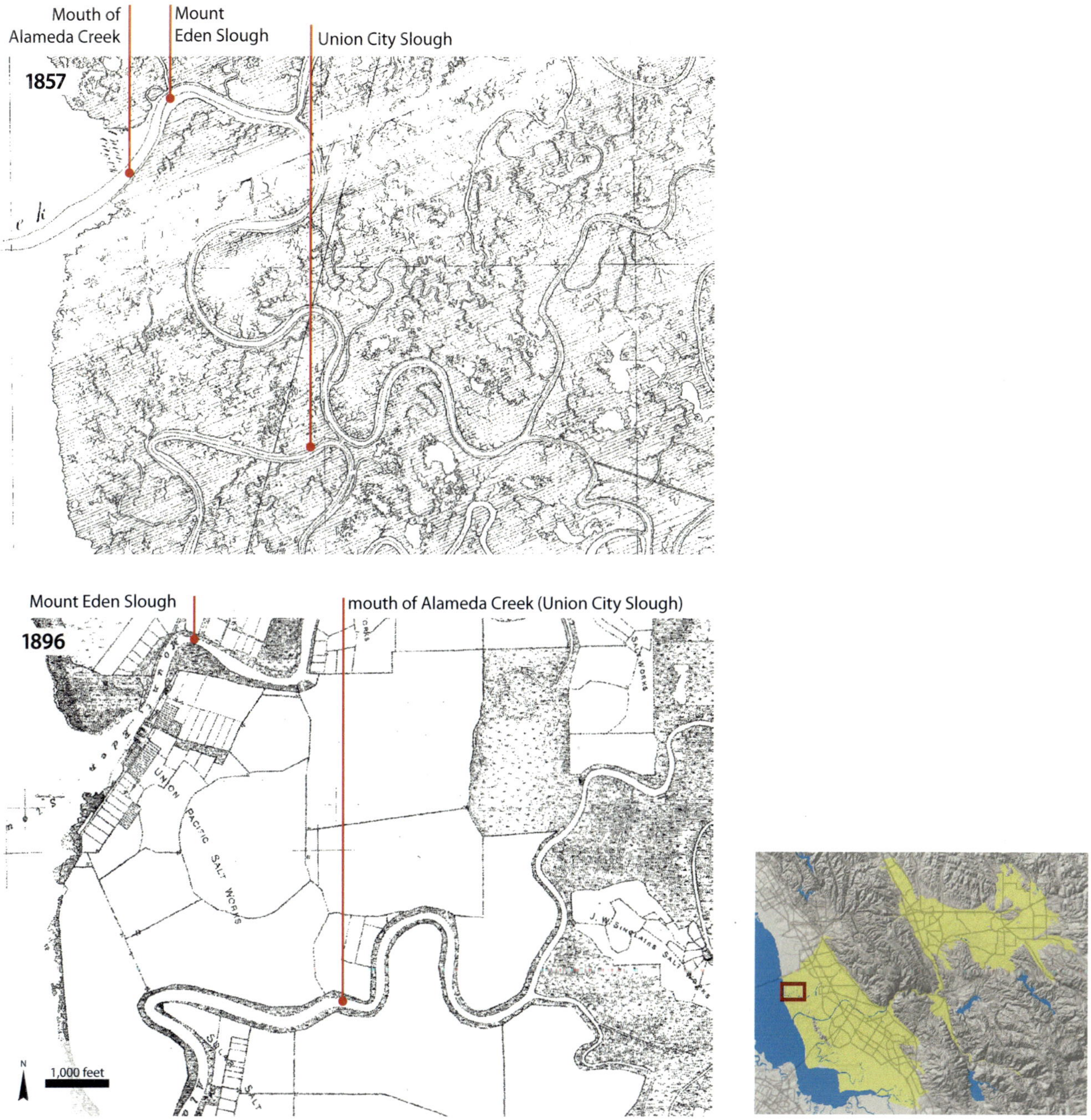

Figure 6.41. Paired images of the mouth of Alameda Creek. In the 39 years between 1857 and 1896, the mouth of Alameda Creek shifted 1.4 miles south, from what is now Mount Eden Slough to Union City Slough. This diversion of the creek was reported by Westdahl (1896c), and may have been a partially natural transition. The 1857 T-sheet shows deep water flowing through both Union City Slough and Mount Eden Slough, and the two channels come together at the point where Alameda Creek was later diverted entirely into the southern channel. Natural channel processes may have brought these two channels close together, allowing for the eventual capture of the main flow of Alameda Creek by the southern branch. However, given the complete transformation of the surrounding tidal marsh to salt ponds and the importance of Alameda Creek for shipping, it seems likely that people had a role in actively reshaping this channel. Throughout the marshlands, constructed levees caused a loss of tidal prism, which resulted in sedimentation. (Kerr 1857b, Westdahl 1896a, both courtesy NOAA)

Figure 6.42. Several early diseño maps (ca. 1840) provide evidence for riparian forest along Alameda Creek. These depictions of trees along Alameda Creek suggest that they were a prominent feature. (USDC ca. 1840c, courtesy The Bancroft Library, UC Berkeley)

Riparian cover along Alameda Creek

Alameda Creek was historically surrounded by a mixed riparian forest as it crossed the Niles Cone (fig. 6.42). The riparian trees were a striking and notable feature in the otherwise largely treeless plain. Spanish explorers described how the upper portion of Alameda Creek across the cone in 1776 was bordered by "many sycamores, cottonwoods, and some live oaks and other trees, and it appears to flow west to empty into the estuary … toward which runs a thick growth of trees" (Font and Bolton 1933:356). Dantí described the stream "heavily overgrown with willow, cottonwood, and some laurels" (Dantí 1795 in Clarke 1952). Further downstream, where the creek entered its lower perennial reach, Hermenegildo Sal, a member of the same expedition, noted that there was substantial tree cover: "Here it is thickest with timber of many large cottonwoods, sycamores, and small willows" (Sal 1795). Other accounts are more general: an 1876 county history described "the river, which when first discovered, as now, was lined

Table 6.3. Textual riparian data. The following table summarizes historical data describing riparian conditions along the creek from Niles Canyon to the marsh.

Location	Source	Year	Quote
General summary of lower Alameda Creek	Dantí in Clarke 1952	1795	"willows"
	Sal 1875	1795	"populated with alders, some willow, madrono and laurel … [after reappearing] more populated … cottonwood, alders, and small willows"
	Halley 1876	1876	"lined with willow and sycamore trees, while the rest of the valley was bare"
	Hilgard 1884	1884	"banks are mostly timbered with sycamore and willows"
	Enos 1887	1887	"its banks being bordered, then as now, with cottonwood and willow trees in the midst of an otherwise scarcely wooded plain"
	Township Register 1910	1910	"its banks were once linked with willow, sycamore, and other trees"
	Clarke 1952	in past	"sycamores, cottonwoods, laurels, willows, live oaks, alders"
mouth of canyon	Henion 1901	1856-1873	"above that there were a good many trees and the creek was narrower and deeper"
near Niles	Dantí in Clarke 1952	1795	"heavily overgrown with willow, cottonwood, and some laurels"
near Niles	Fages 1770	1770	"thickly covered with alisos, laurels, and other trees"
upper Niles Cone	Font and Bolton 1933	1776	"many sycamores, cottonwoods, and some live oaks and other trees"
near Shinn's Island	Shinn 1901	1870s	"lone sycamore tree"
near Shinn's Island	Ellsworth 1901	1871	"the willows growing there were 12 feet, while the present willows are growing almost as high as the tops of the willows that were in the bed when I knew the place"
train from Niles to Decoto	Williams et al. 1878	1878	"trees mark the Alameda Creek"
Decoto Road	Whipple 1901	1901	"the banks are wooded there"

with willow and sycamore trees, giving it the appearance of an alameda or road lined with trees" (Halley 1876:10). Others described the creek with "a great many trees on its bed" (Crespí 1769 in Stanger and Brown 1969:106), and "heavily overgrown with willow, cottonwood and some laurel" (Dantí in Cook 1957:140).

Other sources support this depiction. Images of the creek indicate a tree corridor as late as the 1910s and aerial imagery shows remnant valley oaks in 1939 (USDA 1939). Riparian trees depicted in early maps along Alameda Creek include sycamore (*Platanus racemosa*), cottonwood (*Populus* spp.), willow (*Salix* spp.), live oak (*Quercus agrifolia*), and maple (*Acer* spp.) (Tracy 1852a, Dyer 1861a, Dyer 1862d, Dyer and Hopkins 1863, Cash 1870, Nusbaumer 1889b, Poathon 1895). Textual accounts describe a mix of species including sycamore, cottonwood, willow, alder (*Alnus* spp.), laurel (*Umbellularia californica*), and oak (*Quercus* spp.) (table 6.3). This corridor was particularly notable in contrast to the surrounding grassland.

Although most sources are not spatially explicit enough to document local changes in riparian vegetation along the length of the creek, some shifts can be detected. At the top of the Niles Cone, before surface water had percolated into the fan, riparian vegetation was noted to be particularly water tolerant. Jepson described that "Black cottonwood ... inhabits banks of valley or mountain streams or moist bottoms ... [it] is scattered ... along Alameda Creek near Niles" (Jepson 1909:143). Live oaks, alders, willows and buckeyes (*Aesculus californica*) were also mentioned in the reach where the stream spread from a confined channel across a fan (Brotherhood of Locomotive Engineers 1875:547, Nusbaumer 1889b). As the creek exited Niles Canyon it spread to over 2,000 feet wide, creating a broad zone of riparian influence.

Toward the Bay, the creek narrowed and riparian cover was more dominated by willows. The T-sheet of the estuarine reach of Alameda Creek shows a 50 foot wide corridor of willows lining each side of the creek (Westdahl 1896a). Near the intersection of Whipple Road and Union City Boulevard, Alameda Creek overflowed into a grove of trees which was described as a "swamp" (Westdahl 1896c), likely a willow swamp. (Historian Shinn described Union City "in the willow-swamps at the edge of salt water"; Shinn 1907:23-24.) The grove was at least seasonally flooded (Westdahl 1896c).

Persistent pools

A number of persistent pools were recorded in the late 1800s within the upper perennial and intermittent reaches of Alameda Creek. These pools were recorded by the earliest European visitors to the Niles Cone. The early explorers described "an arroyo with little water, most of it in very deep pools" (April 1776; Font and Bolton 1933:356) towards the top of the cone, and described that the water of Alameda Creek "in various parts ... makes pools" (October; Sal 1795).

As the water recedes from the Alameda creek at Niles, pools are left in various places from which a number of fine specimens of the salmon trout have been taken, some of them measuring two feet or more in length.

—*DAILY ALTA CALIFORNIA* 1889A

Residents documented five pools in particular as notable for their size, depth, and persistence through the late 1800s. Major flood events would have caused some pools to fill and others to develop; Shinn (1901:824) described the pools shifting "periodically; [in] flood periods." However, despite some changes, these pools remained substantially the same: "those holes continued there from 1871 to 1888 in substantially the same form. They were always there. In high water would sometimes fill up and change a little but not very much" (Tyson 1901:844). A large portion of testimony from the Clough case centered on these pools and their historical shape and form; the pools were well known by local residents.

The intermittent reach was particularly notable for the presence of smaller stretches of dry-season flow and pools. The pools would have allowed sediment to settle and would have create thermal stratification (see fig. 8.10). Deeper, cooler water provides fish habitat and refuge after the rest of the creek has become dry. Three of the five documented pools were located near the beginning of the intermittent reach of Alameda Creek, while a fourth was further downstream within the intermittent reach. Witnesses described pools ranging from 50 feet to hundreds of feet long, 20 to 60 feet wide, and 6 to 10 feet deep at the deeper points. The pools were used for swimming and watering livestock through the dry season, so local residents were able to estimate depth based on how it compared to their own height, and could accurately recall when water had been available towards the end of the dry season.

On our way we passed through a fresh water marsh, formed by the overflow of a creek near Alvarado and just in passing saw a Bittern, a Song sparrow, Tree swallow, mud hen ... A grand place for collecting except for the deep water which would necessitate a boat or swimming.

—KELLOGG 1909

In the intermittent reach, the pools remained wet well into the dry season, but surface flow between the pools dried up through the summer in most places (Shinn 1901). The pools were able to maintain cool water in the dry season because they were fed by water flowing through the gravels of the stream bed under the surface. Residents described the pools as fed by "living water" and "always fresh, cool, and nice ... something like well water" (Ford 1901:1198, Overacker 1901:775). Some portions within the intermittent reach of the creek were described with flowing water, but in the "deep places" (the pools) there was "no movement, no current of water" (Tyson 1901:838).

Even these pools could dry up by the end of the summer, particularly in dry years. Some of the pools dried annually; Shinn (1901:786) described that "in ordinary seasons in the dry season, the pools dried up" opposite his property, which bordered one of the large pools recorded by him and others. Many of the pools were used for swimming through mid-summer, but not through the fall: "The water would stay in pretty late, June and July, would be a good bathing pond ... Occasionally in deep holes there would be water left for a few weeks, and finally disappear" (Overacker 1901:1055). Whipple also suggested that the pools did not last through the summer:

> I think there was one hole the last one to dry up was there by Peterson's. That is pretty near in the middle of the dry part. It was a very deep hole. In July, I think, it was deep enough, for the boys went in swimming, as much as 3 or 4 feet, and lasted about a month, and it was entirely dry in September. (Whipple 1901:741)

However, some water could persist in pools through the summer. Shinn gave a detailed description of which pools dried at what point in four dry years (1861, 1863, 1877, and 1882), and two wet years (1861 and 1864):

> During those 4 years water ceased running in the late fall absolutely, and the holes dried up excepting the very deep hole upon the Clough place [near the railroad bridge]. There was water in that two or three years out of those four, but not on the surface. That was an excessively deep hole. That dried up, got absolutely dry; that I think was in 1877, I should say in 1861 and '64 there was water in that hole. (Shinn 1901:869)

It seems likely that some or most of the pools dried towards the end of the dry season, at least in the late 1800s, but that water may have persisted in the deepest pools in years of average or above-average rainfall.

> *It was a regular flood plain, the whole thing in there. The whole works was all the flood plain. I don't know how much it covered, well the flood plain all together, I'd say, was about eight to nine hundred acres, all filled with water out there.*
>
> —FURTADO 1987, DESCRIBING LAND EAST OF COYOTE HILLS

Overflow

During high flows, water overflowed the creek banks and spread west across the Niles Cone, inundating a large area (fig. 6.43, fig. 6.44). Frequent overflows characterized the lower perennial reach of Alameda Creek in particular (downstream of Decoto Street Bridge; Westdahl 1896a, SVWC 1915, Bailey 1917:2). The cone was described during the flood of 1876 as "one vast inland sea" (*The Oakland Tribune* 1876a). Water could remain for hours or days, depositing silt and saturating the soil (Patterson 1977). Overflows from the creek could be powerful, carrying vast quantities of water and debris, including trees, onto the floodplain.

Land use changes upstream, particularly the drainage of the Pleasanton marsh complex and channelization of streams in the late 1800s, likely increased fine sediment loads and peak flood flows in the creek. Larger amounts of fine sediment flowing downstream would have resulted in more fine deposits across the cone and would have caused the lower creek to fill in, resulting in increased flooding. Although flooding was likely a

Figure 6.43. Flooding. This image shows flooding across the Alvarado-Niles Road on February 21, 1914. (AD-162, © San Francisco Public Utilities Commission)

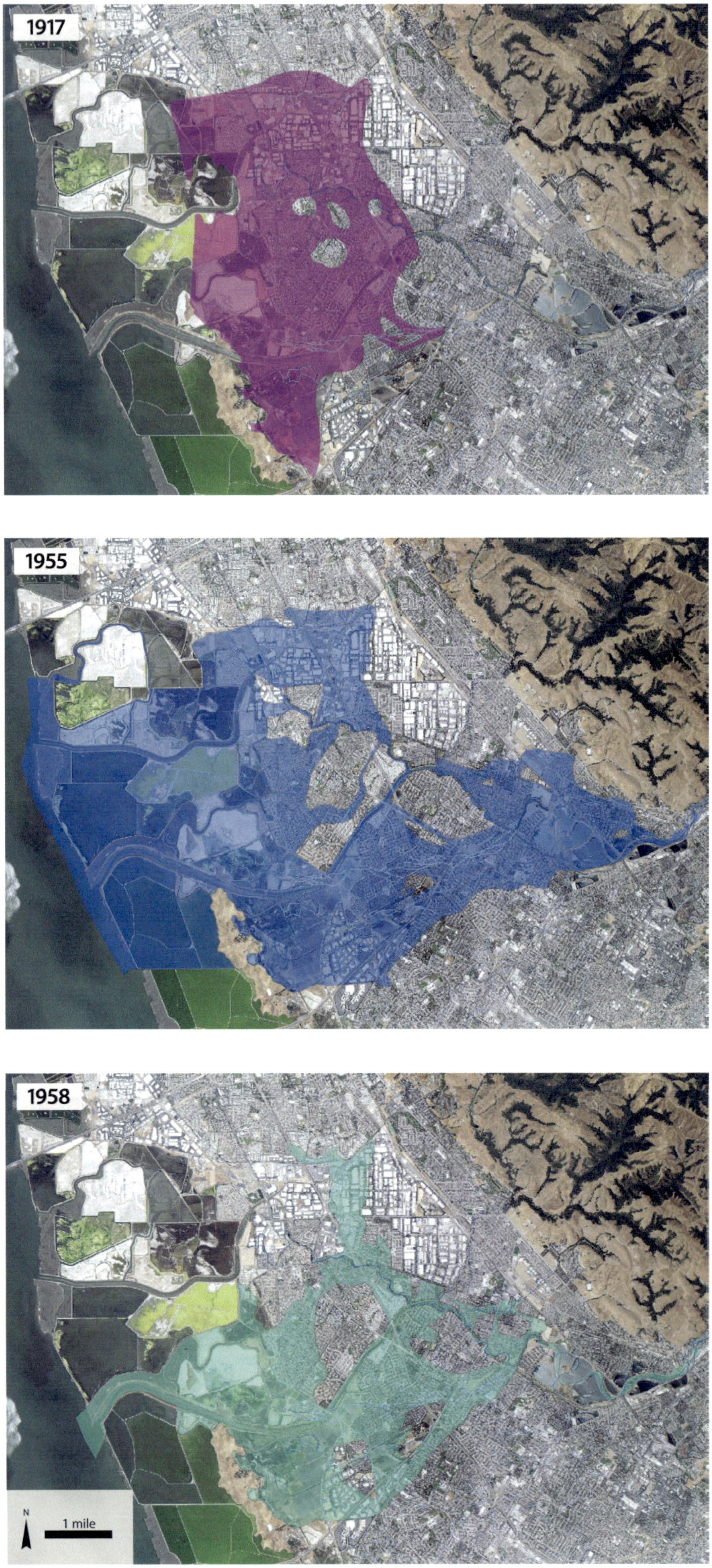

Figure 6.44. Flood extent in 1917, 1955, and 1958. These three maps show flooding extent across Niles Cone during three different flood events. The record-setting flood of 1955 covered 16,560 acres, while the two other floods inundated just under 8,000 acres. Although not every year had the same dramatic flood extent, annual overflow was part of the regular pattern of Alameda Creek. The 1950s floods directly motivated the construction of the Alameda Creek Flood Control Channel. The straight lines in the 1917 image are formed by levees along the edge of the salt ponds. (flood extent digitized from Bailey 1917 and USGS 1962; USDA 2009)

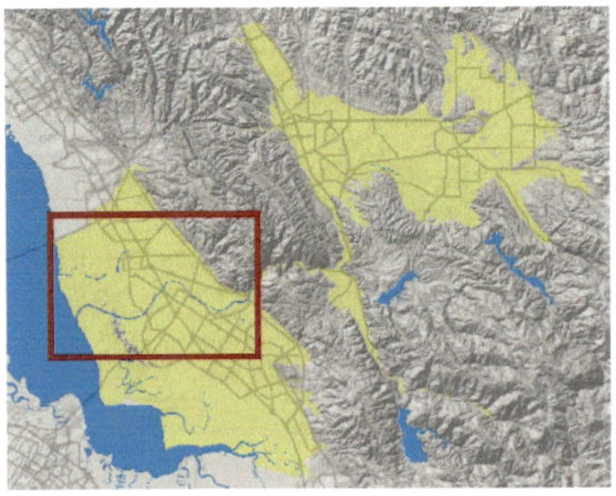

Figure 6.45. Flooding along Alameda Creek on January 4, 1916. A: "Washout." Sandbags were placed to control the flooding of Alameda Creek across the Southern Pacific Railroad tracks. B: "Flooded, in Alvarado west of main street." Note the people paddling a raft and the flooding almost to the top of the fence in the background. (A: AD-136; B: AD-549, both © San Francisco Public Utilities Commission)

regular occurrence prior to these upstream land use changes, these changes certainly exacerbated flooding problems.

Flooding appeared to be a particular problem around Alvarado, near the western border of present-day Union City (fig. 6.45). The town of Alvarado was built on the edge of the tidal marsh, and historical accounts repeatedly describe floods in Alvarado, with water that would remain for days (*Oakland Tribune* 1909, Borghi 1987b), and fields "underwater ranging in depth from a few inches to several feet" (*Oakland Tribune* 1921). The annual floods could also pose serious problems for farmers, and ditches and levees to restrict flow of water were common in the days before the flood control district was formed (Allardt 1888). Residents described water four or five feet deep near Alvarado (McKeown 1975b, Goold 1986, Furtado 1987). Others recalled placing sandbags along the natural levees of the creek to stop flooding; one described being towed as a child for a quarter of a mile in a floating bathtub across the fields (Patterson 1977, Patterson 1987).

It was a perfect ranch ... because the Alameda Creek flooded every year and left at least a quarter of an inch of sediment every year fresh land, fresh soil. It would produce anything. The land out there would produce anything in the world.

—TILLIE GOOLD 1986

Although flooding could cause problems, overflows from Alameda Creek were also seen as beneficial, at least through the early 20th century. The creek overflow historically restricted farmers along the western edge of the cone to farming one crop during the dry season, as crops grown at other times of the year were likely to be drowned or washed away by floods (Brown 1912, Goold 1986, Campbell 1986, Borghi 1987a). But some early farmers appear to have adapted to this pattern and seemed well aware of the benefits that the water and silt deposits brought to their land. Residents recalled floods providing irrigation and an input of new soil to farmlands on the edge of the cone (Patterson 1977). Donald Patterson recalled flooding in the early 1900s:

> In my grandfather's time and in my father's time as well, they depended on the winter floods ... a dry winter, where there was no flooding, meant that the ranch productivity would be substantially less than in the wet years. (Patterson 1977)

One farmer recalled "they kind of wanted that land to go under water; with all the silt that would come in" (Williams 1986; fig. 6.46). Farmers directed flood waters across their land in hopes of building up sediment in areas that were swampy or had poor quality soil (McKeown 1975b, Patterson 1977).

Within the large flood-prone area, two distinct overflow channels persisted through much of the historical period (figs. 6.47, 6.48). The nature of these two channels was ambiguous as early as the 1920s, when a court case attempted to determine whether they were "natural watercourses." It was determined that Crandall Slough, being a slough rather than a well-defined channel, "did not have the physical characteristics of a watercourse," and that the Splits was a constructed channel (Supreme Court of California 1929). These two channels, Crandall Slough and the Splits, are discussed below (fig. 6.48).

Figure 6.46. "Silt dunes ... The ground is not cultivated this year on account of these dune-levee silt deposits." Although in many cases farmers valued deposits of fresh soil, here the large quantity of fine deposits made cultivation impossible. This photo was taken May 30, 1916. (AD-852, © San Francisco Public Utilities Commission)

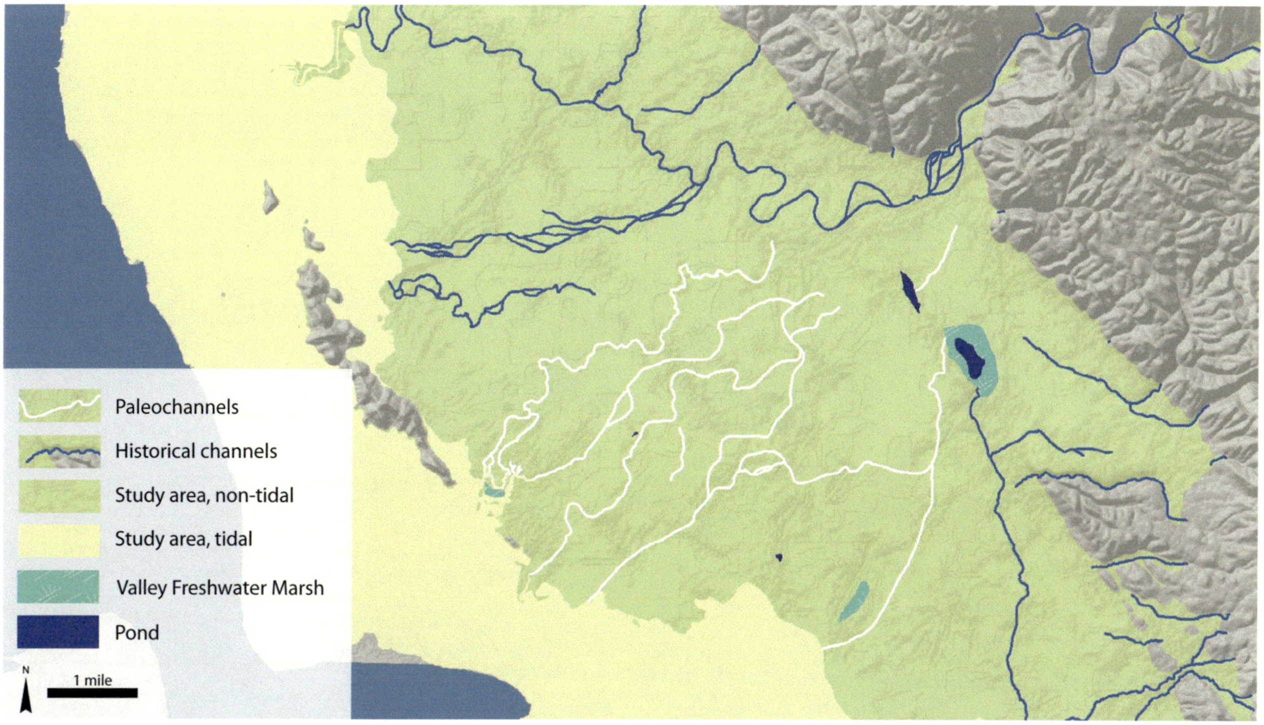

Figure 6.47. Paleochannels across the Niles Cone. In addition to overflow channels, a number of paleochannels across the cone were mapped in 1915 (Stoner et al.), and are still visible in 1939 aerials. These channels likely represent former overflow channels or former courses of the creek, although some may be remnants of constructed ditches (see Shinn [1889]1991). The northern-most of these paleochannels is Sanjón de los Alisos, likely the most recent former course of Alameda Creek. These channels were mapped as sloughs on the habitat map.

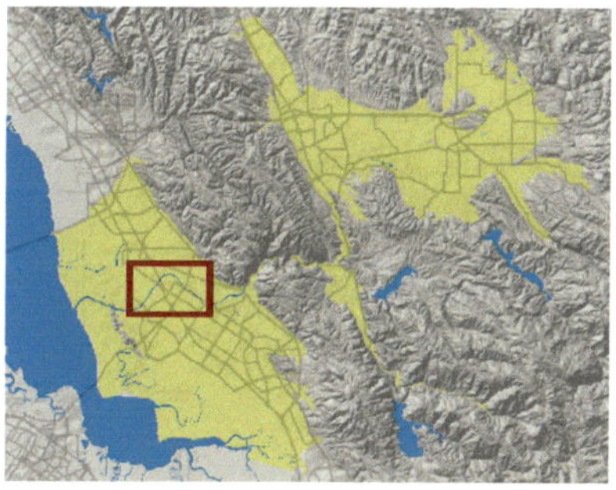

Figure 6.48. Cyril Williams map. This detailed 1915 map shows both Crandall Slough and the Splits flowing from Alameda Creek west to the tidal marsh. Note the many branches of the two channels and the words "Old Channel" along Crandall Slough. (Stoner et al. 1915, courtesy ACWD)

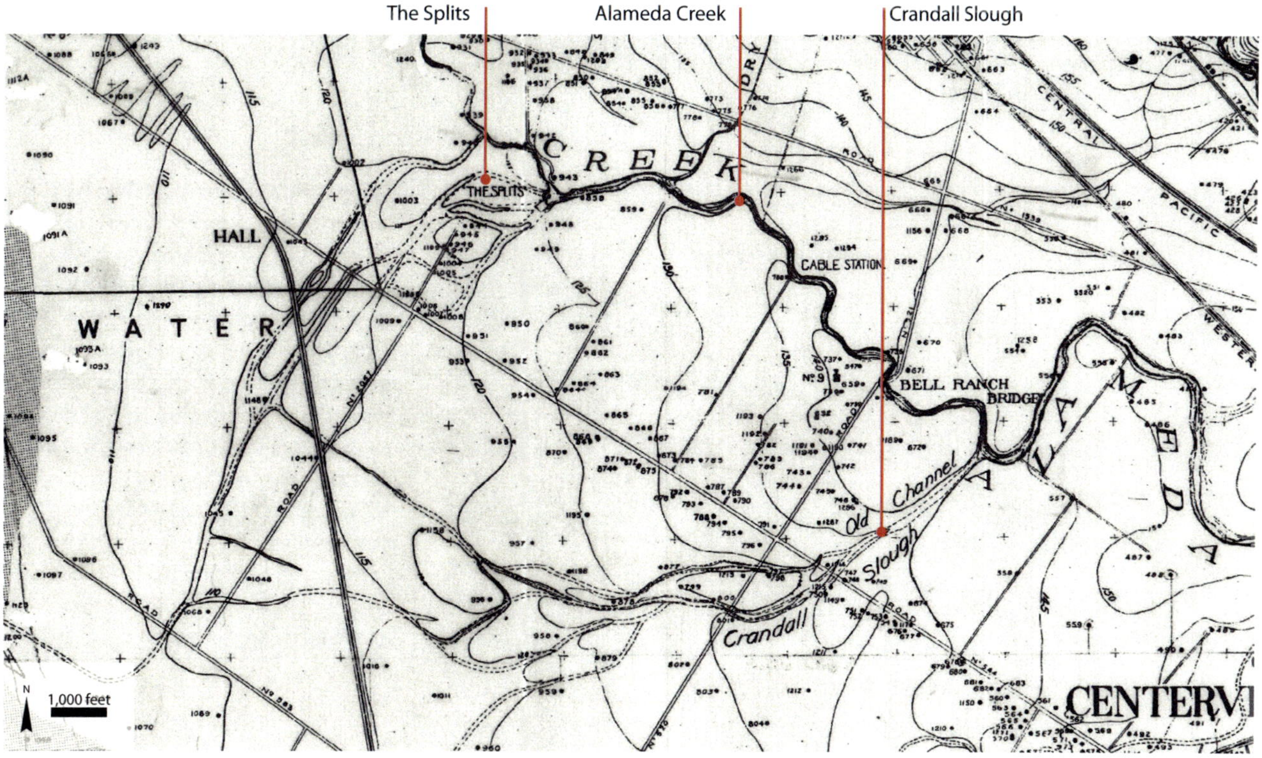

Figure 6.49. Overflow along Crandall Slough, 1955. The 1955 flood spread west across the Niles Cone. Flows spilled from Alameda Creek to flow more directly west along Crandall Slough towards the Coyote Hills, which are visible in the distance as islands within the zone of overflow. Note Coyote Hills Slough connecting the overflow zone to the Bay, and the large overflow zone stretching north to Alvarado, where Alameda Creek met the tides. (courtesy ACFCWCD)

Figure 6.50. "First branch of Crandall's Slough." In 1916, farmers planted orchards and constructed fences across Crandall Slough. A court summary described the slough as follows: "in many places it is cultivated the same as the adjoining lands. In most places there are no evidences of running water" (Burroughs 1925). However, a survey report noted that "these channels, although cultivated like the neighboring fields, are liable to be filled with freshet overflows during the wet season" (Westdahl 1896d), as can be seen in this image. (image A-563, January 4, 1916, © San Francisco Public Utilities Commission)

CRANDALL SLOUGH Crandall Slough split from Alameda Creek about a half mile upstream of Decoto Road, within the perennial, overflow reach. Crandall Slough carried only flood flow (Bailey 1917), with water reported in 1925 as present "only a few days in each year, and usually but a few hours on those days ... this flow usually occurs only at such times as the surrounding country is overflowed by the waters of Alameda Creek overtopping its banks" (Burroughs 1925:2; fig. 6.49).

The slough was broad and shallow, without clearly defined channels over much of its length (fig. 6.50) and "naturally spread widely on each side... because of the flatness of the country" (Supreme Court of California 1929). There were "a number of channels at the mouth of the Crandall Slough ... caused by the currents of the different floods that flowed through" (Lawrence 1920). A detailed account of the overflow into Crandall Slough described conditions in 1925:

> It is reached by the waters of Alameda Creek when the waters have risen in said Creek to a point seventeen and one half feet above the bed of that creek. At this point the slough has a depth of about seven feet and is practically one hundred and twenty five feet wide. ..there is no distinct channel. In many places it is lost in the lands of the surrounding country, with a depth of from two to four feet, and in many places from one to two hundred feet in width. (Burroughs 1925)

Crandall Slough's history is complex. The traces visible on the 1939 aerials almost align with a ditch that is recorded leading from Alameda Creek to the tidal marsh as early as 1852, approximately along the course of Decoto Road (Lewis 1852). Local historian Robert Fisher described this ditch as "'The Old Pacheco Ditch' a cattle boundary dug on the rancho by Indian labor" (Fisher n.d.).

The ditch became the border of the Potrero de los Cerritos land grant between Pacheco and Alviso, and was referred to as "an artificial ditch from the Alameda Creek to the Willows" in land grant testimony (Dyer 1861b). No other accounts were located documenting whether the ditch was dug by local labor or was the product of natural overflow from the creek. The ditch appears to have held perennial water in its lower reaches due to high groundwater (Patterson 1977).

Records of a feature named Crandall Slough appear as early as 1880 (*Morning Oregonian* 1880). Witnesses for the 1901 Clough case refer to Crandall Slough as a local landmark (Shinn 1901, Barry 1901), and the feature is clearly mapped in 1915 and labeled "old channel" (Stoner et al. 1915, see also Clarke 1924b). Although the ditch and Crandall Slough did not follow exactly the same course, they were very similar, and a number of sources use the 1852 date as evidence for the historical presence of Crandall Slough, apparently conflating the two (Supreme Court of California 1929, Superior Court 1920:4).

If the ditch and the slough are the same feature, then it is possible that Crandall Slough was an artificial channel later maintained by natural

The levee at Crandall's Slough broke this morning, giving vent to a large body of water from the Alameda Creek. Where it crosses the country road Alvarado and Centerville it is five feet deep, and a large amount of grain will be a total loss. About a thousand acres are under water.

—THE MORNING OREGONIAN
4/22/1880

processes. However, it is also possible that the "ditch" represents a natural feature that was enhanced, or that the two features were in fact distinct. We included Crandall Slough in our mapping, as it seems to have been present through the historical period and likely could represent an old course of Alameda Creek. Further investigation might help definitively establish whether or not this feature was natural.

THE SPLITS The Splits, also known as Patterson Creek, was an artificial overflow channel that connected with Alameda Creek two miles downstream from Crandall Slough (see fig. 6.48). The Splits channel was created in the early 1900s to control the path of water and sediment across the plain. The Patterson oral history records the process by which the Splits was constructed. Local landowners plowed the land to loosen soil so that water flowing across their property would scour a channel and deposit sediment to fill low marshy areas. McKeown (1975b) remembered "going up there and digging the ground out with a team of horses just loosening it up to wash out." A levee was also constructed to direct water into the developing channel (McKeown 1975b). Donald Patterson described how they "diverted the Alameda Creek by plowing in the overflow channels in the summer and letting it wash out in the winter, until the full flow of Alameda Creek came through the ranch" (Patterson 1977).

He was aware of Mr. Patterson's operations in constructing the canal known as the Splits and that the property owners should get together and prevent him from making this diversion of water.

—LOCAL RESIDENT MR. GALLIVAN, IN LAWRENCE 1920

This channel was later modified to create the present-day Flood Control Channel, which follows an almost identical path across the cone (fig. 6.51). In 1938, the Pattersons made a deal with the SFPUC to allow diversion of Calaveras Dam water releases across their property through the Splits channel (Unknown 1938). This agreement allowed for the construction of a diversion dam to force water through the Splits rather than allowing it to flow down Alameda Creek, where it caused frequent flooding at Alvarado. After the large floods of the 1950s, this diversion channel was replaced by the Alameda Creek Flood Control Channel, completing the transformation.

Figure 6.51. The Splits and the Alameda Creek Flood Control Channel. This 1915 map shows the newly created Splits channel (labeled Patterson Slough), which lines up with the present-day Flood Control Channel, shown in blue. The multi-thread pattern of the Splits channel reflects the semi-natural processes that created it. See also figure 6.48. (Unknown 1942, courtesy ACWD)

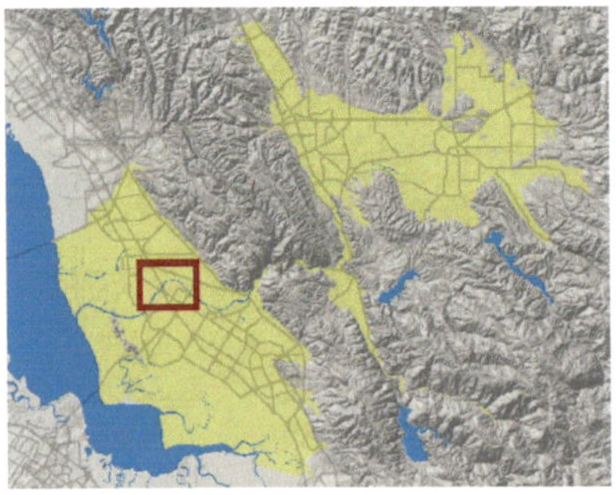

Change in channel form

Although incision is often caused by anthropogenic drivers, Alameda Creek appears to have been down-cutting through its alluvial fan before the arrival of Europeans. Incised channels commonly occur naturally on the upper portion of alluvial fans (Blair and McPherson 1994), and many sources over the 18th and 19th centuries noted how surprisingly deep Alameda Creek was on the Niles Cone. In the fall of 1795 at the top of the Niles Cone fan, along Mission Boulevard, explorer Dantí noted that "in all this stretch the bed of the river or arroyo, is deep and the removal of water impossible" (Cook 1957). In a similar location in 1796, a visitor remarked that "although it contains a creek, still that it affords but little water, and that the channel is so deep that it is difficult to obtain water therefrom for irrigating the extensive plains" (Alberni in Halley 1876). As late as 1888, surveyor George Allardt described that "the bed of the creek after it leaves the canon [canyon], lies from 15-30 feet below the general level of the surrounding country" (Allardt 1888).

Naturally incised conditions on lower Alameda creek may have been a function of climatic conditions, periods of increased rainfall, variation in rainfall intensity, and changing sediment supplies, along with tectonic uplift (Leopold 1951, Montgomery 1999). When Europeans first visited the Niles Cone in the 1760s, Alameda County was beginning a 60 year period of below-average rainfall following a 130 year period of high to normal rainfall (Fritts and Gordon 1980; see also Lamarch 1974, Stockton and Meko 1975). The wet years may have caused an imbalance of water and sediment, first introducing more sediment into the system with increased runoff and then decreasing sediment through longer term establishment of sediment-slowing vegetation and continued runoff (Leopold 1951, Shumm 1973, Balling and Wells 1990). Furthermore, increased periods of high intensity rainfall may have created landslides that acted as temporary sediment dams in the upper watershed (Collins 2005). By the time the explorers reached the Niles Cone in 1769 it seems plausible that Alameda Creek would already have been forced to abandon its old channel (Sanjón de los Alisos) in the period of increased sediment, and then had begun to downcut through its relatively new channel under reduced sediment loads.

A wetter period during the mid-19th century coincided with the period of extensive modifications and "plumbing" of the channel and marsh network, causing an increased sediment load and causing the channel to aggrade. Land use changes upstream resulted in increased erosion and sediment loads at the same time that diversions began to limit the amount of water flowing through the lower Alameda Creek channel, thus creating an opposing imbalance of water and sediment in the system. Increased piping and draining of the Pleasanton marsh may have led to sedimentation of the deep channel on the Niles Cone as sediment was flushed downstream from the marsh bed, which had acted previously as a sediment sink (Fritts and Gordon 1980). Construction of the Calaveras Dam in early 1900s may have had a similar effect, releasing fine sediment downstream.

Figure 6.52. Southeastern drainages. A number of these small drainages spread out onto the cone well before they reached the tidal marsh, including Sabre Cat Creek, Canada del Aliso, Agua Fria, and Toroges Creek. Agua Caliente is the only creek among these drainages that maintained a direct connection to the tidal marsh.

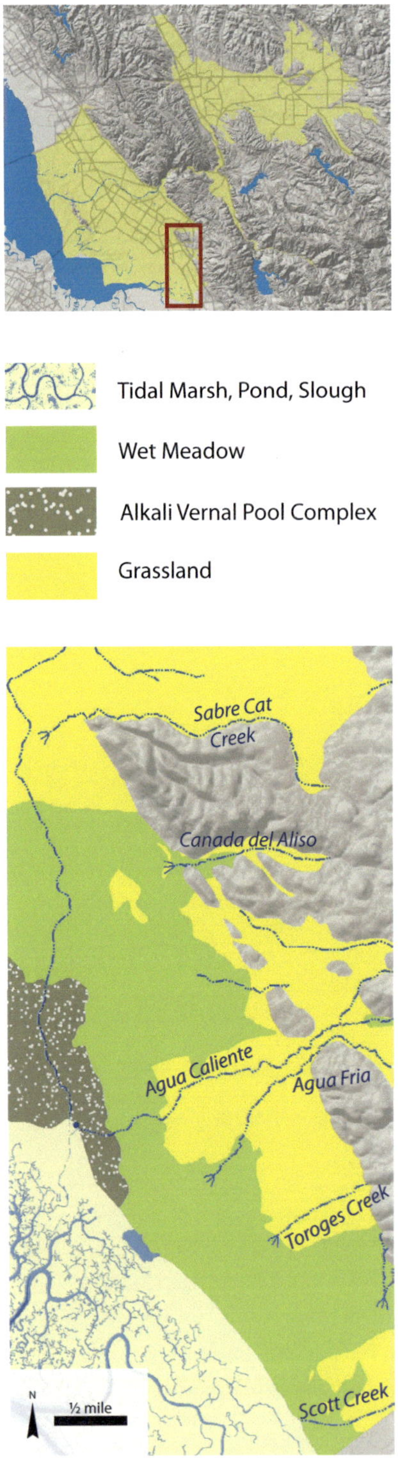

By 1900, local residents began to comment that the creek was filling in. A 1916 report declared that "the channel is silting up and getting shallower" (SVWC 1916). In 1917, engineer Bailey reported, "The main channel diminishes in size very rapidly westward from here until it is practically non-existent beyond Alvarado" (Bailey 1917:2). Landowners that gave depositions for the Clough case repeatedly described gravels and other sediments building up along specific reaches of lower Alameda Creek (see box 6.1). Shinn (1901:782) reported that "when I first knew it distinctly the gravel beds and flowing water were lower in relation to the surface … the banks … were noticeably steeper than now, and the gravel has been washed in."

Other drainages

South of the Niles Cone, a series of small and largely discontinuous drainages flowed from the hills above Fremont to the Bay (including creeks such as Mission Creek, Cañada del Aliso, Agua Caliente Creek, and Lone Tree Creek; fig. 6.52). The creeks were largely fed by a series of flowing springs in the hills, and may have supported perennial flows in at least upper Mission Creek (Sal 1795, Shinn [1889]1991). These springs were noted by the earliest explorers and surveyors to visit the area (Howe 1851, Bryant [1848]1985, Dantí in Cook 1957).

As the creeks flowed from their confined drainages in the hills out across the plain many of them sank into the ground or spread to form wet meadows (fig. 6.53). The exact endpoints of these discontinuous streams would have varied depending on rainfall and time of year, so the mapped extents represent only an approximation. Over time, the creeks were ditched to facilitate drainage and prevent flooding.

Mission Creek provides a well-documented example of modification over time (fig. 6.54). The creek that we know today as Mission Creek was historically two distinct drainages north of the Mission (Higley 1857a, Allardt 1874), which were ditched to flow together by the late 1870s (Thompson and West 1878, USGS 1906). Further downstream, these creeks historically spread into distributary channels before reaching the Lagoon (Tracy 1852, Higley 1857a, Allardt 1874). The 1906 USGS map of this area shows a distinct braided pattern beginning at the same place that Allardt shows the creek forming a distributary. It seems likely that at this point the well-defined single thread channel of Mission Creek ended, spreading across an area that may have had seasonal overflow and connection with the Lagoon.

One exception to the historical pattern of disconnected hillside drainages was Agua Caliente Creek, which flowed from hot springs in the hills and is consistently depicted with a direct connection to Mud Creek Slough in the tidal marsh (see Higley 1857a, Unknown 1865, Allardt 1874, Morse and Westdahl 1896). This creek had a relatively large watershed and short distance to the tidal marsh, which may have helped it maintain a defined

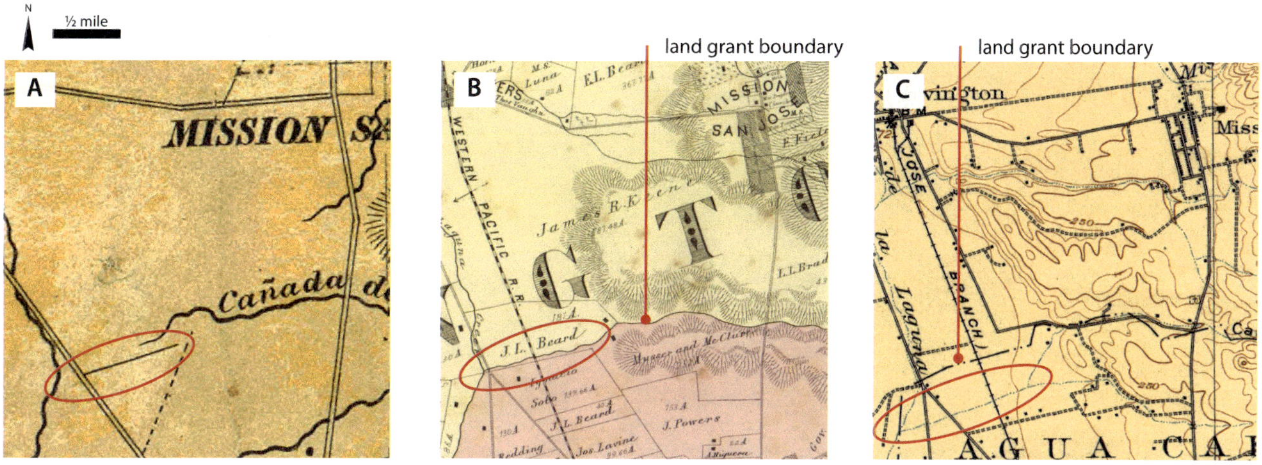

channel. Arroyo de la Laguna, which flowed from the Lagoon, also maintained a channel to the tidal marsh (Clarke et al. 1852).

A series of smaller, undefined swales also occasionally carried water towards the tidal marsh. The swales are shown in a 1915 map, labeled "old channel" (Stoner et al. 1915). The swales were likely either abandoned channels or ditches created as cattle fences (Shinn [1889]1991:50). Water would have only occupied these swales during high rainfall events.

DRY CREEK The major tributary to Alameda Creek on the Niles Cone was Dry Creek, also known as *Arroyo Segunda* [sic], which flowed from the hills towards Alameda Creek along a course similar to the present-day course. However, Dry Creek is depicted consistently on a number of historical maps sweeping to the west of Decoto, along a different path than it follows today (Boardman 1884, Nusbaumer and Boardman 1899). Unlike other drainages, it appears to have flowed in a defined channel directly into Alameda Creek (Boardman 1867, Allardt 1874, Thompson and West 1878, Boardman 1884, USGS 1899a).

Dry Creek was bordered by a band of riparian cover, which is depicted in both maps and imagery (fig. 6.55; USDC ca. 1840c). A number of fine-scale survey maps record survey trees along the bank of Dry Creek, including maples (*Acer* spp.) and a sycamore (*Platanus racemosa*).

Creek summary

Table 6.4, below, summarizes the morphology, flow, and riparian vegetation characteristics for each of the reaches of Alameda Creek described above, and includes a summary of Dry Creek, a major tributary (see fig. 6.33). We developed broad classes applicable to streams throughout the study area, and focused on the large stream systems.

The data presented below were synthesized from a variety of sources, including narrative records, historical aerials, historical maps, digital elevation datasets, and geology and soils data (both historical and modern). For further discussion of riparian width classes and geomorphic processes, see Chapter 9.

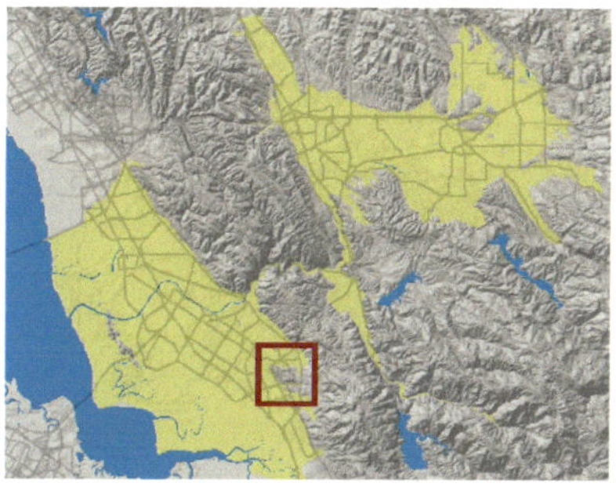

Figure 6.53. Early surveyors provide an unusually detailed description of the distributary historically formed by Cañada del Aliso near the intersection of current-day I-680 and Auto Mall Parkway/Durham Road. In January of 1857 (A), Surveyor Higley crossed "Aliso Creek at the point where it begins to spread" and then, continuing west beyond the distributary, observed the "low ground covered by water from the Aliso Creek" (Higley 1857b). By the 1870s (B), maps show Cañada del Aliso connecting directly with Arroyo de la Laguna, and 30 years later (C) this confluence appears to have been re-ditched and moved further south, although the old course is preserved in the dotted line of the land grant boundary. (A: Higley 1857, courtesy Bureau of Land Management; B: Thompson and West 1878, courtesy David Rumsey Map Collection; C: USGS 1906).

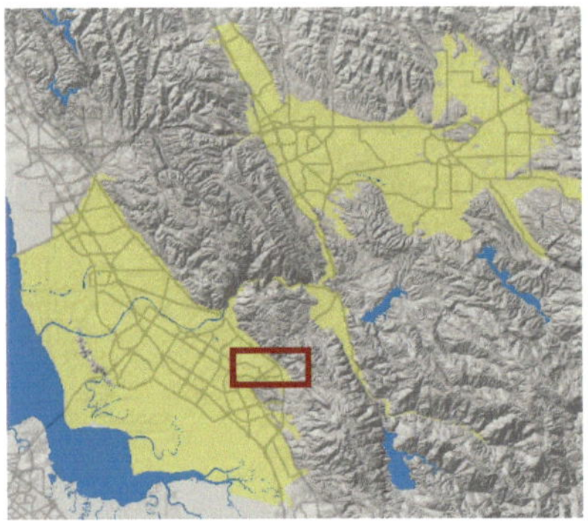

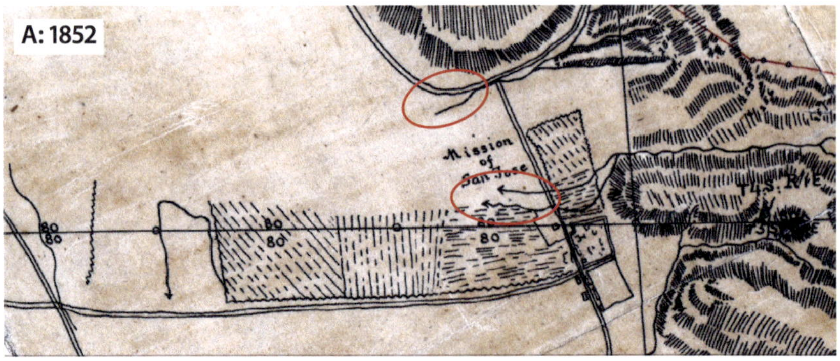

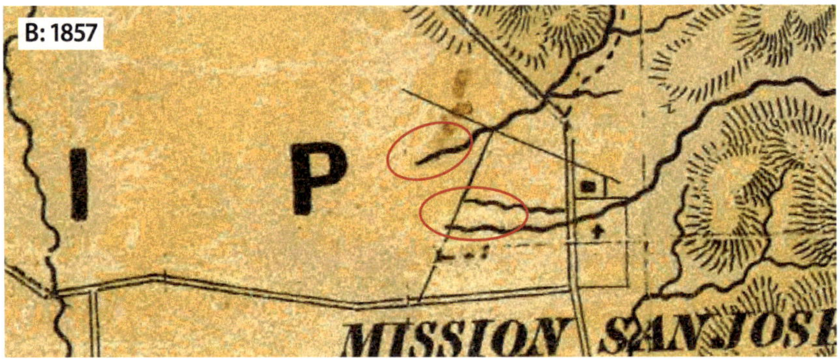

Figure 6.54. Mission Creek. The construction of Mission San José in 1797 created a zone of early modification in the lands immediately surrounding the Mission. Several creeks that historically drained the hills behind Mission San José, including Mission Creek, were used for irrigation of Mission fields. Mission Creek was dammed as early as 1814 to provide water for Mission gardens, and as early as 1795 recently converted native groups were diverting water from the creek for irrigation (Dantí in Cook 1957). Note how Mission Creek is shown spreading and ending in A and B, but connecting to follow its current path in C. (A: Tracy 1852, courtesy The Bancroft Library, UC Berkeley; B: Higley 1857, courtesy Bureau of Land Management; C: Thomspon and West 1878, courtesy David Rumsey Map Collection)

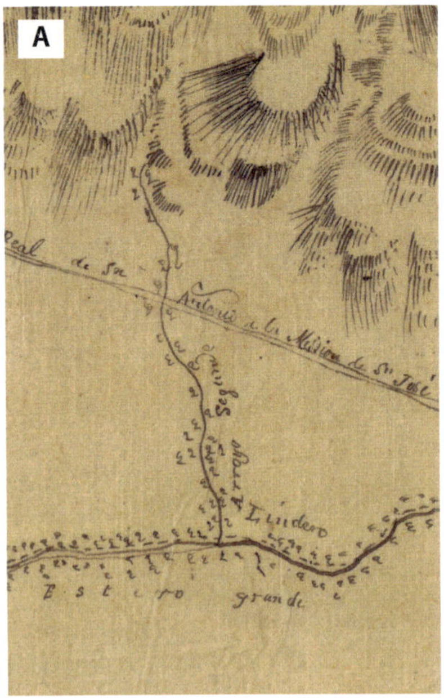

Figure 6.55. Riparian cover along Dry Creek. Both this ca. 1840 *diseño* map (A), and this photograph from May 31, 1915 (B) show well developed riparian tree cover along Dry Creek. (A: USDC ca. 1840c, courtesy The Bancroft Library, UC Berkeley; B: AD-282, © San Francisco Public Utilities Commission)

Table 6.4. Historical stream reach characteristics. This table describes stream characteristics within the study area, and so is restricted to the valley floor.

Creek	Reach	Watershed Area (sq. miles)[1]	Dominant Morphology[2]	Dominant Process [2] (Geomorphic)	Substrate[3]	Dry Season Flow	Riparian Corridor Width Classes[4]	Riparian Vegetation[5]
Alameda Creek (Niles Cone)	Mouth of Niles Canyon *(mouth of Niles Canyon to railroad crossing)*	645	Braided	Depositional/ Transport	Boulders, gravels, sand	Perennial	>1320 ft (>400 m)	Mixed Riparian Forest
	Perennial, gravel-bedded reach *(Mission Blvd to Bart weir)*	648	Braided	Depositional/ Transport	Boulders, gravels, sand	Perennial	660-1320 ft (200-400 m)	Mixed Riparian Forest
	Intermittent with pools reach *(Bart weir to Decoto Rd)*	653	Single-stem, meandering	Transport	Gravels, sand	Intermittent with pools	220-660 ft (60-200 m)	Mixed Riparian Forest
	Perennial, overflow reach *(Decoto road to Alvarado)*	684	Single-stem, meandering	Transport	Sand, silt, clay	Perennial	220-660 ft (60-200 m)	Mixed Riparian Forest
	Estuary reach *(Alvarado to Bay)*	717	Tidal distributary channels	Depositional	Bay mud	Tidal	<200 ft (<60 m)	Willow-Cottonwood
Dry Creek		10	Single-stem, meandering	Transport	Gravels, silt, clay	Intermittent	<200 ft (<60 m)	Mixed Riparian Forest

[1]Watershed area was calculated at the downstream endpoint of each reach using contemporary USGS Streamstats, but is likely representative of the historical drainage area.

[2]Dominant morphology and process were determined from the mapped form of the creek, our understanding of fluvial geomorphic processes, and geology data.

[3]Substrate classes were developed from historical descriptions, soils and geology mapping, and historical photographs.

[4]Riparian corridor width measures from the outer edge of stream-associated vegetation on one side of the stream to the other, including the stream bed. These widths were based largely on the width of a representative reach in the historical aerial imagery and current geomorphic features in the LiDAR data (USDA 1939-40, LiDAR 2007).

[5]Broad riparian vegetation classes were developed from those species data that exist, and describe the inner corridor of riparian vegetation. Further from the creek, riparian vegetation would have included valley oaks and/or sycamores.

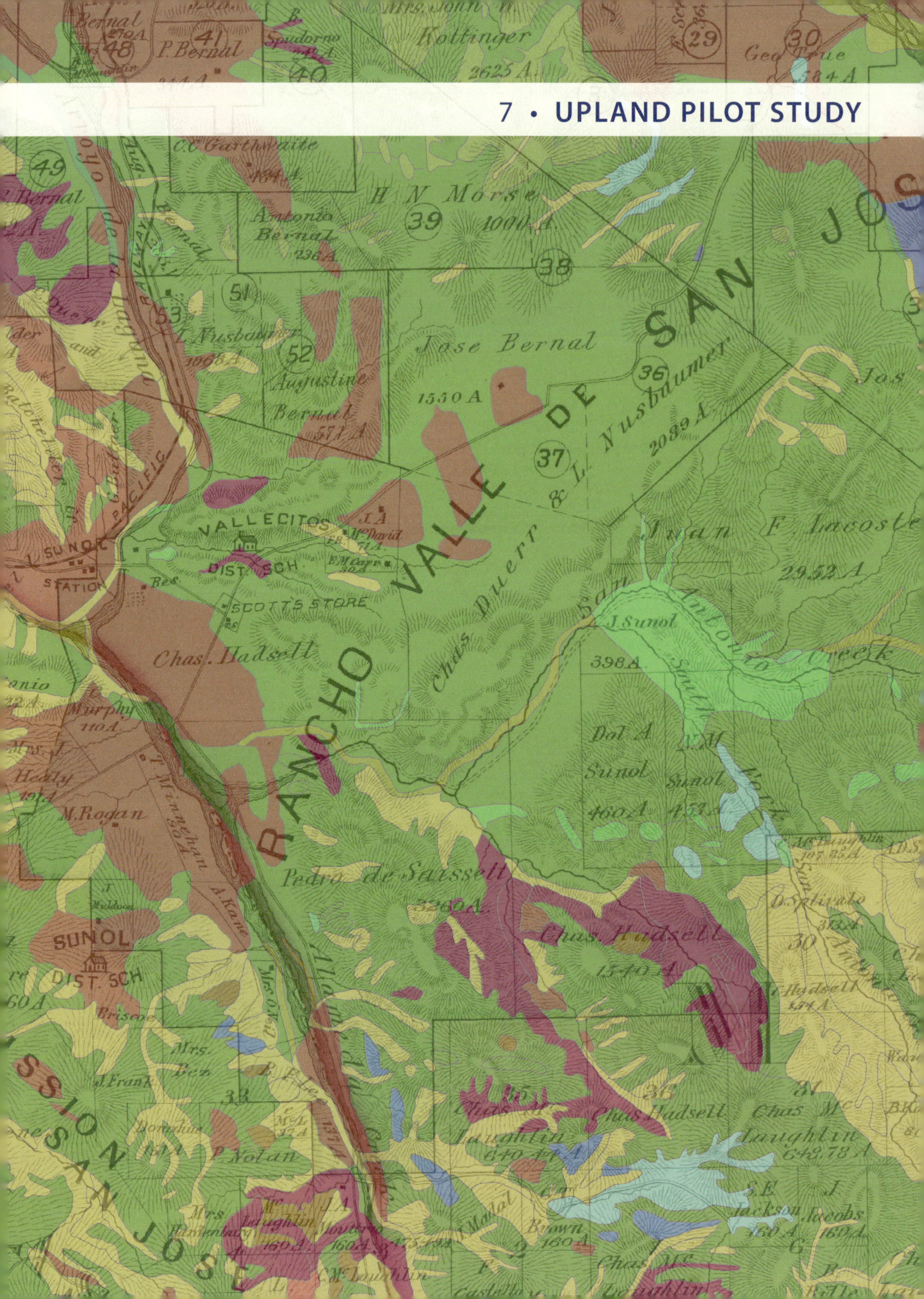
RANCHO VALLE DE SAN JOS
Jose Bernal
1550 A
H. N. Morse
1000 A
C.C. Garthwaite
Kottinger
2625 A
P. Bernal
Antonio Bernal
296 A
Augustine Bernal
577 A
T. Nusbaumer
Chas. Duerr
& L. Nusbaumer
2089 A
Juan F. Lacoste
2952 A
VALLECITOS
DIST. SCH
SCOTT'S STORE
SUNOL
STATION
PACIFIC
Chas. Hadsell
J. Sunol
398 A
San Antonio Creek
Pedro de Saissett
Chas. Hadsell
1540 A
Murphy
110 A
M. Rogan
SUNOL
DIST. SCH
Chas. Hadsell
Chas. Mc Laughlin
643.78 A
P. Nolan

Introduction

Most SFEI historical ecology studies have focused on valley floors, waterways, wetlands, and coastlines, where modifications and current environmental changes are generally most extreme. These areas also tend to have extensive recoverable data about early conditions because they were more intensively settled, managed and explored. As the underlying form and function of our valley ecosystems has become obscured by urbanization, agriculture, landscaping, and hydro-modification, the upland areas (the steeper, bedrock-based hillslopes of the watershed above the valleys and plains) have become even more important in a number of respects. Resource managers and planners are increasingly looking to watershed uplands for solutions to downstream resource challenges. Uplands analyses may have important implications for habitat restoration and conservation. As expanding human populations and development pressure further encroach on lowland open spaces, undeveloped uplands must be managed carefully to balance recreation, farming, and hunting with wildlife needs for refugia and habitat corridors. In addition, conservation efforts are increasingly emphasizing the importance of upland areas to downstream functions and overall watershed health (Bay Area Open Space Council 2011).

While they appear more "natural," many upland areas have also experienced intense modification. Changes in the uplands are often more subtle than those in the valleys, and may include more gradual transitions such as change in vegetation from native to non-native grasses, changes in the density and impact of cattle grazing, or changes in the density of roads. Altered vegetation patterns due to development and changing land management have affected the survival of certain species. For example, the expansion of woody vegetation may affect the range of the kit fox (*Vulpes macrotis mutica*) by reducing the extent and patch size of available open grassland. Alternatively, conversion of chaparral to either woodland or grassland would affect the Alameda whipsnake (*Masticophis lateralis euryxanthus*) by removing essential scrub habitat. Shifts such as these were not deemed to have been adequately documented to provide a basis for management strategies in the Alameda Creek Habitat Conservation Plan (ACHCP). To identify primary areas for protection and restoration, a better understanding of historical habitat conditions in upland environments may be useful for directing management decisions.

The increased attention on conserving, enhancing, and restoring the biological diversity of upland habitats highlights the need for an accurate historical understanding of the recent past. While upland areas may look unmodified—especially compared to urbanized or agricultural valleys—dramatic changes may have taken place in the last 150-200 years, and restoration targets without this context may be less effective, and less

Figure 7.1. Upland vegetation mapping. This excerpt shows vegetation patterns mapped near Sunol by Wieslander in the 1930s overlaid on an 1878 map of the area. See figure 7.6 for the full map of land cover patterns in the upland pilot study area with a legend. (Thompson and West 1878, courtesy David Rumsey Map Collection) resilient to new or ongoing perturbations (Hamilton 1997, Holstein 2001, Foster and Motzkin 2003).

In order to address this information gap we focus this chapter on the following questions:

1. CHANGES IN VEGETATION COVER

Have there been identifiable changes in terrestrial vegetation cover, including expansion/contraction of woodland, savanna, grassland, and chaparral, in relation to grazing, agricultural clearing or more recent infrastructure changes such as reservoirs?

Have changes in fire regimes affected endangered species populations, oak savanna/woodland health or regeneration, and/or dominant vegetation cover types?

2. LAND MANAGEMENT CHANGES

How have land management and infrastructure change impacted upland habitats?

What were the condition and configuration of historical stream, riparian, and off-channel wetland habitats of now-inundated valleys (especially the lands now covered by the San Antonio Reservoir)?

We focus on a selected target area to the south of Livermore-Amador Valley to apply several methods to assess their value for regional applications. As a pilot study, this chapter explores several techniques that can be used to study the historical ecology of the uplands. We incorporated methods including interviews with long-time residents of the area and mapping land use changes, and we use these to reconstruct a picture of historical land cover and change in the uplands of Alameda County.

Study area

The pilot study area is a 30,000 acre expanse centered on San Antonio Reservoir (fig. 7.2). It is bounded by Sunol Valley to the west, Livermore-Amador Valley floor to the north, and Lake Del Valle to the east. To the south, it includes the watersheds draining into San Antonio Creek. The area was chosen because of its varied land ownership (including land holdings of the East Bay Regional Park District and San Francisco Public Utilities Commission), as well as its variety in aspect, slope, and hillslope formations. San Antonio Reservoir was historically known as La Costa Valley and was drained by Apperson Creek and Indian Creek, which flowed from headwaters in the Maguire Peaks and Valpe Ridge, respectively. Welch Creek drains the south side of Maguire Peaks towards Sunol Valley. To the north of La Costa Valley, Vallecitos Valley remains open rangeland with rural residential development.

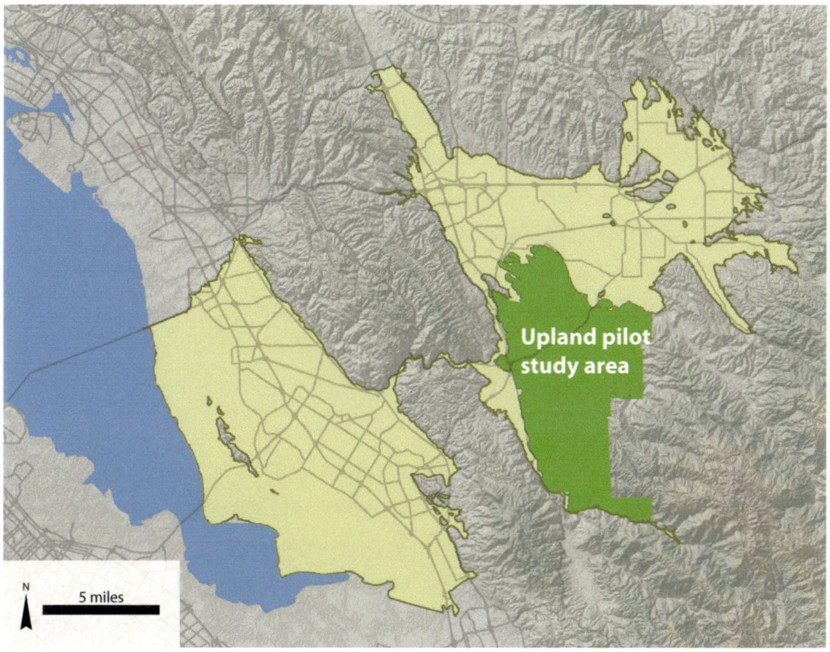

Figure 7.2. Upland Pilot Study Area, shown here in relationship to the Alameda Historical Ecology project boundary (shown in light green).

GEOLOGICAL CONTEXT The study area can be broken into two distinct geologic regions. The northern part of the study area, centering on La Costa and Vallecitos valleys, is dominated by Pleistocene alluvial surficial deposits. The average slope is less than 30%. The southern part, south of Vallecitos and La Costa valleys, is primarily underlain by a mix of Briones and Neroly Sandstone formation and Franciscan mélange bedrock and forms a tightly folded pattern of northwest to southeast trending Coastal Range ridges (with slopes greater than 30%) associated with the Calaveras Fault (Helley and Graymer 1997).

There are different natural vegetation patterns associated with each combination of underlying geological formation, slope, and aspect. These differing landscape forms were conducive to different land uses, which shaped their land cover and development trajectories.

Vegetation change analysis

Today, the uplands of Alameda Creek watershed are composed of woodlands, oak savanna, chaparral, and grassland communities, as well as open water, riparian zones, and rural residential land uses. The uplands provided diverse ecological functions and historically supported home ranges for species of concern including the San Joaquin kit fox *(Vulpes macrotis mutica)* and the Western burrowing owl *(Athene cunicularia hypugaea)*, both of which are associated with open grassland or savanna communities (ICF International 2010b). The Alameda whipsnake *(Masticophis lateralis euryxanthus)* is dependent on chaparral for cover and foraging. A number of threatened and endangered amphibian species, such as the California tiger salamander *(Ambystoma californiense)* and the California red-legged frog *(Rana aurora draytonii)*, depend on upland communities connected to aquatic habitat for foraging, dispersal, and cover.

In order to assess the change in overall distribution, abundance, and composition of terrestrial vegetation over time, we analyzed datasets from the 1850s, 1930s, and 2010, as well as 1940 aerial photography. Each of these sources supported different types of analyses, discussed below. To more thoroughly investigate the changes that occurred, we also examined two example areas in depth: a hillside and a small valley.

Vegetation change: methods 1930s–2010

The earliest comprehensive effort to map California vegetation occurred in the 1920s and 30s, when Alfred Wieslander led the Vegetation Type Mapping (VTM) project for the U.S. Forest Service. The effort produced a set of vegetation maps, as well as plot data, species inventories, and landscape photography, the protocols for which were documented in a detailed field manual (Wieslander et al. 1933). In this area, the vegetation types were mapped in the field directly onto 15-minute U.S. Geological Survey (USGS) quadrangles without the benefit of aerial photography. Vegetation was classified as either complex vegetation mosaics or as pure or mixed stands, with the dominant as well as subdominant species listed.

We analyzed the VTM data for this area in conjunction with two other datasets. In the late 1930s and early 1940s, shortly after the Wieslander effort, the first air photography became available for the area. This imagery provided a useful comparison with the VTM data. For the 2010 Habitat Conservation Plan, Jones and Stokes (now ICF International) used 2010 aerial imagery to map land cover patterns in the uplands of southern Alameda County, providing a comparable modern dataset (ICF International 2010b). However, it is important to note that this dataset was based on aerial photography with limited ground truthing and was not fully field-verified.

We compared the extent of easily discernible classes such as woodland and grassland mapped by the VTM effort to the 1940 USDA aerial photos (fig. 7.3). These two datasets are separated by only a ten year gap, so they should show a relatively similar picture of vegetation patterns. However, we found that the spatial accuracy of the VTM maps was too low to allow them to be meaningfully compared. For this study area, the VTM dataset is most useful as a source describing the composition of vegetation alliances and showing the general proportion of different vegetation types in the area. It is not spatially accurate enough to be used to calculate vegetation change without a substantial effort to adjust this dataset to more closely match the 1940 aerials (e.g., Stanford et al. 2011).

Rather than adjusting the Wieslander polygons, we were able to use the 2010 ICF mapping and attribute it with the VTM vegetation information (box 7.1). We found that the 2010 mapping and the 1940 aerials were a very close spatial match, so we reattributed the 2010 polygons with the 1930s Wieslander habitat/alliance information in order to reconstruct a dataset representing vegetation patterns in the 1930s/1940s. We left the vegetation polygons at the same extent, but removed rural residential, roads, and reservoir polygons if they were not present in the 1940 aerial. We replaced these features with the land cover described by Wieslander. This resulted in a spatially accurate polygon layer describing the extent and composition of 1930s-40s vegetation patterns.

Vegetation change: methods pre-1930s

As the VTM dataset dates back only to the 1930s, and few early maps extend into the hills, we were unable to create a detailed map representing earlier vegetation patterns. We attempted to approximate earlier conditions using three methods: selectively replacing agriculture from the 1930s dataset with native habitat types; interpretation of textual data; and analysis of General Land Office (GLO) data in selected areas (see hillslope analysis p. 232).

Using the 1940 USDA aerial photos, we digitized areas that appeared cultivated, cleared, or grazed, using Wieslander as a guide since his mapping calls out some cultivated areas, but lacks spatial accuracy (see fig. 7.3). These areas were often identifiable on the 1940 aerials by white

Figure 7.3. Comparison of Wieslander and ICF mapping with 1940 aerials. Wieslander digitized mapping overlaid on 1940 aerial imagery is shown at left, while ICF mapping of the same area overlaid on the 1940 aerials is shown at right. The footprint of upland vegetation patterns in this area has changed little enough that the polygons from the 2010 mapping are a more accurate match for for 1940 conditions than the 1935 Wieslander mapping. (USDA 1940; Wieslander 1935; ICF International 2010b)

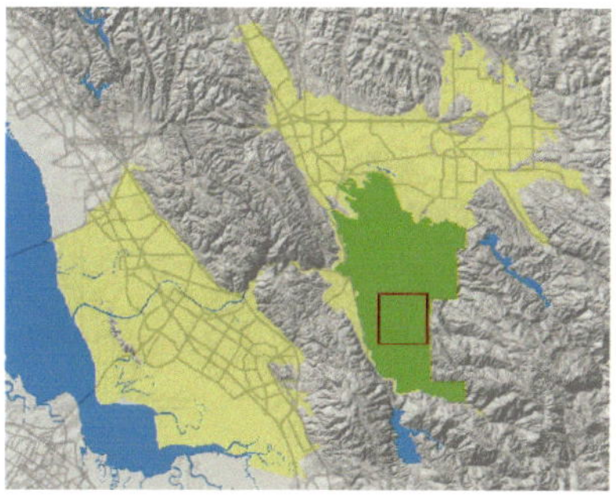

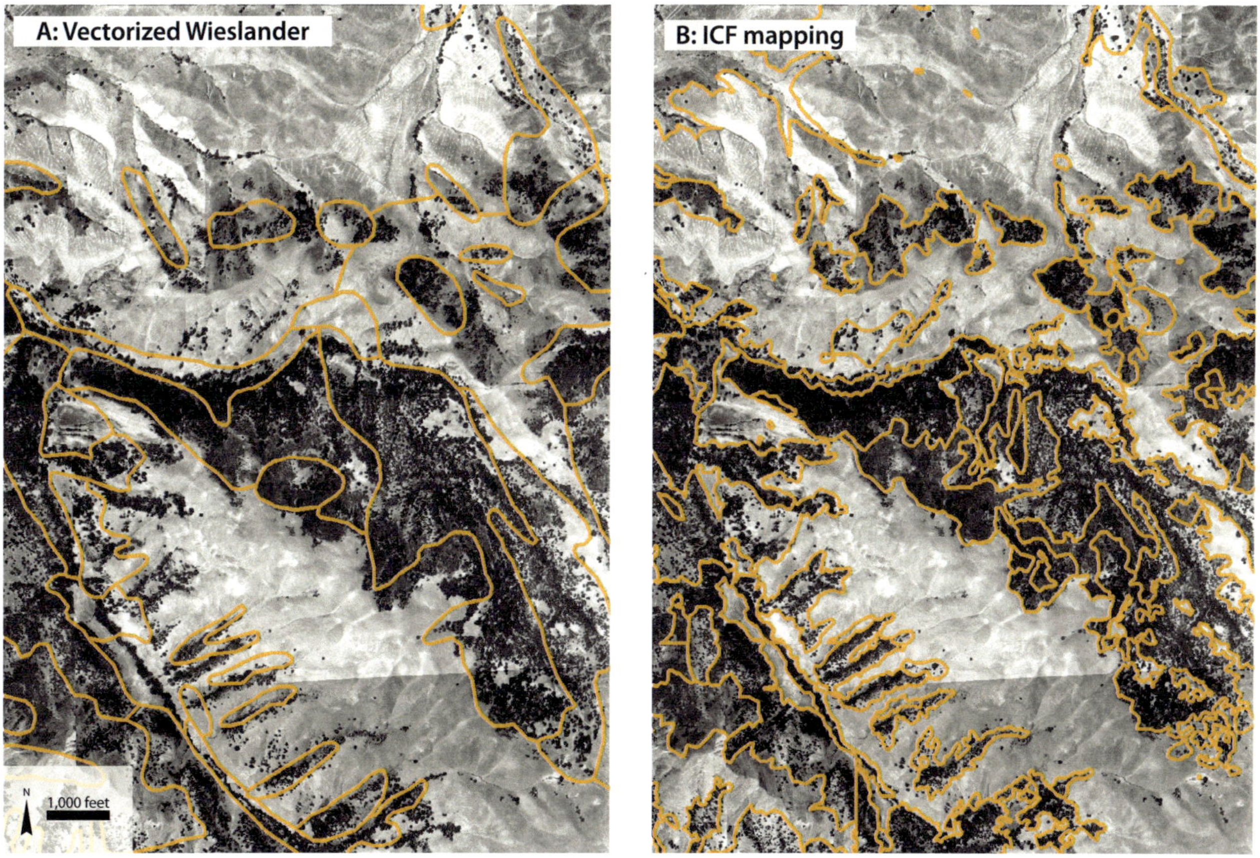

BOX 7.1. VEGETATION AND HABITAT TYPES USED IN THIS CHAPTER

We used the vegetation classes developed by ICF for our analysis. The following habitat types occur today in the upland pilot study area, and are arranged from lowest to highest elevation.

CALIFORNIA ANNUAL GRASSLAND (non-native) is the most common land cover type in the study area. This herbaceous plant community includes species such as wild oat *(avena)*, brome *(bromus)*, Italian ryegrass *(Festuca perennis)* and annual fescue species *(Festuca)* (Sawyer et al. 2009, ICF International 2010b). Other species such as clover, lupin, and California poppy occur as well. Grasslands occur often in areas of alluvial deposits as well as wherever soils exist with clay horizons thick enough to hold significant water near the surface (Holstein 2000).

SERPENTINE BUNCHGRASS GRASSLAND is uncommon in the pilot study area, and occurs only on soils underlain by serpentinite bedrock. Few species thrive in these areas, often called barrens, but purple needlegrass *(Stipa pulchra)* is the dominant species (ICF International 2010b).

NORTHERN COASTAL SCRUB / DIABLAN SAGE SCRUB is relatively common in the pilot study area. It is most often found on south-facing slopes on rocky, well drained soils. Subject to periodic fires, it responds quickly to burning. Common species include bush monkeyflower *(Mimulus auranticus)*, coyote brush *(Baccharis pilularis)*, California matchweed *(Gutierrezia californica)*, poison-oak *(Toxicodendron diversilobum)*, California broom *(Acmispon glaber* var. *glaber)*, and bush lupine *(Lupinus albifrons)*. Coastal and sage scrub is most often used by gopher snakes, rattle snakes, coyote, and bobcat. Lesser goldfinch (*Carduelis psaltria*) and Alameda whipsnake (*Masticophis lateralis euryxanthus*) also occur in this habitat (ICF International 2010b).

MIXED WILLOW RIPARIAN SCRUB occurs along the banks of intermittent and perennial streams. Red willow *(Salix laevigata)* and arroyo willow *(Salix lasiolepis)* are the most common species.

MIXED RIPARIAN FOREST AND WOODLAND is similar to mixed willow riparian scrub and is found flanking major streams in the study area. The major canopy species throughout the study area are California sycamore *(Platanus racemosa)*, valley oak *(Quercus lobata)*, coast live oak *(Quercus agrifolia)*, red willow *(Salix laevigata)*, and California bay *(Umbellularia californica)*. Associated trees and shrubs include California black walnut *(Juglans hindsii)*, other species of willow *(Salix* spp.), California buckeye *(Aesculus californica)*, Fremont cottonwood *(Populus fremontii)*, and bigleaf maple *(Acer macrophyllum)*.

SYCAMORE ALLUVIAL WOODLAND is uncommon within the study area, and occurs mainly along San Antonio Creek. The dominant canopy tree is western sycamore *(Platanus racemosa)*. This habitat type generally occurs along broad braided channels of intermittent streams (Keeler-Wolf et al. 1996).

BLUE OAK WOODLAND is a common feature of the upland pilot areas, mainly occupying north or northeast facing slopes and typically occurring between 100 ft and 230 ft in elevation (Allen 1991, Allen-Diaz et al. 2007, ICF International 2010b). The understory varies from shrubby to open non-native grassland (Allen-Diaz 2007) and blue oaks *(Quercus douglasii)* are often found on rocky, well-drained soils. Blue oaks are relatively slow growing, and grow in stands of mature trees, often with relatively few young saplings. This regeneration problem may be attributed to land management, including fire regime management (Allen-Diaz 2007). Blue oak woodlands make up 13% of the current land over of the upland pilot area, and are the most extensive hardwood rangeland type found in both the Coast Range and the Sierra Nevada (Holzman and Diaz 1991).

COAST LIVE OAK FOREST AND WOODLAND occurs on steep slopes, in mosaics with grasslands, shrublands, and riparian habitats, and is characterized by a dense canopy dominated by coast live oak *(Quercus agrifolia)*; however, other oaks and bay trees are often present (Barbour et al. 2007, ICF International 2010b). Mixed evergreen forest/oak woodland is characterized by a diverse overstory often dominated by coast live oak with a mix of co-dominant oaks such as coast live oak, blue oak *(Quercus douglasii)*, and valley oak *(Quercus lobata)*. Broad-leafed evergreen and deciduous trees are present, including California bay, madrone, California buckeye, and black oak (Holland 1986; Sawyer and Keeler-Wolf 1995).

FOOTHILL PINE - OAK WOODLAND occurs only in three small places in the pilot study area, along higher elevation drainages. The canopy is dominated by foothill pine mixed with oaks, typically with a dense understory of scattered shrubs.

COULTER PINE WOODLAND also was mapped in only a few places in the study area in patterns similar to foothill pine-oak woodland vegetation type. Coulter pine *(Pinus coulteri)* is typically dominant in these closed canopy stands, and this is the very northern extent of its range. Other tree species that are commonly associated with Coulter pine woodlands include bigcone Douglas-fir *(Pseudotsuga macrocarpa)*, black oak *(Quercus kelloggii)*, canyon live oak *(Quercus chrysolepis)*, coast live oak *(Quercus agrifolia)*, interior live oak *(Quercus wislizeni)*, foothill pine *(Pinus sabiniana)*, and ponderosa pine *(Pinus ponderosa)*. Coulter pine woodlands can occur throughout upland environments on steep slopes with shallow soils.

Figure 7.4. Agriculture. In the 1940 aerial imagery, tracks left by haying are easily discernable and are a good indicator of the historical presence of agricultural land uses. (USDA 1940)

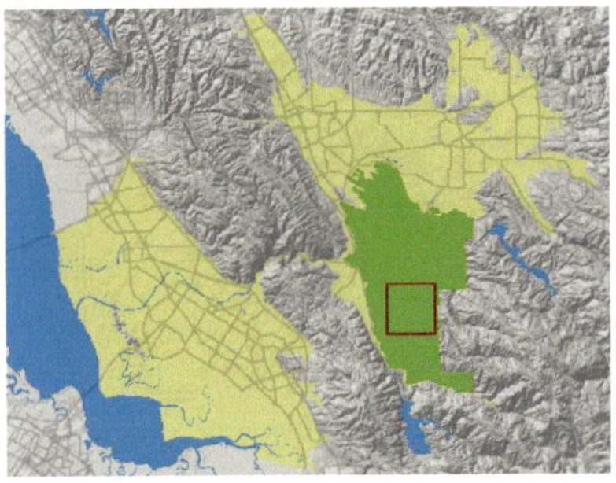

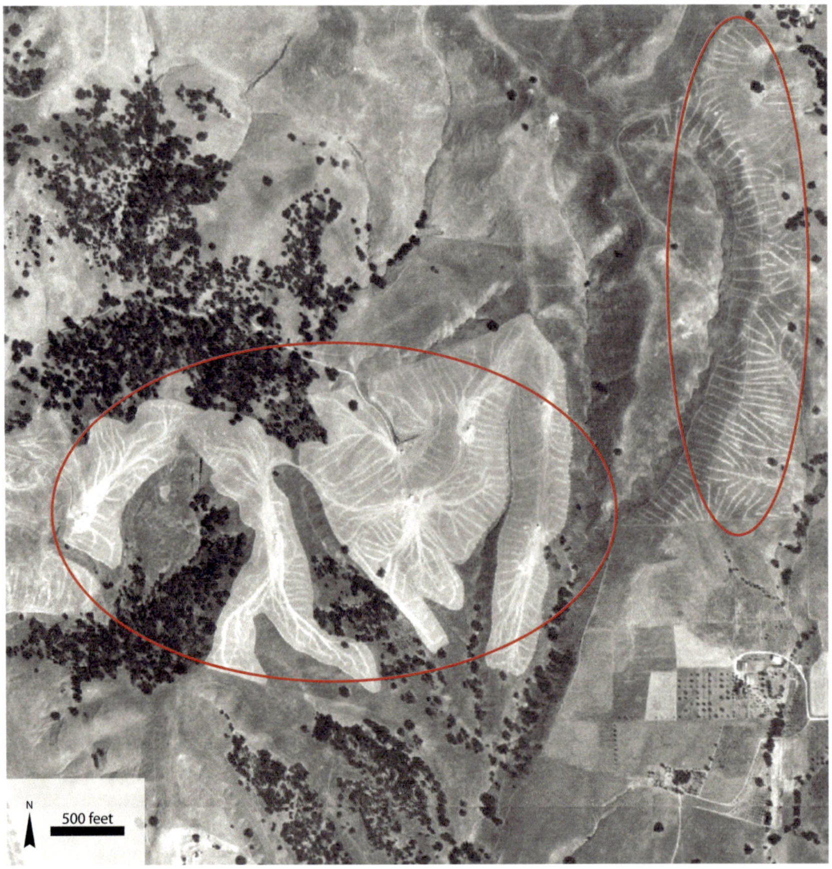

cow trails, tractor lines, or straight lines cutting across an otherwise smooth signature of grasslands (fig. 7.4). This approximation allowed us to calculate a minimum number of acres in agricultural production (including hay and grain, orchards, truck farms, and cattle ranching) in the upland pilot area during the mid-20th century. We replaced these agricultural areas with the surrounding vegetation type, resulting in a rough estimate of vegetation patterns in the 1800s.

We also used descriptions to estimate the early impacts of Europeans on woodlands in particular. Early records suggest that most hillsides in the area had relatively little tree cover in the mid-19th century. In 1853, surveyor Sherman Day surveyed a transect moving east from Sunol Valley across the Welch Creek watershed, and then worked north, skirting the eastern edge of La Costa Valley. He described the landscape as having "very few trees, chiefly oaks" (Day 1853:259), and "surface mountainous, with very few trees. Hills, especially the mountain spurs, gravelly. A little water in the ravines" (Day 1853:258). He also described a "good crop of wild oats on the hills. Very few scattered oaks" (Day 1853:254).

The low tree density may have been a result of the actions of early settlers. Traveler Titus Cronise (1868) suggested that tree clearing was responsible for the low densities:

> With the exception of a belt of evergreen-oak, *quercus agrifolia*, which margins the bay, and gives name to the several encinals [*encinal* being the Spanish word for an oak grove], a few groves of deciduous oak, *quercus*

sonomensis, and a small number of redwood trees in the mountains south of Suñol valley and east of Fruitvale, the county is at present poorly timbered. It was in a much better condition, in this respect, a few years ago. The redwood at one time grew to an enormous size in the mountains about five miles east from San Antonio.

However, it is also possible that early European clearing in the uplands may have been relatively limited. According to longtime resident, Tim Koopman, early settlers did not cut trees for most of the local hillside farming:

> If they had flats [flatlands], dotted with oak stands, they would cut those trees and then remove the stumps to make these open fields. But in the hillsides, they farmed around the trees, I can remember that, they farmed around the trees. You take a look at the Altamont, there wasn't an Altamont forest there at one time, but even in our home country here, there are reasons there aren't trees there ... mother nature's plan perhaps. We've got this riparian land here, with water and so the trees grow there, but for some reasons, whether its wind exposure, there wasn't tree growth there that was removed. (Koopman pers. comm.)

Figure 7.5. Use of photography to understand historical upland vegetation patterns. In addition to textual sources and survey data, we were able to use photographs such as this ca. 1910 image of the future Calaveras Dam site to explore historical vegetation patterns. Note the variety of vegetation types and densities visible in this image. (courtesy The Bancroft Library, UC Berkeley)

Vegetation change: results 1930s–2010

The most significant changes between the 1930s and 2010 were an increase in open water and decrease in sycamore alluvial woodland extent, both due to the construction of San Antonio Reservoir and the drowning of much of La Costa Valley (fig. 7.6). Blue oak woodland extent decreased by just over 10%, and coast live oak forest and woodland alliance extent decreased by 16%. Canopy tree cover maintained roughly

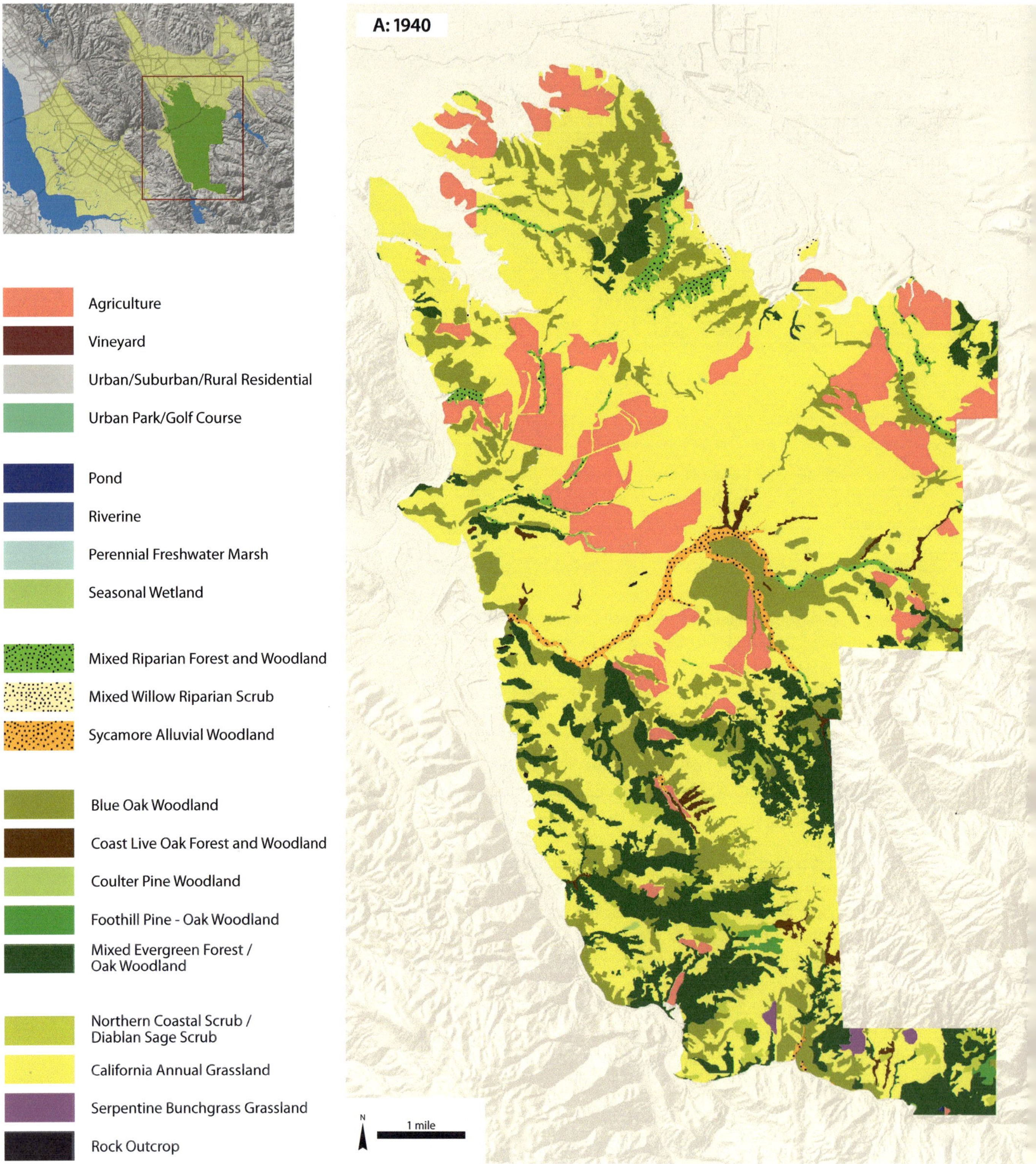

Figure 7.6. Adjusted land cover map: 1940-2010. The 1940 map was created using ICF vegetation mapping adjusted to 1940s aerials. We attributed the polygons with Wieslander VTM field data from the 1930s.

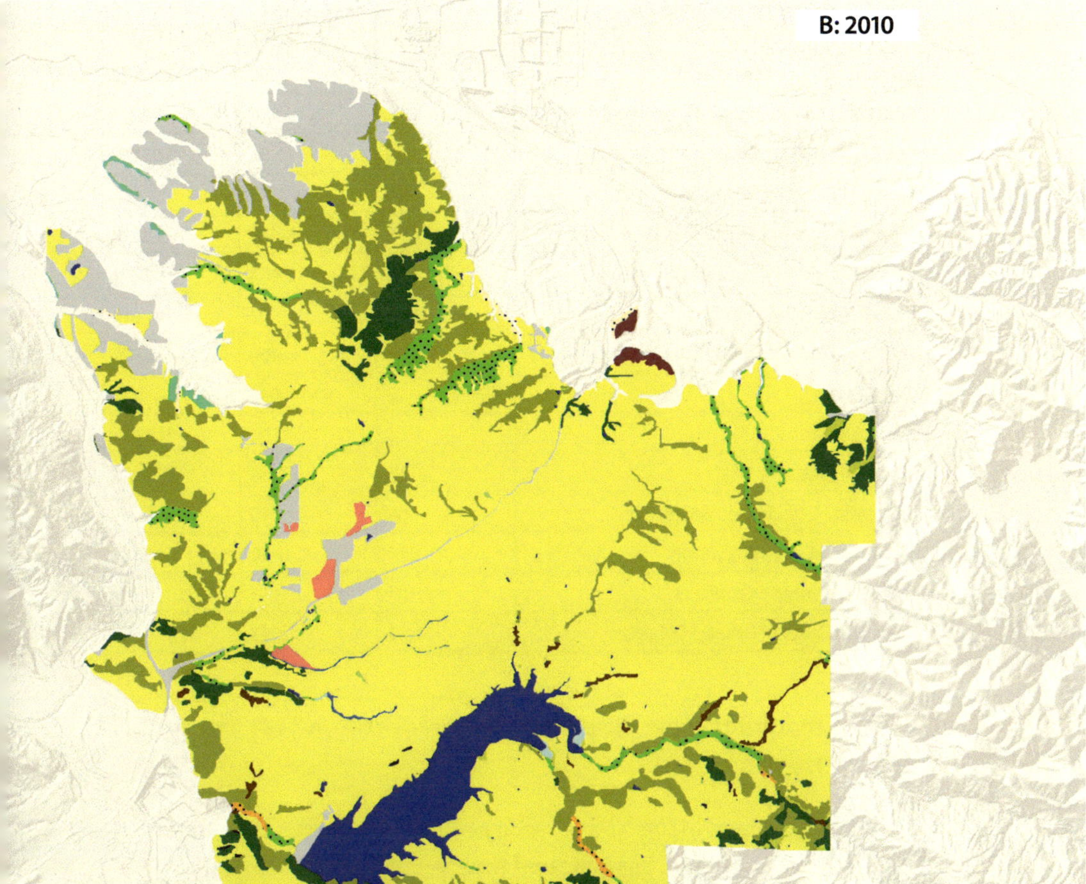
B: 2010

Table 7.1. Changes in land cover extent between 1930s and 2010.

Type	1930s (ac)	2010 (ac)	Percent remaining
Blue Oak Woodland	4,072	3,651	89.7%
California Annual Grassland	13,071	17,499	133.9%
Coast Live Oak Forest and Woodland	232	196	84.3%
Coulter Pine Woodland	56	56	100.0%
Foothill Pine - Oak Woodland	73	73	100.0%
Mixed Evergreen Forest / Oak Woodland	3,462	3,418	98.7%
Mixed Riparian Forest and Woodland	500	462	92.4%
Mixed Willow Riparian Scrub	27	27	100.0%
Northern Coastal Scrub / Diablan Sage Scrub	460	460	100.0%
Open Water	3	844	--
Rock Outcrops	8	10	114.2%
Seasonal Wetlands	12	4	29.6%
Serpentine Bunchgrass Grassland	65	65	100.0%
Sycamore Alluvial Woodland	311	52	16.7%
Dryland Farming / Cropland	3,105	116	3.7%
Rural Residential / Urban / Suburban	<10	1,367	--

the same footprint within the pilot project area between 1940 and 2010 (table 7.1).

While vegetation on higher slopes, such as Mixed Evergreen Forest and Foothill Pine Woodland, has appeared to stay constant both in spatial extent and composition, some lower elevation land that was farmed in the 1930s has since been returned to grassland with some grazing. This return to grasslands is particularly common in regional parklands.

While rural residential land uses and subdivisions increased from less than 10 acres to over 1,000 acres between 1940 and 2010 (fig. 7.7), dryland farming land uses decreased dramatically, and open space/grazing land increased by 33%. Interviews with longtime residents of the area suggested that the 1940s represented the height of agriculture in the region, and that after World War II, the diverse, small-scale agriculture practiced in the Livermore-Amador Valley was unable to keep up with the new globalizing market (Banke pers. comm.).

Some of the most noteworthy changes to upland vegetation may be the least visible. As in the rest of California, dramatic changes in grassland species composition occurred in the Livermore upland area shortly after Spanish contact (Mensing and Byrne 1999). Non-native wild oats, mustard, and other exotic grassland species were commonly found on California hillsides by the early 1800s. Species composition has continued to change with the more recent introduction of medusa-head *(Elymus caput-medusae)*, yellow star-thistle *(Centaurea solstitialis)*, and other invasive plants. The introduction of these invasives has caused native grasslands to become one of the more

threatened ecosystems in California (Holmes 1990, D'Antonio et al. 2000). These introduced taxa also benefit from reduced fire frequency, with dramatic growth and reproduction in the wet season, paired with long-lived, hardy seeds able to weather long periods of drought in a senescent state. Several residents reported observing rapid expansion of these exotics as well as fast-growing native shrubs upon the elimination of both fire and regular grazing.

We also found some dramatic changes in wetland extent. Acres of open water increased from under 10 acres to almost 850 acres. This increase was due in part to the construction of many small stock ponds developed for cattle ranching but was largely due to the construction of the San Antonio Reservoir (Turner Dam) in 1965 (fig. 7.8). The flooding of the La Costa Valley resulted in the loss of 84% of the sycamore alluvial woodland habitat in the study area (fig. 7.9). The San Antonio Dam blocks 37 square miles of watershed upstream of the USGS gage near Sunol.

In addition to inundating stands of sycamore alluvial woodland, the construction of the dam and flooding of the valley had many other effects on native species throughout the area. Longtime resident Paul Banke described the changes in the Arroyo del Valle gorge before and after the construction of the Lake Del Valle (fig. 7.10). We can infer similar wildlife losses in other flooded valleys:

> The ecological changes in terms of the different species was so remarkable when they put the [Del Valle] dam in … The quail all but disappeared. There used to be lots of dove in that area. The dove disappeared. The deer herd was never like it was before. There are still a lot of deer up there relative to other properties around, but nothing like it was when the creek was there, and I'd often wonder about that … They just disappeared. And my theory is that the creek was such a great area for the does to fawn in, and that their survival rate when they moved up into the hills really dropped. I don't know why, but it sure seemed that way. (Banke pers. comm.)

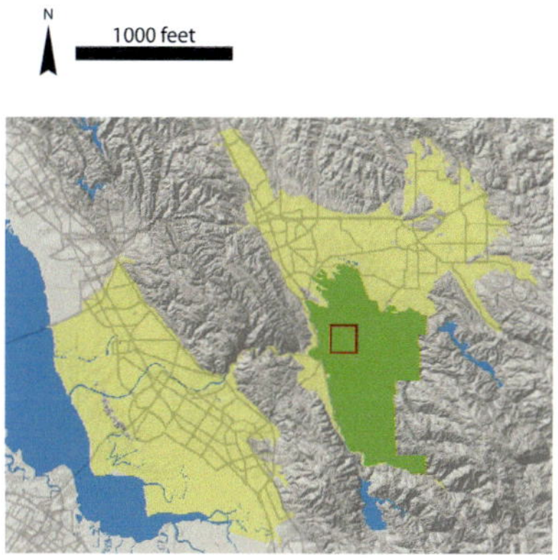

Figure 7.7. Example of the spread of rural residential development. These two aerials show an example of an area where residential development has grown since 1940. (USDA 1940; USDA 2005)

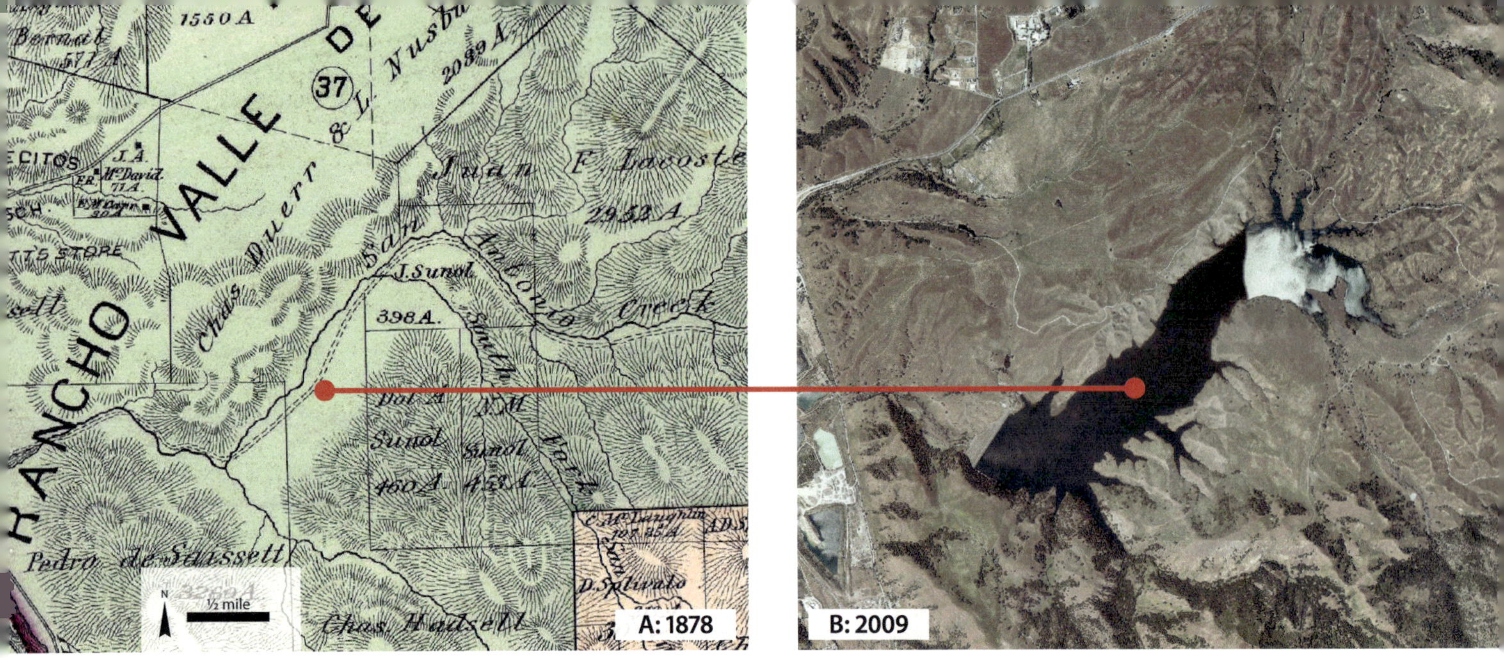

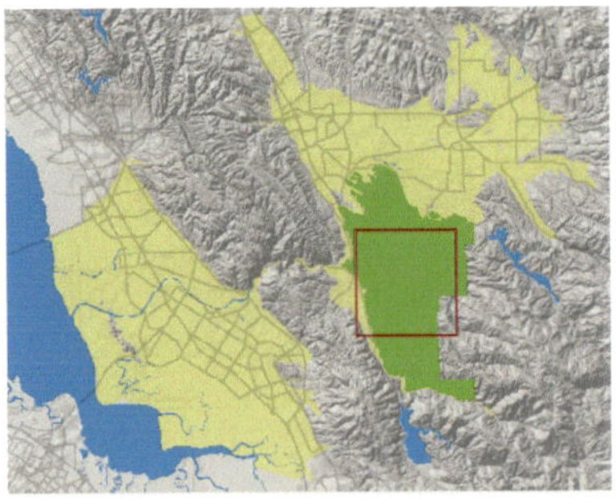

Figure 7.8. La Costa Valley and San Antonio Reservoir. (top) San Antonio Reservoir covers the area once known as La Costa Valley, shown on this 1878 map. (A: Thompson and West 1878, courtesy David Rumsey Map Collection; B: USDA 2009)

Figure 7.9. Sycamores on San Antonio Creek. (bottom) This ca. 1910 photo shows some of the large sycamores that lined San Antonio Creek prior to the construction of Turner Dam. (courtesy The Bancroft Library, UC Berkeley)

Figure 7.10. Oaks in former upland valleys. These ca. 1910 photos were taken to evaluate future dam sites. Both the site of San Antonio Reservoir (A) and Lake Del Valle (B) supported valley oaks and blue oaks. (courtesy The Bancroft Library, UC Berkeley)

Figure 7.11. GLO transect across a hillside. GLO survey notes overlaid on 1940 aerial imagery. These survey points indicate the presence of vegetation paterns different from those visible in 1940. The mismatched width of chaparral in particular may indicate a change in sucession patterns brought on by the end of managed fire. Surrounding survey points help confirm the placement of the points, which appears to be shifted slightly west compared to the aerials. (USDA 1940; Day 1853)

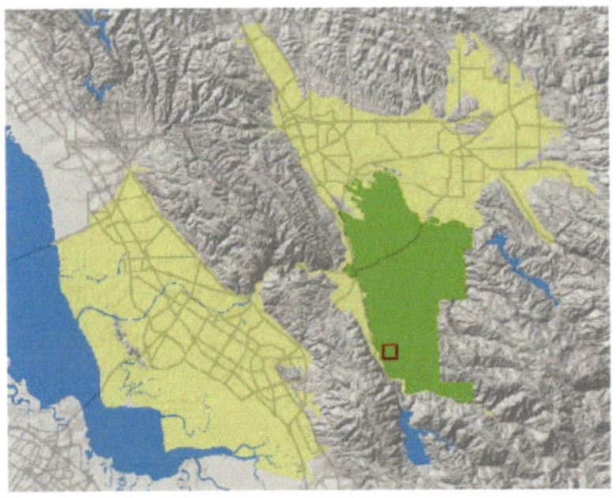

Hillslope change analysis

In addition to these broad analyses of vegetation change, we also conducted more focused analyses of localized change in two selected areas. These analyses allowed us to explore whether there were shifts in vegetation prior to 1930.

The first location we chose was a single north-facing hillside. Several GLO surveyors crossed this hillside, making it a promising site for pre-1930s analysis. GLO data is sparser in the upland areas than in the valleys: only three section lines traverse the upland pilot study area.

GLO data for this hillside provided an unusually detailed description of 1850s conditions (fig. 7.11). In May of 1853, GLO surveyor Sherman Day described the steep north-facing slope bordering Welch Creek as covered with chaparral. He described a "slope of mountain in midst of wild field of chapparal [sic], sand and shells … No bearing trees near" (Day 1853:220; May 23). He traversed the hill slope until he noted, "Cross creek 21 links [14 feet] wide in bottom of deep ravine, running northwest. Chaparral ends about 10 chains [660 feet] from section stake" (Day 1853:220; May 23).

We compared this GLO data with our two later datasets. Wieslander, in 1935, described this same north-facing hillslope as coast live oak woodland,

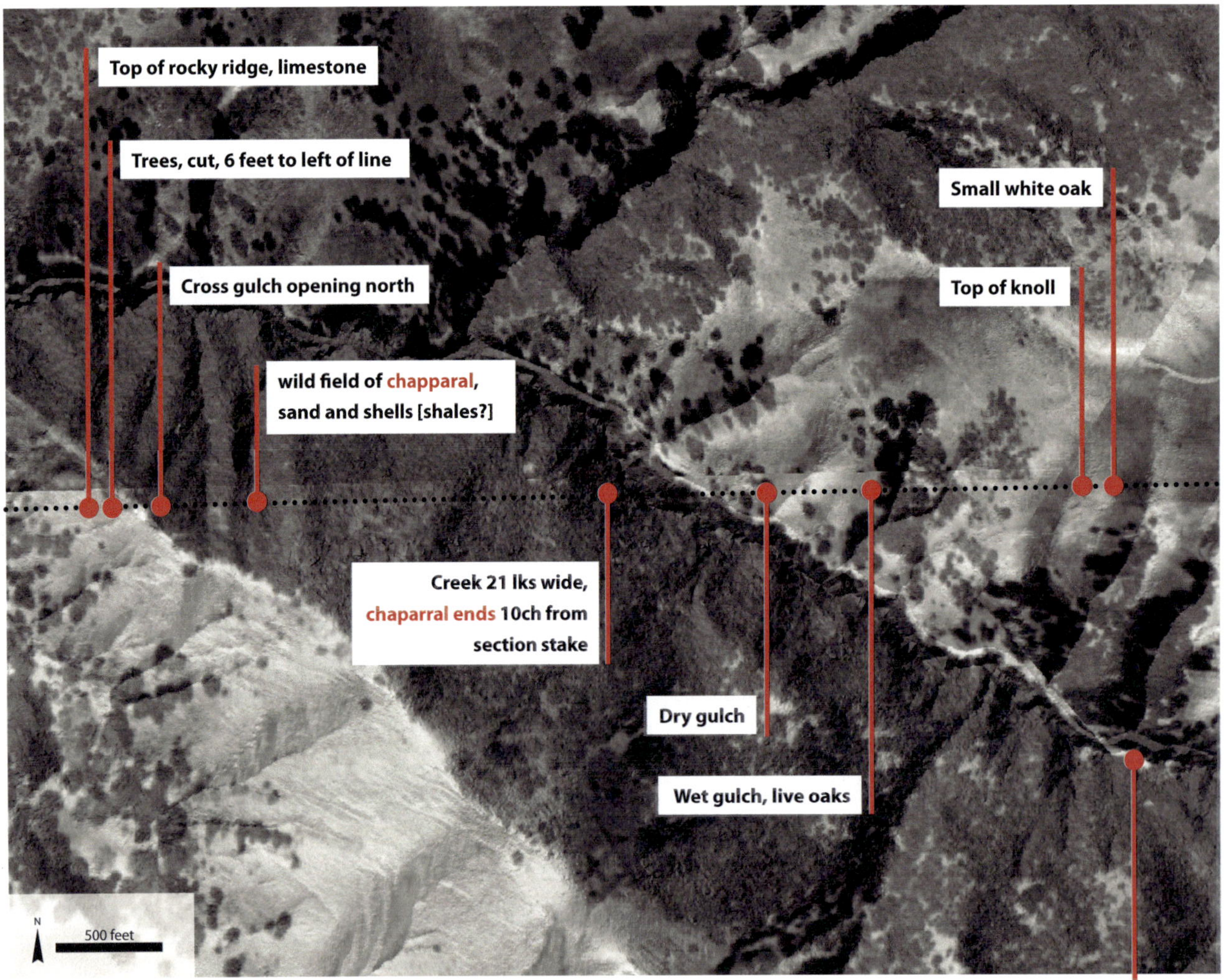

while the modern ICF mapping shows this area as mixed evergreen and oak woodland (Wieslander 1935, ICF International 2010b).

These data appear to describe linear plant succession. With the introduction of fire suppression, grassland/forb communities may be replaced by chaparral over time, which may in turn give way to oak woodland and evergreens (e.g., gray pines) (Haidinger and Keeley 1993, Russell and McBride 2003). However there is some debate over the pre-European condition of much of the East Bay hills, and whether their botanical character was dominated, in fact, by grasslands or by woody shrubs (Hopkinson and Huntsinger 2005).

Tribal peoples made widespread use of fire to modify plant succession and control large swaths of coastal California's terrestrial ecosystems, and as far back as the early to mid-Holocene may actually have been responsible for the proliferation of grasslands in much of the state (Keeley 2005). Tribes used fire to control the distribution of chaparral, maintain grassland cover and forage for wildlife, control pathogens, improve access to seeds and acorns, and aid in hunting rabbits, insects, and other small game—among a great many other purposes (Kroeber [1925] 1976, Keeley 2002, Anderson 2005, Lewis 1973). The complex patterns of fire management first observed by Europeans in the late 1700s had assuredly evolved over the millennia. However, our understanding of this practice and how it was applied—by which tribes, in which environments, at what frequency, for how long, and to what effect—is still very much a topic of continuing research and investigation.

Keeley (2005) and others (e.g., Hopkinson and Huntsinger 2005) have made the case that due to a paucity of grass-produced opal phytolith soil samples in the East Bay hills, those hills were more likely dominated by scrub rather than grasslands. However, it is also possible that grasslands may have had a high proportion of forbs (which tend not to produce phytoliths in abundance), a hypothesis supported by ethnobotanical records of forb use by tribes (Bocek 1984).

It is likely that the chaparral cover observed by GLO surveyors on these hillsides was the result of centuries of careful management. This specific north-facing hillside may have been a "wild field of chapparal, sand and shales" (Day 1853:220; May 23) as late as the 1850s, and may be representative of other north-facing slopes (now mainly mixed evergreen and oak woodlands) that have been managed under a more suppressed fire regime during the latter half of the 19th century and the 20th century and have also made this transition. However, our conclusions are based on only a single location in the study area, and more research would have to be done in order to determine whether other north-facing slopes presently covered with dense oak and evergreen woodland historically cycled through a chaparral/oak woodland burn cycle as well. If in fact this were the case, there may be many implications for managing these areas for species of concern that favor chaparral, such as the Alameda whipsnake *(Masticophis lateralis euryxanthus)*.

Though our hillslope analysis focused on just a single location, we hypothesize that the end of native burning allowed oak woodland to replace chaparral in the steep upland terrain, and provided an opportunity for succession to the evergreen/mixed oak woodland that we find today. This change might also have been due in part to the decline of dryland farming on the steep hillsides.

Valley change analysis

The second site we chose for a detailed local analysis was La Costa Valley. The northern half of the pilot study area is dominated by low-sloped alluvial or downdropped valleys, including La Costa and Vallecitos (fig. 7.12a). In the aerial imagery from 1940, differences in vegetation cover between the northern and southern parts of the study area are great (see figure 7.12). The lower slopes of these valleys have very few trees compared to the densely wooded north-facing slopes of the steeper southern part of the study area.

The timber is very scattering, being composed of live oak and pine, with generally a very sparing growth of underbrush. Over a large area the hills are perfectly bare of any timber growth, the timber and brush being the most plentiful upon the north slopes … From a consensus of opinion; it seems that the soil is not as good as it was years ago. The effect of the cattle and sheep tramping over the steep hills is to cut up the soil. The winter rains wash the loose matter off, thereby preventing the old time growth of grass. To these causes may also be attributed the fact that the growth of young timber and brush is very small.

—WAGGONNER 1906:1-2

For this analysis, we used aerial photography and other supporting sources to explore historical tree density. Using individual trees visible in the 1940 aerials, we calculated oak and sycamore densities, focusing on areas near or under the current San Antonio Reservoir (La Costa Valley), as this was the portion of the study area that experienced the greatest loss of valley oak habitat. We examined 1940 tree density and its relationship to soil type, geology, and slope in order to estimate pre-modification oak density in heavily farmed and grazed areas. Based on these calculations, we found that of the 3,884 acres that make up La Costa and Vallecitos valleys, 700 acres with remnant oaks had an density of 2.12-2.53 trees/acre in 1940 (USDA 1940; fig. 7.12b,c).

There were few data sources pre-1940. GLO surveyor Sherman Day described the hills north of La Costa Valley in July 1853 as "covered with square broken stone of granwacki [greywacke], quartz and porphyry … steep gorges. Timber sparse" (Day 1853:261; July 16). We found little other evidence describing which plant communities were established in La Costa and Vallecitos valleys. Early maps show wide alluvial valleys with multi-threaded channels coursing through the lowland areas (e.g., USGS 1906).

Wieslander (1933) mapped vegetation cover on the upstream and downstream ends of the wider part of La Costa Valley as valley oak alliance with sycamore alluvial woodland. ICF International (2010b) mapped the few oaks that remain on the edges of the reservoir today as blue oak woodland. These areas have the same soil type and underlying geology as small alluvial basins. We hypothesize that the density of blue and valley oaks in other small alluvial depressional valleys in this area (including the rest of La Costa and Vallecitos valleys) might have been similar to densities shown in the 1940 aerial (which shows valley oaks established closer to

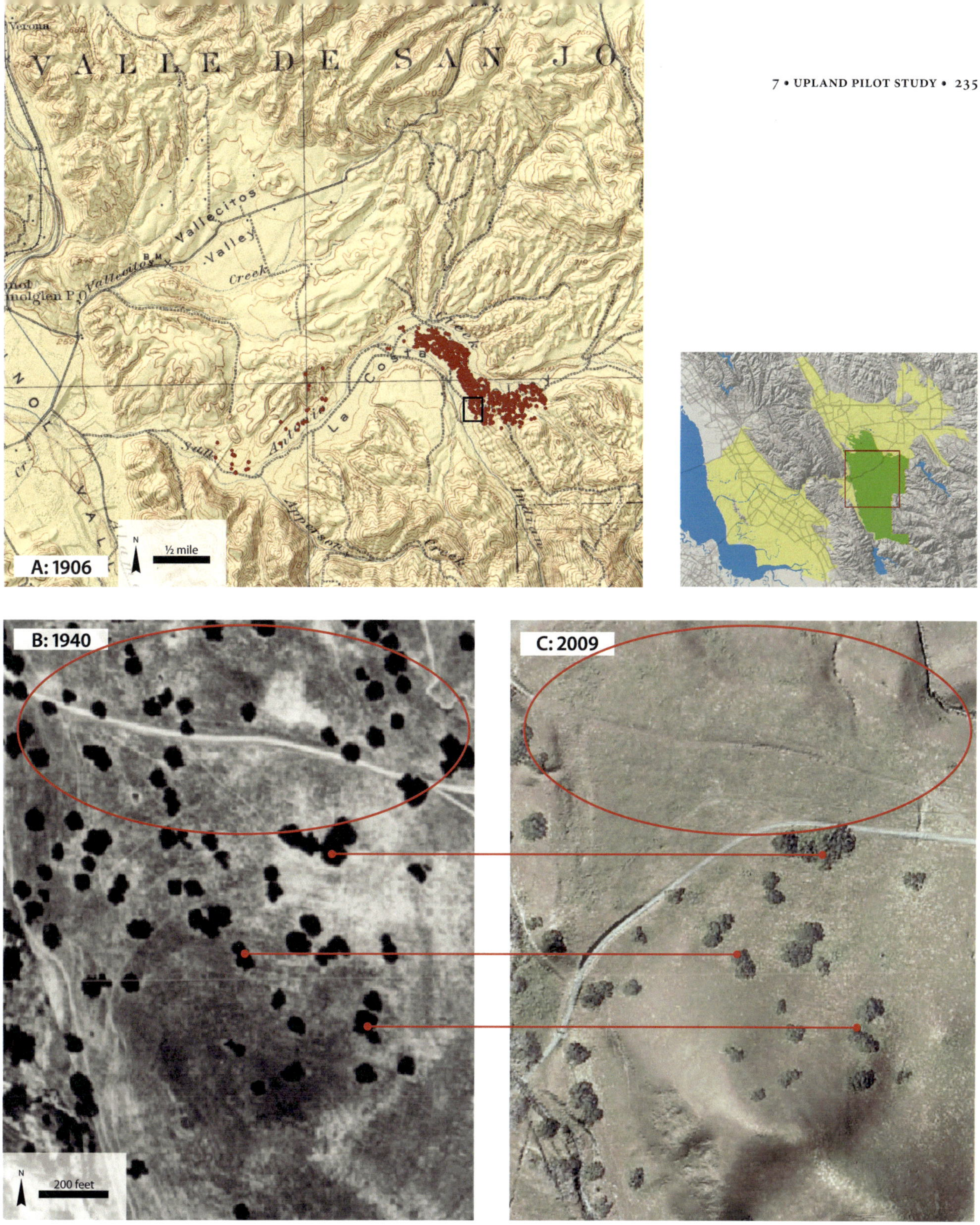

Figure 7.12. Oaks in La Costa and Vallecitos valleys. (A) USGS map showing the two valleys in 1906. The red dots indicate valley oaks visible on the floor of La Costa Valley in the 1940 aerial. (B) A zoom in to the historical aerial in the area within the black box on (A) shows these oaks in 1940. Compare the same area in 2009 (C). Even outside of the area now covered by the reservoir, many oaks present in 1940 have been lost today. For those trees that remain, canopy size clearly increased between 1940 and 2009. (A: USGS 1906; B: USDA 1940; C: USDA 2009)

OAK REGENERATION. RICH FLETCHER, LONGTIME RESIDENT:

One thing we've noticed about oak trees is that you seldom see successful germination and growth, except in brush. A major reason for the lack of oak regeneration is that acorns are highly desirable as food and many animals seek them out. Therefore, acorns in open grassland seldom sprout before something finds them. Everything seems to eat acorns, from insects and birds to deer and cattle. When trees successfully sprout in open grassland, they are also highly desirable as food for many animals. Grasshoppers, squirrels, rabbits, deer and cattle all feed on young oaks and in the open the young trees are very vulnerable. Trees that manage to grow to a few feet in height, are also vulnerable to rubbing and thrashing by deer and cattle reducing their chance of making it to maturity even more. On the other hand, some acorns are deposited in brushy areas by birds or other critters, where they are more likely to sprout and grow while being protected from the dehydrating effects of direct sunlight. The small oak trees are protected and sheltered by the brush, which allows them to reach a substantial height without being eaten or knocked down by rabbits, deer and/or cattle.

channels than blue oaks). However, historical patterns may have differed even from this 1940 picture. By 1940, some oaks present historically may have already been cleared for cropland.

Land management analysis

In this section, we explore the history of road development, farming, and land management dating back to the earliest European settlers of the area. We quantify changes in road density and describe impacts that roads, population, grazing, and fire would have had on the upland pilot area, with a focus on implications for future land management. An increase in farming in the late 19th century and early 20th century, along with higher density rural residential populations and active dryland farming, decreased fractal dimensions and patch sizes in the upland areas which would have had implications for species and plant communities across the landscape (Boren et al. 1997).

Road network analysis

Roads and paths can act as barriers to animal passage and disturb aquatic and upland environments by increasing habitat fragmentation and wildlife mortality (Case 1978). The decline of many populations of species of concern is directly linked to habitat encroachment and fragmentation due to road construction and urban development (ICF International 2010b).

Roads are often the "pioneer vector" of land management, leading the way for farming, ranching, subdivisions and other development. Road construction results in more edge conditions and disturbed areas, and has effects on habitat connectivity, hill slope processes, and vegetation. Road crossings of streams can cause increased erosion both downstream and upstream from the crossing, and often act as a barrier to wildlife using the riparian corridor for migration. Road networks, especially ungravelled roads, can function as conduits to creeks during rain events, conducting fine sediment and excess water directly to the creek and changing the timing and intensity of the hydrologic system (Wemple et al. 1996). This can destabilize stream banks and increase erosion. In low flow events, certain types of road crossings, especially culverts, can act as a barrier to fish passage and thereby limit upstream habitat access (Warren and Pardew 1998, Gibson et al. 2005).

We used a number of current and historical sources to quantify the change in road density in the upland pilot study area. We digitized roads shown on early land ownership maps drawn by Thompson and West (1878), mentioned in GLO surveys (1850s), depicted on an early USGS quadrangle (1906), and indicated on USDA aerial photos (1940). Roads present today were mapped by the SFPUC (fig. 7.13). We then used spatial analysis tools in a GIS to calculate the change in road density over time, as well as quantify the change in the number of road stream crossings over time (fig. 7.14).

Earlier transportation corridors, such as Indian trails, stage roads, and skid trails were not included in the scope of this analysis for lack of available data sources. While pre-colonial Native populations in the area were relatively high, the trails and paths they created and used over centuries functioned very differently from modern transportation corridors, and did not likely create similar stresses or disturbances to landscape functionality.

The analyses show that a major increase in road density in the upland pilot area occurred between 1878 and 1906 (table 7.2). Density of roads and stream crossings continued to increase into the 20th century, but the initial large changes in habitat fragmentation and patch size may have occurred during the initial phase of population increases in the late 19th and early 20th centuries, when many roads were built. We can guess that the early roads were built as major throughways, and that as the area became more densely populated and more heavily used as farm land or ranch land, roads were constructed to provide access to individual land holdings.

We also found some remarkable persistence in road location. As early as 1853, Sherman Day (1853:260) described that "A large trail runs through the hollow into Sunol's Valley." In 1874, surveyor George Allardt (1874) labeled it "the road to Patterson farms." This road was in the same location as the main road crossing La Costa valley in 1878 (Thompson and West 1878), and is the current location of Highway 84 (Vallecitos Road).

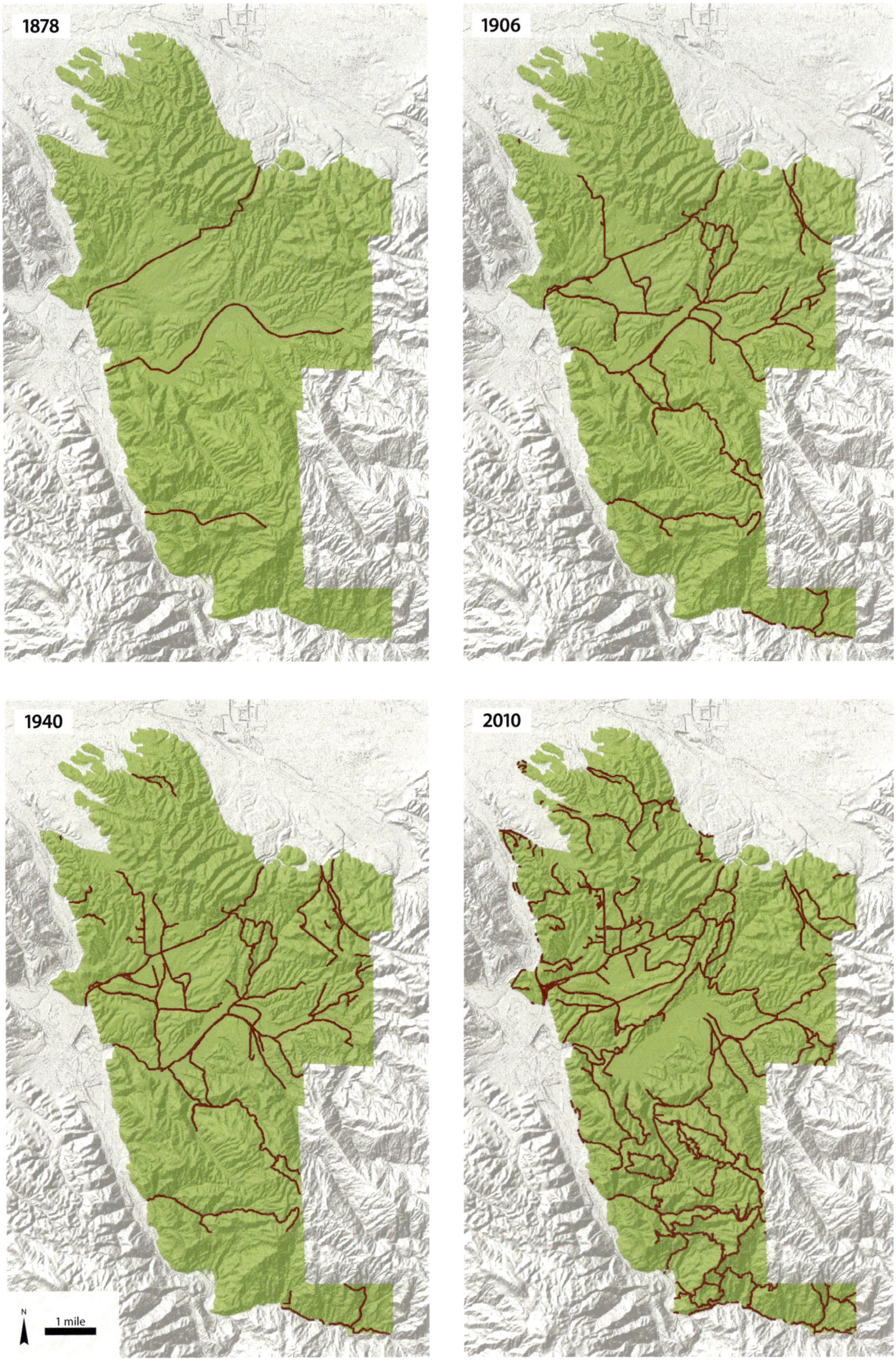

Figure 7.13. Changes in road network density over time. These maps show the road network over time, tracking the major increase in roads due first to farming at the turn of the 20th century and then to the boom of population and suburban development in the late 20th century.

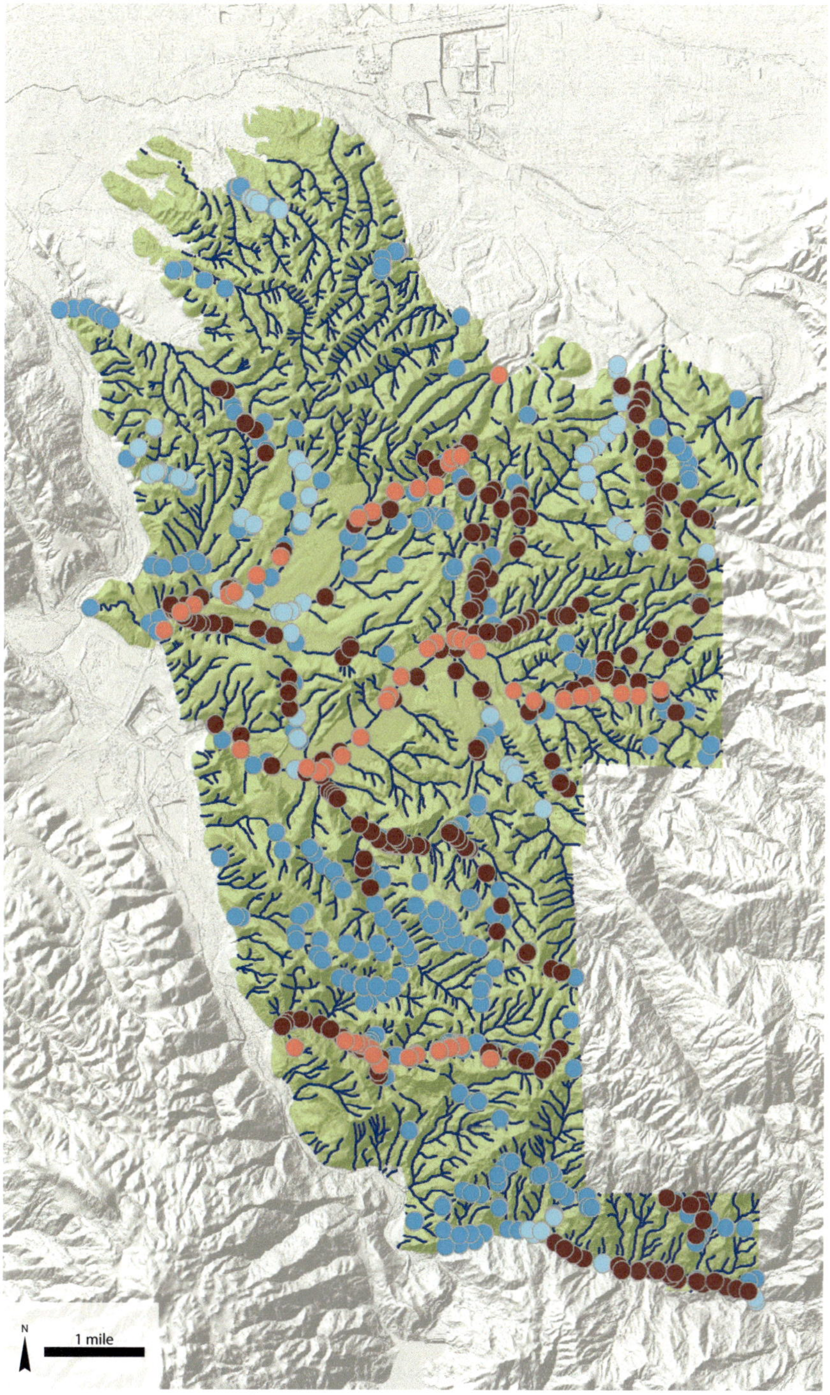

Figure 7.14. Spatial display of stream crossings in the upland pilot. This map summarizes the number of intersections at which roads crossed streams through the late 1800s and 1900s. Although a number of roads were constructed very early, more road crossings exist today than even during the peak of farming in the 1940s. Many of these roads may be unused today, although they likely still affect stream processes. Road crossings can impact stream dynamics and aquatic habitat in a number of ways, including causing channel bed and bank erosion, fish barriers,

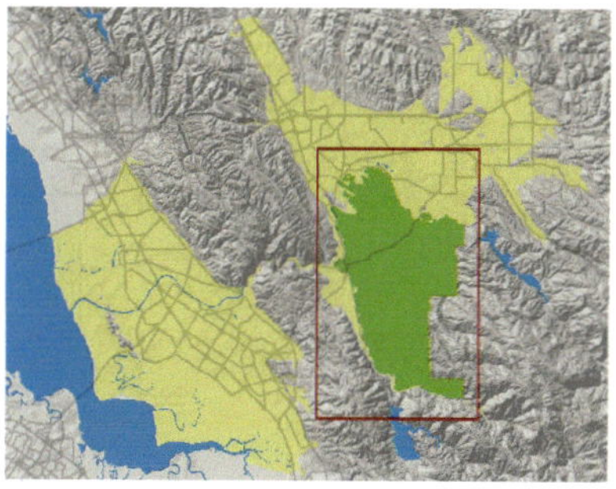

Table 7.2. Summary of road construction: 1878-2009.

Year	Miles	Density (linear road/sq mile)	Intersections with streams	Source
1878	14	0.32	54	Thomson and West 1878
1906	53	1.20	222	USGS 1906
1940	68	1.54	275	USDA 1940
2009	129	2.92	395	SFPUC data set

DIVISION OF THE UPLANDS

I've done some reading about the early history of our ranch, and as the gold rush era ended, a lot of people were looking for a place to live and a way to make a living. Some of these people took their animals to the hills and squatted on government owned land where there was free range. The Homestead Act of 1862 provided an opportunity for many of those same people to legally acquire title to 160 acres of this land. Our ranch, which now has two houses, and only one full-time residence, at one time had as many as eight separate homesteaders. The number of people living on the land up there was a lot higher 125 years ago than it is today.

—RICH FLETCHER , PERS. COMM.

A further step in this analysis would be to perform landscape ecology metric evaluations on these road networks and compare them to species home ranges. Such analysis could help to determine whether endangered species are being affected by the fragmentation of habitat in these upland areas and if so, whether there are unused or unnecessary roads or road crossings that can be removed and restored, thus increasing corridor length or patch size for upland habitat. The existence of a dense but unused road network (and associated stream crossings) could have an adverse effect on fish habitat quality, as well as upland species home ranges.

Farming and ranching history through interviews

Upland areas were rarely visited by GLO surveyors, explorers, and other early observers. As a result, maps, historical survey notes, and old photographs are few and far between. However, some descendants of the families that settled the area in the late 1800s are still living in the area, and their memories can supplement physical records. In partnership with the Alameda County Resource Conservation District, we interviewed four long-time residents of Livermore-Amador Valley, focusing on individuals with a long family history of occupation or landholding in the upland pilot area. We discussed their family histories, their memories of the area, and their grandparents' recollections of the landscape, as well as changes they have noticed over time.

The residents remarked on the type of farming that was done in the upland areas of Alameda County around the turn of the 20th century, noting that there was a lot of farming in the upper hills, mainly done with teams of horses. For example:

> Even where I am on the Patterson ranch, as steep and remote as this … there were whole areas farmed with horses, and this is all country that I

Figure 7.15. Abandoned cabin. This cabin belonged to John Rodrigues. (courtesy Rich Fletcher)

> would never take a tractor on. I mean, it's steep. (Banke pers. comm.)
>
> One of the things that did go on in the early days was that a lot of the hills were farmed. Hills that are very topographically challenging, pretty steep, but there was a lot of hill land farming, mostly done with teams of horses at that time. You can trust a team of horses to pull equipment better than you can trust a tractor. There's a little sense about a horse that a tractor doesn't have. I know that there was an awful lot of farming going on in the dryland hills. (Koopman pers. comm.)
>
> There was a time when all of the Altamont and all that area north of 580, it was all dryland farming. All the way to the Contra Costa border. It's all pasture now, it used to be dryland farming. You see a little bit out there, but not how it used to be. (Sachau pers. comm.)

In the 1940s, at the time of the earliest aerial photography in the study area, grazing and agricultural land uses comprised a minimum of 16% of the upland pilot study area (over 3,000 acres), including on steep hillsides and ridgetops. Many people agreed that the 1930s constituted the height of agriculture and dryland farming in the region (fig. 7.15). It was also a time when subsistence farming was integral to the development of local economies, which decreased through the 20th century as "houses became more a valuable crop than vegetables," and grain grown in Livermore-Amador Valley could not compete with the cheaper prices of imported goods (Banke pers. comm., Koopman pers. comm., Sachau pers. comm.).

ON UPLANDS POPULATIONS

Evidence on our ranch indicates that early ranchers used convenience fences of barbed wire and natural barriers to control their grazing animals. We have several sites where cabins either fell down in disrepair or burned down. They were small buildings, maybe 15 feet by 20 feet with one room, a bed and stove build near a spring for a water source. Typically there was one of these homes per quarter section of land. Accessed by the trails that are today's roads, the routes are most often along paths of the least resistance.

—RICH FLETCHER, PERS. COMM.

Conclusion

Through this upland pilot, we explored data sources such as early GLO surveys, textual descriptions, land use history, aerial photographs, road maps, Wieslander mapping, and interviews with longtime residents. Using these sources, we were able to document evidence of possible shifts in vegetation community makeup, although overall patterns have remained remarkably constant since the 1930s. Prior to the 1930s, sources are relatively limited, and we were not able to develop a comprehensive picture of land cover in the pilot study area. However, we were able to use some data rich areas to speculate about succession, land management, and the impact of fire pre-1930s. Given the history of agriculture and land use, most of the significant recent changes in vegetation composition and distribution likely occurred between the 1850s and the 1930s, during the height of agricultural expansion and population growth in the upland areas of Alameda County.

Fire management in these hills may have contributed to a recurring cycle between grassland, scrub, woodland, and evergreen forest. One GLO survey transect hinted at the different composition of upland vegetation historically, and the dynamic nature of the uplands under native management. The recent expansion of shrubland in the Bay Area may be the result of the combined effects of fire suppression and the expansion and then contraction of high density grazing (Hamilton 1997, Davis and Borchert 2006).

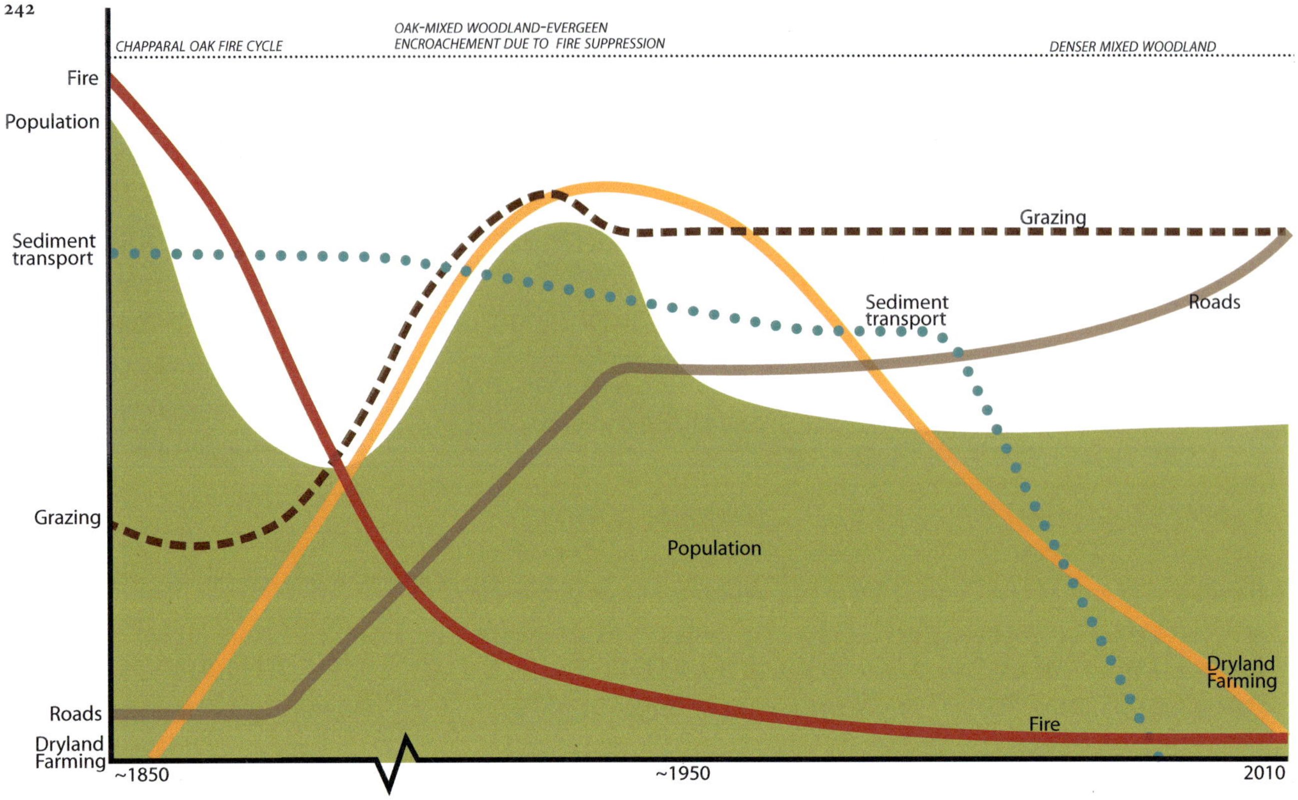

Figure 7.16. Land use changes over time. This conceptual diagram estimates relative intensities of land uses, management regimes, populations, and natural drivers in the upland pilot area.

More recent land use changes have also impacted the uplands. Population, road building, and the influence of dryland farming peaked in the 1940s and have been on the decline since, though ranching has continued to help maintain open grasslands in the absence of fire. These changes in land use have varying implications for management and species recovery (fig. 7.16).

Implications for species of concern

Changes in land use and vegetation patterns impact the species that depend on upland habitats. Some species benefit from open grasslands, while others need scrub or woodland. For example, sycamore alluvial woodland, which has been reduced to only 16% of its 1930s extent in the upland pilot study area (due largely to the construction of the San Antonio Reservoir), thrives in summer-dry, flashy streams, a habitat type that may be less beneficial for over-summering salmonids. Favoring any one of these habitat types risks protecting one suite of species at the expense of others.

Here we focus on impacts to a few of the species of concern listed in the Eastern Alameda County Conservation Strategy: whipsnake *(Masticophis lateralis euryxanthus)*, badger *(Taxidea taxus)*, and San Joaquin kit fox *(Vulpes macrotis mutica)*. All of these species have suffered habitat fragmentation associated with roads and urban development. We can use our understanding of the changes to upland land cover patterns to explore the potential effects of management decisions on these species of concern. How have these species been impacted by fire suppression, the conversion of chaparral to oak woodland, or other land use and management patterns?

The Alameda whipsnake *(Masticophis lateralis euryxanthus)*, listed as threatened at the state and federal level, occurs in coastal scrub and chaparral communities, but also forages in other communities such as grasslands and woodlands (ICF International 2010b). Fire suppression negatively impacts the whipsnake by creating a closed scrub canopy which often reduces the diversity of microhabitats that whipsnakes require (ICF International 2010b).

The American badger *(Taxidea taxus)* is considered a special-status species in California. Badgers occurred in short grass and dry pasture, as well as some scrub habitat near Del Valle Reservoir. Badgers thrive on open or sparsely vegetated ground amenable to digging burrows (Grinnell et al. 1937:373). Badgers would require relatively open, grassy expanses.

The state and federally endangered San Joaquin kit fox *(Vulpes macrotis mutica)* has experienced the almost total elimination of its valley floor habitat (ICF International 2010b). The kit fox prefers foothill grasslands and valley oak savanna.

How we choose to manage the upland areas of watersheds directly impacts the fate of these species. Prior to current management, these upland systems were dynamic, responding to periodic burning that would favor each of woodlands, grasslands, and scrub in turn. The static system present today risks favoring one of these types at the expense of the others. Determining historical levels or populations of species of concern and further exploring early vegetation patterns would help us fully understand the upland environment under Native management.

Future directions for study

This pilot project merely scratches the surface of the types of ecological analyses that could be useful in piecing together the story of the Alameda County uplands. Several types of analyses could be done using data sources we have already uncovered, and several new methods of data collection and analysis could be added to enrich the historical narrative.

1. FURTHER EXAMINATION OF VEGETATION CHANGE.

While we were able to detect change between the 1940 aerial attributed with Wieslander data and the 2010 vegetation mapping, we were fairly unsuccessful in mapping the habitat composition of upland areas in the 1800s. We were able to extrapolate to estimate vegetation patterns in areas with agriculture and cattle grazing, but we are less certain of what the grasslands and south-facing slopes might have looked like before European contact. While it is possible that these descriptive data do not exist, more time could be spend in data collection with the goal of reconstructing habitat types and extent from the 1800s. If an 1800s era map could be created, spatial analysis of fractal dimensions could be performed, including patch size, edge-to-area ratios, and other analyses comparing habitat structure across time frames. This would give current managers a way to set goals for current species management given the existing conditions within a historical context.

2. FURTHER EXAMINATION OF ROAD IMPACTS

With the information about road impacts gathered from this pilot study, several management tools could be implemented or explored. An assessment of the impact of road crossings on stream incision, degradation, and erosion could help managers target areas that might support increased habitat area for resident fish populations and other aquatic organisms. Landscape metrics describing changes in patch sizes due to road and fence construction would help prioritize areas for conservation. Strategically closing unused roads, or removing unused or degraded culverts to regain sufficient corridors or habitat patches could greatly help species conservation in the area.

3. FURTHER EXAMINATION OF FIRE MANAGEMENT IMPACTS

Field data collection of tree cores and phytoliths could help managers understand the role of fire and the historical extent of grasslands. This could help managers evaluate how to best manage grazing and fire for species of concern.

4. FURTHER INTERVIEWS WITH LONG-TIME RESIDENTS

Finally, some of the richest information we gathered in this upland pilot study was from the time spent with residents of the Livermore Valley whose families have been in the area since the mid-19th century. Additional oral interviews with these and other "old timers" would flesh out the picture of the 19th and 20th century eastern Alameda County uplands to inform our understanding of the present.

Upland pilot study summary

Vegetation footprints within the upland pilot have not changed significantly since the 1930s. Oak dominated hillsides have remained relatively stable in area since the 1930s, but some previously agricultural areas have been taken over by grasslands. The height of agriculture in the uplands of Alameda Creek watershed was in the late 19th and early 20th centuries. The most significant changes between the 1930s and the present are an increase in open water and decrease in sycamore alluvial woodland area, both due to the construction of reservoirs.

The vegetation make-up of hillside communities in the upland pilot area has changed since the 1930s. Chaparral on the north-facing slopes transitioned to oak woodlands by the 1930s. As fire management became infrequent these areas have continued the process of succession to evergreen/oak woodland.

Roads and fences have dissected upland habitats, even in publicly owned lands and parks. Smaller patch sizes and fragmented habitats threaten several endangered species, including the Alameda whipsnake and the San Joaquin kit fox.

Creek culverts and stream crossings can interrupt habitat corridors and limit the movement of aquatic species. Steps could be taken to evaluate roads and close those that are no longer in use. Additionally, examining road crossings for fish passage barriers and removing those roads that are not useful could decrease fragmentation.

The majority of upland areas within the upland pilot have been dominated by low herbaceous cover since at least the Holocene, and perhaps earlier. Since the arrival of the Spanish, native bunchgrasses and forbs have been replaced by European annual grasses, which continues today with the invasion of such species as yellow star thistle, and other noxious weeds. Efforts should be taken to control these invasive weeds, or manage them with fire or grazing.

Native bunchgrasses and forbs likely each dominated specific habitats of the historical landscape. Forbs and woody shrubs may be more suited to certain hilltops or south-facing slopes. Little is known about the composition of grasslands in the early 1800s and before. Species composition likely shifted over time—the arrival of the California Indians in the late Pleistocene may have initiated a shift to open grasslands. This is an ongoing field of study and has implications for managing certain endangered species.

Understanding of historical conditions should be used to direct grazing and burning as management tools. Currently, in much of the uplands, grazing is the primary management tool and there is little burning compared to native management practices. Land managers are under pressure to manage the uplands for endangered species, which may benefit from different combinations of burning and grazing practices. When managing for certain species, the desired ratio of woody plants to open grassland should be considered to determine the appropriate burning/grazing regimes. For example, the expansion of woody vegetation from lack of burning affects the range of the kit fox by reducing the extent and patch size of available open grassland. In contrast, conversion of chaparral to either woodland (due to decreased burning) or grassland (due to increased burning or grazing) may impact Alameda whipsnake *(Masticophis lateralis euryxanthus)* by removing essential habitat.

8 • FISHES OF THE ALAMEDA CREEK WATERSHED

by Robert A. Leidy and Bronwen Stanford

Native fish assemblages

We combined historical evidence describing habitat conditions and historical and recent records of fish within the Alameda Creek watershed to reconstruct probable native fish assemblages associated with major habitat types. Streams, lakes, ponds, wetlands, and other aquatic habitats within the Alameda Creek watershed are part of the Sacramento-San Joaquin Fish Province (Moyle 2002). The Alameda Creek watershed historically supported many of the species of freshwater and saltwater dispersant, as well as endemic, fishes found in the Great Central Valley and Sacramento-San Joaquin River Delta (Leidy 2007; table 8.1, see summary table 8.4).

At least 41 native fishes likely historically characterized the Alameda Creek watershed and its adjoining tidal marshes, slough channels, and shallow near shore bay environments. Fishes that historically would have been found in freshwater habitats within the Alameda Creek watershed include Pacific lamprey (*Lampetra tridentata*), Pacific brook lamprey (*L.*cf. *pacifica*), river lamprey (*L. ayresii*), thicktail chub (*Gila crassicauda*), hitch (*Lavinia exilicauda*), California roach (*L. symmetricus*), Sacramento blackfish (*Orthodon microlepidotus*), Sacramento splittail (*Pogonichthys macrolepidotus*), Sacramento pikeminnow (*Ptychocheilus grandis*), speckled dace (*Rhinichthys osculus*), Sacramento sucker (*Catostomus occidentalis*), coho salmon (*Oncorhynchus kisutch*), Chinook salmon (*O. tshawytscha*), rainbow trout/steelhead (*O. mykiss*), threespine stickleback (*Gasterosteus aculeatus*), Sacramento perch (*Archoplites interruptus*), tule perch (*Hysterocarpus traski*), prickly sculpin (*Cottus asper*), and riffle sculpin (*C. gulosus*)(Buchan et al. 1999, Gobalet et al. 2004, Leidy et al. 2005a, Leidy 2007).

Native fishes that would have characterized tidally influenced aquatic environments include Pacific lamprey, river lamprey, leopard shark (*Triakis semifasciata*), brown smoothhound (*Mustelus henlei*), bat ray (*Myliobatis californica*), white sturgeon (*Acipenser transmontanus*), Pacific herring (*Clupea pallasii*), longfin smelt (*Spirinchus thaleichthys*), northern anchovy (*Engraulis mordax*), coho salmon, Chinook salmon, steelhead, jacksmelt (*Atherinopsis californiensis*), topsmelt (*Atherinopsis affinis*), threespine stickleback, Pacific staghorn sculpin (*Leptocottus armatus*), white croaker (*Genyonemus lineatus*), shiner perch (*Cymatogaster aggregata*), longjaw mudsucker (*Gillichthys mirabilis*), arrow goby (*Clevelandia ios*), bay goby (*Lepidogobius lepidus*), bay pipefish (*Syngnathus leptorhynchus*), plainfin midshipman (*Porichthys notatus*), English sole (*Parophrys vetulus*), speckled sanddab (*Citharichthys stigmaeus*), diamond turbot (*Hypsopsetta guttalata*), California halibut (*Paralichthys californicus*), and starry flounder (*Platichthys stellatus*) (Stevenson et al. 1987, Gobalet 1992, Moyle 2002, Leidy 2007, Hobbs 2011, Moyle et al. 2012).

Many of these fishes are either entirely missing from the watershed today, or are present in very low numbers (see Leidy 2007 and Leidy et al. 2011 for discussions of the status of freshwater fishes in the Alameda Creek watershed). Comparing historical records with recent fish collections shows

Figure 8.1. Paired views of creek habitat and native fishes. Top photo: Lower Alameda Creek between Alvarado and Decoto roads (May 1915; courtesy San Francisco Public Utilities Commission). Inset: Sacramento perch (CAS 26724), collected in 1957 from lower Alameda Creek near Niles, opposite California Nursery Co. (specimen photographed by Jon Fong, California Academy of Sciences.) Bottom photo: Western sycamores adjacent to San Antonio Creek, a larger intermittent stream in the upper Alameda Creek watershed (ca. 1910; courtesy The Bancroft Library, UC Berkeley). Inset: Steelhead pictured with San Francisco Water Department pipeline surveyors within the upper Alameda Creek watershed, circa 1935 (photo courtesy East Bay Regional Park District).

a change in the species distribution of freshwater fish within the Alameda Creek watershed from a dominance by native species to dominance by non-native species. For example, until the 1950s native fish species accounted for over 90% of total fish species within freshwater environments of the Alameda Creek watershed. By 1953-1969, the percentage of total species represented by native species had dropped to 61%, and by 1972-1987 the percentage of native species had further dropped to 46% of total species. Currently, native fish species comprise about 46% of the total fish species found in the watershed (19 of 41 total fish species)(table 8.1). Interestingly, the dramatic increase in the number of non-native fish species in the watershed is correlated with the completion of Del Valle and San Antonio reservoirs and the South Bay Aqueduct in the mid-to-late 1960s. These water projects resulted in water inputs from the Central Valley and the stocking of reservoirs with non-native fishes for recreational purposes, both of which appear to have greatly facilitated the introduction of non-native fishes into the Alameda Creek watershed.

The thicktail chub is extinct, and the Sacramento splittail and coho salmon have been extirpated from the Alameda Creek watershed. A number of other species have also been severely impacted by human modifications. Sacramento perch and tule perch, while present in the watershed, persist in relatively small numbers and have very restricted distributions. The status of river lamprey, riffle sculpin, and speckled dace is uncertain, as these fishes have not been recorded in the watershed in recent years. As further discussed below, Alameda Creek and its tributaries historically supported anadromous rainbow trout, or steelhead (Leidy et al. 2005a). While resident rainbow trout are locally abundant in the headwater streams of the Fremont Hills and Diablo Range, steelhead are currently unable to access suitable historical habitats because of migration barriers in the lower watershed (Leidy et al. 2005a). In recent years, a small number of adult Chinook salmon have been observed periodically during their fall upstream migration below barriers in the lowermost watershed. The origins (i.e., hatchery produced vs. naturally produced fish) and overall population status of Chinook in the watershed are not well understood.

Steelhead

The Alameda Creek watershed historically supported significant numbers of steelhead although there are no reliable quantitative estimates for the number of adult fish or spawning run size (Daily Alta California 1889b, Welch 1931, Shapovalov 1938a,b,c, California Department of Fish and Game 1953, Evans 1954, Fisher 1959, Smith 1998, Leidy et al. 2005a, Becker et al. 2007, Alameda Creek Alliance 2012). Steelhead remains have been recovered from Native American archaeological sites adjacent to Alameda Creek (Gobalet et al. 2004). In addition to steelhead, Alameda Creek also historically supported resident rainbow trout in headwater streams inaccessible to steelhead, typically in stream reaches above physical barriers such as waterfalls and cascades (Leidy 2007). Leidy et al. (2005a)

Table 8.1. Changes in the fish fauna of the Alameda Creek watershed, 1865-2012, Alameda and Santa Clara counties (table adapted from Leidy 2007). Changes to the system occurred when dams, reservoirs, and flood control projects were built (Sunol Dam, 1900; Calaveras Dam, 1925; Flood Control Channel, 1960s; James H. Turner Dam, forming the San Antonio Reservoir, 1965; and Del Valle Dam, 1968).

Common Name	Period of Record				
	1855-1860 (1)	1895-1948 (2)	1953-1969 (3)	1972-1987 (4)	1992-2012 (5)
Native Species					
Sacramento splittail	X	P	X		
River lamprey	P	P	X		
Coho salmon	P	P	X		
Chinook salmon	X	?	?	?	X
Speckled dace	P	X	?	?	?
Riffle sculpin	P	X	?	?	X
Hardhead	P	X	X	P	X
Tule perch	X	X	P	X	X
Shiner perch	P	P	P	X	X
Pacific brook lamprey	P	P	P	P	X
Pacific lamprey	X	X	X	X	X
Hitch	P	X	X	X	X
California roach	X	X	X	X	X
Sacramento blackfish	P	X	X	X	X
Sacramento pikeminnow	P	X	X	X	X
Sacramento sucker	X	X	X	X	X
Rainbow trout	P	X	X	X	X
Threespine stickleback	X	X	X	X	X
Sacramento perch	X	X	X	X	X
Prickly sculpin	X	X	X	X	X
Staghorn sculpin	P	P	P	X	X
Longjaw mudsucker	P	P	P	X	X
Starry flounder	P	P	P	P	X
Non-native Species					
Smallmouth bass		X	X	X	X
Brown trout		X	X		
Common carp			X	X	X
White catfish			X	X	X
Brown bullhead			X	X	X
Mosquitofish			X	X	X
Black crappie			X	X	X
Green sunfish			X	X	X
Bluegill			X	X	X
Largemouth bass			X	X	X
Goldfish			X	X	X
Golden shiner			X	X	X
Rainwater killifish			X	X	X
Threadfin shad				X	X
Channel catfish				X	X
Black bullhead				X	X
Inland silverside				X	X
Striped bass				X	X
Redear sunfish				X	X
Bigscale logperch				X	X
Yellowfin goby				P	X
Redeye bass					X
Tui chub					X
Total number of species	23	24	33	37	41
Number Native Species	23	22(+ 1?)	20(+ 3?)	17(+ 3?)	19(+ 1?)
Percent native species	100	92	61	46	46

Abbreviations: X, present; P, not recorded but likely present; ?, status uncertain;
shading denotes species with reproducing populations primarily restricted to elevations below 100 m.

For sources, see Leidy 2007, Table 13.

documented the historical existence of a spawning run or reproducing population of rainbow trout/steelhead in 19 streams within the Alameda Creek watershed, and a probable spawning run or reproducing population in another two streams. Prior to the construction of dams and other barriers, steelhead would likely have had complete or partial spawning access to at least 16 of the 21 (76%) assessed streams (Leidy et al. 2005a). Bjorkstedt et al. (2005) reviewed historical information and modeled physical habitat suitability for steelhead in local watersheds in order to predict which streams historically contained viable populations. Alameda Creek steelhead populations were identified as exhibiting viability-in-isolation (i.e., a population having a low (<5%) probability of going extinct over a 100-year time frame under conditions where exchanges with other populations have a negligible influence on its extinction risk (Bjorkstedt et al. 2005)). By reviewing historical evidence of stream habitat conditions assembled for this report in the context of currently available suitable steelhead habitat, we have inferred probable levels of historical steelhead life history support functions (i.e., spawning migration, spawning, juvenile rearing, and out-migration) and their significance by aquatic system and habitat type within the Alameda Creek watershed (table 8.2).

Steelhead life history

Central California coastal rainbow trout/steelhead exhibit considerable life history variation (McEwan 2001). Rainbow trout within a stream below migration barriers are facultatively anadromous and, as such, may be weakly or strongly anadromous, resident, or diverse mixtures of these various forms depending on environmental conditions (Satterthwaite et al. 2009a,b). Diverse migratory life history tactics would have allowed these populations of steelhead/rainbow trout to persist under the extremely variable environmental conditions within the Alameda Creek watershed.

The National Marine Fisheries Service lists steelhead within the Alameda Creek watershed as part of the Central Coast Distinct Population Segment (NMFS 2007). Alameda Creek steelhead are ocean-maturing or winter steelhead that spend one to four growing seasons in the ocean before initiating their spawning migration during the fall and winter (typically December through March depending on streamflow conditions), and then spawn (from December to April, but primarily January to March) within a few weeks to a few months after they enter freshwater (McEwan and Jackson 1996, Moyle 2002). Steelhead may return to the ocean after spawning and then return again to spawn the following year, and potentially a third or fourth time (McEwan and Jackson 1996). Steelhead typically rear in freshwater for one to four years and migrate downstream once they reach 130 to 250 mm total length (Moyle 2002).

Under historical conditions steelhead within the Alameda Creek watershed spawned and reared primarily within mid-elevation and headwater streams of the Diablo Range, within Niles Canyon, and in several small intermittent and perennial tributaries draining the Fremont-Livermore Hills (Smith

1998, Leidy 2007). Presumably, during years of above normal precipitation the geographic extent of potentially suitable spawning and rearing habitat for steelhead increased, especially within the warmer, lower elevation streams of the northern Alameda Creek watershed. Historically, headwater streams within the Diablo Range would have exhibited flow and temperature conditions similar to conditions currently found in the most undisturbed streams that remain today. As such, headwater streams would have been characterized by small-to-moderate sized pools separated by short, shallow riffles, permeable sand-gravel-cobble substrates, high levels of dissolved oxygen, relatively cool summer water temperatures, and high water clarity (Leidy 2007, Leidy et al. 2011). Steelhead rearing habitat within the streams of the lower watershed (e.g., Niles Canyon, lower Alameda Creek downstream of Niles Canyon) would presumably have been available in clear, thermally stratified pools and in warmer stream reaches with abundant, high-quality food (i.e., aquatic and terrestrial invertebrates, see Keith (1995) for an analysis of rainbow trout diets within upper Alameda Creek).

It is not possible to accurately reconstruct historical thermal regimes within the Alameda Creek watershed for the probable full range of habitats utilized by steelhead. However, it is reasonable to infer that historically water temperatures that did not regularly exceed 21°C (69.8°F) were suitable for juvenile steelhead growth, although local adaptation of steelhead to warmer water temperatures would have been probable (Spina 2007, Atkinson et al. 2011). Thermal refugia likely played a critical role in juvenile steelhead rearing; under current land use conditions and water management practices the recovery of these historical thermal conditions on a large geographic scale is likely no longer practicable (Aktinson et al. 2011).

McBain and Trush (2008) describe a very plausible conceptual model of historical steelhead life history tactics within the Alameda Creek watershed:

> Historic steelhead life history tactics within the Alameda Creek watershed likely occurred in two broad categories: (a) fry born in the upper tributaries reared for one or two years, then migrated rapidly to San Francisco Bay and (b) following emergence, the fry moved downstream and reared in the mainstem and/or Niles Cone before entering San Francisco Bay. Historically, headwater tributaries likely contributed large smolts directly to San Francisco Bay, especially during consecutive wetter years, but many additional large smolts were likely produced by slower migrating juveniles that grew on their way downstream through the mainstem channels, before smolting and entering Alameda Creek Estuary and then San Francisco Bay.

The diverse suite of life history tactical options available historically to steelhead within the Alameda Creek watershed was a local ecological adaptation to the highly variable precipitation, streamflow, and fluvial geomorphology which characterized the Alameda Creek watershed (box 8.1). Presumably local ecological adaptations allowed steelhead to spawn, rear, and grow under extremely variable climatic and geomorphic conditions before entering San Francisco Bay as smolts. The twelve historical steelhead life history tactics identified by McBain and Trush (2008) require

Table 8.2. Steelhead life cycle function and inferred level of support by habitat type.

System	Habitat Type	Example(s)	In-Migration
Tidal Features *(Marine and Estuarine)*	Shallow bay	Estuary Reach	Significant
	Tidal sloughs and flats	Estuary Reach, Alvarado to Bay	Significant
	Tidal marsh	Estuary Reach	Significant
	Panne/Salina	Estuary Reach	-
Non-tidal Features *(Riverine)*	Mainstem perennial streams	Alameda Creek, Little Yosemite to Upper Sunol Valley	Essential
		Alameda Creek, Niles Canyon	Essential
		Alameda Creek, Gravel-bedded Reach on Niles Cone	Essential
		Alameda Creek, Floodplain Reach from Decoto Road to Alvarado	Essential
	Mainstem intermittent streams	Alameda Creek, Gravel-bedded reach from BART weir downstream to Decoto Road	Essential
	Large perennial tributary creeks	Arroyo de la Laguna, Calaveras Creek, Arroyo Hondo	Essential
	Large intermittent tributary creeks	Lower-to-Middle Arroyo del Valle and Arroyo Mocho, San Antonio	Essential
	Small perennial-to-intermittent tributary creeks	Dry, Stonybrook, Sinbad, Upper Alameda, Calaveras, Arroyo Hondo, Isabel, Smith, Upper Arroyo Del Valle, Trout, Upper Arroyo Mocho	Essential
	Discontinuous creeks and distributaries	Lower Arroyo Mocho, Lower Arroyo del Valle	Essential
Non-tidal Wetland *(Palustrine and Laustrine)*	Perennial freshwater ponds	Pleasanton Marsh Complex Tule Pond (Fremont)	Significant for tributaries to ponds (e.g., Arroyo Mocho, Arroyo del Valle)
	Perennial valley freshwater marsh	Pleasanton Marsh Complex	Significant in association with perennial freshwater ponds and tributaries to ponds (e.g., Arroyo Mocho, Arroyo del Valle)
	Perennial valley freshwater marsh willow grove/Sausal (Type 1)	Pleasanton Marsh Complex	Significant in association with perennial freshwater ponds and tributaries to ponds (e.g., Arroyo Mocho, Arroyo del Valle)
	Estuary fringe willow grove/ Sausal (Type 2)	Lower Alameda Creek Floodplain – Upper Tidal Marsh Transition Zone	-
	Large playa	Frick Lake (Livermore Valley)	-
	Springs and seeps	Widespread	-
	Wet meadow	Lower Arroyo Mocho, Lower Alameda Creek Floodplain – Upper Tidal Marsh Transition Zone	Potential surface hydrologic link between discontinuous streams during floods
	Sycamore alluvial woodland	Alameda Creek above Sunol, San Antonio Creek, Arroyo del Valle, Arroyo Mocho	-

Habitat Support Function

Spawning Migration = Immigration of adult steelhead to their spawning grounds

Spawning = Habitat used by steelhead for spawning

Freshwater Juvenile Rearing = Spring/Summer or Over-Winter rearing of juvenile steelhead or smolts in non-tidal, freshwater habitats

Estuary Juvenile Rearing = Rearing of juvenile steelhead or smolts in tidally influenced portions of the estuary

Outmigration = Emigration of juvenile steelhead, pre-smolts, or smolts to rearing habitat or to the ocean

Spawning	Freshwater Juvenile Rearing Spring/Summer	Freshwater Juvenile Rearing Over-winter	Estuary Rearing	Out-migration
-	-	-	Potentially significant	Significant
-	-	-	Potentially significant	Significant
-	-	-	Potentially significant	Significant
-	-	-	-	--
Significant	Significant	Potentially significant	-	Essential
Potentially significant	Essential	Essential?	-	Essential
Potentially significant	Potentially significant	Potentially significant	-	Essential
Likely insignificant	Potentially significant	Potentially significant	-	Essential
Potentially significant	Potentially significant	Potentially significant	-	Essential
Significant to essential	Significant to essential	Potentially significant to significant	-	Essential
Potentially significant	Potentially significant	Likely insignificant	-	Essential
Potentially significant to essential	Potentially significant to essential	Potentially significant to critical	-	Essential
Insignificant	Insignificant	-	-	Potentially significant surface hydrologic link between discontinuous streams
-	Potentially significant	Potentially significant	-	Significant for tributaries to ponds
-	Potentially significant	Potentially significant	-	Potentially significant
-	Potentially significant	Potentially significant	-	Potentially significant
-	-	-	Potentially significant	Potentially significant
-	-	-	-	-
-	Essential contributions of cool water during summer/fall for maintenance of stream rearing habitats	-	-	-
-	-	-	-	Potential surface hydrologic link between discontinuous streams during floods
-	Significant allochthonous contribution in support of food web	-	-	-

Importance of Habitat Support Function

Essential = Critical or vital to the maintenance of viable steelhead populations

Significant = Of or likely to be of major consequence to the maintenance of viable steelhead populations

Potentially Significant = Exhibiting the possibility or capacity to be of major consequence to the maintenance of steelhead populations

Insignificant = Of minor consequence to the maintenance of steelhead populations when considered alone

BOX 8.1 STEELHEAD LIFE HISTORY TACTICS

These proposed steelhead life history tactics within the Alameda Creek watershed require that adult steelhead have both spawing access to headwater tributaries and suitable conditions for rearing, and smolt outmigration in the mainstem of Alameda Creek from Niles Canyon to San Francisco Bay.

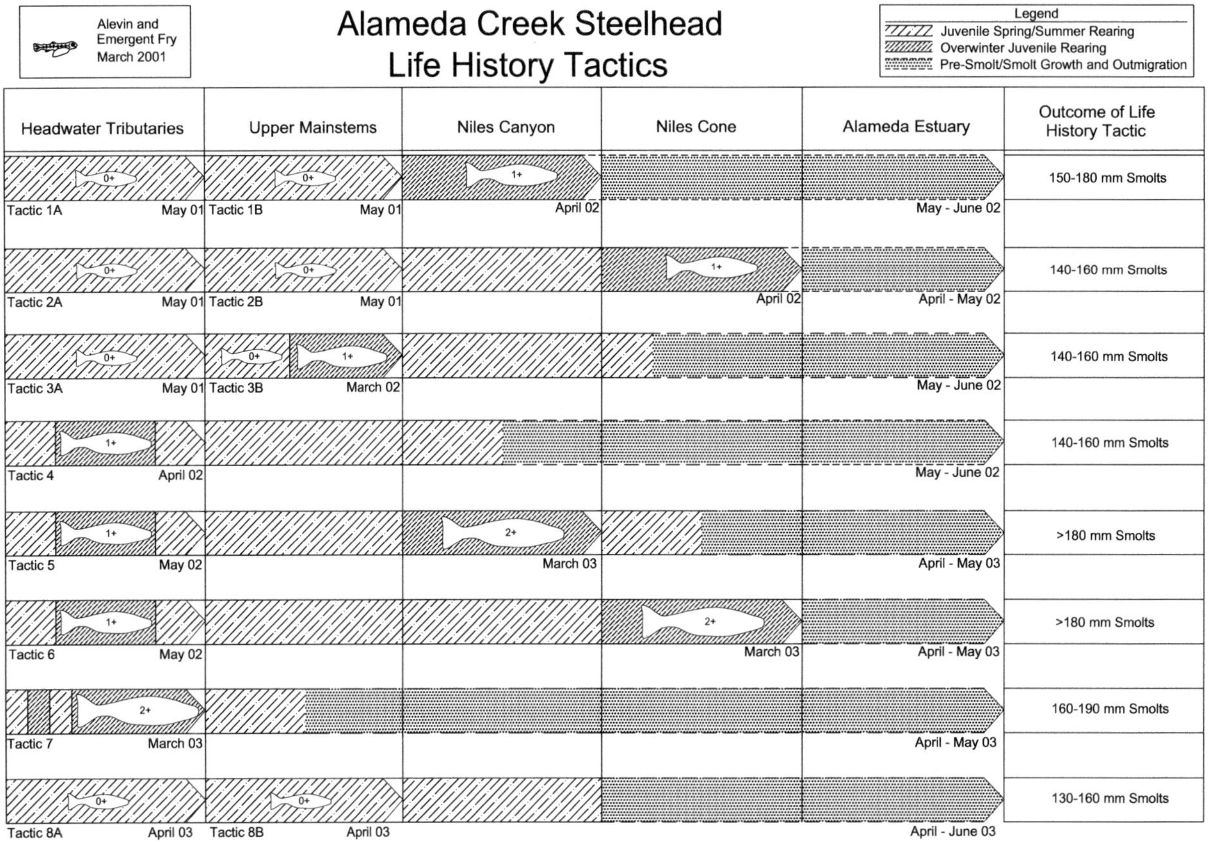

Figure 8.2. Steelhead life history tactics. Table and text from McBain and Trush (2008). For more details see full report.

Tactic 1A and Tactic 1B. Steelhead fry emerge from headwater tributaries in Tactic 1A or the upper mainstem in Tactic 1B (e.g., just below Little Yosemite Canyon). The fry then migrate into Niles Canyon within a few months and rear in Niles Canyon throughout their first summer and autumn. Over-wintering in Niles Canyon is followed by pre-smolt outmigration in spring (now as 1+ fish) and eventually entering San Francisco Bay in late-spring or very early-summer as 1+ smolts. Tactic 1A may be prevalent in Wet years when adult access is best, while Tactic 1B may be prevalent in Dry years when adult access into smaller tributaries is restricted and the window for successful spawning is very narrow.

Tactic 2A and Tactic 2B. Steelhead fry emerge from headwater tributaries in Tactic 2A or the upper mainstem in Tactic 2B. The fry then migrate through Niles Canyon by early summer and spend the remaining summer and autumn in Niles Cone, either in the backwater pools or farther downstream. Over-wintering in Niles Cone is followed by rapid pre-smolt movement farther downstream in Niles Cone by mid-spring, then entry into San Francisco Bay by late-spring as 1+ smolts.

Tactic 3A and Tactic 3B. Steelhead fry emerge from headwater tributaries in Tactic 3A or the upper mainstem in Tactic 3B. Fry from the headwaters in Tactic 3A soon travel downstream and spend the summer and autumn with fry from Tactic 3B in an upper mainstem channel. Over-wintering in the upper mainstem channel is followed by rapid presmolt movement farther downstream, perhaps spending some time in Niles Canyon, before entering San Francisco Bay by late-spring as 1+ smolts.

Tactic 4. Steelhead fry emerge from a headwater tributary and remain in the tributary (though likely moving downstream) through their first winter, then migrate downstream in early spring or late winter and enter San Francisco Bay by mid-spring as 1+ smolts. This tactic might rely on back-to-back Wet years for adult access, high spawning success, tolerable summer rearing, and downstream access the following spring.

Tactic 5. Steelhead fry emerge from a headwater tributary and remain in the tributary (though likely moving somewhat downstream) through their first winter, then migrate downstream in early spring or late-winter to Niles Canyon where they spend their second summer and autumn. In early spring they would continue downstream as pre-smolts, entering San Francisco Bay in early- or mid-spring as 2+ smolts.

Tactic 6. Steelhead fry emerge from a headwater tributary and remain in the tributary (though likely moving downstream) through their first winter, then migrate downstream in early spring or mid-spring eventually to Niles Cone where they spend their second summer and autumn. In the following early spring they would enter San Francisco Bay in early or mid-spring as 2+ smolts. This tactic might apply to later downstream migrating 1+ presmolts that experience a temperature threshold preventing smoltification and forcing them to 'wait-out' a second winter before smolting.

Tactic 7. Steelhead fry emerge from a headwater tributary and remain in the tributary (though likely moving somewhat downstream) through their second winter, then migrate downstream in early spring or late-winter and enter San Francisco Bay by mid-spring as 2+ smolts. This tactic might rely on a Wet year (for adult access and spawning success), followed by a Dry year and then a Normal/Wet year (the Dry year preventing downstream migration as 1+ juveniles and forcing a second summer and winter).

Tactic 8A and Tactic 8B. Steelhead fry emerge from headwater tributaries in Tactic 8A or the upper mainstem in Tactic 8B. The fry then migrate through Niles Canyon and Niles Cone by early summer and enter San Francisco Bay as 0+ smolts. This tactic would rely on wetter years with good growth potential. These 0+ smolts could have spent the summer in the estuary, and then have migrated to the ocean in fall.

that adult steelhead have both spawning access to headwater tributaries (primarily in the southern Diablo Range, or the mainstem of Alameda Creek just downstream of Little Yosemite Canyon) and suitable conditions for rearing and out-migration in the mainstem of Alameda Creek from Niles Canyon to San Francisco Bay (fig. 8.2). However, even this model simplifies the probable diversity of tactical options available to steelhead within the Alameda Creek watershed under historical conditions, as the likely contributions of small intermittent and perennial tributaries flowing from the Fremont and Livermore hills and northern Diablo Range are largely omitted from the model. Smaller intermittent and perennial streams outside of the southern Diablo Range probably were important contributors to overall steelhead production, particularly during years of above average precipitation when flows created conditions suitable for spawning, rearing, and out-migration. In addition, non-natal, off-main-channel (e.g., floodplains, perennial-intermittent tributaries) habitats also probably played significant roles historically as contributors to steelhead growth and survival (refer to following discussion). The construction of large dams (i.e, Calaveras, San Antonio, and Del Valle reservoirs) in the upper watershed, construction of the Alameda Creek Flood Control Channel below Niles Canyon, changes to streamflow patterns within mainstem Alameda Creek and tributaries in Livermore Valley, and widespread declines in the surface water quality of most urban valley streams has severely reduced the number of practicable life-history tactical options available to steelhead.

One factor identified as critical in determining whether steelhead will survive to reproduce is fish length at the time of smolting: the greater the size of steelhead at smolting, the greater the likelihood that fish will return as spawning adults (McBain and Trush 2008). Presumably historical conditions provided suitable habitat for the growth of pre-smolts within a range of habitats, including Niles Canyon, the mainstem of Alameda Creek below Niles Canyon, non-tidal floodplain wetlands, non-natal seasonal tributaries, and estuarine channels and wetlands at the mouth of Alameda Creek. It is also likely that a much larger reproducing population historically meant that low survival rates were more acceptable than they are today. Even populations that had extremely low return rates due to small size at smolting may have historically been able to persist due to higher production overall. Historically, the relatively widespread availability of suitable spawning and rearing habitat and life history tactical options adapted to these variable habitat conditions would increase the total number of young fish produced and thus the number of smolts becoming adults, and this would contribute to overall population persistence within the watershed. Currently, far fewer steelhead are produced, largely because of limited suitable habitat. A logical consequence of the current limited availability of suitable habitat is that maximizing the number of large-bodied steelhead smolts takes on greater significance in maintaining viable populations. Clearly, restoration of steelhead within the Alameda Creek watershed will depend largely on the success of management actions to restore multiple steelhead life history tactical options (i.e., increasing pre-smolt/smolt survival by increasing the amount and quality of spawning and rearing habitat and by creating suitable conditions for out-migration to San Francisco Bay).

Many critically important historical habitats for steelhead within the Alameda Creek watershed have been degraded, isolated, or destroyed. While resident rainbow trout are still locally abundant in several headwater streams of the Diablo Range, steelhead currently are not able to access suitable historical spawning habitats or return to the ocean as smolts because of migration barriers in the urbanized lower watershed (Leidy et al. 2005a, Leidy et al. 2011). For example, the BART weir in Fremont blocks adult steelhead access to the upper watershed. The Alameda Creek Flood Control Channel as currently designed and managed also acts as an impediment to steelhead pre-smolts or smolts that attempt to migrate to the San Francisco Estuary (Robinson et al. 2012).

Migration barriers in the upper watershed have blocked access to some of the highest quality steelhead spawning and rearing habitat (Leidy et al. 2011). Calaveras Dam (completed in 1925) prevented steelhead from reaching high quality spawning habitat on perennial Arroyo Hondo and its tributaries. Calaveras Reservoir flooded other important steelhead spawning and rearing habitat on perennial canyon reaches of lower Calaveras Creek. The reservoir and dam also blocked the downstream movement of juvenile steelhead that used these perennial streams for rearing. Completion of San Antonio (1965) and Del Valle (1968) reservoirs blocked steelhead access to additional high functioning spawning and rearing habitat found in their small perennial-to-intermittent headwater tributaries. Additionally, operation of these three reservoirs and other water conveyance facilities altered natural flow regimes in streams below the dams, further degrading suitable steelhead spawning and rearing habitat in the lower watershed (e.g., Alameda Creek in Niles Canyon).

Other anthropogenic changes in the watershed further degraded the quality of important steelhead habitat. Access by steelhead to several small headwater creeks flowing from the Fremont Hills that likely provided high quality spawning and rearing habitat was either completely blocked by downstream barriers (i.e., as on Dry Creek) or degraded through sedimentation, drawdown of groundwater levels from the pumping of wells, or the creation of migration barriers at road crossings (e.g., Stonybrook and Sinbad creeks). The diking and filling of large expanses of tidal marsh and sloughs historically connected to Alameda Creek within the lower floodplain likely eliminated significant estuary rearing habitat used by steelhead and possibly Chinook salmon prior to smolting. Subsequent channelization of lower Alameda Creek further isolated the creek from its productive floodplain and eliminated shaded perennial and intermittent stream reaches beginning at the mouth of Niles Canyon and continuing downstream to the tidal marshes. Construction and management of the Flood Control Channel also created conditions unsuitable for rearing and smolting steelhead, primarily by creating migration barriers, increasing water temperatures and turbidity, and creating habitat favorable to fishes that prey on juvenile steelhead.

The potential significance of off-channel habitats to steelhead

In addition to the mainstem perennial reaches and headwater tributaries of Alameda Creek, steelhead may have also reared historically in non-natal, off-channel aquatic habitats within the Alameda Creek watershed. Potential habitat in tidal marsh and sloughs (see fig. 6.20), floodplain lakes (see fig. 8.23), perennial valley freshwater marsh and ponds associated with the Pleasanton marsh complex (see fig. 8.22), and perennial and intermittent tributaries (see fig. 6.55) historically covered thousands of acres and many stream miles. These features likely provided significant habitat for juvenile steelhead growth, thereby improving overall survival. The importance of off-channel aquatic habitats to salmon for feeding and growth has been documented elsewhere in California (Sommer et al. 2001, Moyle et al. 2007, Limm and Marchetti 2009, Benigno 2011). For salmon, non-natal seasonal tributaries may provide warmer water temperatures and higher prey densities compared to main-channel habitats and this may lead to increased growth rates and improved juvenile survival (Limm and Marchetti 2009). Limm and Marchetti (2009) also point out that off-channel habitats may 1) be less turbid than main-channel areas during high flows which could improve juvenile salmon feeding efficiency, 2) increase prey availability because overall wetted area is increased in floodplain wetlands, and 3) decrease the density of potential predators and their direct and indirect effects on juvenile salmon, again because the wetted channel area has increased. These studies emphasize the likely historical and potential future importance of off-channel habitat diversity to rearing salmonids, including steelhead.

Coho salmon

Leidy et al. (2005b) used direct historical evidence and an assessment of current habitat conditions to conclude that coho salmon likely occurred historically within the Alameda Creek watershed. The number of adult coho salmon that entered the watershed is unknown. Given the limited amount of potentially suitable coho habitat in the watershed, the number of adult fish would be expected to be relatively low. Leidy et al. (2005b:243) further concluded that:

> The most likely location of historically suitable habitat for coho salmon was in the small, perennial tributaries to Alameda Creek (i.e., Dry, Stoneybrook, Pirate), mainstem Alameda creek at Little Yosemite, headwaters to San Antonio Creek (i.e., La Costa and Indian creeks), and Calaveras Creek and its primary tributary Arroyo Hondo Creek, below migration barriers.

By the 1920s the construction and operation of dams and other water diversion and conveyance structures would have destroyed, degraded, and isolated significant reaches of potentially suitable coho salmon habitat.

Sightings of coho salmon after the 1920s may have represented straying adult fish (Leidy et al. 2005b).

Chinook salmon

There are few reliable records for the occurrence of Chinook salmon in small streams tributary to the San Francisco Estuary watershed prior to the mid-1980s, and only two definite historical records, from the San Leandro and San Mateo creek watersheds (Leidy 2007). Chinook salmon remains have been recovered from an archeological site adjacent to lower Alameda Creek, but the origins of these remains are attributed to San Francisco Bay, as Native Americans were known to transport salmon captured in the estuary to inland sites (Gobalet et al. 2004). Beginning in the mid-1980s, a few Chinook salmon have been fairly regularly sighted or captured below the salmon migration barrier formed by the BART weir in the Alameda Creek Flood Control Channel (Alameda Creek Alliance 2012). The origins of these Chinook salmon are not well documented, but the general consensus is that most adult Chinook salmon are related to fish hatchery stocks (Leidy 2007). However, successful natural spawning, hatching, and juvenile survival of Chinook salmon has been documented from other Bay Area watersheds, and smolts have been recorded in at least three watersheds (Leidy 2007). This opens the possibility that following modification of the BART weir in lower Alameda Creek to allow for fish passage, Chinook salmon and steelhead may establish permanent spawning runs within the Alameda Creek watershed.

Tidal features (marine and estuarine)

Shallow bay, tidal flats, and sloughs

Tidal habitats adjacent to San Francisco Bay contained the greatest diversity of native fishes in the Alameda Creek watershed (table 8.3). The abundance and diversity of fishes using shallow embayments, tidal flats, and tidal sloughs were controlled by seasonal changes in water levels, salinities, and temperatures, and the spatial arrangement and connectivity between different habitats. The amount and timing of freshwater discharges through the Delta and from Alameda Creek influenced local tidal water salinities and controlled seasonal accessibility by fishes to various tidal habitats. For these reasons, fish assemblages in tidal environments likely varied considerably in species composition and abundance both seasonally and interannually. Probable fishes associated with shallow open bay and tidal flat environments included Pacific lamprey, bay pipefish, leopard shark, brown smoothhound, bat ray, white sturgeon, Pacific herring, longfin smelt, northern anchovy, Chinook salmon, steelhead, jacksmelt, topsmelt, threespine stickleback, white croaker, shiner perch, barred surfperch, arrow goby, bay goby, plainfin midshipmen, staghorn sculpin, English sole, California halibut, speckled sand dab, diamond turbot, and starry flounder (table 8.4; Goals Project 1999, Hobbs 2011, Moyle et al. 2012).

Lower Alameda Creek was directly connected to an extensive network of tidal sloughs and channels (see figs. 6.20, 6.21), estimated at approximately 170 linear miles (fig. 8.3). Fishes associated with tidal sloughs varied by slough size, accessibility, and connectivity with freshwater streams, but likely included Pacific lamprey, white sturgeon, Chinook salmon, steelhead, topsmelt, threespine stickleback, shiner perch, longjaw mudsucker, arrow goby, bay goby, Pacific staghorn sculpin, and starry flounder (Stevenson et al., 1987, Goals Project 1999, Moyle 2002, Leidy 2007, Moyle et al. 2012). Sloughs also functioned as migration corridors for adult Pacific lamprey, steelhead, and coho and Chinook salmon to suitable habitats in the perennial headwater reaches of Alameda Creek watershed. Historical documentation for the occurrence of coho and Chinook salmon are limited, but it is likely that streams within the watershed supported these species historically (Leidy et al. 2005b,c). Tidal sloughs in conjunction with tidal marsh may have functioned as potentially significant rearing habitat for juvenile steelhead.

Tidal marsh

The estuarine portions of Alameda Creek formed a dense and extensive interconnected network of periodically flooded tidal sloughs, marshes, and marsh pannes (see figs. 6.17, 6.18). A seasonally variable saline-to-brackish-to-fresh water salinity gradient likely characterized tidal marshes. At its landward extent, tidal marsh connected to lowland freshwater wetlands in areas of artesian springs. This diverse mosaic of interconnected, variable

Tidal Marsh, Pond, Slough
Tidal Flat
Bay/Deep Channel
Wet Meadow
Grassland

Figure 8.3. Extensive networks of tidal sloughs, marshes and pannes were connected to the Alameda Creek channel and supported diverse assemblages of marine and euryhaline fishes. In conjunction with adjacent freshwater wetlands, tidal channels and marshes likely functioned as critical rearing habitat for juvenile steelhead and possibly Chinook salmon prior to their migration to the Pacific Ocean as smolts.

salinity wetlands probably functioned differently than tidal sloughs as fish habitat. Tidal marsh supported similar but less diverse assemblages of native fishes compared to tidal channels, in part because dense marsh vegetation restricted access and movement by large-bodied fish species (e.g., sturgeon). Other smaller-bodied pelagic fishes such as topsmelt would not be expected to regularly occur in densely vegetated tidal marshes. Significantly, tidal salt and brackish marshes in conjunction with adjacent freshwater wetlands may have functioned as important rearing habitat for juvenile steelhead and possibly Chinook salmon prior to their ocean migration; however the extent of tidal marsh use by steelhead adjacent to Alameda Creek is unknown. Other probable native fishes that occurred in tidal marshes adjacent to Alameda Creek include jacksmelt, Pacific staghorn sculpin, threespine stickleback, longjaw mudsucker, and arrow goby (Moyle et al. 2012).

Marsh panne/salina

Pannes or salinas (i.e., salt-pans; see fig. 6.22) were characterized by intermittent, seasonal ponding and were embedded within dense tidal marsh vegetation, which restricted access by most fishes. Pannes were typically covered by shallow, warm, high salinity water relative to nearby tidal channels, although local environmental conditions were likely highly dependent on tidal conditions and the amount and timing of freshwater outflows. Pannes probably supported threespine stickleback, arrow goby, or bay goby depending on opportunities for access and water salinities.

Non-tidal features (riverine)

Mainstem perennial streams

Mainstem perennial streams were typically associated with areas of groundwater discharge from seeps and springs and were underlain by impervious formations such as bedrock or clay soils, often in areas characterized by extensive faulting or a high water table. Examples of large perennial stream reaches include the lower floodplain reach of Alameda Creek (i.e., Decoto Road to Alvarado) (fig. 8.4), the gravel-bedded reach of Alameda Creek below the mouth of Niles Canyon, Alameda Creek within Niles Canyon (fig. 8.5, see figs. 5.16, 5.18, 5.19), Alameda Creek from just upstream of the confluence with Calaveras Creek within Little Yosemite to upper Sunol Valley, and Alameda Creek canyon downstream from the Alameda Creek Diversion Dam (fig. 8.6). Greater fish species diversity likely would be found in the low-to-mid-elevation reaches of large perennial streams (i.e., lower floodplain and Niles Canyon reaches of Alameda Creek) characterized by deep, permanent pools, compared to smaller, upper-elevation reaches (i.e., upper mainstem Alameda Creek below Little Yosemite).

Archeological evidence from Alameda Creek and Arroyo de la Laguna confirms the presence of a diverse fish assemblage dominated by lowland forms (Gobalet et al. 2004). The perennial mainstem reaches of Alameda Creek supported a fish assemblage similar to that found historically in

Figure 8.4. Views along lower Alameda Creek. View of mainstem perennial reaches of Alameda Creek (A) between Alvarado and Decoto roads (May 1915), and (B) within the gravel-bedded reach below the mouth of Niles Canyon (June 1915). These mainstem perennial reaches functioned as essential migration corridors and significant freshwater rearing habitat for steelhead. Deep permanent pools within these reaches would also have supported diverse assemblages of native fishes. (AD-275 and AD-289, © San Francisco Public Utilities Commission)

Figure 8.5. Mainstem perennial reaches of Alameda Creek through Niles Canyon (A) 1908 and (B) June 2009. Niles Canyon under natural flow conditions likely functioned as important spring/summer and winter rearing habitat, as well as spawning habitat for steelhead. (A: courtesy California Historical Society CHS2O12.930; B: photo by Robert A. Leidy)

Figure 8.6. Upper Alameda Creek canyon downstream of the Alameda Creek Diversion Dam, summer 2009. This high-gradient, upper mainstem, perennial reach is characterized by small-to-medium-sized pools fed by cool groundwater. Pacific lamprey, rainbow trout, California roach, and prickly sculpin characterize the fish assemblage within this reach. (photo by Robert A. Leidy)

similarly situated lowland riverine environments within the Central Valley, and in tributaries to the San Francisco Bay Estuary (Moyle 2002, Leidy 2007). Species collected from or likely to have occurred within mainstem perennial streams before significant Euro-American environmental modifications include Pacific lamprey, thicktail chub, hitch, California roach, Sacramento blackfish, hardhead, Sacramento pikeminnow, Sacramento sucker, steelhead, threespine stickleback, prickly sculpin, Sacramento perch, and tule perch (table 8.3, fig 8.7; Gobalet et al. 2004, Leidy 2007).

In addition to functioning as steelhead migration corridors, some mainstem perennial streams likely contained significant spawning and rearing habitat for steelhead and/or resident rainbow trout, especially within the mid- to upper-elevation reaches of these streams (Leidy et al. 2005a, Leidy 2007).

Figure 8.7. Fishes of perennial reaches. (A) Tule perch (SU 5929) collected circa 1890s from Alameda Creek, and (B) Sacramento perch (CAS 26724) collected March 25, 1957 from Alameda Creek near Niles, opposite California Nursery Co. Tule perch and Sacramento perch historically were common within perennial stream reaches such as Niles Canyon and in pools below the mouth of Niles Canyon. Both species are now extremely rare within the Alameda Creek watershed. (specimens photographed by Jon Fong, California Academy of Sciences).

For example, prior to human modifications to spring and summer surface flows, Niles Canyon may have supported significant steelhead spawning and juvenile rearing habitat (figs. 8.8, 8.9). Pools in Niles Canyon under natural spring and summer flow conditions were likely thermally stratified, because natural low spring-summer flows limited mixing of the water column. Contributions of cool, shallow groundwater and/or water from adjacent seeps and springs combined with extensive shading from riparian vegetation would likely create cool water temperatures in pools. This created suitable habitat for juvenile steelhead rearing within the deeper portions of pools, which could remain relatively cool and clear through the warmest summer months (fig. 8.10).This effect can be observed today in pools in upper Alameda Creek that receive low summer surface inputs (Hanson Environmental 2002). More broadly, pools in summer-warm streams in northern California and the Pacific Northwest are known to

Figure 8.8. Niles Canyon habitat. (A) Extensive pool habitat within Niles Canyon, 1885. Note the pool shading by willows and alder on the left side of the photo and the shaded pool on the far right of the photo. Natural summer low flows combined with the shading effects of riparian vegetation and groundwater and hyporheic flow likely resulted in thermal stratification of pools containing coldwater patches and low turbidity that would have provided high quality summer rearing habitat for steelhead. (see figure 8.10). (B) Riffle transitioning to pool, Alameda Creek, upper Niles Canyon, July 2009. Current water management practices augment summer flows through Niles Canyon, which likely reduces thermal stratification of pools. (A: courtesy California Historical Society, CHS2012.931; B: photo by Robert A. Leidy)

Figure 8.9. Shaded pools and riffles associated with large boulders, as seen in this 1885 image. Species such as torrent sedge (*Carex nudata*), white alder (*Alnus rhombifolia*), and various willows (*Salix* spp.) provided shading that moderated summer water temperatures, helping to sustain high quality rearing habitat for steelhead through Niles Canyon. (courtesy California Historical Society, CHS2OT2.932)

display significant thermal heterogeneity often characterized by discrete cold water patches that form in response to multiple environmental influences on water temperature, including riparian shading, hyporheic and groundwater flow, and channel bedform (Ebersole et. al 2003a, Wondzell 2012). Thermally stratified pools containing cold water patches may function as thermal refuges for salmonids, including steelhead and Chinook salmon (Nielsen et al. 1994, Ebersole et. al 2003b).

Beginning in the early 20th century, water management practices (including constructed flood control channels, dams, and the draining of the Pleasanton marsh) had altered the natural flow regime in Niles Canyon and had increased spring and summer flows through the canyon. These increased warm water flows during spring and summer likely reduced thermal stratification of pools through increased mixing of the water column, reducing or eliminating suitable rearing habitat for steelhead (see fig. 8.10). Warmer water temperatures would also improve conditions for predatory native minnows, as well as non-native predatory basses and sunfish. Population increases of piscivorous fish such as the native Sacramento pikeminnow, non-native largemouth bass,

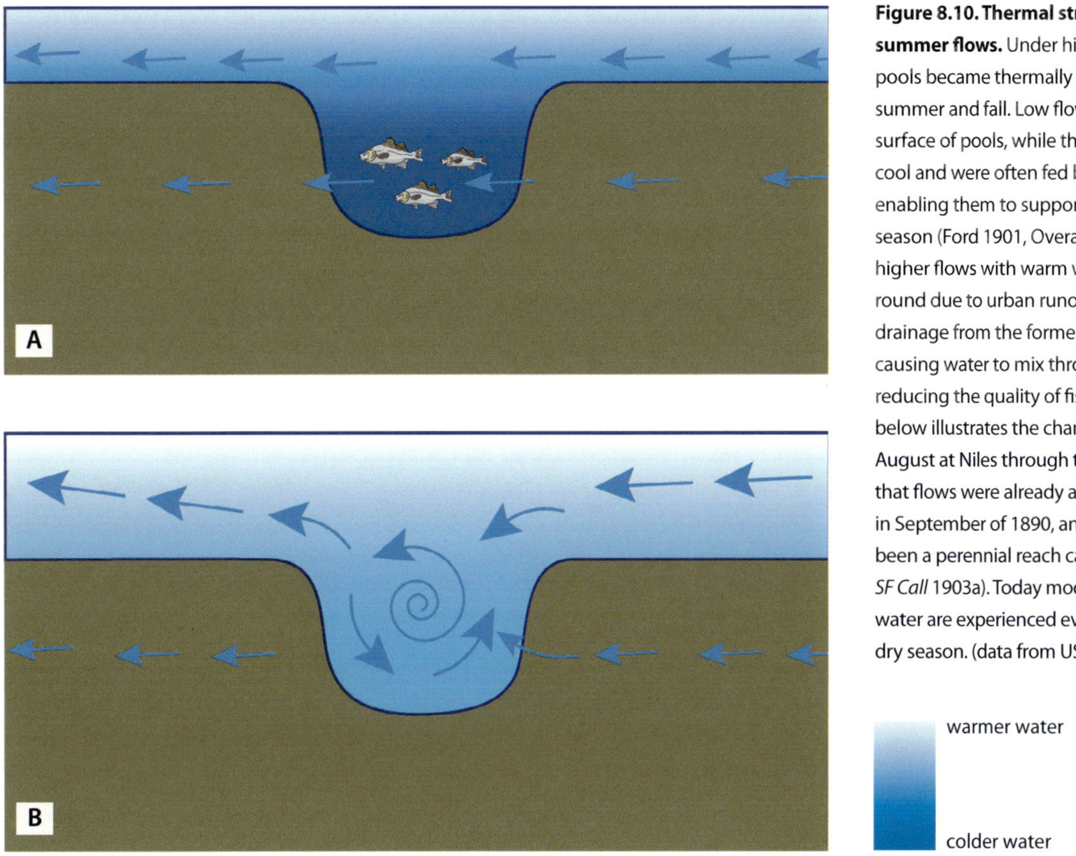

Figure 8.10. Thermal stratification and summer flows. Under historical conditions (A), pools became thermally stratified during the summer and fall. Low flows passed across the surface of pools, while the depths remained cool and were often fed by groundwater, enabling them to support fish through the dry season (Ford 1901, Overacker 1901). Today, (B), higher flows with warm water continue year round due to urban runoff, water imports, and drainage from the former Pleasanton marsh, causing water to mix throughout the pool and reducing the quality of fish habitat. The graph below illustrates the change in flow patterns in August at Niles through the 20th century. Note that flows were already altered by diversions in September of 1890, and this would have been a perennial reach ca. 1800 (Hilgard 1889, *SF Call* 1903a). Today moderate flows of warm water are experienced even at the end of the dry season. (data from USGS gage at Niles)

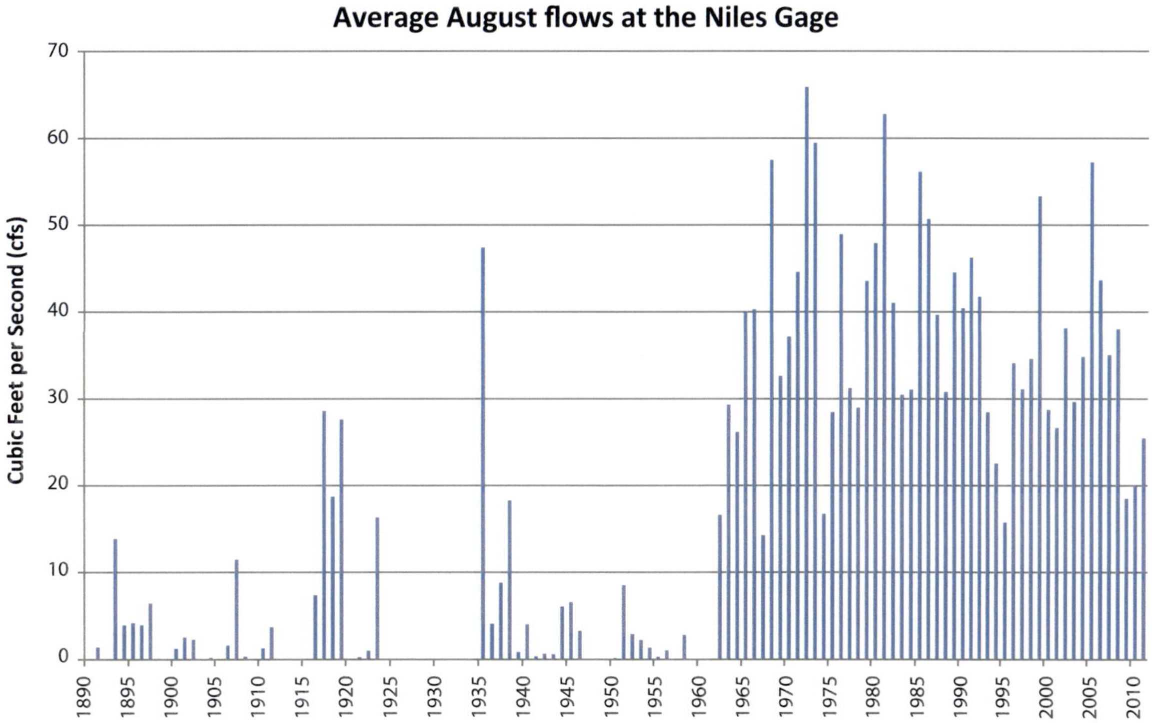

and several other sunfish and catfish species may also have drastically decreased the suitability of Niles Canyon for steelhead rearing and migration through predation.

Concomitant increases in stream turbidity would have also had negative effects on stream ecosystem primary and secondary productivity. Development of agriculture and urbanization within the Livermore-Amador Valley over the past century resulted in substantial increases in suspended sediment which in turn affected the feeding efficiency and growth of juvenile steelhead. The combination of warm water temperatures and high turbidity has been identified as a stressor that likely has significant deleterious effects on the growth and survival of juvenile and adult steelhead in Alameda Creek (Atkinson et al. 2011).

Mainstem intermittent streams

Mainstem intermittent streams would support fish species similar to those found in mainstem perennial streams, albeit for more restricted periods (see table 8.3). Examples of mainstem intermittent streams include Alameda Creek (1) within Sunol Valley, (2) downstream from the mouth of the upper Alameda Canyon to Little Yosemite, and (3) from the BART weir to Decoto Road. Large mainstem intermittent streams would function as critical migration corridors and as potentially significant rearing habitat for steelhead. For example, there were large deep pools between the BART weir and Decoto Road that historically contained steelhead during the summer (Shinn 1901, Tyson 1901). Mainstem intermittent reaches of Alameda Creek below Niles Canyon likely supported juvenile steelhead into at least the late 1950s as evidenced by the collection of three halfgrown rainbow trout in a large pool near Niles, opposite the California Nursery Company (see bottom image in fig. 6.37), during late March of 1957; these specimens may have been rearing or smolting steelhead (Follett 1957). Other fish species would use intermittent reaches seasonally, as movement corridors to suitable upstream or downstream intermittent or perennial habitats, or would remain in deeper pools that persisted throughout the summer and fall. Probable fishes from mainstem intermittent reaches include Pacific lamprey, thicktail chub, hitch, California roach, Sacramento blackfish, hardhead, Sacramento pikeminnow, Sacramento sucker, steelhead, threespine stickleback, prickly sculpin, Sacramento perch, and tule perch (table 8.3, fig. 8.11; Gobalet et al. 2004, Leidy 2007).

Large perennial tributary creeks

Examples of large perennial tributary creeks include Arroyo de la Laguna, Arroyo Hondo, and Calaveras Creek (fig. 8.12). Permanent flow, cool water temperatures, and high quality spawning and rearing habitat for steelhead and/or resident rainbow trout along much of their lengths characterized Calaveras Creek and Arroyo Hondo. As large perennial tributaries, these creeks were critical to the production and persistence of steelhead within the Alameda Creek watershed (fig. 8.13). The construction of Calaveras Dam

Figure 8.11. Pool-associated fishes. (A) Sacramento pikeminnow and (B) Sacramento sucker (from San Antonio Creek) were two native fishes commonly associated with pools within mainstem intermittent streams. (A: courtesy SFPUC; B: photo by Robert A. Leidy)

and Reservoir in 1925 isolated steelhead spawning and rearing habitat that likely occurred in Calaveras Creek, and also blocked access by steelhead to many miles of high functioning spawning and rearing habitat in Arroyo Hondo. In addition to steelhead/resident rainbow trout, Calaveras Creek and Arroyo Hondo would have supported Pacific lamprey, California roach, Sacramento sucker, threespine stickleback, and prickly sculpin.

Arroyo de la Laguna was a low gradient, valley floor stream with higher summer water temperatures than Arroyo Hondo (fig. 8.14). Arroyo de la Laguna would have been an important steelhead migration corridor for steelhead migrating into and out Arroyo del Valle and in wet years

Figure 8.12 Large perennial tributaries in the Alameda Creek watershed. These images show historical Calaveras Creek (A) and recent Arroyo Hondo (B). Both creeks provided important steelhead spawning and rearing habitat prior to the completion of Calaveras Dam in 1925. (A: courtesy The Bancroft Library, UC Berkeley; B: photo by Robert A. Leidy)

Figure 8.13. Steelhead pictured with San Francisco Water Department pipeline surveyors within the Alameda Creek watershed, circa 1935. (photo courtesy EBRPD)

Figure 8.14. Pool, riffle, and gravel bar along lower Arroyo de la Laguna. Arroyo de la Laguna functioned as a migration corridor for steelhead utilizing streams such as Arroyo del Valle and Arroyo Mocho. This is a low gradient, valley floor stream with higher summer water temperatures compared to either Calaveras Creek or Arroyo Hondo, due to its lower elevation and generally open riparian canopy. (photo by Robert A. Leidy)

Arroyo Mocho. Lowermost Arroyo de la Laguna may have contained steelhead spawning and rearing habitat depending on the annual and seasonal distribution of rainfall in the northern portions of the Alameda Creek watershed. Other fishes that would have used Arroyo de la Laguna include Pacific lamprey, thicktail chub, hitch, California roach, Sacramento blackfish, Sacramento pikeminnow, Sacramento sucker, steelhead, threespine stickleback, tule perch, Sacramento perch, and prickly sculpin (Gobalet et al. 2004, Leidy 2007).

Large intermittent tributary creeks

The lower-to-middle reaches of Arroyo del Valle and Arroyo Mocho, and mainstem San Antonio Creek are examples of large intermittent tributary creeks (fig. 8.15). Intermittent reaches characterized by alluvial soils were often entirely dry by summer or fall. Under conditions where the stream reach completely dried, fishes would use intermittent reaches seasonally, particularly as a migration corridor to other suitable intermittent or perennial habitats with year-round pools found within the watershed. Large intermittent tributary creeks would function as critical migration corridors and potentially significant rearing habitat for steelhead depending on the amount and distribution of annual precipitation.

Fish assemblages in intermittent streams would be highly variable depending on local environmental conditions. Generally, intermittent streams with large, deep permanent pools would be expected to support greater fish species diversity than reaches with small pools. Fishes likely associated with large intermittent streams include Pacific lamprey, Pacific brook lamprey, thicktail chub, hitch, California roach, Sacramento blackfish, Sacramento pikeminnow, Sacramento sucker, steelhead, threespine stickleback, and prickly sculpin (fig. 8.16). Steelhead adults and smolts utilized these intermittent reaches as migration corridors to other suitable intermittent or perennial habitats. For example, adult steelhead likely would migrate through the lower-to-middle elevation reaches of San Antonio Creek to reach high quality spawning habitat and perennial tributaries in the upper watershed such as La Costa and Indian creeks.

Small perennial-to-intermittent headwater creeks

Small perennial and intermittent headwater creeks flowed from the Fremont and Livermore hills and Diablo Range. Examples of headwater streams include Upper Dry, Sinbad, Upper Alameda, Isabel, Smith, La Costa, Indian, Upper Arroyo del Valle, Trout, Tassajara, Upper Arroyo las Positas, and Upper Arroyo Mocho creeks (fig. 8.17, see fig. 6.55). Headwater creeks flowing south and west from the Fremont Hills directly into Alameda Creek (e.g., Dry and Sinbad creeks) and creeks within the mountainous Diablo Range (e.g., Upper Alameda, La Costa, and Smith creeks) would have been characterized by shallow riffles interspersed with pools of small-to-moderate size and depth, relatively cool summer water temperatures, and coarse sand-gravel-cobble-boulder substrates. Perennial

Figure 8.15. Examples of large intermittent tributaries. Historical photograph (A) shows riparian cover along either San Antonio Creek or Arroyo del Valle (it was found labeled with both locations). Photograph (B) shows riparian cover along Arroyo del Valle in 1897. Both Arroyo del Valle and San Antonio Creek were large tributaries that were summer-dry in their lower reaches but contained persistent summer pools in their upper reaches that were likely used by steelhead and other native fishes. (A: courtesy The Bancroft Library, UC Berkeley; B: courtesy Livermore Heritage Guild)

Figure 8.16. Hitch is a native cyprinid that is locally common within large intermittent tributaries. This fish was collected from Arroyo de la Laguna. (photo by Robert A. Leidy)

Figure 8.17. Four examples of perennial headwater creeks within the upper Alameda Creek watershed. (A) Smith Creek, a tributary to Arroyo Hondo that supports healthy populations of rainbow trout; (B) Indian Creek, a tributary to San Antonio Reservoir; (C) Upper Alameda Creek; and (D) Dry Creek, which flows from the Fremont Hills to lower Alameda Creek. Perennial and intermittent headwater creeks were historically essential as spawning and rearing habitat for steelhead. (photos by Robert A. Leidy)

headwater creeks likely functioned as critical spawning and rearing habitat for steelhead and/or resident rainbow trout. The close proximity of Dry, Stonybrook, and Sinbad creeks to Niles Canyon and lower mainstem Alameda Creek compared with headwater streams within the Diablo Range would have reduced the out-migration distance, thereby improving the overall survival rate for steelhead smolts emigrating to San Francisco Bay. These small intermittent streams characterized by cool spring and early summer water temperatures may have played a significant role in steelhead production (Follett 1955a, 1955b). Rainbow trout were documented in mid-to-late October 1955 in Stonybrook Creek up to 1.5 miles upstream from its confluence with Alameda Creek in Niles Canyon. This indicates that this small intermittent stream was important for steelhead into at least the late fall in some years (Follett 1955a,b). Juvenile rainbow trout are known elsewhere in California to emigrate downstream from intermittent tributaries into suitable permanent streams as water levels drop (Erman and Leidy 1975). Such a life history tactic may have been especially important in small intermittent tributaries to Niles Canyon (i.e., Stonybrook and Sinbad creeks) and lower Alameda Creek (i.e., Dry Creek) where downstream

migration distances to suitable, permanent habitat in mainstem Alameda Creek were relatively short (fig. 8.18).

Other native fishes found in these creeks included Pacific lamprey, hitch, California roach, speckled dace, Sacramento sucker, prickly sculpin, and possibly riffle sculpin (fig. 8.19; Leidy 2007, Leidy et al. 2011). Headwater streams within the northern Alameda Creek watershed that flowed from the drier Fremont-Livermore-Altamont hills into the Livermore Valley were typically intermittent (e.g., Tassajara Creek, Altamont Creek) and would have likely produced fewer steelhead than those draining the wetter portions of the Diablo Range. Small, intermittent tributary creeks that maintained permanent pools embedded in otherwise summer-dry stream reaches supported fishes such as thicktail chub, California roach, hitch, Sacramento sucker, threespine stickleback, and prickly sculpin (Leidy 2007, Leidy et al. 2011).

Large discontinuous creeks and distributaries

Large discontinuous creeks and distributaries such as those on the lowermost Alameda Creek floodplain were characterized by surface hydrologic connections with tidal wetlands and large intermittent or perennial streams during periods of flooding. These connections would allow fish move into these temporary habitats (fig. 8.20). Presumably dry creeks and distributaries embedded within floodplains would be characterized by fish species residing in adjacent perennial/intermittent stream or tidal marsh habitats. Persistence of these fishes would depend largely on the extent and duration of connectivity to other permanent aquatic habitats and the permanence of surface water in distributaries.

Figure 8.18. Step-pool sequence on Stonybrook Creek, a small tributary to Niles Canyon, 1897. This small intermittent stream would provide high quality spawning and rearing habitat for steelhead. Downstream migrating smolts and pre-smolts would have to travel a relatively short distance to reach Alameda Creek and then the San Francisco Estuary compared to headwater populations spawned in the Diablo Range. (courtesy The Bancroft Library, UC Berkeley)

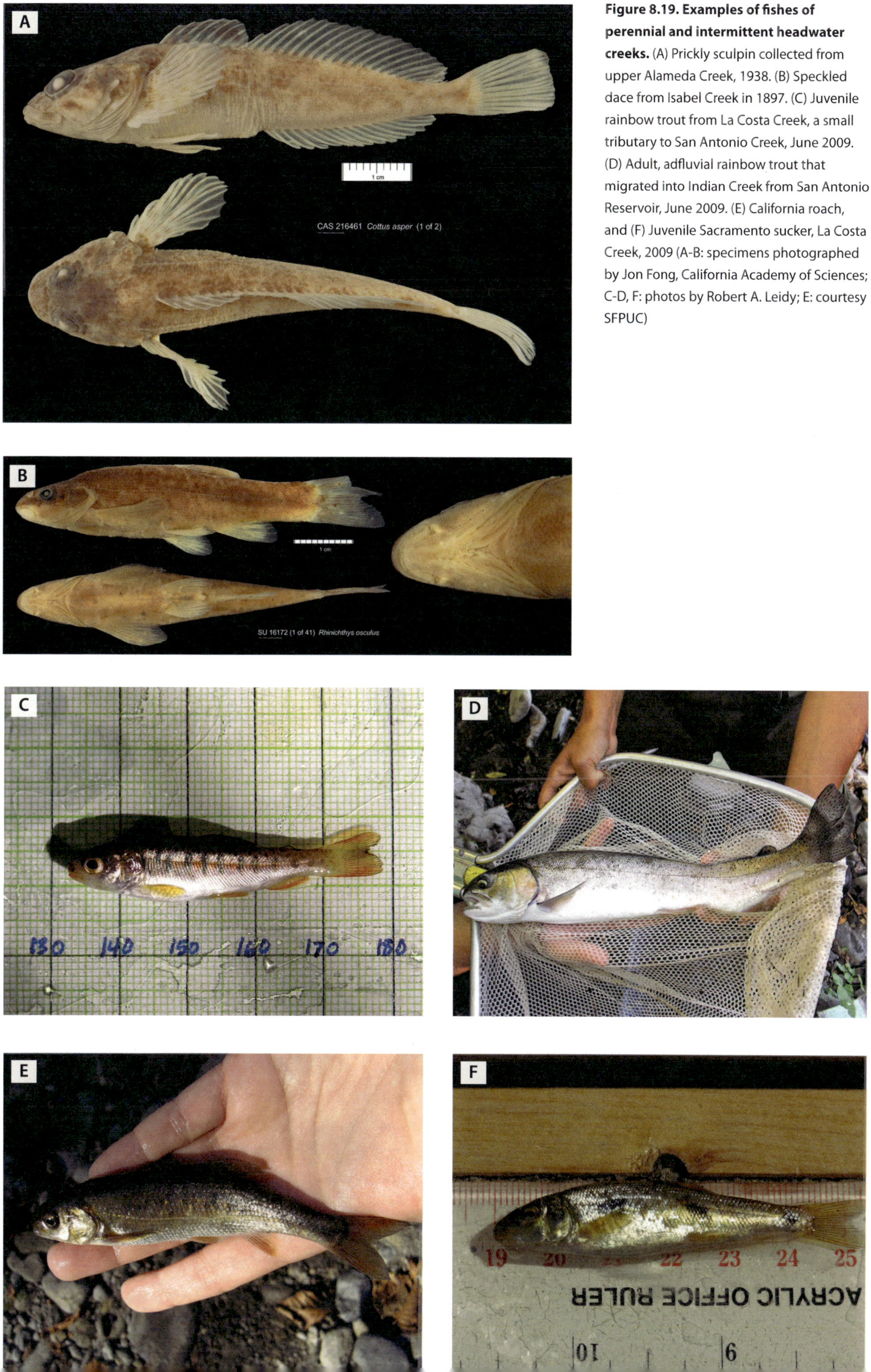

Figure 8.19. Examples of fishes of perennial and intermittent headwater creeks. (A) Prickly sculpin collected from upper Alameda Creek, 1938. (B) Speckled dace from Isabel Creek in 1897. (C) Juvenile rainbow trout from La Costa Creek, a small tributary to San Antonio Creek, June 2009. (D) Adult, adfluvial rainbow trout that migrated into Indian Creek from San Antonio Reservoir, June 2009. (E) California roach, and (F) Juvenile Sacramento sucker, La Costa Creek, 2009 (A-B: specimens photographed by Jon Fong, California Academy of Sciences; C-D, F: photos by Robert A. Leidy; E: courtesy SFPUC)

Figure 8.20. Crandall Slough, lower Alameda Creek floodplain reach. This shallow, ephemeral distributary channel would have created temporary connectivity for fishes between Alameda Creek and the tidal marsh during flood flows. (AD-563, © San Francisco Public Utilities Commission)

Some fishes likely would be trapped as distributaries dried following separation from other more permanent aquatic habitats. This pattern of seasonal fish use is consistent with similar small, discontinuous creeks (and adjacent watersheds) of the San Joaquin Valley and San Francisco Bay Estuary (Leidy 2007, Leidy pers. observ.).

The lower-to-middle reaches of Arroyo del Valle and Arroyo Mocho, two large intermittent tributary creeks in the Livermore-Amador Valley, were characterized by distributary channels on the valley floor where water often sank into porous alluvial soils before reaching downstream wetlands (Sowers 2003; see fig. 4.3). As a result, in years of below average precipitation, fish (including steelhead) may not have been able to migrate into the upper watersheds of these streams. However, during years of average to above average precipitation, periodic flooding of wet meadows adjacent to the lowermost reaches of Arroyo del Valle and Arroyo Mocho would have provided a temporary surface water connection between the lower and upper watershed of sufficient duration to allow steelhead and other fishes to migrate through this area.

Small discontinuous creeks and distributaries

The lower reaches of many small creeks within the Alameda Creek watershed descended from the foothills in defined channels before dissipating into undefined or multiple small channels on their alluvial fans (see fig. 4.49). These creeks were typically characterized by ephemeral or intermittent flows within their upper reaches (fig. 8.21). Small

Figure 8.21. Upper channelized reach of Alamo Creek. Alamo Creek flows from the hills north of Livermore-Amador Valley to join South San Ramon Creek, flowing into the Alamo Canal and eventually Arroyo de la Laguna. Historically, Alamo Creek was an intermittent, discontinuous stream that may have been fishless (see fig. 4.2a). The channelization of Alamo Creek and urbanization of its watershed has created perennial flow conditions that allow lowland fishes such as the native hitch and non-native sunfish species to colonize the stream. (photo by Robert A. Leidy)

discontinuous creeks and distributaries that rarely or never had surface connections with other larger permanent streams or tidal wetlands were probably fishless.

Non-tidal features (palustrine and lacustrine)

Perennial freshwater ponds

Fish bones recovered from Native American archeological sites in the vicinity of the Pleasanton marsh complex (see fig. 4.25) confirm the use of this marsh by several native fish species (Gobalet et al. 2004). The Lagoon and Tule Pond (see fig. 6.16) near Fremont also likely contained native fishes that colonized these ponds when Mission Creek or Arroyo de la Laguna, tributaries that flowed into and out of the Lagoon, flooded and connected the two ponds through adjacent perennial marsh and wet meadow. Probable fishes associated with perennial freshwater ponds, depending on seasonal hydrologic conditions, include thicktail chub, hitch, Sacramento pikeminnow, Sacramento sucker, threespine stickleback, juvenile steelhead, Sacramento perch, tule perch, and prickly sculpin (Gobalet 2006).

Perennial valley freshwater marsh

Perennial valley freshwater marsh was typically associated with perennial and seasonal freshwater ponds. Perennial freshwater marsh and associated large ponds with seasonal surface water connections to larger perennial streams or tidal marsh likely supported fishes such as Sacramento blackfish, hitch, Sacramento splittail, thicktail chub, Sacramento pikeminnow, Sacramento sucker, threespine stickleback, Sacramento perch, and tule perch. Perennial freshwater marsh and ponds embedded within grassland landscapes with no surface water connections to other permanent aquatic features were likely fishless. As with alkali wetlands and ponds, the areal extent of flooding and persistence of ponding water would vary from year to year depending on regional and local precipitation and runoff patterns and therefore would be an important determinate of fish assemblage membership and distribution.

Willow grove/sausal

TYPE 1 - PLEASANTON MARSH COMPLEX Seasonally or permanently flooded willow groves were found as wetland types in association with perennial valley freshwater marsh and perennial-to-intermittent ponds within the Pleasanton marsh complex (see fig. 4.19). Fish bones recovered from Native American archeological sites in the vicinity of the Pleasanton marsh complex confirm the use of the marsh by a number of native species, including thicktail chub, hitch, Sacramento blackfish, Sacramento pikeminnow, Sacramento sucker, prickly sculpin, threespine stickleback, tule perch, and Sacramento perch.

The Pleasanton marsh complex potentially functioned as seasonal habitat; it may have provided rearing habitat for steelhead during winter and early spring before water temperatures became too warm for juvenile trout to survive. In addition to the potential to function as seasonal rearing habitat for steelhead, the open water and flooded willow wetlands associated with the marsh likely provided a seasonal surface hydrologic connection for juvenile and adult steelhead migrating between the upper reaches of Arroyo Mocho and Arroyo del Valle through Arroyo de la Laguna to lower Alameda Creek. Finally, marsh wetlands may have functioned as sources of nutrients and organic matter that supported downstream food webs important to native fishes, including rearing steelhead (Trush pers. comm.).

TYPE 2 - WILLOW GROVE/SAUSAL – LOWER ALAMEDA FLOODPLAIN – UPPER TIDAL MARSH TRANSITION ZONE Seasonally or permanently flooded willow groves also formed in areas of significant fresh surface groundwater discharge within the lower Alameda Creek floodplain (see fig. 6.14). These sausals, in combination with contiguous tidal marsh, had the potential to provide significant estuarine rearing habitat for steelhead and Chinook salmon. Other fishes that may have been associated with sausals include threespine stickleback, prickly sculpin, Sacramento blackfish, tule perch, and Sacramento perch.

Large alkali playa

Frick Lake (see fig. 4.17) in eastern Livermore Valley was thought to be an intermittent alkali lake and therefore it was likely fishless. However, if Frick Lake retained water throughout the year and/or maintained a seasonal surface water connection with tributary creeks, then it possibly supported fishes such as threespine stickleback. Threespine stickleback is often associated with shallow, seasonal alkaline waters (Moyle 2002, Stanford et al. 2011).

Springs and seeps

Springs and seeps were geographically widespread throughout the Alameda Creek watershed and were characterized by shallow subsurface to surface water saturation, and/or small permanent pools with variable discharges. Seeps were fishless but likely played an important role in supporting native fishes, most notably rearing steelhead, through the discharge of cool groundwater into summer pools embedded within otherwise summer dry stream reaches. Springs were likely fishless; however, perennial springs in close proximity with surface hydrologic connections to other permanent aquatic and wetland habitats could have been colonized by fishes such as California roach, rainbow trout, and threespine stickleback (Leidy pers. observ.)

Wet meadows

Wet meadows were seasonally flooded herbaceous wetland communities often adjoining tidal marshes and some stream floodplains. Seasonally flooded wet meadows functioned as a potential surface hydrologic link for fishes between discontinuous streams and/or the floodplains within the upper tidal marsh transition zone. For example, periodic flooding of wet meadows adjacent to lower Arroyo Mocho Creek in Livermore-Amador Valley may have provided a temporary surface water connection for steelhead and other fishes such as California roach, Sacramento sucker, and threespine stickleback to move between the upper and lower watershed through areas characterized by poorly defined, discontinuous channels (fig. 8.22). Similarly, when flooded, the wet meadows bordering the lower Alameda Creek floodplain could have provided temporary surface

Figure 8.22. Flooded meadows along Mocho Canal could have provided seasonal migratory connections for fishes during overflows as well as temporary feeding opportunities. Even after the construction of Mocho Canal, seasonally wet meadows such as the one shown in the left side of this image could have provided seasonal access to rearing habitat for fishes during overflows. (A-1357 from Gutmann 1919, MB 067, courtesy SFPUC Archives)

hydrologic connections and feeding habitat for fishes moving between Alameda Creek and adjacent distributary channels and tidal marshes (fig. 8.23). The importance of floodplain wetlands to the movement of small fishes between watersheds has been documented for other streams tributary to the San Francisco Bay Estuary (Snyder 1905, Leidy 2007).

Sycamore alluvial woodland

Sycamore alluvial woodland growing on higher floodplain benches above the low channels of large intermittent streams such as Arroyo del Valle, San Antonio Creek and Alameda Creek upstream from Sunol would be fishless (fig. 8.24). Intermittent, low flow channels adjoining sycamore woodlands would function as critical migration corridors for anadromous fishes such as steelhead. Persistent pools would support other native fishes tolerant of high water temperatures and low levels of dissolved oxygen such as hitch, California roach, Sacramento blackfish, Sacramento pikeminnow, Sacramento sucker, and prickly sculpin.

Figure 8.23. Floodplain lake by Coyote Hills, January 1916. (top) Floodplain lakes such as these provided seasonal connectivity for fishes between various aquatic habitats. Depending on the duration of ponding, flooded areas may also have provided temporary rearing habitat for fishes. (AD-681, © San Francisco Public Utilities Commission)

Figure 8.24. Sycamores along San Antonio Creek (bottom), a large intermittent stream, ca. 1910. Leaf litter from sycamores likely was an important energy source for aquatic invertebrates upon which fishes would feed. (courtesy The Bancroft Library, UC Berkeley)

A-681

Table 8.3. Historical wetland habitats and associated probable native fish assemblages, Alameda Creek watershed.

System	Habitats	Example(s)
Tidal Features *(Marine and Estuarine)*	Shallow bay	Open water, permanently flooded embayments of San Francisco Bay bordering sloughs and wetlands near Union City, Newark and Fremont
	Tidal sloughs and flats	Periodically exposed mudflats bordering San Francisco Bay and sloughs interconnected with the Bay and Alameda Creek channel
	Tidal marsh	Periodically flooded marsh bordering San Francisco Bay and sloughs and landward wetlands and uplands
	Panne/Salina	Pannes embedded within tidal marsh adjacent to San Francisco Bay
Nontidal Features *(Riverine)*	Mainstem perennial streams	Alameda Creek, Little Yosemite to Upper Sunol Valley Alameda Creek (Niles Canyon) Alameda Creek, floodplain reach from Decoto Road to Alvarado
	Mainstem intermittent streams	Alameda Creek, Sunol Valley Alameda Creek, canyon mouth downstream to Little Yosemite Alameda Creek, gravel-bedded reach downstream from BART weir to Decoto Road
	Large perennial tributary creeks	Arroyo de la Laguna Upper Calaveras Creek Arroyo Hondo
	Large intermittent tributary creeks	Lower-to-middle Arroyo del Valle and Arroyo Mocho San Antonio Creek
	Small perennial-to-intermittent tributary creeks (e.g., headwaters)	Upper Dry, Stonybrook, Sinbad, Upper Alameda, Isabel, Smith, La Costa, Indian, Upper Arroyo del Valle, Trout, Upper Arroyo Mocho, and many other Diablo Range headwater streams
	Large discontinuous creeks and distributaries	Lowermost Alameda Creek floodplain
	Small discontinuous creeks and distributaries	Many small creeks descending from foothills onto alluvial fans

Probable historical native fish assemblage[1]	Notes on habitat use
Bat ray, leopard shark, Pacific herring, northern anchovy, white sturgeon, longfin smelt, jacksmelt, topsmelt, white croaker, Chinook salmon, steelhead, shiner perch, Pacific staghorn sculpin, longjaw mudsucker, arrow goby, bay goby, starry flounder, California halibut	Shallow, permanently flooded embayments supported various combinations of resident and migratory marine and euryhaline fish species typically found in south San Francisco Bay. Assemblage would have varied depending on local and regional environmental conditions, particularly changes to freshwater discharge through the Delta.
Bat ray, leopard shark, northern anchovy, white sturgeon, longfin smelt, jacksmelt, topsmelt, white croaker, Chinook salmon, steelhead, shiner perch, Pacific staghorn sculpin, longjaw mudsucker, arrow goby, bay goby, starry flounder	Tidal sloughs and adjoining mudflats likely supported various combinations of resident and migratory species found in south San Francisco Bay and the estuarine environment of lower Alameda Creek. Assemblage would have varied depending on local and regional environmental conditions, particularly changes to freshwater discharge through the Delta and resulting South Bay water salinities, and slough location, configuration, accessibility, depth, and size. Sloughs connected to upland streams functioned as migration corridors for adult Pacific lamprey and steelhead to upland streams. Juvenile steelhead and Chinook salmon used tidal sloughs for emigration to the bay and potentially in conjunction with tidal marsh for rearing. Tidal flats and sloughs were likely important rearing and feeding environments for many other marine and euryhaline fishes.
Chinook salmon, steelhead (?), jacksmelt, topsmelt, Pacific staghorn sculpin, threespine stickleback, longjaw mudsucker, arrow goby	Tidal marshes supported similar but less diverse assemblages of native fishes compared to tidal sloughs, mud flats and shallow embayments, in part because areas characterized by dense wetland vegetation restricted access and movement by large-bodied fishes. Tidal marsh may have provided significant estuarine rearing habitat for juvenile steelhead and Chinook salmon.
Threespine stickleback, arrow goby, bay goby	Pannes typically formed as isolated, intermittent, seasonally flooded habitat patches surrounded by dense tidal marsh vegetation, which restricted access by most fishes. Pannes were likely shallow and warm, exhibiting high salinities relative to tidal channels and marsh. Local environmental conditions would control whether pannes supported fishes or were fishless.
Pacific lamprey, Pacific brook lamprey, thicktail chub, hitch, California roach, Sacramento blackfish, hardhead, Sacramento pikeminnow, Sacramento sucker, steelhead (both as migration corridor for adults and smolts and as rearing habitat), threespine stickleback, prickly sculpin, Sacramento perch, tule perch	Perennial streams were typically associated with areas of groundwater discharge from seeps and springs along reaches underlain by impervious surfaces, in areas with active faulting, or zones with high or shallow groundwater tables. The greatest fish species diversity was found in mainstem perennial reaches characterized by deep, permanent pools and short, shallow riffles. Mainstem perennial reaches functioned as critical migration corridors and significant freshwater rearing habitat for steelhead.
Pacific lamprey (migration corridor), rainbow trout/steelhead (primarily as migration corridor for adults and smolts), thicktail chub, hitch, California roach, Sacramento pikeminnow, hardhead, Sacramento sucker, threespine stickleback, prickly sculpin, Sacramento perch, tule perch	Large intermittent mainstem streams would have functioned as critical migration corridors and potentially significant rearing habitat for steelhead. Other fish species would have used intermittent reaches seasonally, as a migration corridor to suitable upstream perennial and intermittent habitats, or would have remained in deeper pools that persisted throughout the summer and fall.
Pacific lamprey, Pacific brook lamprey, rainbow trout/steelhead, hitch, California roach, Sacramento pikeminnow, Sacramento sucker, threespine stickleback, riffle sculpin	Large perennial tributary creeks were particularly critical to the persistence of steelhead.
Pacific lamprey (migration corridor), rainbow trout/steelhead (primarily as migration corridor for adults and smolts), hitch, California roach, Sacramento blackfish, Sacramento pikeminnow, Sacramento sucker, prickly sculpin	Large intermittent tributary creeks would have functioned as critical migration corridors and potentially significant rearing habitat for steelhead. Other fish species would have used intermittent reaches seasonally, as a migration corridor to suitable upstream perennial and intermittent habitats, or would have remained in deeper pools that persisted throughout the summer and fall. Fish trapped as the stream dried would have perished.
Pacific lamprey, Pacific brook lamprey, rainbow trout/steelhead, California roach, Sacramento sucker, threespine stickleback, riffle sculpin	Small perennial and intermittent streams were located in the hilly or mountainous headwater reaches of watersheds. Pools would have varied from small to moderate in size and depth, and would have been characterized by cool water temperatures and coarse sand-gravel-cobble-boulder substrates. Perennial and intermittent tributaries would have served as critical spawning and rearing habitat for steelhead and/or resident rainbow trout.
Typically fishless unless associated with other streams supporting fishes.	Assemblage likely highly variable depending on local environmental conditions. Intermittent channels likely would have been characterized by various species characteristic of mainstem perennial/intermittent stream habitats (see above). Persistence of fishes would have depended largely on extent and duration of connectivity to mainstem streams and permanence of surface water in distributaries. Fishes likely were often stranded and perished as distributaries desiccated following separation from mainstem streams.
Typically fishless unless associated with wetland habitat.	Small streams and channels with ephemeral or intermittent flow and no continuous channel connections to downstream waters often contained little or no suitable habitat for fishes.

[1]The probable fish assemblage was derived from historical and recent records and accounts of fish occurrences assessed within the context of historical environmental conditions. Historical fish assemblage diversity and species abundances for each habitat type would likely exhibit significant temporal and spatial variability, and would be contingent on local topography, soils, geology, and annual rainfall patterns, among other environmental factors. For example, habitat suitable for rainbow trout/steelhead immigration, spawning, rearing, and emigration would vary temporally along several environmental axes, most notably the amount and distribution of annual precipitation, water temperature, dissolved oxygen, and the availability of food and cover.

Table 8.3 (continued). Historical wetland habitats and associated probable native fish assemblages, Alameda Creek watershed.

System	Habitats	Example(s)
Non-tidal Wetland *(Palustrine and Laustrine)*	Perennial freshwater ponds	Pleasanton marsh complex The Lagoon, Fremont Tule Pond, Fremont
	Perennial valley freshwater marsh	Pleasanton Marsh Complex The Lagoon, Fremont
	Perennial valley freshwater marsh - willow grove/sausal (Type 1)	Pleasanton Marsh Complex
	Estuary fringe willow grove/sausal (Type 2)	Lower Alameda Floodplain – Upper Tidal Marsh Transition Zone
	Large playa	Frick Lake, Livermore Valley
	Springs and seeps	Widespread
	Wet meadow	Lower Arroyo Mocho Floodplain Lower Alameda Creek Floodplain – Upper Tidal Marsh Transition Zone
	Sycamore alluvial woodland with gravel channels	Alameda Creek, Sunol Valley San Antonio Creek Arroyo del Valle Arroyo Mocho

Probable historical native fish assemblage[1]	Notes on habitat use
Thicktail chub, hitch, Sacramento blackfish, Sacramento pikeminnow, Sacramento sucker, prickly sculpin, three spine stickleback, Sacramento perch, tule perch, steelhead	Fish bones recovered from Native American archeological sites in the vicinity of the Pleasanton marsh complex confirm the use of the marsh by several native species. The Lagoon and Tule Pond in Fremont likely contained native fishes that colonized the ponds when Mission Creek and/or Arroyo de la Laguna, tributaries that flowed into and out of the Lagoon, flooded and connected the two ponds through adjacent perennial marsh and wet meadow.
Thicktail chub, hitch, Sacramento blackfish, Sacramento pikeminnow, Sacramento sucker, prickly sculpin, threespine stickleback, Sacramento perch, tule perch, steelhead	Fish bones recovered from Native American archeological sites in the vicinity of the Pleasanton marsh complex confirm the use of the marsh by several native species. In addition to the potential to function as seasonal rearing habitat for steelhead, open water and flooded wetlands associated with the marsh likely provided a seasonal surface hydrologic connection for juvenile and adult steelhead migrating between the upper reaches of Arroyo Mocho and Arroyo del Valle, and lower Alameda Creek.
For seasonally or permanently flooded willow groves with access to perennial-to-intermittent ponds: steelhead (intermittent rearing), thicktail chub, hitch, Sacramento blackfish, Sacramento pikeminnow, Sacramento sucker, prickly sculpin, threespine stickleback, Sacramento perch, tule perch	Fish bones recovered from Native American archeological sites in the vicinity of the Pleasanton marsh complex confirms the use of the marsh by several native species. In addition to the potential to function as seasonal rearing habitat for steelhead, open water and flooded willow wetlands associated with the marsh likely provided a seasonal surface hydrologic connection for juvenile and adult steelhead migrating between the upper reaches of Arroyo Mocho, and possibly Arroyo del Valle, and lower Alameda Creek.
For seasonally or permanently flooded willow groves with access to tidal channels: Chinook salmon (potential rearing), steelhead (rearing), threespine stickleback, prickly sculpin, Sacramento blackfish, tule perch, Sacramento perch, prickly sculpin	Flooded willow groves embedded within the lower Alameda Creek floodplain may have worked in combination with tidal marsh to provide potentially significant estuarine rearing habitat for steelhead and/or Chinook salmon.
Frick Lake would be fishless if it completely dried annually.	Frick Lake was an intermittent lake containing water seasonally. However, if Frick Lake retained water throughout the year and/or maintained a seasonal surface water connections with streams, then it possibly supported fishes such as threespine stickleback, a species often associated with shallow, alkaline waters.
Typically fishless	Springs with permanent surface connections and in close proximity to streams may have supported California roach, rainbow trout, and threespine stickleback. Surface and shallow subsurface discharges from springs and seeps to nearby streams likely made critical contributions of cool water for the maintenance of suitable fish habitat, most notably rearing habitat for steelhead.
Typically fishless. Possibly California roach, Sacramento sucker, steelhead, and/or threespine stickleback	When seasonally flooded, wet meadow functions as a potential surface hydrologic link for fishes between discontinuous streams and floodplains and/or upper tidal marsh transition zone. For example, infrequent, periodic flooding of wet meadows adjacent to lower Arroyo Mocho Creek in Livermore-Amador Valley may have provided a temporary surface water connection for steelhead between the upper and lower watershed through areas typically characterized by dry, discontinuous channels.
Pacific lamprey (migration corridor), rainbow trout/steelhead (primarily as migration corridor for adults and smolts), hitch, California roach, Sacramento blackfish, Sacramento pikeminnow, Sacramento sucker, prickly sculpin	Sycamore alluvial woodland growing on higher floodplain terraces above the low flow channel would have been fishless except during large floods. Adjoining, intermittent, low flow channels would have functioned as critical migration corridors for anadromous fishes. Persistent pools would have supported other native fishes tolerant of high water temperatures and low levels of dissolved oxygen.

[1]The probable fish assemblage was derived from historical and recent records and accounts of fish occurrences assessed within the context of historical environmental conditions. Historical fish assemblage diversity and species abundances for each habitat type would likely exhibit significant temporal and spatial variability, and would be contingent on local topography, soils, geology, and annual rainfall patterns, among other environmental factors. For example, habitat suitable for rainbow trout/steelhead immigration, spawning, rearing, and emigration would vary temporally along several environmental axes, most notably the amount and distribution of annual precipitation, water temperature, dissolved oxygen, and the availability of food and cover.

Table 8.4. Historical status and evidence for native fish stream assemblages in the Alameda Creek watershed (bold script indicates taxon is endemic to the Sacramento-San Joaquin Zoogeographic Province).

Family	Species	Zoogeographic type	Life history status	Current distributional status	Current estimated abundance
Petromyzontidae/ Lampreys	*Lampetra tridentata* Pacific lamprey	OBF-SD	M, AND, FWR	P	Low
	Lampetra cf. *pacifica* Pacific brook lamprey	OBF-FD	FWR	P	Unknown
	Lampetra ayresii River lamprey	OBF-SD	M, AND, FWR	UR	Unknown
Acipenseridae/ Sturgeons	*Acipenser transmontanus* White sturgeon	OBF-SD	M, AND, EST	UR	Low
Cyprinidae/ Minnows	***Gila crassicauda* Thicktail chub**	OBF-FD	FWR	EX	Extirpated
	***Lavinia exilicauda* Hitch**	OBF-SD	FWR	W, LC	High
	***Lavinia symmetricus* California roach**	OBF-FD	FWR	W, LC	High
	***Orthodon microlepidotus* Sacramento blackfish**	OBF-FD	FWR	LC	Moderate
	***Pogonichthys macrolepidotus* Sacramento splittail**	OBF-FD	FWR, EST	AB	Extirpated
	***Ptychocheilus grandis* Sacramento pikeminnow**	OBF-FD	FWR	W, LC	High
	***Rhinichthys osculus* Speckled dace**	OBF-FD	FWR	UR	Unknown
Catostimidae/ Suckers	***Catostomus occidentalis* Sacramento sucker**	OBF-FD	FWR	W, LC	High
Salmonidae/ Salmon & Trout	*Oncorhynchus kisutch* Coho or silver salmon	OBF-SD	M, AND, FWR	AB	Extirpated
	Oncorhynhus tshawytscha Chinook salmon	OBF-SD	M, AND, FWR	P	Low
	Oncorhynchus mykiss Rainbow trout/steelhead	OBF-SD	M, AND, FWR	W, LC	High
Gasterosteidae/ Sticklebacks	*Gasterosteus aculeatus* Threespine stickleback	OBF-SD	M, EST, AND, FWR	W, LC	Moderate

Zoogeographic type: EM = euryhaline marine; OBF-FD = obligatory freshwater dispersant; OBF-SD = obligatory saltwater dispersant

Life history status: M = marine; AND = anadromous; FWR = freshwater resident; EST = estuarine resident; AMP = amphidromous

Current distributional status in watershed: LC = locally common; W= widespread; UR = uncommon/rare; AB = absent; EX = extinct; P = present but current status poorly documented or unknown

Probable population abundance (number of individuals) in watershed: Extirpated = 0; low = <1,000; Moderate= 1,000-100,000 ; High= ≥ 100,000; Unknown

Notable early record(s) from the watershed (year) (source)
Alameda Creek, Alameda co. (1860)(MCZ 8889, 8890)
Alameda Creek (1966)(Hopkirk 1973)
Archeological site CA-ALA-453 adjacent to Alameda Creek, just north of the confluence of Dry Creek, Union City (Schulz 1986)
Archeological site near vicinity of Mission Creek and Lake Elizabeth (site CA-ALA-576 or CA-ALA-509-342)(Gobalet et al. 2004)
Archeological site near confluence of Arroyo de la Laguna and Arroyo Valle (site CA-ALA-555)(Gobalet et al. 2004) Alameda Creek (1898)(Snyder 1905)
Alameda Creek at Sunol (1898) (J. O. Snyder, CAS 115974, and Snyder 1905) Santa Isabel Creek, Mt Hamilton and Arroyo Honda creeks (Snyder 1905) "Arroyo Crista Blanca, Livermore" [Arroyo valle] (1912) (F.S. Curtis, ANSP 38897) "Arroyo Mocho, a few (15) miles S.E. of Livermore" (1939)(Calhoun and Helm, CAS 137848)
Archeological site near confluence of Arroyo de la Laguna and Arroyo Valle (site CA-ALA-555)(Gobalet et al. 2004) Alameda Creek (1934)(San Francisco Aquarium Society, CAS 7406) Alameda Creek between Niles and bridge just east of Sunol (1939)(R.R. Miller and G. Murphy, UMMZ 133179)
"Arroyo Crista Blanca, Livermore" [Arroyo Valle] (1912) (F.S. Curtis, ANSP 38898) Alameda Creek (1955)(W.I. Follett and G.M. Peckham, CAS 26166)
[Alameda Creek], Sunol (1898)(J.O. Snyder, CNHM 2574) Alameda Creek, Niles Canyon (1927)(CAS 159793) Alameda Creek, immediately below dam in Niles Canyon (1927)(Follett 1974)
Santa Isabel Creek, Mt. Hamilton (1898)(J.O. Snyder, CAS 137823 and CAS 116172) Arroyo Honda and Isabel Creeks (1898)(Snyder 1905) Alameda Creek, at junction with Calaveras Creek (1938)(I. Shapovalov, CAS 73763)
Archeological site CA-ALA-483 and -585 adjacent to Arroyo de la Laguna, west of the intersection of Interstate 680 and Bernal avenue, Pleasanton (Gobalet 1990) MR Santa Isabel Creek, Mt. Hamilton (1898)(J.O. Snyder, CAS 137004) Alameda, Arroyo Honda, Smith and Isabel creeks (1898)(Snyder 1905)
Alameda Creek, 1938 or 1939, by I. Shapovalov (Hopkirk pers. comm.) Alameda Creek watershed (prior to 1962), noted at historical migration route and probable distribution for silver salmon and/or steelhead (Skinner 1962)
Archeological site CA-ALA-453 adjacent to Alameda Creek, just north of the confluence of Dry Creek, Union City (Schulz 1986)
Alameda Creek, described as "some trout ... early in the season" (Hallock 1877: 16) "Stoney brook [Stonybrook], near Niles" (Hallock 1877: 16) Alameda Creek, 1800s "ran year round with steelhead trout, salmon" (Amador-Livermore Valley Historical Society newsletter, October 1989) Arroyo Honda, Smith, and Santa Isabel Creek, Mt. Hamilton (1898)(J.O. Snyder, CAS 116872
Archeological site near vicinity of Mission Creek and Lake Elizabeth (site CA-ALA-576 or CA-ALA-509-342)(Gobalet et al. 2004) Alameda Creek, likely collected in late 1850s or 1860 based on other collections by A. Agassiz (MCZ 6760)

Museum Source Acronyms:
ANSP (Academy of Natural Sciences, Philadelphia, PA)
CAS (California Academy of Sciences, San Francisco)
MCZ (Museum of Comparative Zoology, Harvard University)
SU (Stanford University fish collection, housed at CAS)
USNM (United States National Museum - Smithsonian)

Table 8.4 (continued). Historical status and evidence for native fish stream assemblages in the Alameda Creek watershed (bold script indicates taxon is endemic to the Sacramento-San Joaquin Zoogeographic Province)

Family	Species	Zoogeographic type	Life history status	Current distributional status	Current estimated abundance
Cottidae/ Sculpins	*Cottus asper* Prickly sculpin	OBF-SD	AMP, EST, FWR	W, LC	High
	Cottus gulosus Riffle sculpin	OBF-FD	FWR	UR/AB?	Unknown
	Leptocottus armatus Pacific staghorn sculpin	EM	EST, AMP	LC	High
Centrarchidae/ Sunfish	***Archoplites interruptus*** **Sacramento perch**	OBF-FD	FWR	UR	Unknown
	Hysterocarpus traskii **Tule perch**	OBF-SD/FD	FWR	UR	Unknown
	Cymatogaster aggregata Shiner perch	EM	EST	LC	Moderate
Gobiidae/ Gobies	*Gillichthys mirabilis* Longjaw mudsucker	EM	M, EST	LC	Moderate
Pleuronectidae/ Righteye flounders	*Platichthys stellatus* Starry flounder	EM	M, EST	LC	Moderate

Zoogeographic type: EM = euryhaline marine; OBF-FD = obligatory freshwater dispersant; OBF-SD = obligatory saltwater dispersant

Life history status: M = marine; AND = anadromous; FWR = freshwater resident; EST = estuarine resident; AMP = amphidromous

Current distributional status in watershed: LC = locally common; W= widespread; UR = uncommon/rare; AB = absent; EX = extinct; P = present but current status poorly documented or unknown

Probable population abundance (number of individuals) in watershed: Extirpated = 0; low = <1,000; Moderate= 1,000-100,000 ; High= ≥ 100,000; Unknown

Museum Source Acronyms:

ANSP (Academy of Natural Sciences, Philadelphia, PA)
CAS (California Academy of Sciences, San Francisco)
MCZ (Museum of Comparative Zoology, Harvard University)
SU (Stanford University fish collection, housed at CAS)
USNM (United States National Museum - Smithsonian)

Notable early record(s) from the watershed (year) (source)
Alameda Creek (1898)(Snyder 1905) Alameda Creek, immediately below dam in Niles Canyon (1927)(Follett 1974)
Alameda Creek (1934)(San Francisco Aquarium Society, CAS 6840) Alameda Creek, junction with Calaveras Creek (1938)(L. Shapovalov, CAS 24720)
Lower Alameda Creek (Leidy 2004)
Archeological site CA-ALA-483 and -585 adjacent to Arroyo de la Laguna, west of the intersection of Interstate 680 and Bernal Avenue, Pleasanton (Gobalet 1990) Archeological site CA-ALA-453 adjacent to Alameda Creek, just north of the confluence of Dry Creek, Union City (Schulz 1986) Calaveras Reservoir (1943)(C.E. Holladay, CAS 20926) Alameda Creek, dredger ponds at Niles (1953)(Wi.I. Follett, G.M. Peckham and party, CAS 25736)
Archeological site CA-ALA-483 adjacent to Arroyo de la Laguna, west of the intersection of Interstate 680 and Bernal Avenue, Pleasanton (Gobalet 1990) Archeological site near vicinity of Mission Creek and Lake Elizabeth (site CA-ALA-576 or CA-ALA-509-342)(Gobalet et al. 2004) Alameda Creek, Sunol (Snyder 1905, CAS 105929) Alameda Creek, immediately below dam in Niles Canyon (1927)(Follett 1974)
–
Mouth of Alameda Creek (1975)(G. Scoppettone, Sacramento State University field trip) Alameda Creek, flood control channel at crossing of Newark and Union Blvd, west of Fremont (1981)(C. Swift and B. Crabtree, LACM 42661)
–

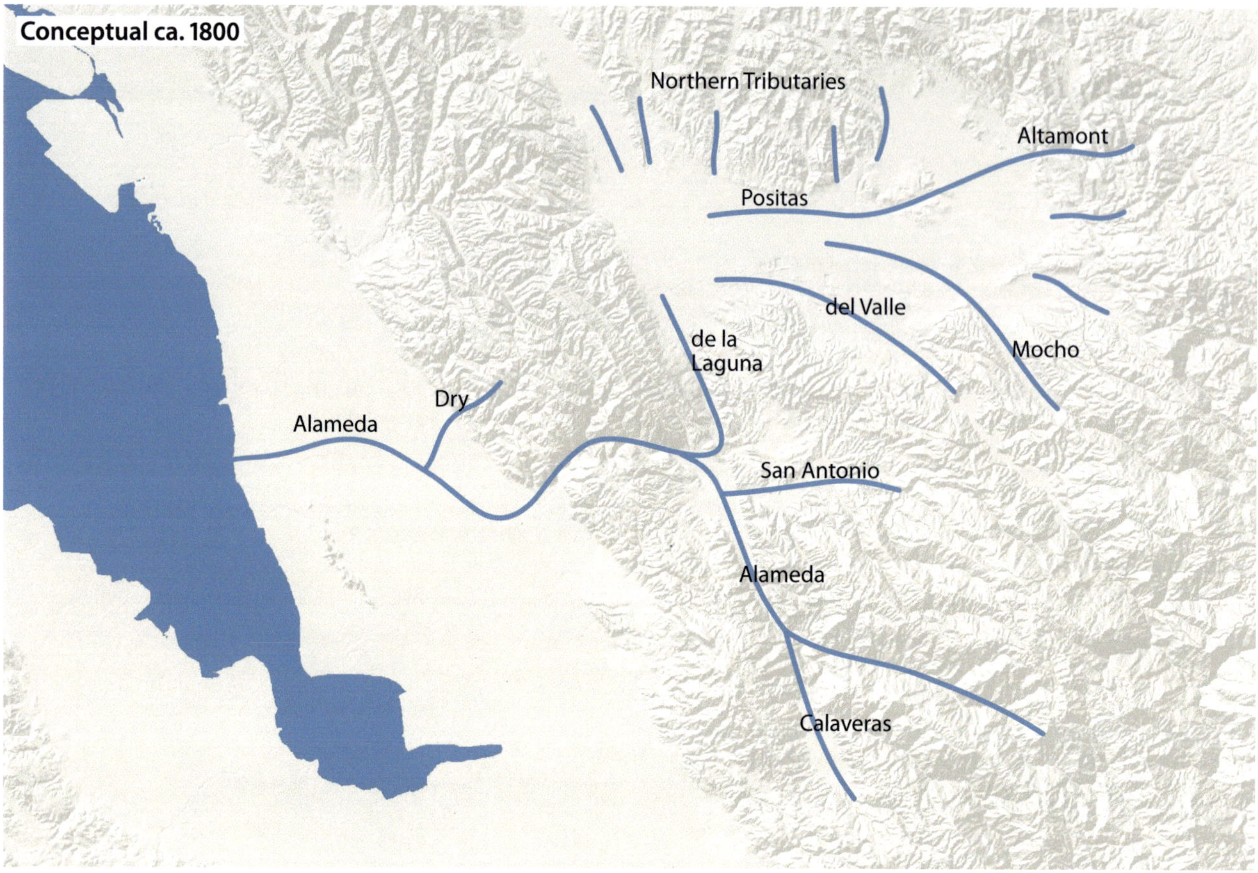
Conceptual ca. 1800
Northern Tributaries
Altamont
Positas
del Valle
Mocho
de la
Laguna
Dry
Alameda
San Antonio
Alameda
Calaveras

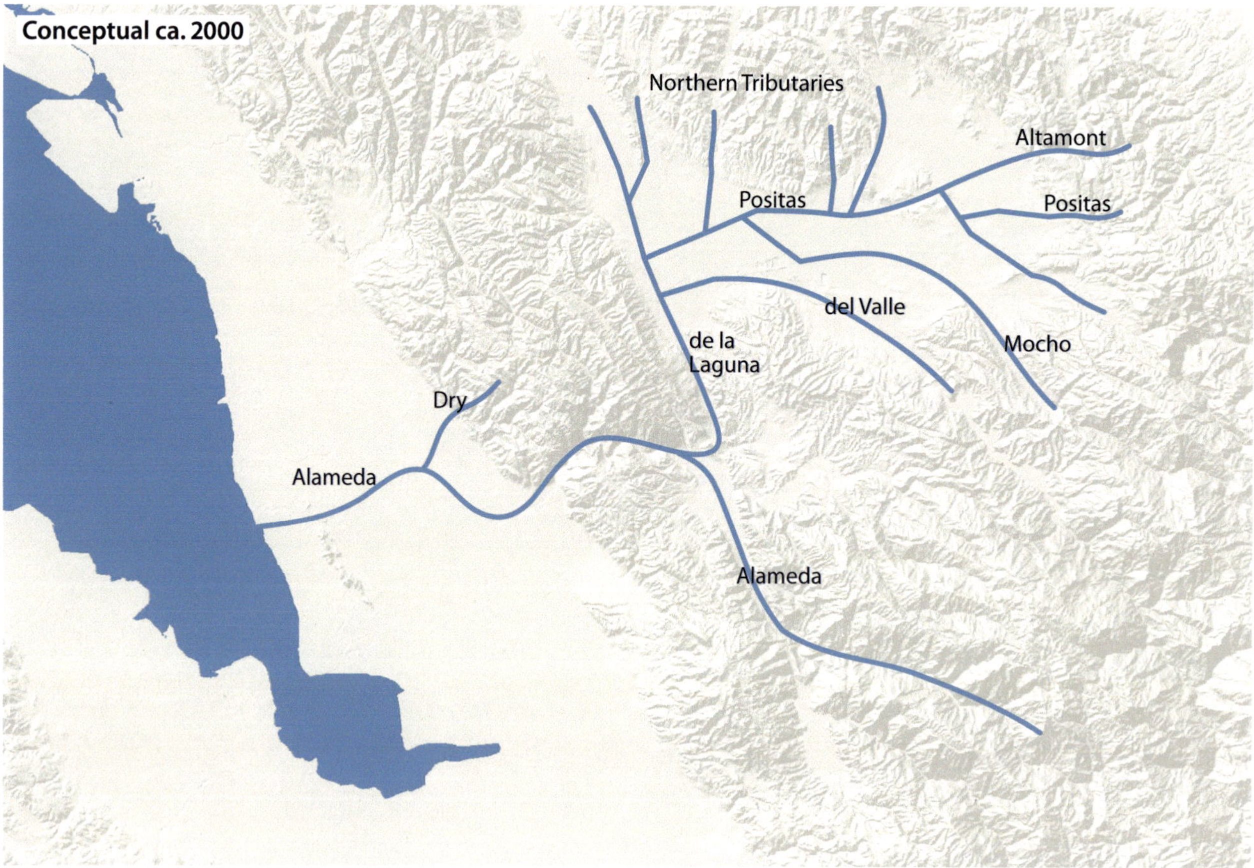
Conceptual ca. 2000
Northern Tributaries
Altamont
Positas
Positas
del Valle
Mocho
de la
Laguna
Dry
Alameda
Alameda

The modifications of the past 200 years have resulted in a profoundly altered Alameda Creek watershed. Rural and urban development now covers 60% of the study area, which is home to over 500,000 people. The large valley wetlands have been drained, and only fragments of the tidal marsh remain. Streams have been straightened, channelized, and connected. Even the species composition of remaining grasslands and riparian corridors has shifted in many cases towards non-native species.

Understanding the scope and character of these modifications can help us recognize the important components of the historical landscape and identify what has been lost. This perspective can help answer questions fundamental to the restoration planning process: How have landscapes changed? How do we make resilient, sustainable systems in the future? What are the priorities, and where are the opportunities?

This chapter initiates that process through a conceptual synthesis that explores differences between the historical and present-day watershed and the impacts that these differences may have had on watershed functions. This chapter begins with an analysis of land cover change and then presents a series of spreads providing a conceptual analysis of historical patterns and functions (fig. 9.1). The chapter ends with some suggested restoration opportunities and future research directions.

We have chosen to analyze the watershed as a whole to emphasize the interconnections between segments of the watershed that today often appear quite distinct or isolated. Many of the problems facing watershed managers can be most effectively approached by considering the range of functions contributed by each portion of the watershed, both historically and today. For example, as illustrated in Chapter 8, steelhead were historically supported by many types of habitat throughout the watershed, including places that may no longer support them today.

Figure 9.1. Conceptual view of Alameda Creek drainage network and watershed. This depiction of the watershed forms the basis for many of the conceptual diagrams used in this chapter. The diagrams focus on the alluvial plains of the Alameda Creek watershed. The channels shown here represent an abstracted network of the major streams ca. 1800 (top) and today (bottom).

Land cover change

To quantify changes in land cover, we combined three contemporary land cover layers. In eastern Alameda County (from Niles Canyon mouth east) we used data from the East Alameda County Conservation Strategy (ICF International 2010a). For the Niles Cone we used the Bay Area Aquatic Resource Inventory (BAARI) for both wetlands and baylands (SFEI 2011), and CalVeg (CalVeg 2004). These data allow a general comparison to be made between historical and contemporary conditions at the scale of the study area, but do not address habitat quality or character (i.e., historical grassland/low herbaceous cover comprised a different suite of species than most present-day annual grasslands, but both are displayed as "grassland"). For more information and a complete crosswalk, see Chapter 2, table 2.3. Some habitat types were excluded from this analysis, including most riparian classes and some small areas that were mapped as woodland but occurred along streams. These were excluded for consistency with the historical mapping (see stream methods discussion p. 32). As a result of these excluded classes, the calculated historical and contemporary total acreages are not exactly the same.

Our analyses document a dramatic shift in land cover type across the study area (figs. 9.2, 9.3). The most obvious change has been the spread of residential and industrial development and agriculture. Developed land has replaced grassland as the dominant land cover type, and together developed and agricultural land today cover 65% of the study area.

Wetlands, once hotspots of ecological diversity and productivity for the watershed, are greatly reduced in area today. Circa 1800, wetlands covered 60% of the Niles Cone and baylands (two-thirds of these wetlands were tidal wetlands). A large proportion of the Livermore-Amador Valley (45%) was also covered by wetlands. Overall, the total extent of non-tidal wetlands decreased to less than 20% of their historical extent. Large, spring-associated willow thickets (or willow swamps), which historically covered over 2,000 acres within the study area, are no longer present.

Much of this wetland area disappeared as a result of purposeful draining, but over the historical period water consumption and groundwater draw-down have also contributed to the reduction in wetland area. A reduction in water supply due to channelization, diversions, or active pumping can affect land cover types such as seasonal wetlands, seeps and springs, and riparian cover. These wetlands performed functions such as denitrification, production and storage of organic nutrients for the watershed, and water and sediment storage and filtration. These functions now must be performed elsewhere, either within the watershed or the Bay, if at all.

There has also been a shift in the type of wetlands, from large seasonal wetlands to perennial open water features such as quarry ponds and stock ponds. Reservoirs occur outside of the study area and were not included in this land cover analysis, but are an additional contributor to this shift in aquatic habitat type for the watershed.

Focusing on a particular area helps illustrate the types of changes that have occurred. Figure 9.4 shows the historical and present-day extent of wetlands within the Livermore-Amador Valley. All wetland types have declined in area with the exception of open water, which expanded with the construction of the quarry ponds. Alkali habitats in Livermore-Amador Valley occupy only a tenth of their historical extent: they have shrunk from 7,500 acres to 785 acres. However, these 785 largely contiguous acres of seasonal alkali wetlands represent a significant feature of the contemporary landscape and provide important remnant habitat. Within the study area, there are only two remnant alkali wetlands: one at Springtown in Livermore Valley and another in the Don Edwards National Wildlife Refuge at Warm Springs near Newark.

The baylands habitats have also been dramatically transformed (fig. 9.5). Today, tidal marshland covers only 23% of its historical extent within the study area and has largely been replaced by salt ponds, many of which are currently inactive. The area of former tidal wetlands is dominated today by perennial lagoons, a remnant of the salt ponds that were constructed through the late 1800s and early 1900s (fig. 9.6). Natural tidal pannes cover less than 7% of their historical area. The upper edge of the tidal marshes has been converted to industrial and residential development, eliminating the ecotone and reducing the overall area of the baylands.

A number of important modifications to the watershed could not be captured through this analysis of change in land cover type. In particular, changes in species composition alter aspects of the watershed including erosion rates, runoff, type of large woody debris in streams, and habitat. Many areas mapped as grassland today are used for grazing and are dominated by non-native species.

Each of the three regions in the study area (Livermore-Amador Valley, Sunol Valley-Niles Canyon, and Niles Cone) shows a distinct historical land cover pattern (fig. 9.7). For simplicity, we created pie charts that group historical land cover classes into grassland, seasonal wetland (alkali habitats and wet meadow), perennial wetland (perennial marsh and ponds), willow thicket, and tidal marsh. Despite differences in historical land cover, the three regions have followed similar trajectories and are dominated by development today, albeit to differing degrees (fig. 9.8). Seventy-seven percent of the Niles Cone region is now covered with development or artificial salt ponds. In Livermore-Amador and Sunol valleys, agriculture and grassland still make up a substantial portion of overall land cover—42% in Sunol and 35% in Livermore.

Figure 9.2a.

HISTORICAL LAND COVER, EARLY 1800s

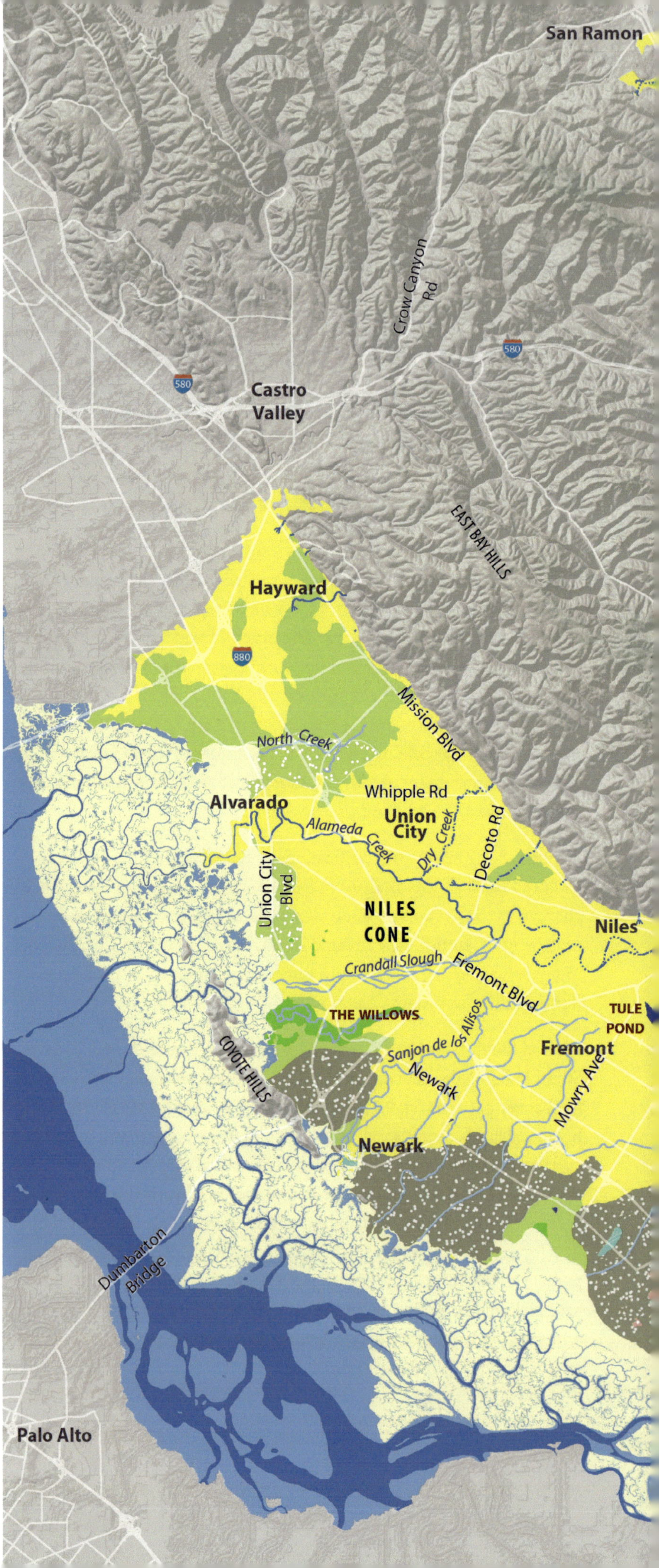

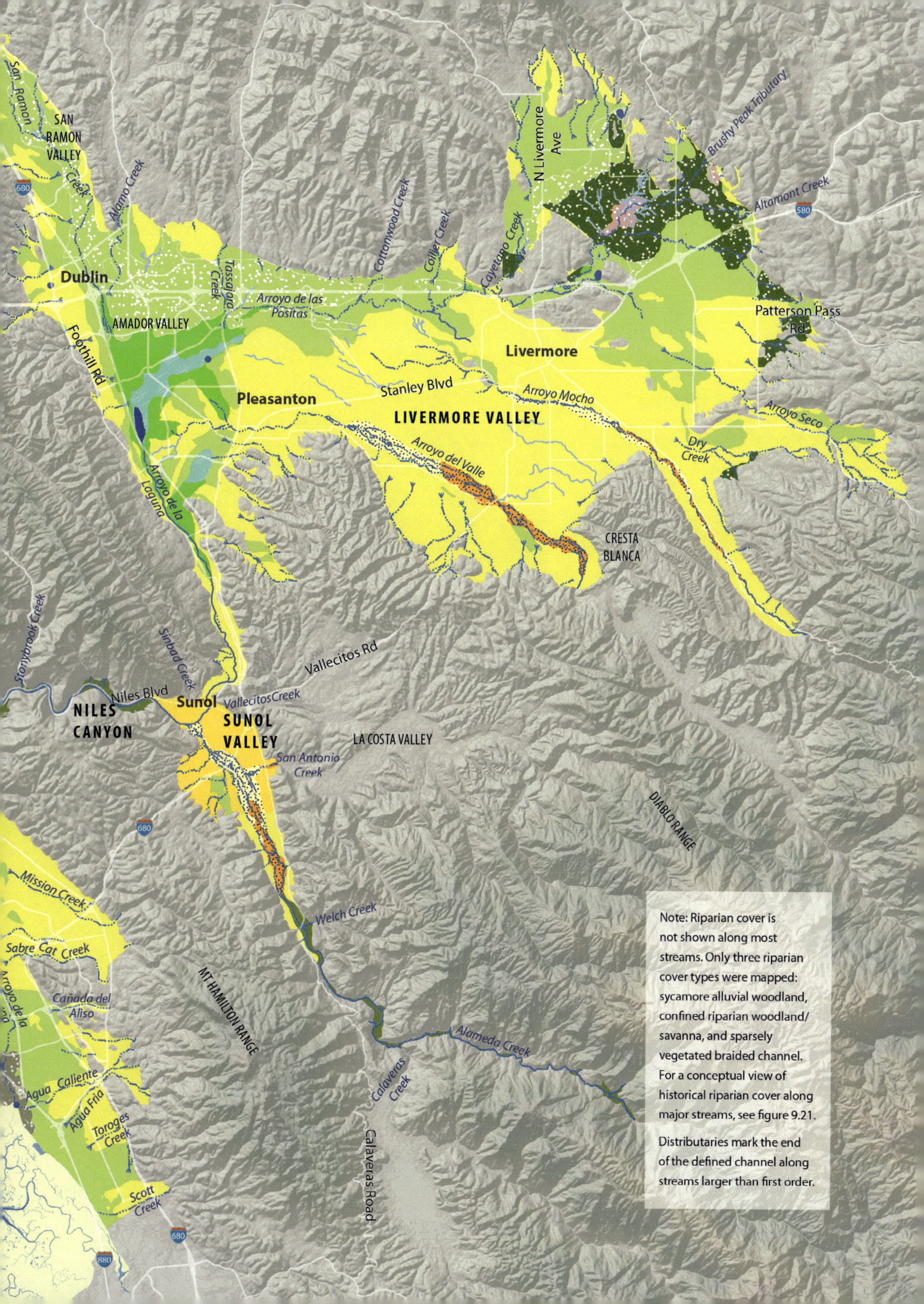

Note: Riparian cover is not shown along most streams. Only three riparian cover types were mapped: sycamore alluvial woodland, confined riparian woodland/savanna, and sparsely vegetated braided channel. For a conceptual view of historical riparian cover along major streams, see figure 9.21.

Distributaries mark the end of the defined channel along streams larger than first order.

Figure 9.2b.

CONTEMPORARY LAND COVER, CA. 2000

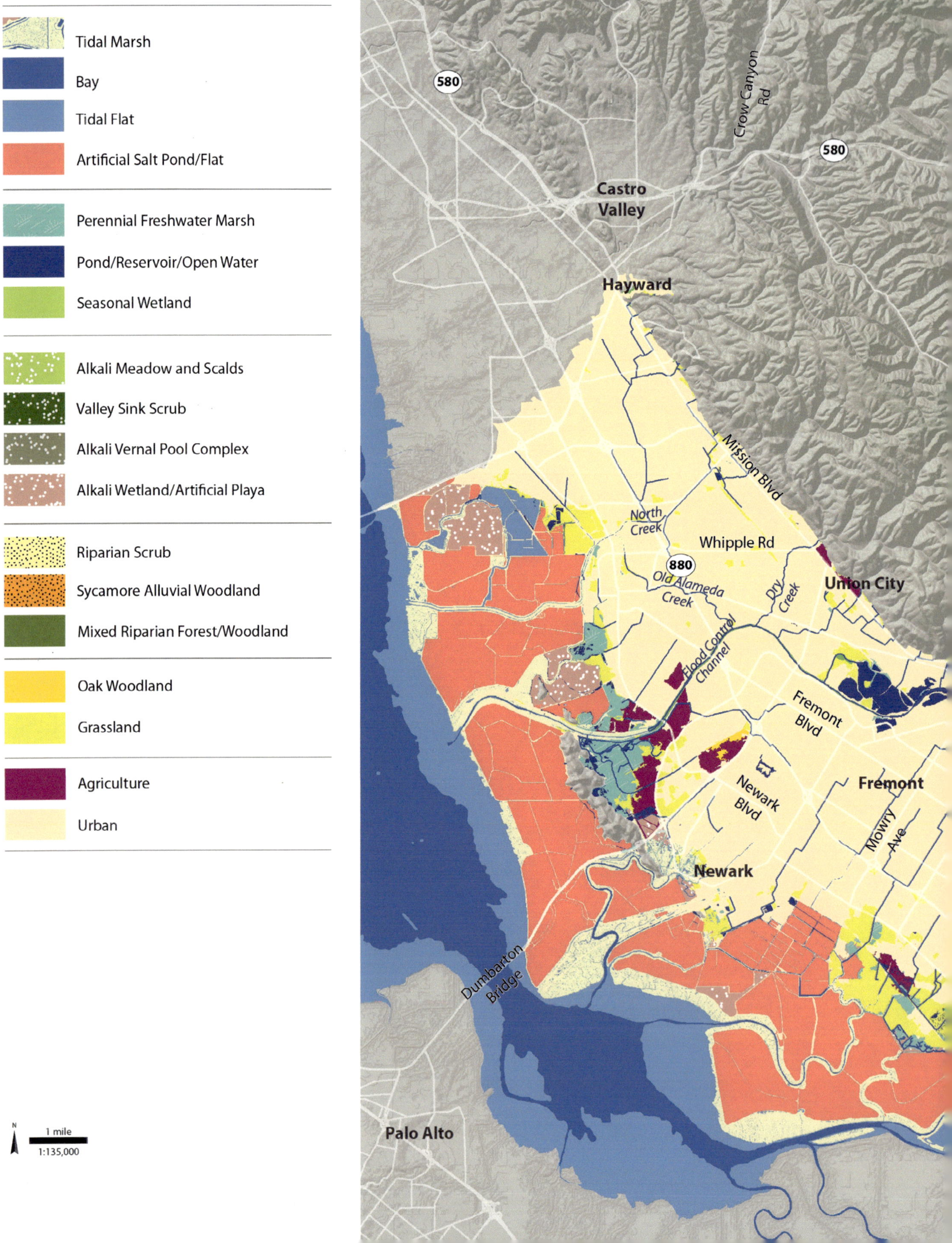

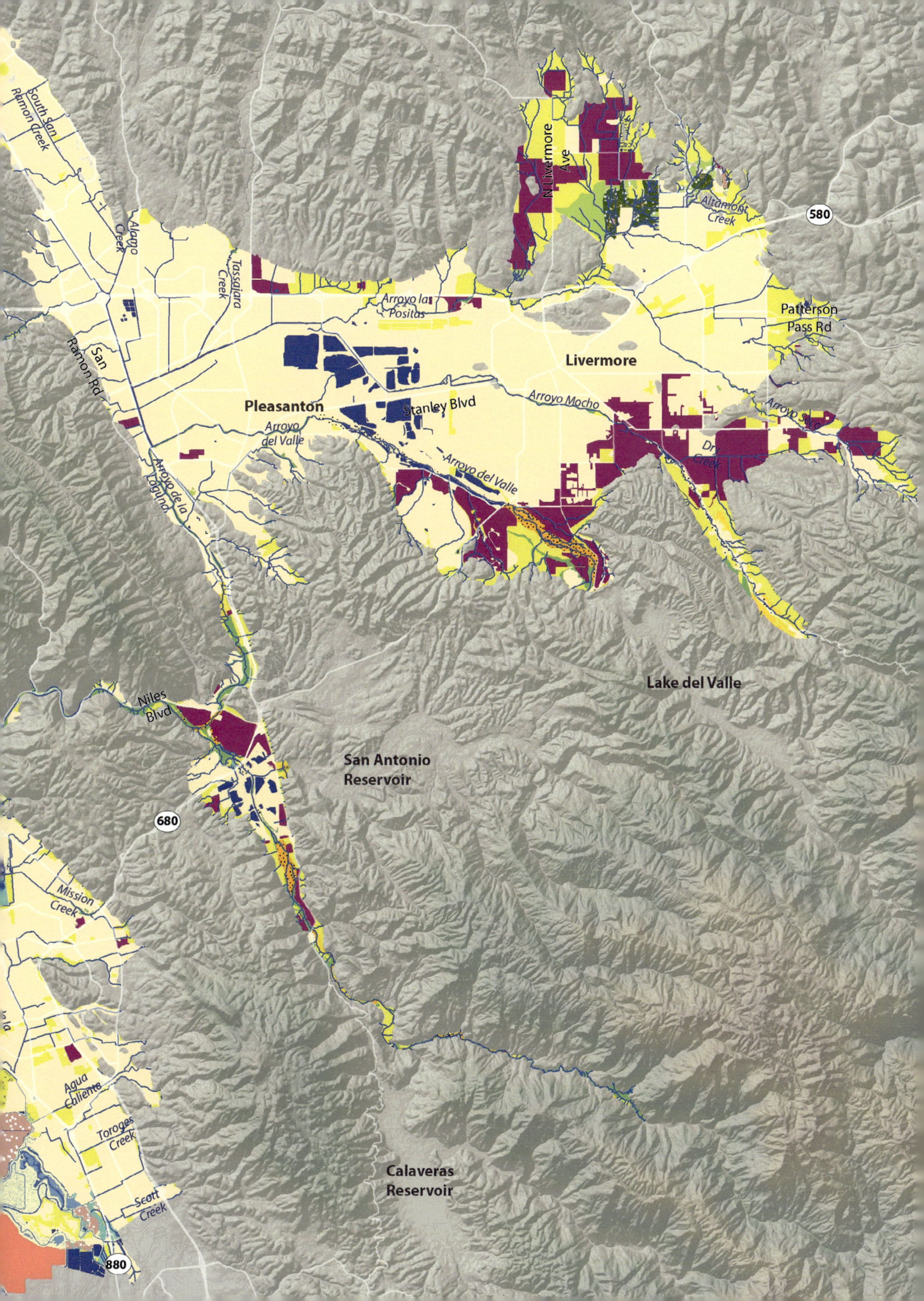
South San Ramon Creek
Alamo Creek
Tassajara Creek
San Ramon Rd
Arroyo las Positas
N Livermore Ave
Altamont Creek
580
Patterson Pass Rd
Livermore
Pleasanton
Stanley Blvd
Arroyo Mocho
Arroyo Seco
Arroyo del Valle
Dry Creek
Arroyo del Valle
Arroyo de la Laguna
Lake del Valle
Niles Blvd
San Antonio Reservoir
680
Mission Creek
Agua Caliente
Toroges Creek
Scott Creek
Calaveras Reservoir
880

Figure 9.2c.

SURFICIAL GEOLOGY MAP WITH FAULTS

Modern

- Open water
- gq - gravel quarries and percolation ponds
- Qhc - stream channel deposits

Later Holocene

- Qhfy - alluvial fan deposits
- Qhly - alluvial fan levee deposits
- Qhty - stream terrace deposits
- Qhay - alluvial deposits, undifferentiated

Holocene

- Qhbm - San Francisco Bay Mud
- Qhb - basin deposits
- Qhfe - alluvial fan-estuarine complex deposits
- Qhf - alluvial fan deposits
  - Qhf1
  - Qhf2
  - Qhf3
- Qhff - alluvial fan deposits, fine facies
- Qhl - alluvial fan levee deposits
  - Qhl1
  - Qhl3
- Qht - stream terrace deposits
  - Qht1
  - Qht2
- Qha - alluvium, undifferentiated

Holocene to Latest Pleistocene

- Qf - alluvial fan deposits
- Qt - stream terrace deposits
- Qa - alluvium, undifferentiated

Pleistocene

- Qpf - latest Pleistocene alluvial fan deposits
- Qpt - latest Pleistocene stream terrace deposits
- Qof - early to late Pleistocene alluvial fan deposits
- Qoa - early to late Pleistocene alluvial deposits, undifferentiated
  - Qoa1
  - Qoa2

Surficial geology mapping adapted from Witter et al. 2006 and Knudsen et a l. 2000. Faults from Jennings 1994.

N

1 mile

1:135,000

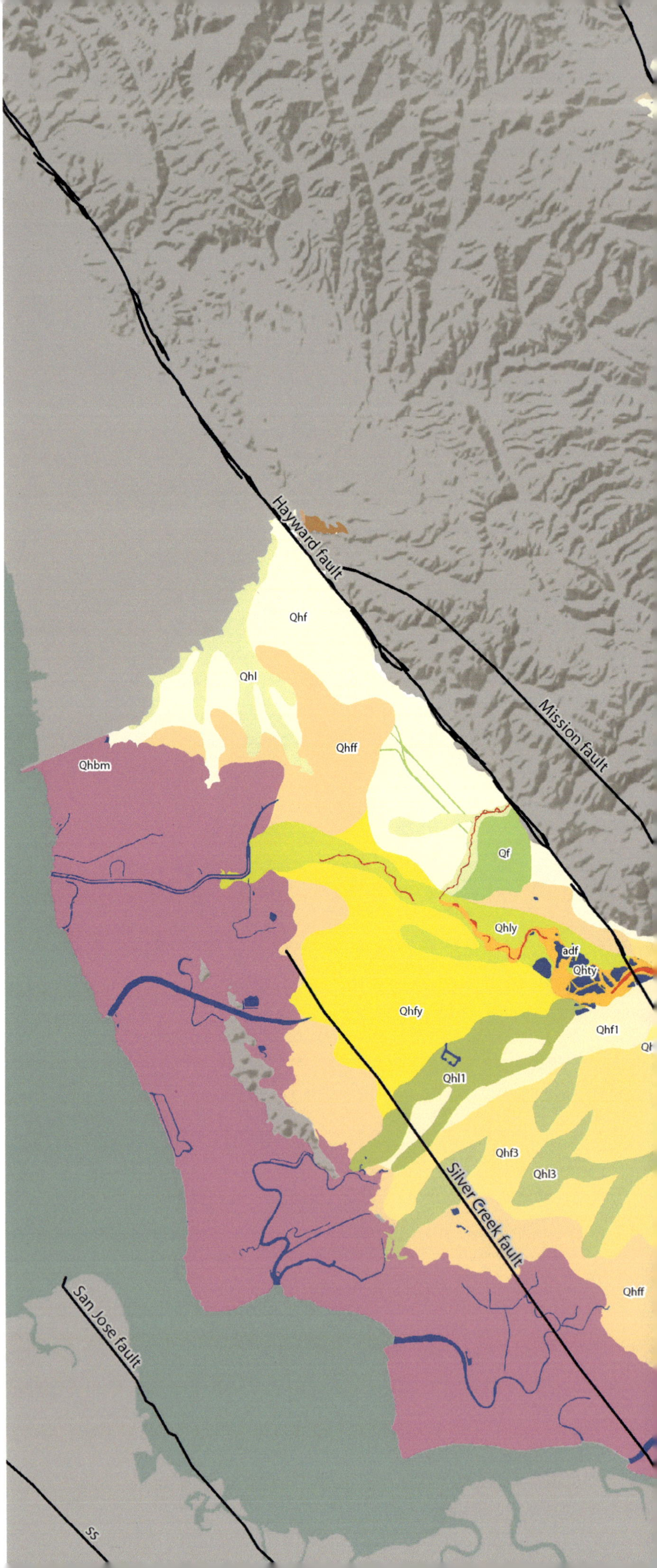

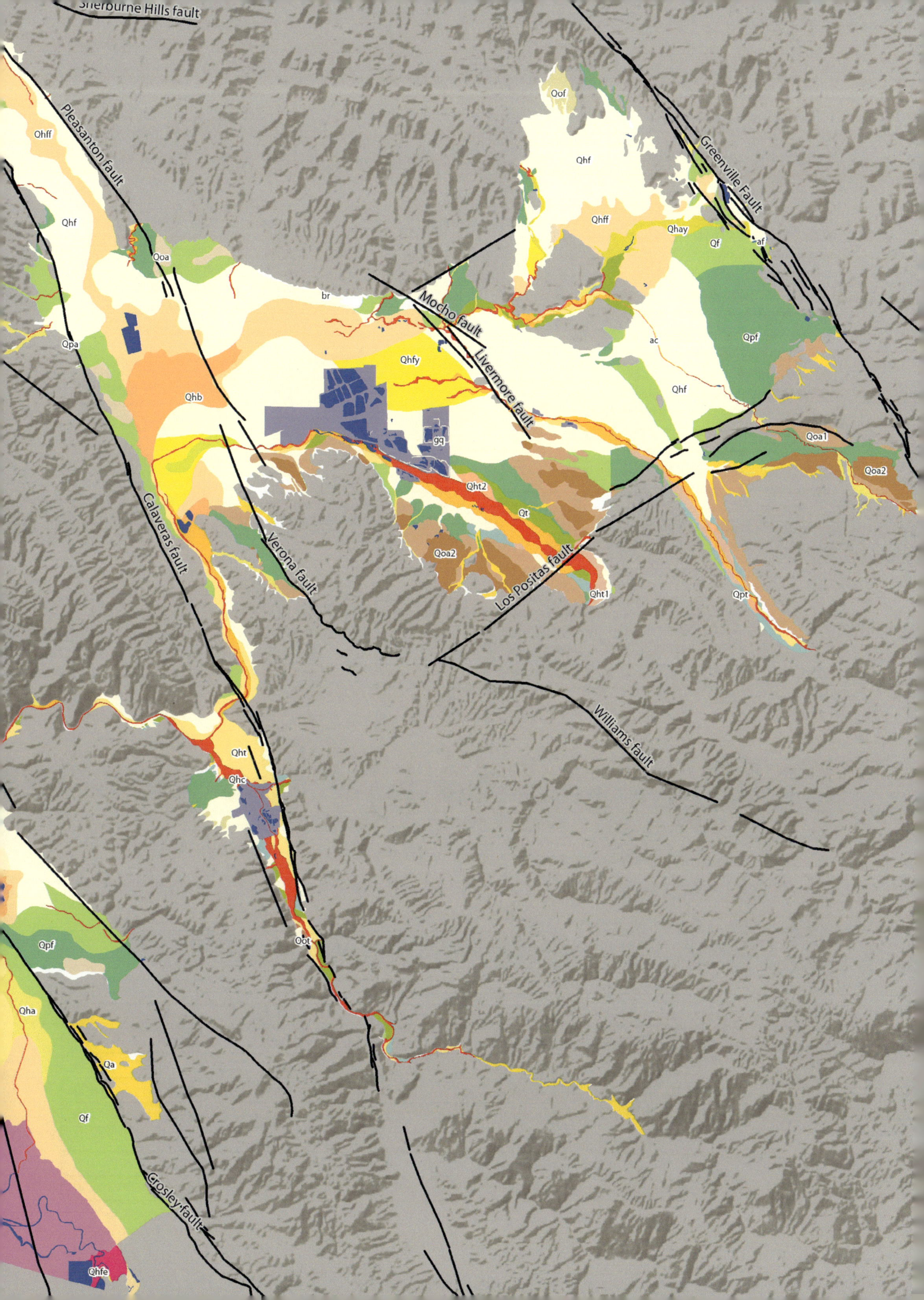

Sherburne Hills fault
Pleasanton fault
Qhff
Qhf
Qoa
Qpa
Qhb
br
Qhfy
gq
Mocho fault
Livermore fault
Qof
Qhf
Qhff
Qhay
Greenville Fault
af
Qf
ac
Qpf
Qhf
Qoa1
Qoa2
Qht2
Qt
Qoa2
Los Positas fault
Qht1
Qpt
Calaveras fault
Verona fault
Williams fault
Qht
Qhc
Qot
Qpf
Qha
Qa
Qf
Crosley fault
Qhfe

Three ways of looking at land cover change

Figure 9.3. Land cover change. The present-day land cover includes mapping completed between 2000 and 2011, so the date ca. 2000 was chosen to reflect this range. The portion of the study area that extends into Contra Costa County (near Dublin) was excluded from the analysis. Refer to Table 2.3 for crosswalk. The stacked bar chart (top) shows the "compression" of native habitats into greatly reduced area and the replacement with new land cover types such as developed lands, agriculture, and salt ponds. Bar charts (center) allow side by side comparison of past and present extent. The table (bottom) gives the numerical data, rounded to the nearest 50 acres. Classes have been lumped for ready comparison between historical and contemporary land cover patterns. (Contemporary data from CalVeg 2004, ICF International 2010a, and SFEI 2011)

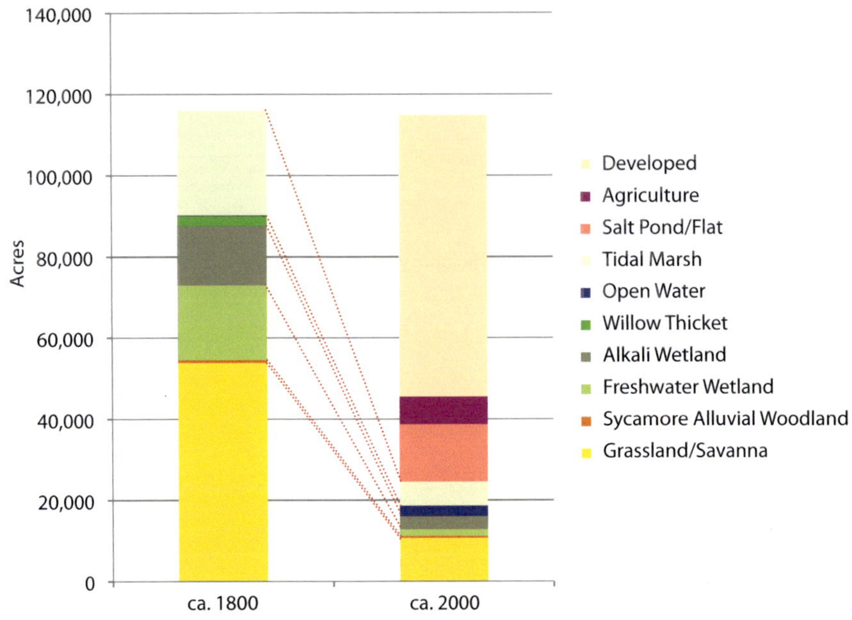

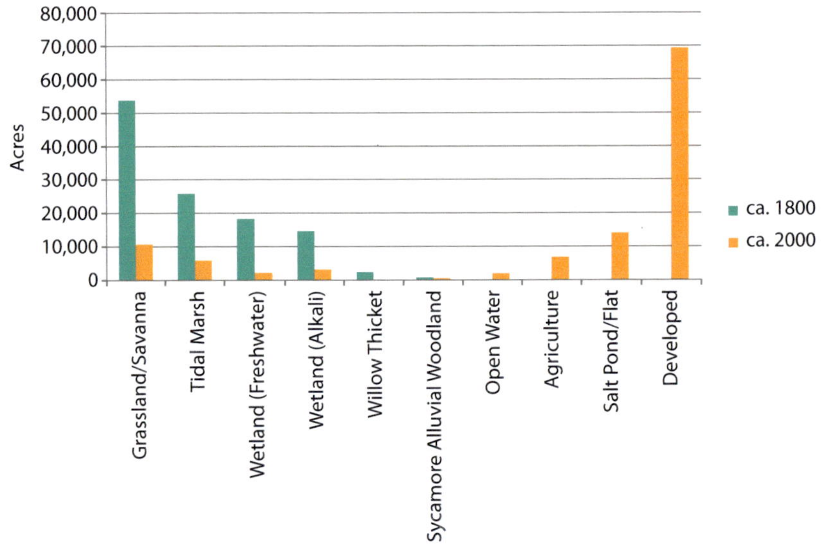

NOTES

- Tidal Marsh includes all natural tidal habitats (flats, pannes, shallow bay, vegetated tidal marsh). The only excluded bayland habitat type is salt pond, which is classified separately.
- Historical grassland and savanna classes have been combined. Grassland ca. 2000 includes all grassland types, including annual and perennial grasslands.
- Riparian habitat types were excluded, with the exception of sycamore alluvial woodland.
- We included historical willow thicket, a wetland type. We excluded the scrubbier present-day mixed willow riparian scrub as it is not equivalent to the historical willow thicket class.

	Acres ca. 1800	Acres ca. 2000	% Change
Grassland/Savanna	54,600	10,750	-80%
Sycamore Alluvial Woodland	800	500	-32%
Wetland (freshwater)	19,500	1,700	-90%
Wetland (alkali)	15,800	3,200	-78%
Willow Thicket	2,400	0	-100%
Open Water	150	2,500	+1,489%
Tidal Marsh	25,850	5,900	-77%
Salt Pond/Flat		14,100	-
Agriculture		6,900	-
Developed		69,400	-

Livermore-Amador Valley wetlands and open water

Figure 9.4. Wetlands and open water. In Livermore-Amador Valley there has been a large reduction in wetland area and a conversion to open water. Most of the non-fluvial water resources today are perennial open water ponds, including quarry ponds, and little remains of the formerly extensive wetland landscape. The numbers in the table have been rounded. The stacked images below show Livermore-Amador Valley through time. (The Contra Costa County portion of San Ramon Valley was excluded from this comparison; contemporary data from ICF International 2010a)

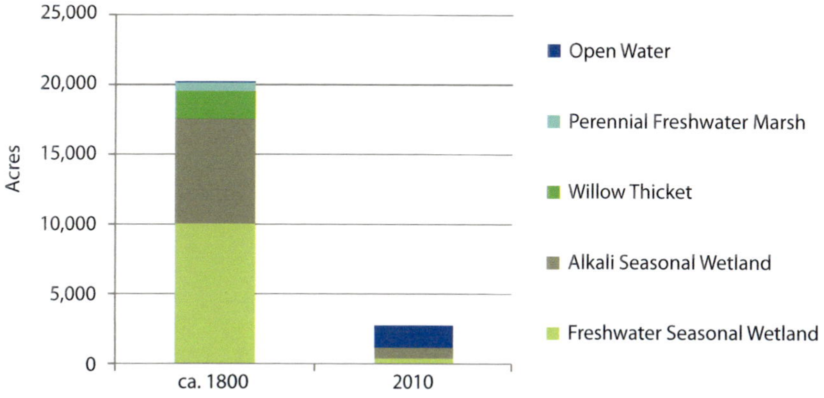

NOTES

- Historical seasonal wetland may be more inclusive than present-day seasonal wetland.
- Open water includes "riverine" in the 2010 mapping.

	Acres ca. 1800	Acres 2010	% Change
Freshwater Seasonal Wetland	10,050	400	-96%
Alkali Seasonal Wetland	7,500	800	-90%
Willow Thicket	2,000	0	-100%
Perennial Freshwater Marsh	600	20	-97%
Open Water	50	1,600	+2,808%

Change in tidal marshland composition

Figure 9.6. Change in tidal marshland composition. This chart and table show the replacement of much of the historical tidal marsh with salt ponds, as well as an overall reduction in area (due to fill). Marsh pannes, historically a common feature, have almost disappeared from the contemporary landscape. Bay-associated tidal flat included in this analysis only extends to the edge of the tidal marsh and does not include the fringe of tidal flat along the marsh edge. Acreages are rounded. (Contemporary data from SFEI 2011)

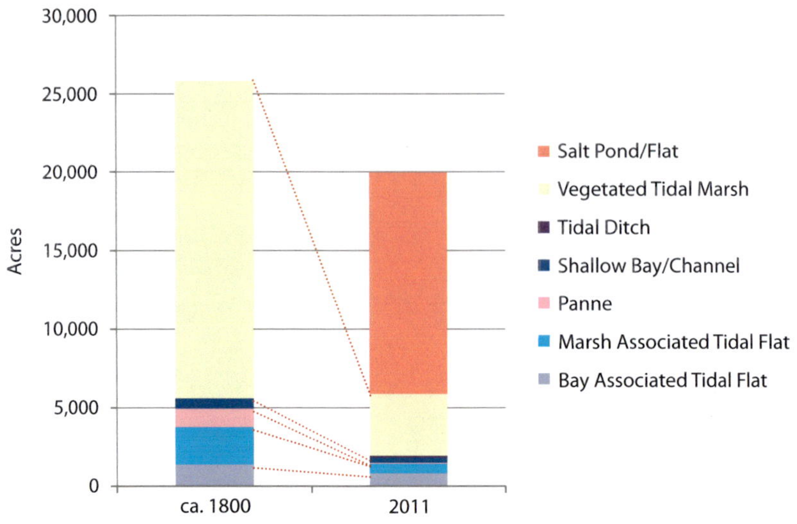

	Acres ca. 1800	Acres 2011	% Change
Bay Associated Tidal Flat	1,400	800	-41%
Marsh Associated Tidal Flat	2,400	600	-75%
Panne	1,150	75	-93%
Shallow Bay/Channel	650	350	-51%
Tidal Ditch	0	100	
Vegetated Tidal Marsh	20,250	3,950	-80%
Artificial Salt Pond/Flat	0	14,100	

Figure 9.5. Spatial comparison of tidal marsh composition. (opposite page) These images show the historical tidal network (bottom) compared with the present-day network (top). Some of the major tidal channels can still be traced through the modified landscape, but many of the smaller channels and the historical connectivity to adjacent wetlands has been lost. The pink lines represent present-day tidal channels, as mapped for BAARI. (Contemporary: USDA 2009, SFEI 2011)

2009/2011

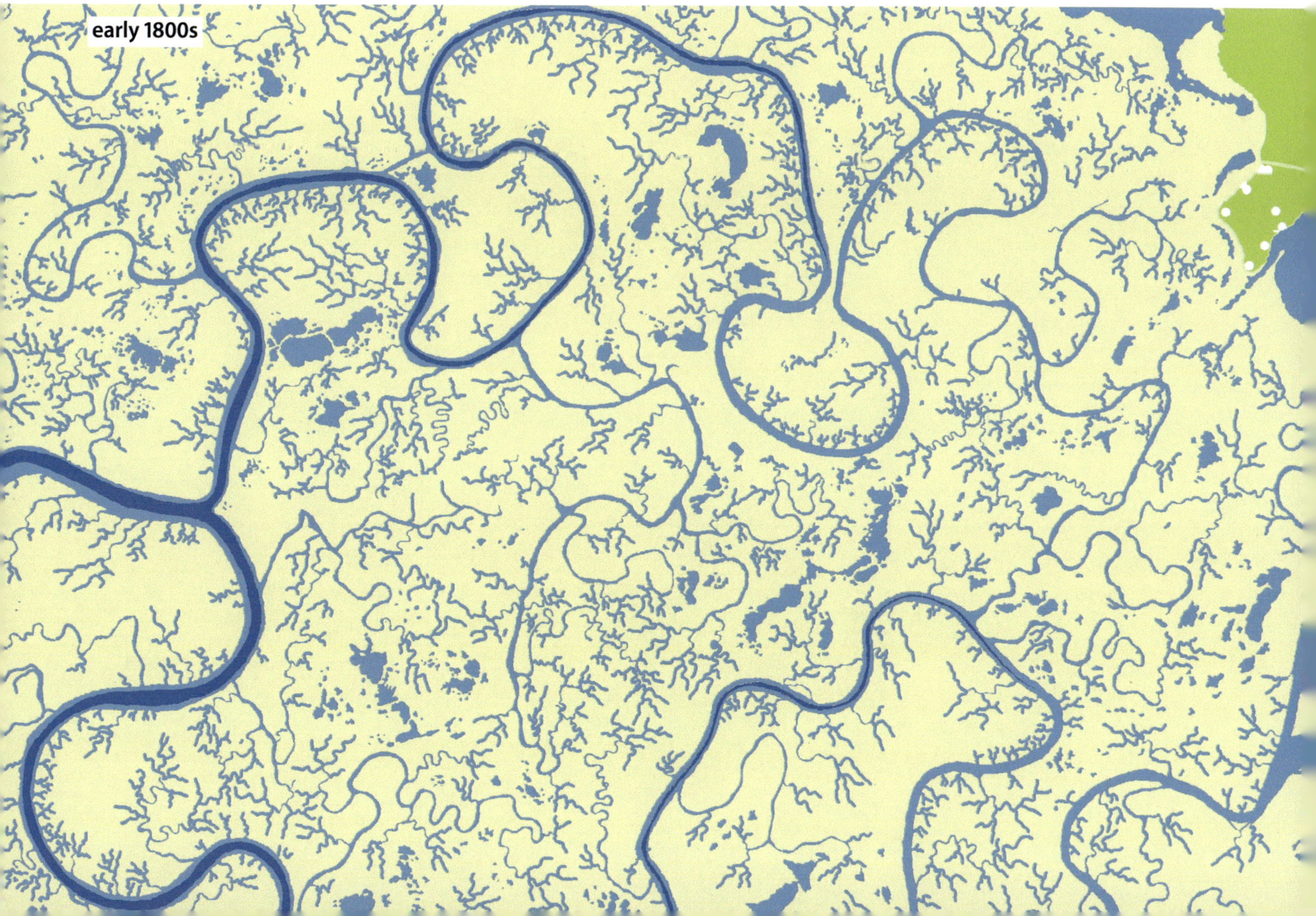
early 1800s

Historical land cover by subregion

Figure 9.7. Historical land cover. These charts show simplified versions of the distinct habitat patterns found in each of the three subregions, ca. 1800. Note the difference in total acres for each region. Riparian cover types were excluded from these charts.

NOTES

- Seasonal wetland includes all alkali wetland types and wet meadows.
- Perennial wetland includes valley freshwater marshes and perennial ponds.

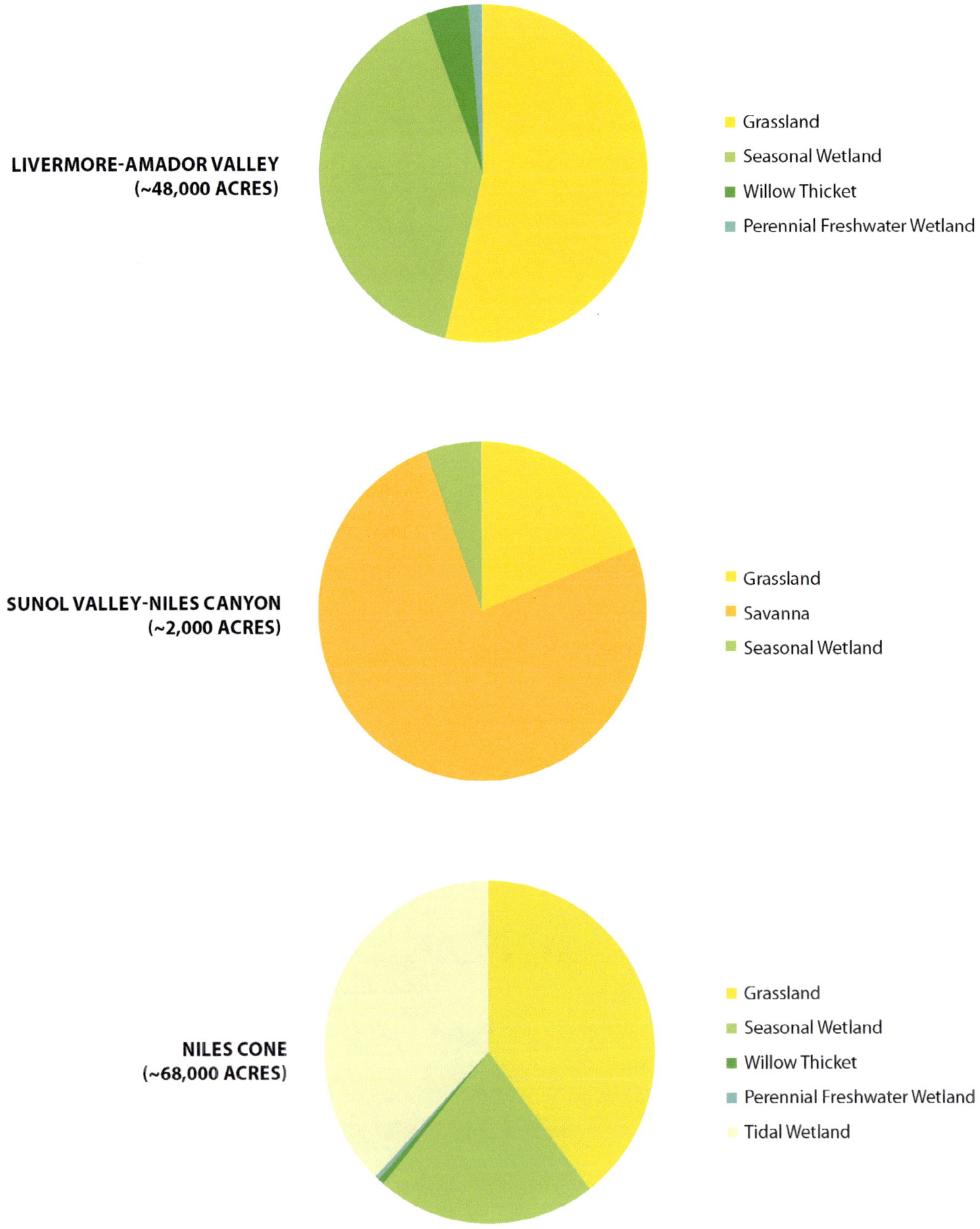

Change in land cover by subregion

Figure 9.8. Changes in land cover. For the same three subregions, these charts compare historical and contemporary land cover. (Contemporary data from CalVeg 2004, ICF International 2010a, and SFEI 2011)

NOTES

- Differences in classes between the three charts reflect differences in the present-day land cover mapping methods and classes for the three regions of the study area, as well as differences in land cover types present. See table 2.3 for more details on the crosswalk.
- Historical seasonal wetland may be more inclusive than present-day seasonal wetland.
- Riparian habitats (with the exception of sycamore alluvial woodland) were excluded.

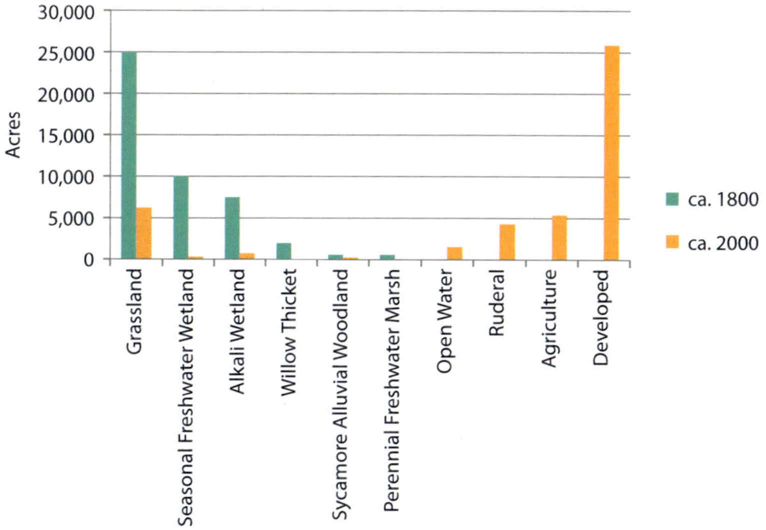

LIVERMORE-AMADOR VALLEY

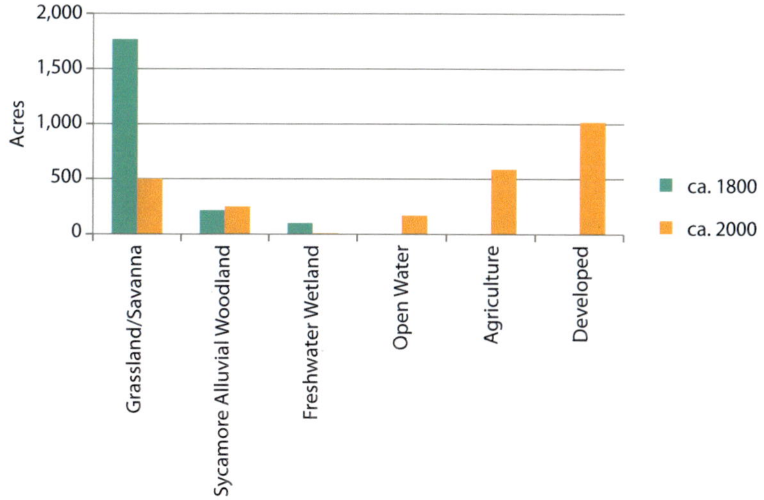

SUNOL VALLEY- NILES CANYON

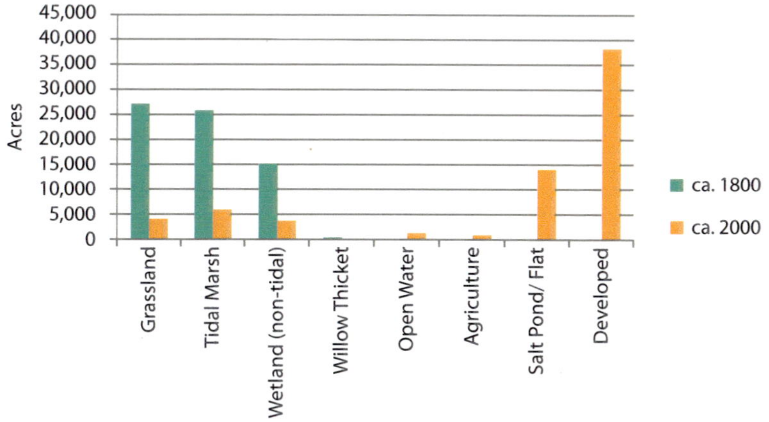

NILES CONE

Summary concepts for land cover change

The Alameda Creek watershed contained a diverse array of land cover types, from the willow groves and freshwater wetlands of the Pleasanton marsh complex to the iodine bush and alkali wetlands of the Springtown sink to the sycamore alluvial woodlands and oak savanna of Sunol Valley.

The distribution of habitats reflected underlying physical processes and characteristics. Topography, soils, geology, surface water, and groundwater availability were primary factors in determining historical habitat distribution.

Grasslands were the dominant vegetation cover across the study area. The grasslands contained wildflowers (forblands) and a mix of herbaceous species. Sparse oaks were found, particularly in northern Sunol Valley and at very low density in Livermore-Amador Valley, but the study area was dominated by low herbaceous cover, generally broken only by riparian tree corridors and willow thickets.

The Pleasanton marsh complex provided a vast amount of habitat that has been lost. It contained a mix of willow thickets, open water ponds, marshes, and seasonal wetlands. Ecosystem functions provided by this wetland complex would have included fine sediment and water storage, summertime water release, organic matter production to support fisheries downstream, and a wide array of wetland habitats.

By contrast, portions of both Springtown alkali sink and the tidal wetlands remain. Potential exists for restoration of additional portions of these once extensive wetland complexes, which may help restore some of the ecosystem functions once carried out by these wetland complexes.

Alkali wetlands contained an array of distinct habitat types, including alkali meadow, alkali sink scrub, alkali vernal pool complex, and alkali playa. These wetlands supported a host of now-rare native species, including vernal pool fairy shrimp (*Branchinecta lynchi*), San Joaquin spearscale (*Atriplex joaquinana*), Congdon's tarplant (*Centromadia parryi* ssp. *congdonii*), and palmate-bracted bird's beak (*Cordylanthus palmatus*). Protection of the remnant alkali wetlands in Alameda County would protect a host of native species that are currently under threat. Protection of the surrounding watersheds will be essential for the persistence of these wetlands.

Non-tidal perennial wetlands (freshwater marsh and ponds) covered fewer than 950 acres, or approximately one percent of the total wetland habitat. Perennial wetlands typically existed within more extensive seasonal wetlands.

Tidal marshlands were one of the most extensive habitat types in the study area, covering almost 26,000 acres and providing a vast area of wetland habitat. Restoration of some of these tidal marshes may help create a buffer against sea level rise and create needed wetland habitat to support steelhead and other species.

Tidal marshlands had a much higher density of channels historically than the remnants do today. These channels contributed to a greater tidal prism and provided more connection between large channels and the tidal marsh.

The connection of Alameda Creek with the tidal marsh (including the overflow connection) historically supported a fresh-brackish ecotone. The variety of habitats available along this ecotone supported a broad array of species.

Only 18% of historical wetland cover remains. The huge loss of wetland area makes those remaining areas highly valuable. Restoration efforts that expand the tidal marshlands and alkali wetlands that remain may be the most effective means to support wetland-dependent species.

Conceptual synthesis

The following conceptual synthesis brings together information from the previous chapters to describe patterns of form and function across the watershed. The section is organized into a series of two-page spreads, each comparing conditions historically and today. The spreads focus on the following topics:

- How has channel connectivity changed?
- How has channel form changed over time?
- How have water transport and storage changed?
- How did sediment move through the watershed?
- Where did wetlands occur?
- How has the seasonality of aquatic resources changed?
- How has riparian habitat diversity changed?

A graphical conceptual model or illustration forms the center of each spread, and the accompanying text guides the reader through the graphic. A page of summary points follows each major topic (channels, transport and storage, wetlands, and riparian cover).

We developed a conceptual view of the watershed, which we use as the foundation for many of the following spreads (see fig. 9.1). This graphical view focuses on major stream systems and those portions of the watershed that are within the study area (alluvial plains/valley floor). We show only an abstracted picture of the smaller tributaries to allow us to focus on the functions that the tributaries as a group perform. The diagrams are designed to show broad conceptual patterns across the study area.

How has channel connectivity changed?

Over the past 150 years, channels in the Alameda Creek watershed have transformed from highly discontinuous to tightly connected. Historically, streams spread into seasonal wetlands in many locations, and there were broad overflow zones and floodplains associated with many of the creeks. Streams that were connected to other streams or to wetlands in wet years would disconnect and dissipate in dry years. Today many miles of artificial channels extend the historical network to control the flow of water (fig. 9.9). Using our historical mapping and BAARI, we calculated that drainage density in the study area (excluding tidelands) increased from 2.3 mi/sq. mi to 2.9 mi/sq. mi.

Channel connectivity influences a host of fluvial processes, including incision and erosion, sediment transport and deposition, flooding, summer baseflow, and groundwater level and recharge, which in turn affect habitat availability for both aquatic and terrestrial species, including invasive species. These large-scale changes to the historical network have shaped the ways that sediment and water flow through the system today and the support of local wildlife.

Figure 9.9. Changes to the stream network. Each region of the study area experienced an increase in total stream miles. Stream alignments also changed, such that many contemporary channels do not match the course of channels in the historical network. These are stream miles that have been lost (shown in light blue). Many of these lost miles have been replaced by new miles of stream in the contemporary network (shown in green) that do not follow historical stream courses. These newly constructed channels may lack some of the functionality of natural stream courses. Storm drains (shown in fig. 9.10) are not included in these calculations. See fig. 9.11 for spatial depiction of channels included in this calculation. (Contemporary data SFEI 2011)

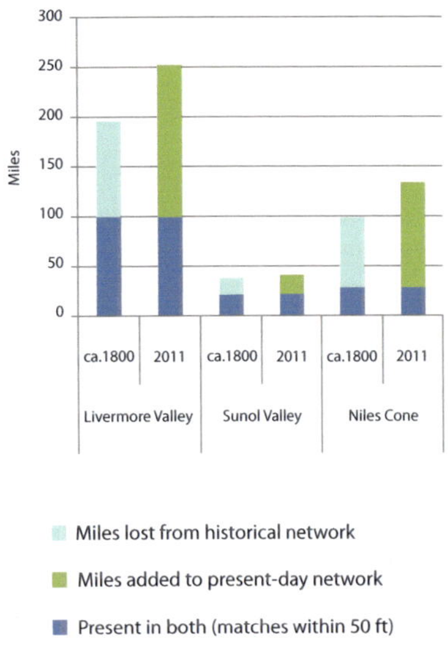

The most striking modification to the drainage network has been the construction of canals across the Pleasanton marsh complex. Over 10 miles of canals were developed through this wetland in the early 1900s in places that previously had no defined drainage. These canals funneled streams directly into Arroyo de la Laguna, rather than allowing them to spread and flood as they had historically (fig. 9.10; Collins and Leising 2003, Bigelow et al. 2008). Similar patterns of increased connectivity can be seen in the Springtown sink to the east, where some broad, shallow channels did exist historically, but have become more defined and numerous today.

Channel connectivity and length have also increased on small, unnamed systems. Most small drainages historically spread and sank into valley alluvium as they flowed from the hills. Today these drainages have almost universally been ditched to connect directly with another stream, helping to move water and sediment rapidly downstream and restricting opportunities for temporary storage that allow water to slow and spread.

At the same time, certain parts of the network were shortened as meanders and swales were removed to permit development, often leading to increased erosion. Stream miles have been lost along reaches that were historically sinuous and have been straightened, notably Alameda Creek on Niles Cone. A series of sloughs on Niles Cone and in Livermore-Amador Valley have been filled and no longer carry surface water.

Our mapped historical stream network measured only 335 miles, in comparison with the 427 (non-tidal) miles mapped by the BAARI dataset. Some of this difference is likely due to differences in mapping detail. However, it is primarily due to channel extension—areas where channels were historically discontinuous but have been extended to form direct connections (fig. 9.11). Only one-third of the contemporary network falls within 50 feet of historical channels. The disparity is greatest in the Niles Cone, where there are more ephemeral and artificial drainages today. The contemporary and historical networks are most similar in Sunol Valley, where the stream network was historically largely continuous. Overall, 44% of the historical network is preserved in the contemporary network (if we do not include sloughs, this percentage increases to 52%).

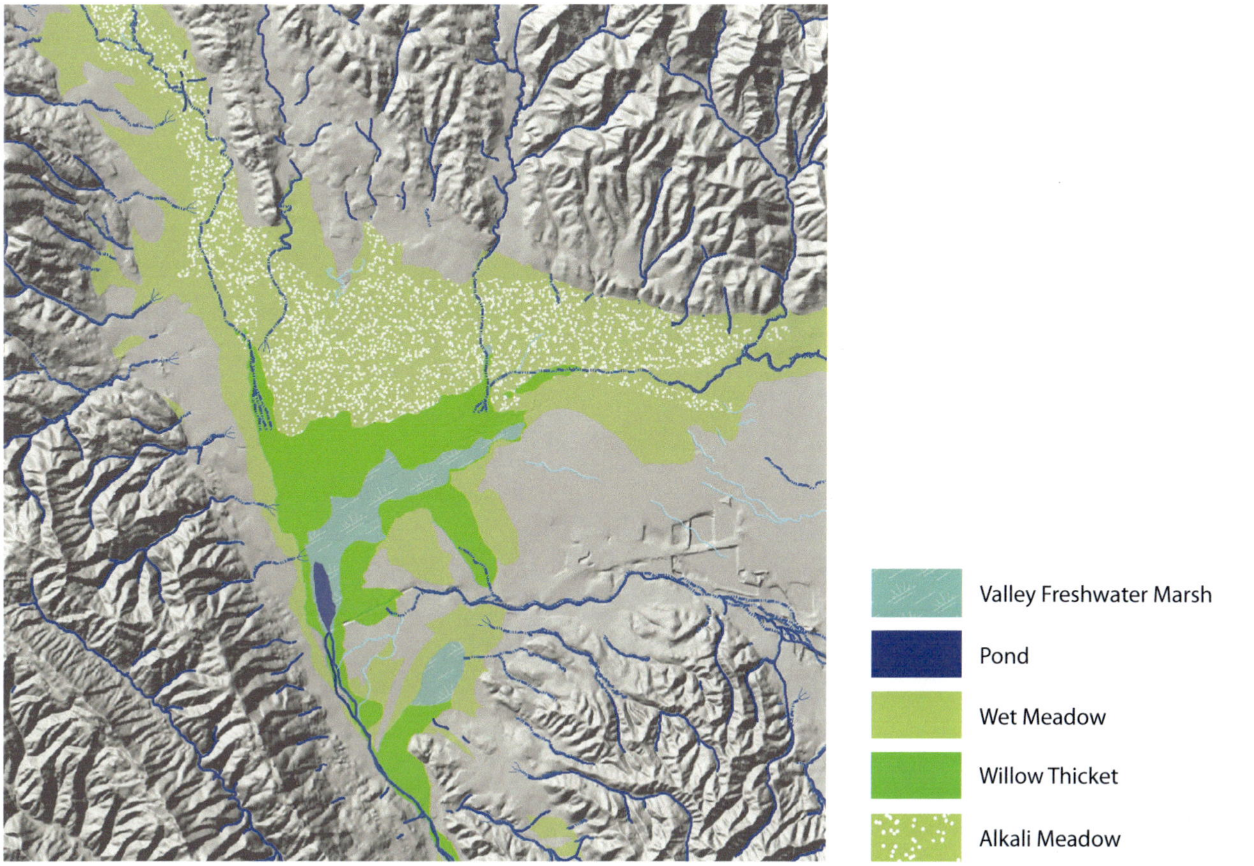

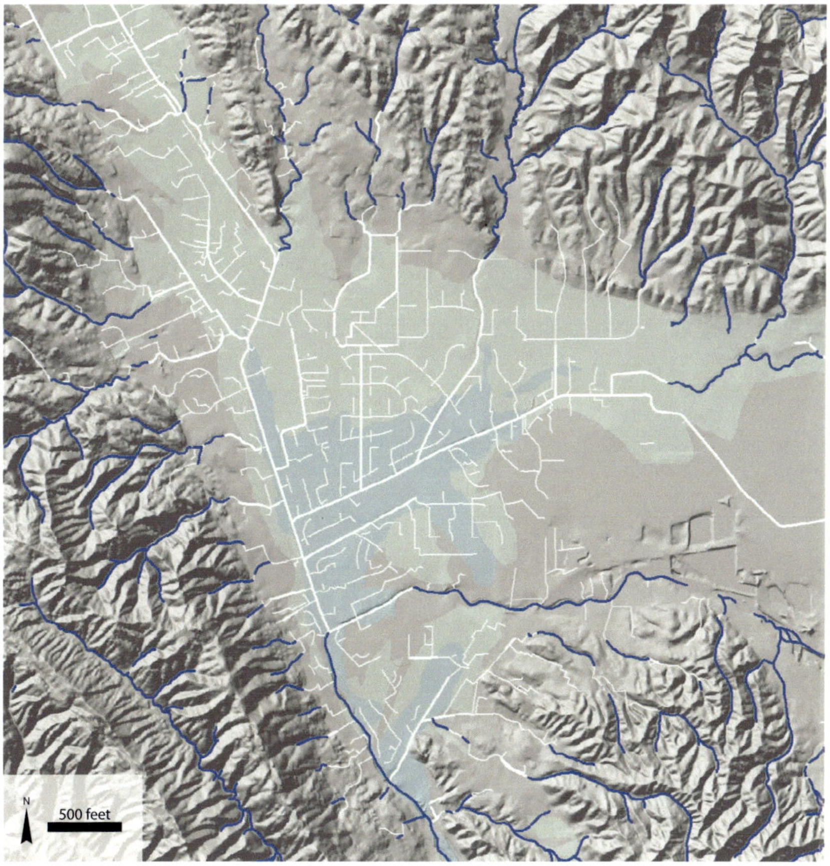

Figure 9.10. Stream extension across the historical Pleasanton marsh. In addition to the major channels mapped for BAARI (see next page) a series of smaller storm drains now connects across the area once occupied by the Pleasanton marsh complex. The white lines represent constructed channels (thicker white lines) and storm drains (thinner lines). The drainage network is overlaid on a depiction of the historical marsh complex. This map only shows storm drains with diameter greater than 24 inches—additional smaller drainages are also present. (Contemporary mapping from Sowers and Richard 2003)

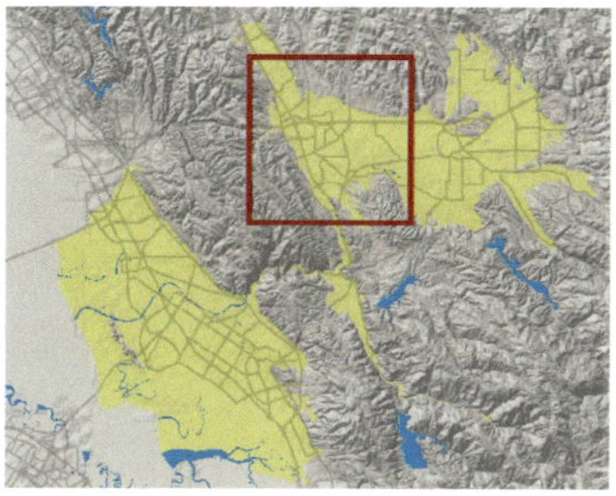

N
1 mile
1:135,000

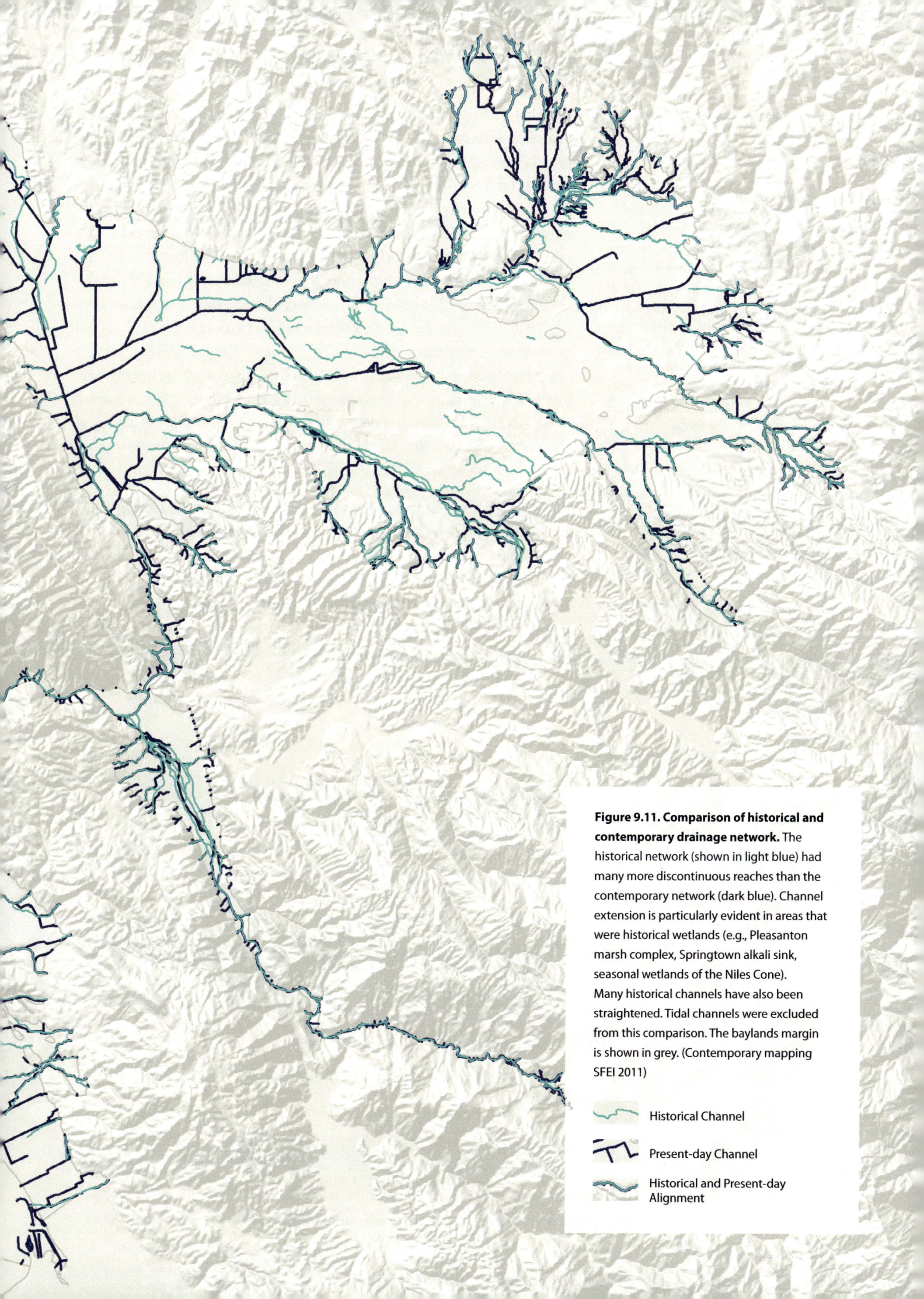

Figure 9.11. Comparison of historical and contemporary drainage network. The historical network (shown in light blue) had many more discontinuous reaches than the contemporary network (dark blue). Channel extension is particularly evident in areas that were historical wetlands (e.g., Pleasanton marsh complex, Springtown alkali sink, seasonal wetlands of the Niles Cone). Many historical channels have also been straightened. Tidal channels were excluded from this comparison. The baylands margin is shown in grey. (Contemporary mapping SFEI 2011)

How has channel form changed over time?

Even before large-scale human modifications to the watershed, natural processes such as landslides, changes in rainfall, and uplift caused shifts in sediment and water transport, causing streams to deepen (incise) or fill (aggrade) over time. The historical watershed had a wide variety of interconnected channel forms in relatively close proximity: sinuous, narrow, clay-bedded channels and broad, gravelly braided channels; multi-threaded and single channel; deep, well-defined streams and shallow sloughs. These patterns have been altered during recent history. Overall, much of the historical diversity of channel form—and associated aquatic habitat—has been lost. Below we show two examples of changes in channel form over time. Arroyo de la Laguna was a multi-thread, marshy creek that incised dramatically with increased flow, resulting in reduced floodplain access and a deep single-thread channel (fig. 9.12). Conversely, lower Alameda Creek was a naturally incised channel at the beginning of the historical period that filled towards the end of the 19th century as a result of upstream erosion (fig. 9.13).

Modifications contributing to these changes include the channelization of streams and wetlands such as the Pleasanton marsh complex and the construction of dams. The confinement of surface waters to narrow channels increased water velocity and sediment transport, flushing water and sediment downstream that had historically been trapped in wetlands, and causing incision along streams such as Arroyo de la Laguna. The construction of impermeable surfaces and storm drains caused a further increase in the speed and volume of flow reaching the channel network, resulting in further incision. Dams trapped coarse sediment in the upper watershed, also

Figure 9.12. Conceptual cross section along Arroyo de la Laguna in 1853 and 2011. This cross section compares the channel pattern in 1853, reconstructed from GLO surveyor Sherman Day's survey notes, with the incised channel today. Day did not record the depth of channels, but he did record the distances and a short description of each channel he crossed—thus, the depths shown below are estimated. This diagram emphasizes the number of smaller channels present historically, in comparison with the one broad channel today. This cross section is approximately 1/3 mile downstream of Verona Bridge. (historical data from Day 1853)

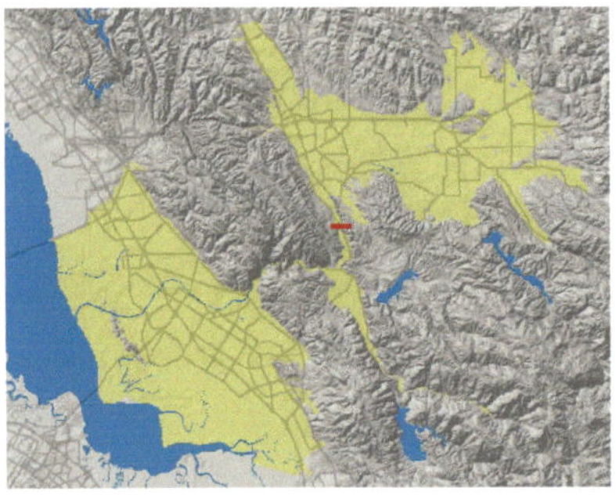

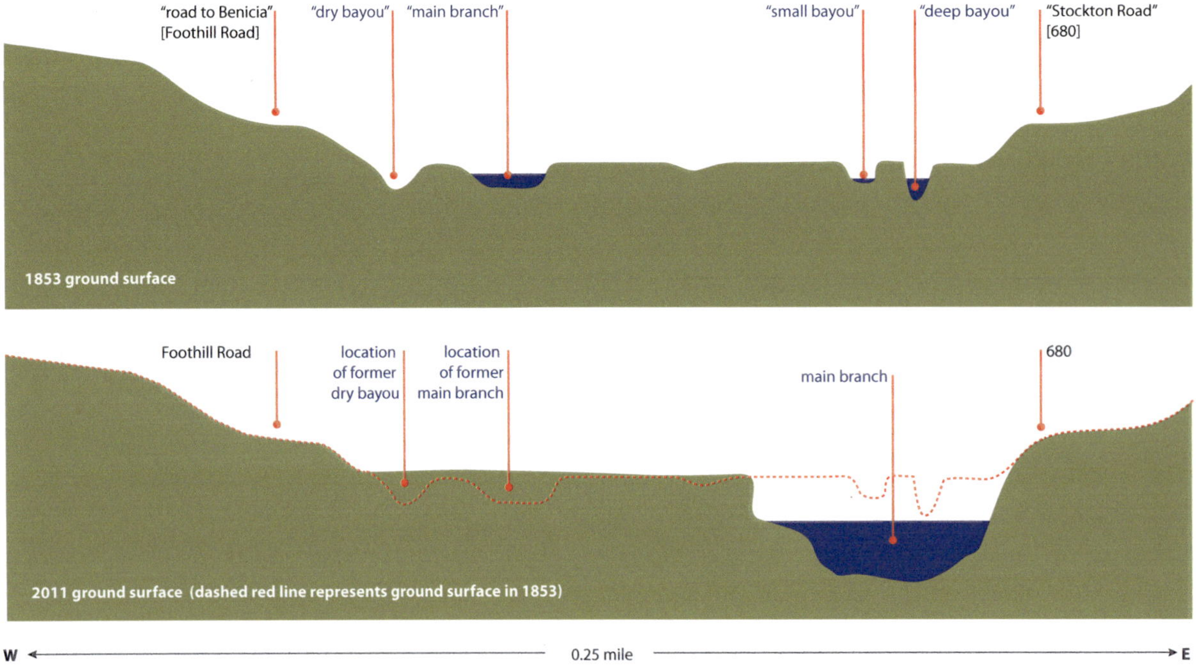

leading to downstream incision. Some channelized streams are cut off from their historical flood plains and now have too much sediment, resulting in sedimentation in channels such as lower Alameda Creek.

These modifications resulted in a loss of in-channel complexity as well. Channel beds were more complex historically, containing pools, riffles, islands, bars, and reaches with variable widths. Channel bed form could change rapidly over a short distance. Channels have been engineered to drain the valleys and increase the amount of land available for agriculture and development, removing much of this complexity. The removal of larger-scale features such as meanders and side channels, as well as some of the historical in-channel features such as bars, pools, and riffles, has reduced both the quantity and diversity of stream habitat.

Together, these modifications radically changed the deposition and erosion patterns of streams through the watershed. Where storm and flood flows once spread across floodplains, slowing and sinking back to groundwater, now peak flows are piped quickly into the flood control channels, increasing shear stress on the bed and banks, and leading to increased bank erosion and incision.

Figure 9.13. Conceptual cross section along Alameda Creek through time. This cross section shows three time points along Alameda Creek near Niles. The first two are based on testimony by Charles Shinn, describing how he crossed the creek in 1871, and then how the channel had filled by 1901. He recorded some depths and angles but not widths, so the widths shown are estimated. (historical data from Shinn 1901)

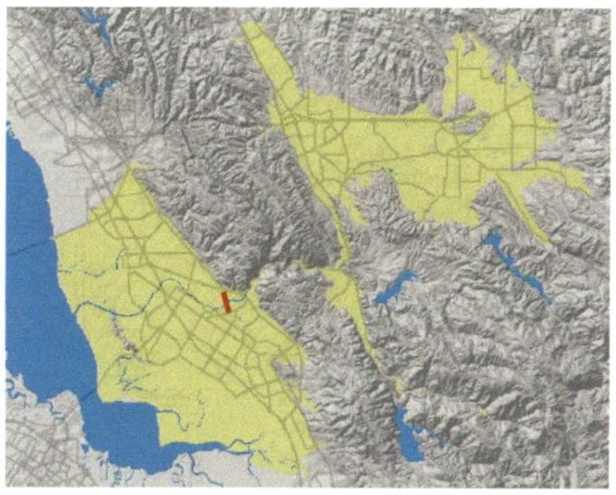

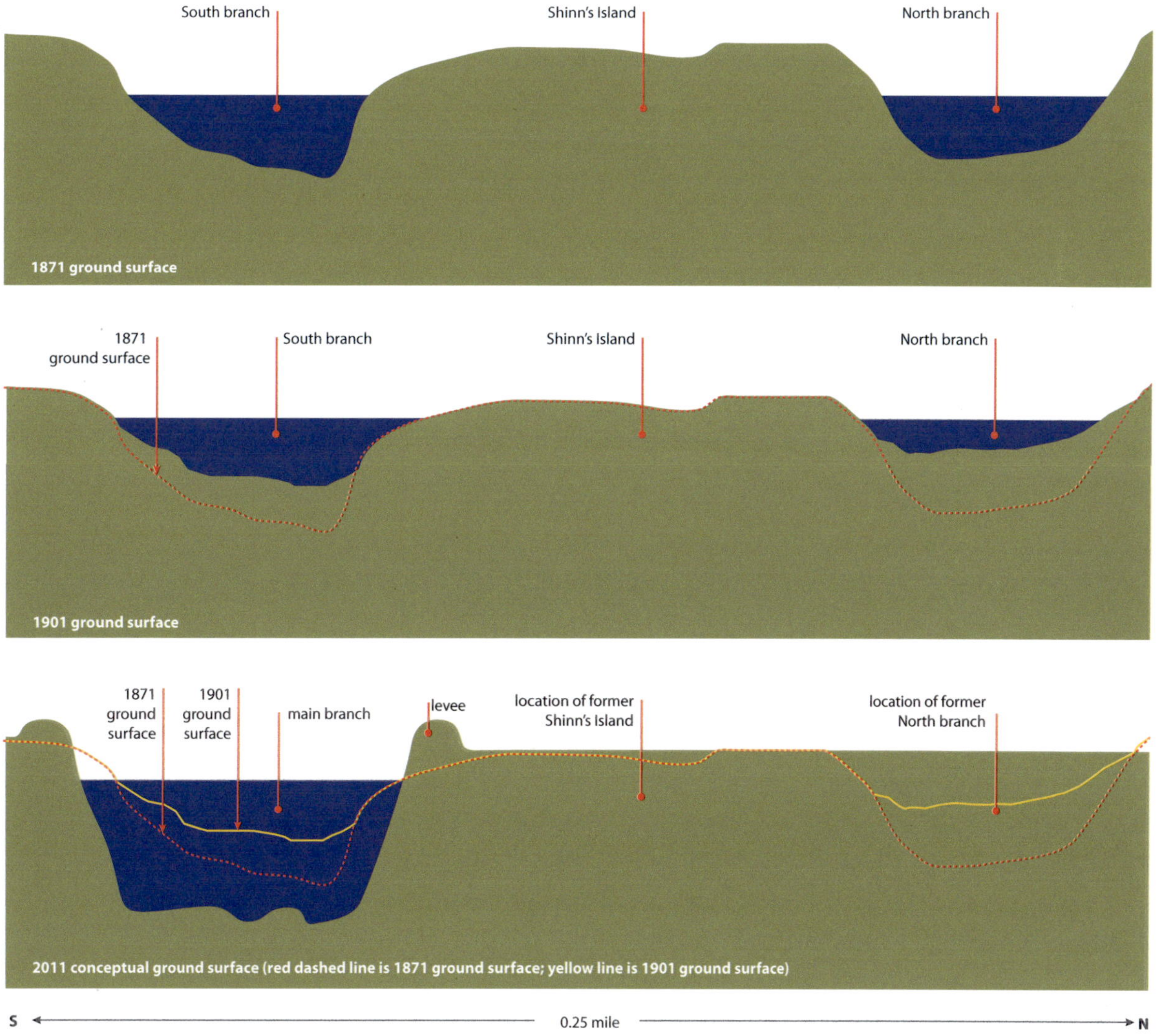

Summary concepts for channel network and morphology

Channel morphology varied widely through the watershed, and different creek forms provided different functions. Creeks varied from broad, braided systems reaching over 1,500 feet wide, to shallow, poorly defined channels and sloughs, to narrow, meandering streams (fig. 9.14a). Braided reaches provided recharge and coarse sediment storage. Swales and sloughs played a role transporting water through relatively flat areas. Narrow, well-defined channels helped to convey water downstream.

Historically, many creek beds were more complex in form. Historically channels had more pools, bars, side channels, riffles, and meanders than most channels do today. The proliferation of artificial canals and flood control channels has shifted the dominant channel form towards smooth, constructed channels, replacing some of the habitat historically available in the complex historical channel form.

Most streams flowed through discontinuous channels. As they flowed from confined canyons to the relatively flat valley floor, streams typically lost definition, sinking into gravels or spreading to form wetlands across less porous substrate. Larger streams (such as Alameda Creek) or streams with a steeper gradient (such as Dry Creek) or less opportunity to percolate (such as Arroyo las Positas) were among the few that were able to maintain continuous channels.

The stream network today is highly connected and streams have limited overflow zones. Streams no longer feed into wetlands or spread to dissipate energy. The Alameda Creek watershed today is a highly engineered system, the result of a series of modifications through the late 1800s and 1900s that were designed to maximize water output for nearby cities. These changes have resulted in more rapid transport of sediment and increased erosion and incision, and have implications for local species populations.

Incision and aggradation patterns began to shift rapidly following the drainage of the Pleasanton marsh. Incision can be largely traced to changes in water and sediment movement through the system as a result of the channelization of wetlands and streams and the construction of dams.

Restoration that allows alluvial creeks some room to meander and shift course might help these creeks more effectively store sediment and alleviate some problems with erosion and incision. Both natural and anthropogenic processes cause incision and sedimentation within the watershed. Many of these are irreversible (bed lowering from large dams, tectonic uplift, mass wasting, etc.) and are intensified by the very rapid movement of water through the system today. Restoration efforts that seek to slow the movement of water through the system, such as allowing inset floodplains to establish, or reconnecting channelized streams to abandoned floodplains, may help slow erosion.

Some relatively natural stream reaches still exist. Protection and restoration efforts should focus on habitats remaining downstream of dams and on improving channelized streams. For example, Old Alameda Creek, a modified version of the historical Alameda Creek channel, still exists on the Niles Cone, north of the Flood Control Channel (fig. 9.14b). This more natural channel may have some potential as habitat for steelhead and other fish. Some stands of sycamore alluvial woodland also remain in the watershed, and should be considered a high conservation priority.

Mid-sized tributaries (Dry, San Antonio, Sinbad, Stonybrook) likely provided important steelhead habitat. Although many streams were discontinuous, these streams connected directly with Alameda Creek and provided easy access to perennial, shaded, upper reaches. These systems may offer high value steelhead restoration opportunities (fig. 9.14c).

Figure 9.14. Contemporary streams. (A) Sycamores and other riparian vegetation along braided Arroyo del Valle within Sycamore Grove Park. (B) A straightened reach of Arroyo Mocho between Livermore and Pleasanton, parallel to Stanley Avenue, which was ditched in the early 20th century. (C) Arroyo Mocho in Livermore, showing a weir in the flood control channel just upstream of Santa Rita Road. (All photos from 2011. A: Ruth Askevold; B&C: Sarah Pearce)

How have water transport and storage changed?

Alameda Creek, its tributaries, and their connected aquifers constitute a complex system of natural water storage and transport. Under more natural flow conditions, certain reaches primarily conveyed water downstream while others recharged groundwater or spread into wetlands during wet periods. The watershed worked as a unit, filtering, storing, transporting, and recharging. Over the past 200 years people have redesigned this system of storage and transport. Some of the changes have enhanced natural processes of the watershed (e.g., groundwater recharge through porous gravels), while others have reversed the natural processes (e.g., a transport reach becomes a storage reservoir). Figure 9.15 illustrates some of the major changes to the movement of water through the Alameda Creek watershed.

Historically, groundwater recharge occurred through alluvial valleys and fans in losing reaches of the stream network. Water percolated rapidly through the coarse gravels of southeastern Livermore Valley, Sunol Valley, and the upper Niles Cone, reaching underground aquifers. Recharge also took place through the channel bed, particularly in broad braided channels, as well as through the surrounding overflow floodplain zone. During the historical period these natural recharge areas were artificially expanded: Sunol Valley was engineered to function as a large subsurface water filter and storage site, and in Livermore-Amador Valley engineers discussed rerouting Las Positas to flow over more porous soils so that it could sink into groundwater. This continues today—Arroyo Mocho and Arroyo del Valle receive water from the Delta (State Water Project) which then percolates underground through permeable Livermore gravels to be stored as groundwater.

In contrast, streams flowing through bedrock canyons (e.g., Niles Canyon, upper Alameda Canyon) and clay soils (northern and western Livermore-Amador Valley) provided little natural recharge. Some of these reaches have changed dramatically in function through conversion to storage reservoirs, while others continue to function as transport reaches.

Figure 9.15. Conceptual depiction of water storage and transport. (opposite page) Much of the historical surface storage that absorbed flood flows (in the form of wetlands and overflow zones) is no longer present in the watershed. The distinction between recharge and transport indicates the primary process for each reach; some recharge may occur within reaches that are shown as transport or storage.

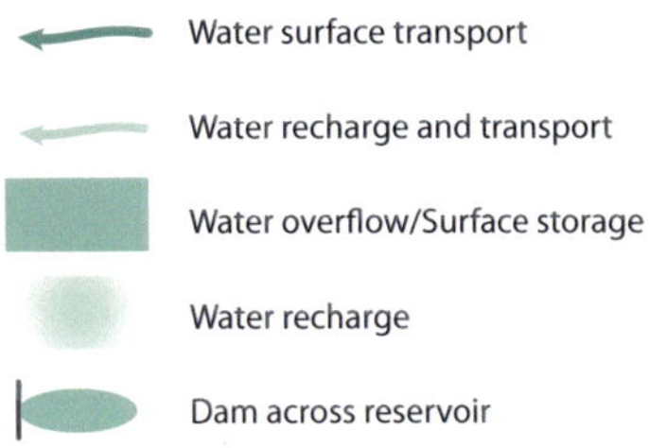

Historically, the primary permanent surface water storage site was the Pleasanton marsh complex, which covered 2,600 acres. The marsh captured flows and served as a detention basin, slowly releasing water downstream through the summer. In addition, vast areas of seasonal wetlands stored water temporarily in the rainy season, reducing peak flows. Today engineers have concentrated recharge into specific reaches and recharge ponds, using former gravel quarries for this purpose at Niles. Most surface storage has been moved out of the valley floor and into large reservoirs. Natural storage patterns have further been altered by water imports from the Delta and Hetch Hetchy, and export of water to San Francisco.

This redesign of the watershed has produced many social benefits, such as increased area for development, flood protection, and drinking water for a growing population, but has also resulted in a series of unintended consequences. In particular, changes to the water transport system have resulted in changes to aquatic resources and the movement of sediment, discussed next.

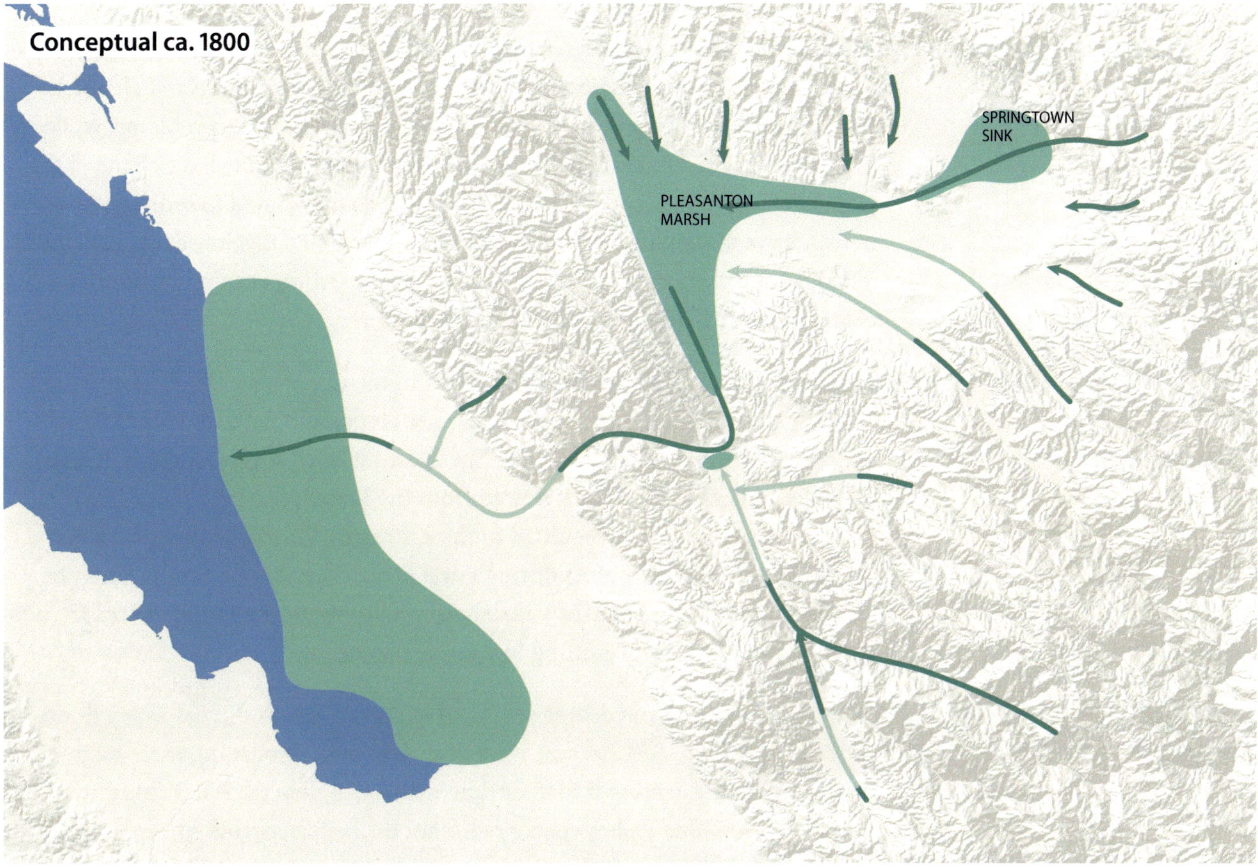
Conceptual ca. 1800
SPRINGTOWN
SINK
PLEASANTON
MARSH

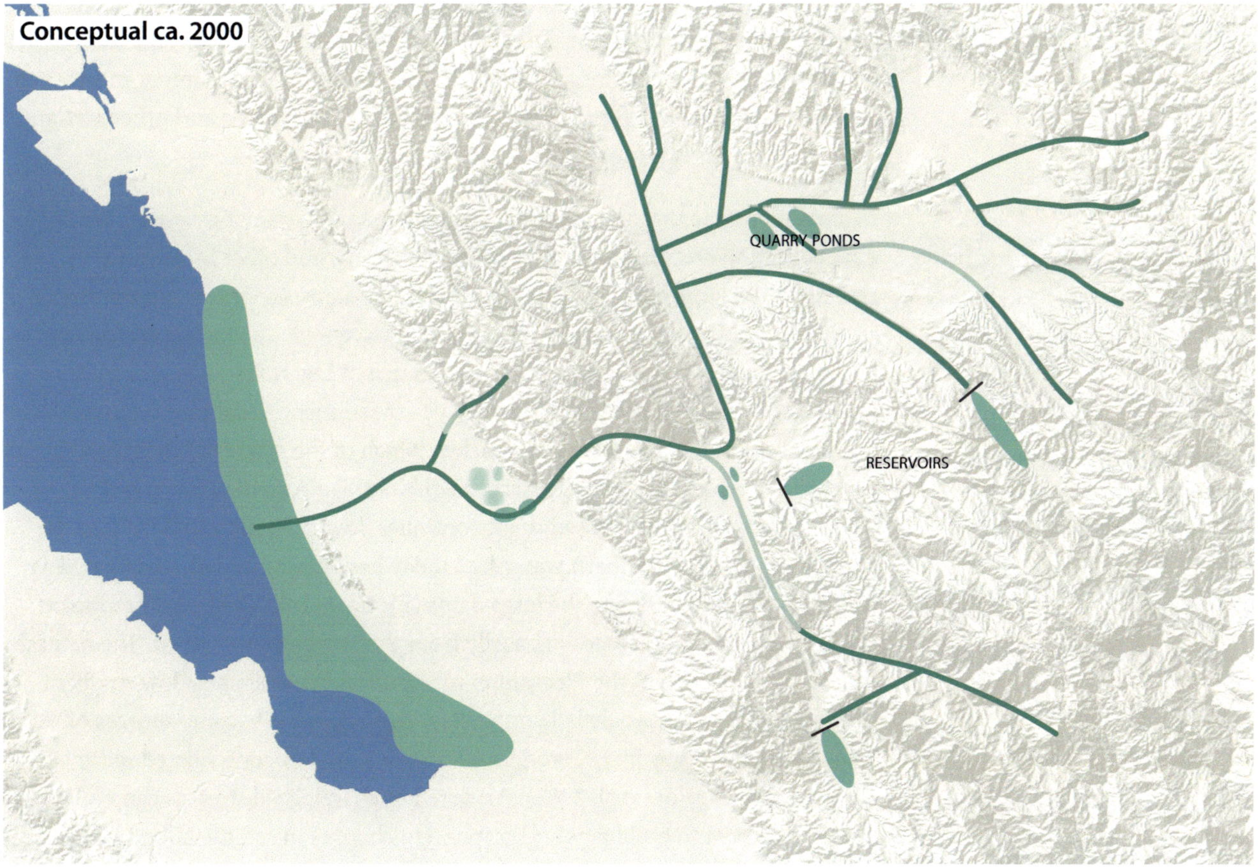
Conceptual ca. 2000
QUARRY PONDS
RESERVOIRS

How did sediment move through the watershed?

Altered sediment patterns have been one of the most dramatic unintended consequences of hydrologic modifications. Straightened, connected channels through the Livermore-Amador Valley have created incision problems, while levees constructed to prevent overflow across the Niles Cone have changed the pattern of sediment deposition and erosion and contributed to sedimentation of the channel (Collins and Leising 2003, Collins 2005, Bigelow et al. 2008, SFEI 2012). Meanwhile, the large dams have withheld sediment supply from the channel network, increasing incision (fig. 9.16).

Historically, most sediment was produced in the upper watershed. The geology of the surrounding hills influenced sediment types and patterns of deposition and storage through the watershed. The headwaters of Arroyo Mocho, Arroyo del Valle, and Alameda Creek flowed from the Franciscan formation, resulting in a comparatively coarse bedload in these streams. Coarse sediment entered the valley primarily in pulses during flood events. Smaller drainages from the northern Livermore-Amador Valley have smaller watersheds and source from the Great Valley Complex, resulting in finer sediment.

Storage of both fine and coarse sediment occurred across alluvial valley floors in times of overflow. Most streams were discontinuous, depositing their sediment in fans rather than carrying it further downstream. Across the Niles Cone and the Livermore-Amador Valley, coarse sediment deposits occurred towards the upper portion of the valley or at the tops of individual fans. By the 1800s, Alameda Creek had incised into its coarse fan at the top of Niles Cone and was creating a newer, finer fan downstream. The Pleasanton marsh complex and the tidal marsh acted as large sediment storage basins, allowing suspended sediment in water to settle and deposit, and facilitating the growth of tules, sedges and other wetland plants, which in turn trapped more sediment.

Sediment passage through the watershed has changed dramatically with the channelization of streams and construction of dams and other in-channel structures. As large flood flows are dampened by dams, scouring and resetting of gravel bars and in-channel forms becomes less frequent. The high connectivity and smooth trapezoidal channels through much of Livermore-Amador Valley and the Niles Cone provide a rapid passage for sediment. Tidal marshes, once sediment sinks, are now transport reaches. Much of the storage of coarse sediment now occurs behind the dams of Arroyo del Valle, San Antonio Creek, and Calaveras Creek, while fine sediment continues downstream to settle within the creek channels. The southern watershed today has been essentially eliminated as a source of sediment due to the large dams (Bigelow et al. 2008), so all sediment (both coarse and fine) comes primarily from the northern watershed. The natural sediment sinks, such as the Pleasanton marsh complex or sinuous low-gradient channels such as Arroyo de la Laguna, have in some cases become sources of fine sediment, as they incise, erode, and release stored unconsolidated sediment downstream (Bigelow et al. 2008). Armored or channelized streams can no longer spread sediment to build fans and respond to changes in sediment load. As we change the ways that water is stored and transported through the system, we need to resolve the resulting problems with erosion and sedimentation.

Figure 9.16. Conceptual depiction of sediment storage and transport. These diagrams show changes in sediment deposition patterns that the watershed has accommodated. Rather than spreading sediment broadly across their floodplains, streams now often store sediment within their lower channels, resulting in aggradation and the need for active management. Much of the historical coarse sediment supply has been trapped behind reservoirs. Almost all reaches transport both fine and coarse sediment—this diagram attempts to show which is dominant. An additional ongoing process today not shown in this diagram is the de-silting of the Flood Control Channel. (Concept after Collins 2005)

Fine sediment transport

Fine sediment storage

Coarse/fine sediment transport

Coarse/fine sediment storage

Dam across reservoir

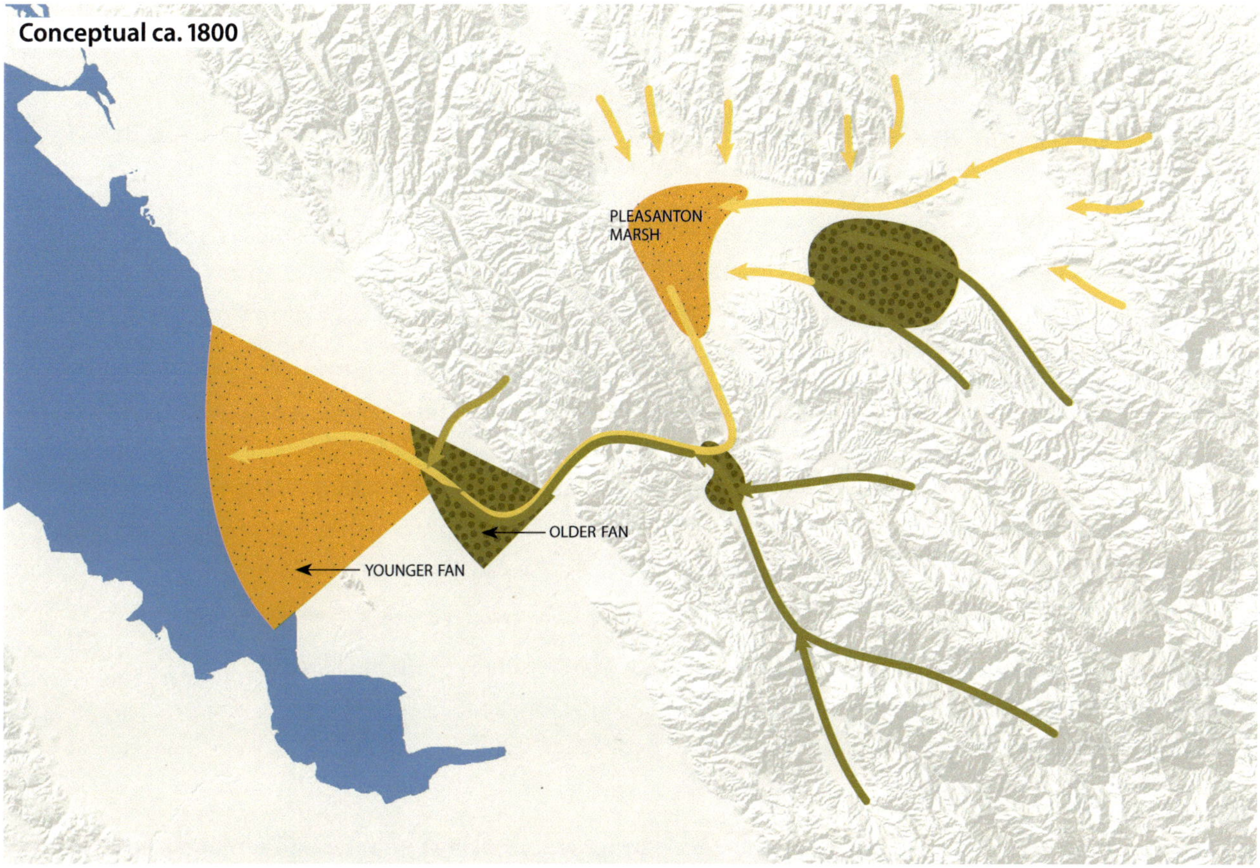
Conceptual ca. 1800
PLEASANTON MARSH
OLDER FAN
YOUNGER FAN

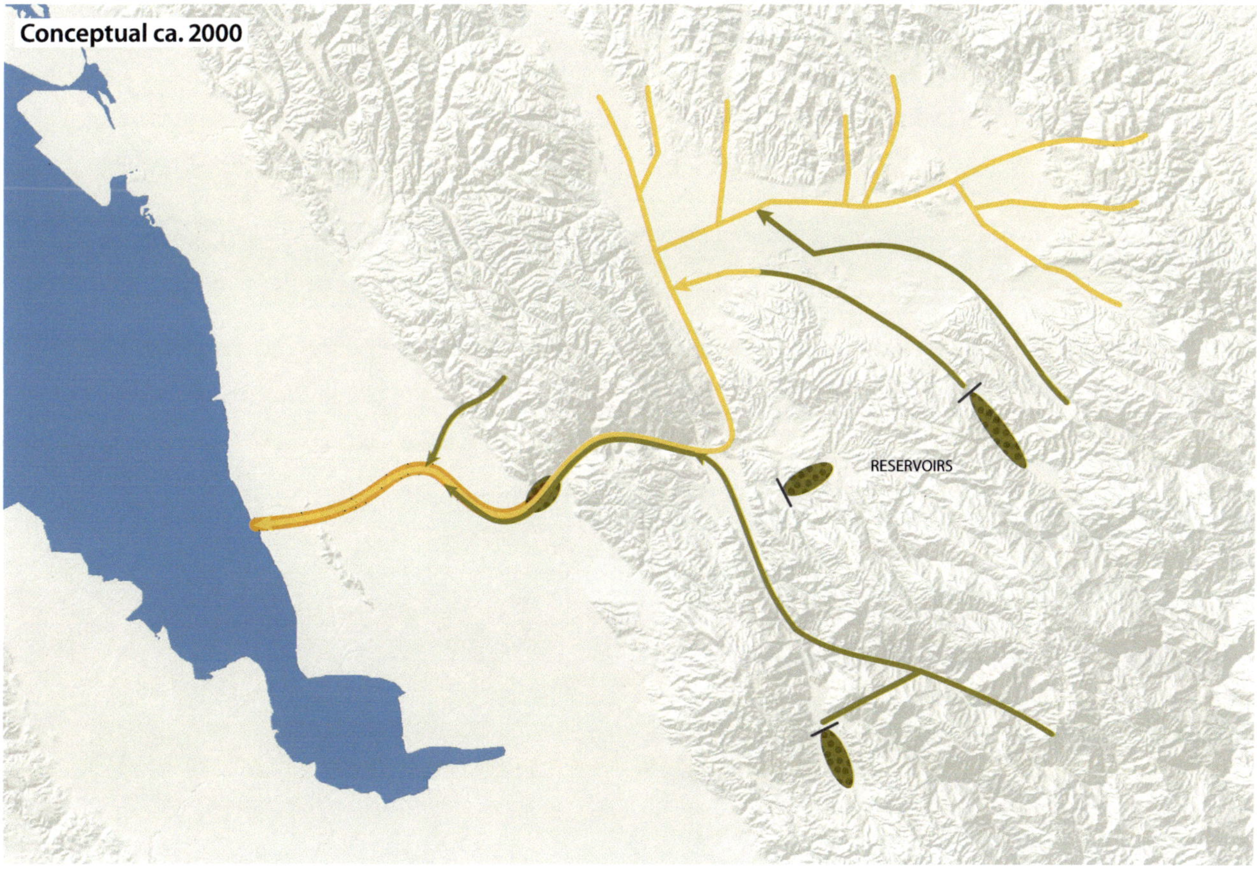
Conceptual ca. 2000
RESERVOIRS

Summary concepts for transport and storage of water and sediment

Natural sediment and floodwater storage was provided by large wetlands, historically drained by slow-moving streams. Wetlands along the edge of the Niles Cone and the western side of Livermore-Amador Valley in particular accommodated fine sediment (fig. 9.17).

Tidal marshes were important natural sediment sinks. The connection between Alameda Creek and the tidal marshes has been eliminated by levees, which deprives the tidal marshes of the sediment they need to grow as sea levels rise. Reconnecting the tidal marshes with Alameda Creek could provide both natural sediment transport and storage, helping sustain the tidal marshes.

During high flow events, much of Amador Valley and the western Niles Cone would have flooded. This flooding created a temporary surface connection from the discontinuous tributary creeks to Alameda Creek and from Alameda Creek to the tidal marshes, providing potential fish passage. Under normal flow conditions, Arroyo Mocho formed a distributary two miles before reaching the Pleasanton marsh complex, while Arroyo del Valle, Arroyo las Positas, Tassajara Creek, and South San Ramon Creek formed distributaries that emptied into the Pleasanton marsh.

Many of the ecosystem services provided by wetlands historically are missing from the watershed. Wetlands provided sediment storage, water filtration, habitat for a broad range of species, and water storage. Effective restoration and support for focal species may require creative thinking to restore some of the missing processes and habitats.

Streams and wetlands functioned as a system for storage and transport of water and sediment through the watershed. Creeks and wetlands were historically much more interconnected than they are today—creeks flowed and distributed water into seasonal wetlands, and then often re-formed at the bottom of wetlands. Today ditching connects creeks directly and allows them to bypass wetlands.

Annual overflow was a feature of many of the historical streams, particularly larger streams such as Alameda Creek, Arroyo del Valle, and Arroyo Mocho (fig. 9.17). Overflow zones and seasonal wetlands performed important roles storing and slowing the movement of water and sediment, as well as replenishing farm soils and providing temporary feeding opportunities for steelhead.

A

B

Figure 9.17. Overflow and sediment storage on the Niles Cone. These images show flooding across a field (A) and through Crandall Slough (B) in the 1914 and 1916. The bottom image (C) shows some of the sediment left in deposits across fields in the wake of flooding. (AD-83, AD-565, and AD-852, © San Francisco Public Utilities Commission)

C

Where did wetlands occur?

In the recent past, the Alameda Creek watershed supported a diversity of wetland types. Wetlands formed in largely predictable patterns shaped by faults, bedrock barriers, alluvial fans, and tidal waters (fig. 9.18). They were controlled by the underlying geology and form of the land, forces which persist today, although their expression is altered by management.

FAULTS. Wetlands controlled by faults included Tule Pond and the Lagoon (Lake Elizabeth). These two wetlands occur along the Hayward Fault, far removed from other wetland complexes. These wetlands are known as sag pond wetlands, and formed in the depression between the two strands of the fault. Throughout the study area, fault-related springs supported additional small fault wetlands.

BEDROCK BARRIERS. Bedrock barriers contributed to the formation of both of the large wetland systems in Livermore-Amador Valley—the Springtown alkali sink and the Pleasanton marsh complex. Natural bedrock barriers helped create high groundwater levels and broad depressions, which received surface waters from the surrounding watershed. Both wetlands also occurred directly upstream of faults, which may have helped disrupt groundwater flows. The Pleasanton marsh complex was confined by the convergence of branches of the East Bay Hills, forming in the low point of the western Livermore-Amador Valley, where water collected before moving slowly downstream to Sunol Valley. In the Springtown alkali complex, the Las Positas anticline created a series of small hills, forming a natural barrier to drainage. The seasonal wetlands occurred at the bottom of this small basin.

BASE OF ALLUVIAL FANS. Broad alluvial fans, which dominated most of Niles Cone and Livermore-Amador Valley, had steep slopes and good drainage, precluding extensive wetlands. At the toe or bottom edge of these fans, however, wetlands formed on the fine-grained, poorly drained clay soils, supported by stream overflows as well as springs and high ground water.

TIDAL WATERS. Extensive tidal marshlands formed at the Bay margin as tidal and fluvial sediment built up over time and then was colonized by halophytic (salt tolerant) plants.

Today the wetlands along the Hayward Fault are open water features, and Lake Elizabeth has replaced the marshes of the historical Lagoon. The confined basin wetland at Springtown still floods seasonally, and the Pleasanton marsh complex flooded as late as the 1950s. The bottomland wetlands at the toe of alluvial fans exist only in small pockets (e.g., near Coyote Hills)—for the most part they have been eliminated by a combination of creek channelization, filling and diking of lowlands, and groundwater pumping. Understanding the forces that created and maintained these wetlands can help present-day managers and inhabitants to understand flooding and ground saturation patterns today, and to recognize opportunities for restoration.

Figure 9.18. Conceptual depiction of wetland forcing features. This diagram shows how landscape-level patterns of topography and soils determine where wetlands occur—and where they do not. The well drained upper fans of Arroyo Mocho, Arroyo del Valle, and Alameda Creek precluded wetlands across much of the study area, instead recharging groundwater. Wetland mosaics formed in the poorly drained distal ends of fans. These processes combined with natural bedrock barriers to create the watershed's most important wetland complexes at Pleasanton and Springtown. Wetlands also formed at the base of Niles Cone and the Bay margin. Finally, the Hayward Fault created another type of wetland—sag ponds. Today most large natural wetlands are remnants of these former features, maintained by many of the same processes. For example, along the edge of Niles Cone fragments of tidal marsh and seasonal wetlands still persist, as does a small wetland near the Willows.

Hayward Fault
Topographic confinement
Alluvial fan
Direction of flow
Wetland
Tidal marsh

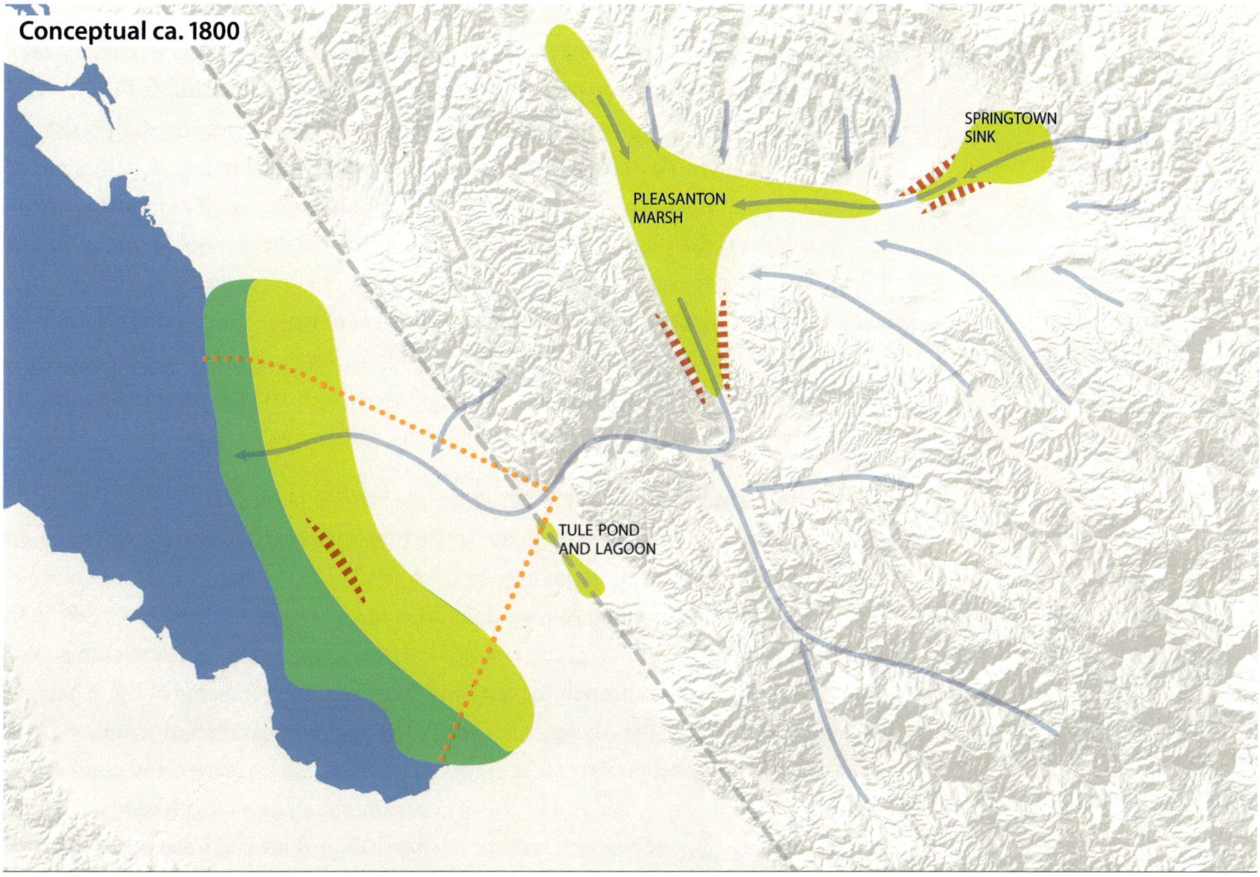
Conceptual ca. 1800
SPRINGTOWN
SINK
PLEASANTON
MARSH
TULE POND
AND LAGOON

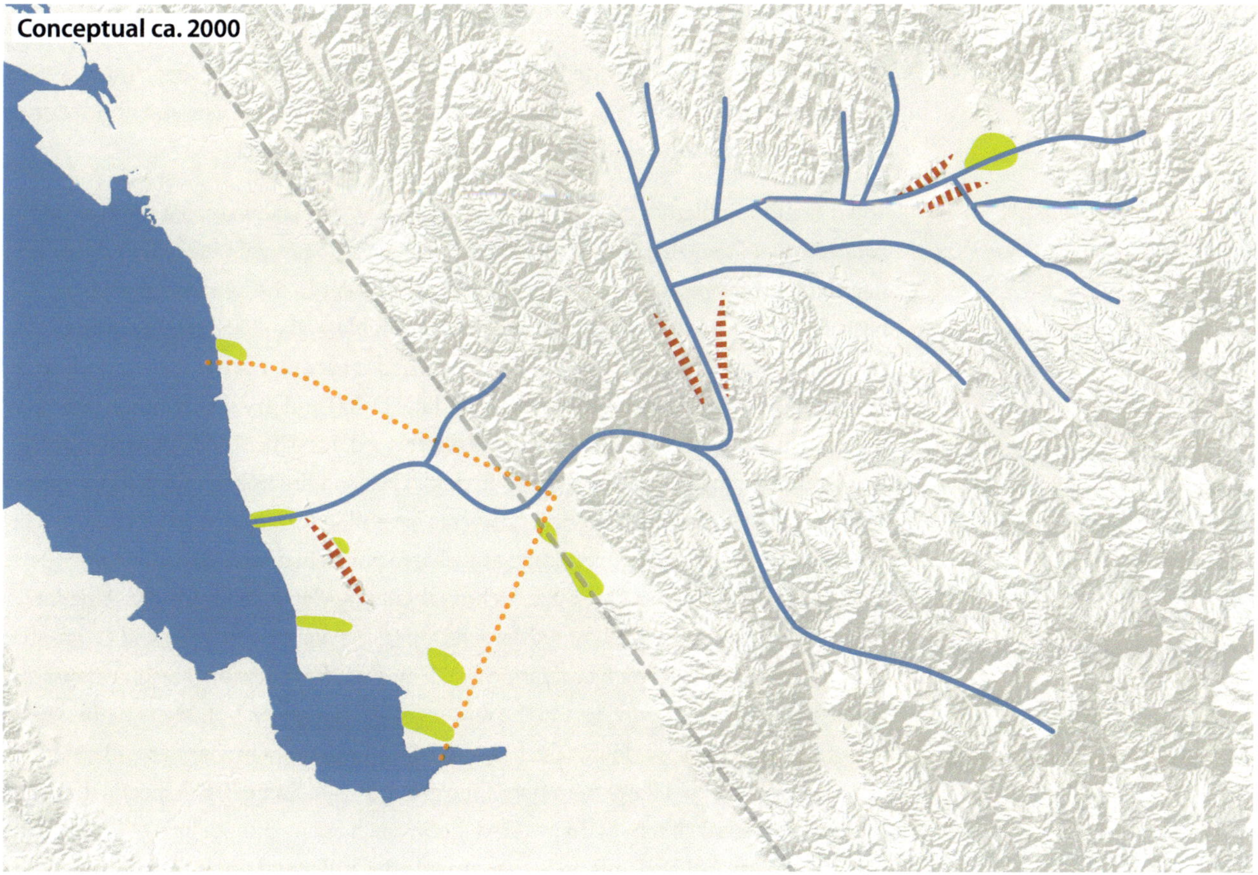
Conceptual ca. 2000

How has the seasonality of aquatic resources changed?

Much of the Alameda Creek watershed ran dry in the summer, typical of streams in Mediterranean climates. Year-round flow occurred primarily in spring-fed upper reaches, away from the alluvial plains. However, some pockets of summer water did persist within the lowlands. Dry season aquatic resources are essential refuges for the survival of over-summering migratory fish or resident fish populations, as well as for other aquatic species such as red-legged frog (*Rana aurora draytonii*). Reliable pools are also important as watering holes for terrestrial species such as kit foxes (*Vulpes macrotis mutica*), deer, and other large mammals. A small but important set of marshes and ponds, perennial stream reaches, and groundwater-fed pools within summer-dry reaches all contributed to the dry season aquatic resources of the valley floors (fig. 9.19). Changes in this resource availability have many consequences for local species population dynamics.

Documented historical summer water resources include pools along lower Arroyo Mocho, Arroyo del Valle, and Alameda Creek; perennial marshes and ponds in the center of the Pleasanton marsh complex; a spring-fed pond and perennial reach on Arroyo las Positas; perennial flows along Arroyo de la Laguna and much of Alameda Creek; water in Tule Pond and the Lagoon; and the vast tidal marshes. The extent of dry season aquatic resources changed both through the year and from year to year. Some of these habitats remained wet through the dry season, such as the perennial ponds and marshes. Others (such as the in-channel pools) would slowly dry through the summer, and some may have dried completely by the end of the dry season. The size and depth of pools within creeks and the extent of perennial reaches depended on short and longer term climatic patterns. Many of these pools were shaded and groundwater-fed and as a result were able to maintain low temperatures through the dry season (see fig. 8.10, p. 267).

Seasonal wetlands also provided important habitat across the study area. These wetlands expanded the wetland area in winter and provided services such as water and sediment storage, filtration, and denitrification, as well as additional habitat.

Today there is similarly a range of summer-available water resources, but the type and connectivity of available aquatic resources has changed. Seasonal wetlands in the valley floor have for the most part been replaced by open water in the hills, and there has been a substantial loss in the total amount of accessible wetland habitat. Some tidal marshlands remain, but most have been diked, and most other perennial wetlands have been drained. Water releases (from dams and the Delta) and dry season urban runoff (from lawn watering and other household water uses), have increased dry season creek flows, resulting in higher temperatures in pools (Hanson Environmental 2002). Reservoirs such as Del Valle, San Antonio, and Calaveras have added large open water resources in the hills. Quarry pits now functioning as recharge ponds in upper Niles Cone provide additional summer water. Stock ponds throughout rangelands in Livermore-Amador Valley are particularly useful for red-legged frog (*Rana aurora draytonii*) and California tiger salamander (*Ambystoma californiense*). However, these new dry season resources are in many cases not equivalent to the resources that have been lost. Many native species depend on cool summer flows, which these artificial resources may not provide; salmonids especially require cool flows from upwelling groundwater. Some native species do benefit from warm water habitats, as do invasive species such as largemouth bass (*Micropterus salmoides*). Seasonal wetlands no longer provide the buffer and water and sediment storage that they once did, resulting in more rapid movement of water through the watershed.

Figure 9.19. Conceptual depiction of aquatic resources. These diagrams compare the wet season (green and blue) and dry season (blue) resources available on the lowlands of the Alameda Creek watershed historically and today. Many historical seasonal wetlands have been removed today. Contemporary streams coded "managed" are streams that have flow patterns that are controlled by releases rather than natural flow patterns. "Managed intermittent" streams are dry for a significant portion of the year.

Perennial (or most of year)

Seasonal

Managed flow regime

Managed intermittent

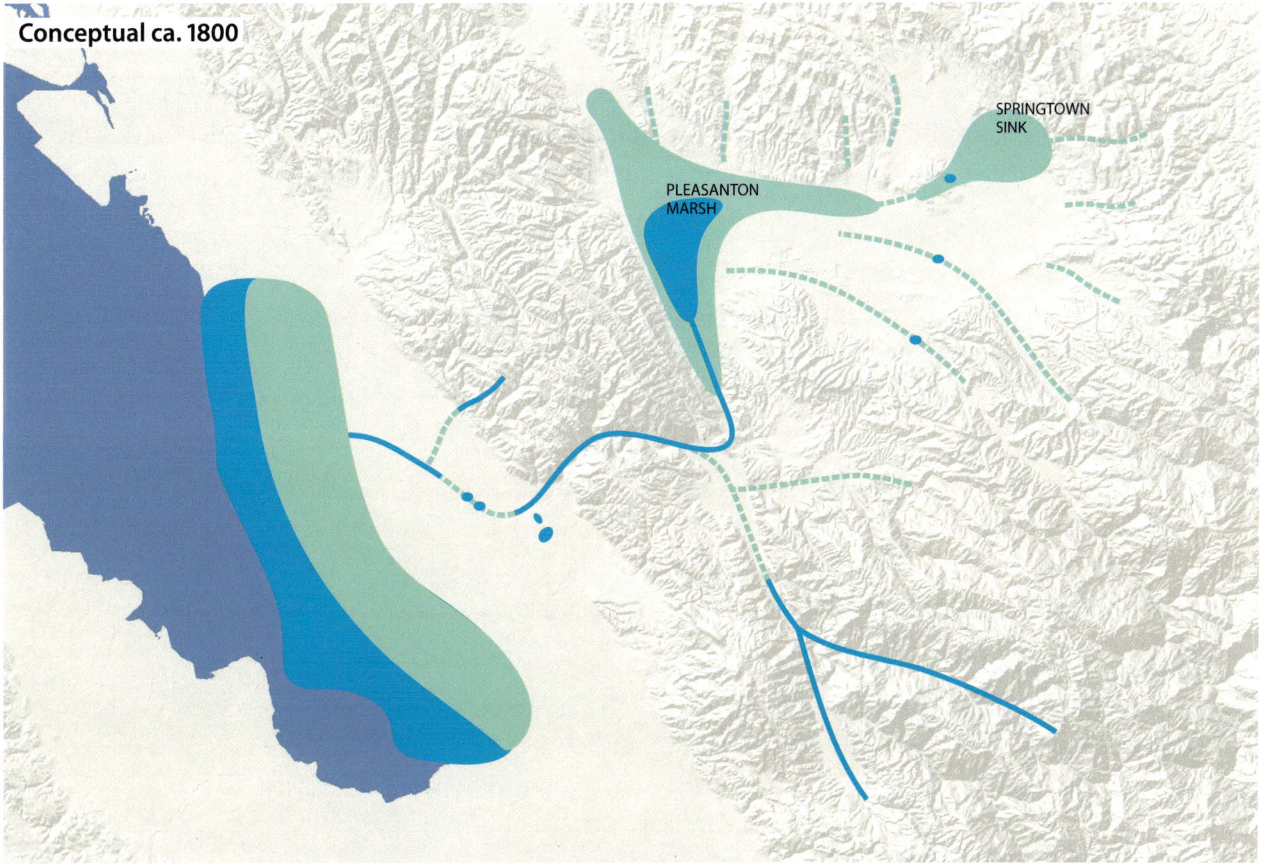
Conceptual ca. 1800
PLEASANTON
MARSH
SPRINGTOWN
SINK

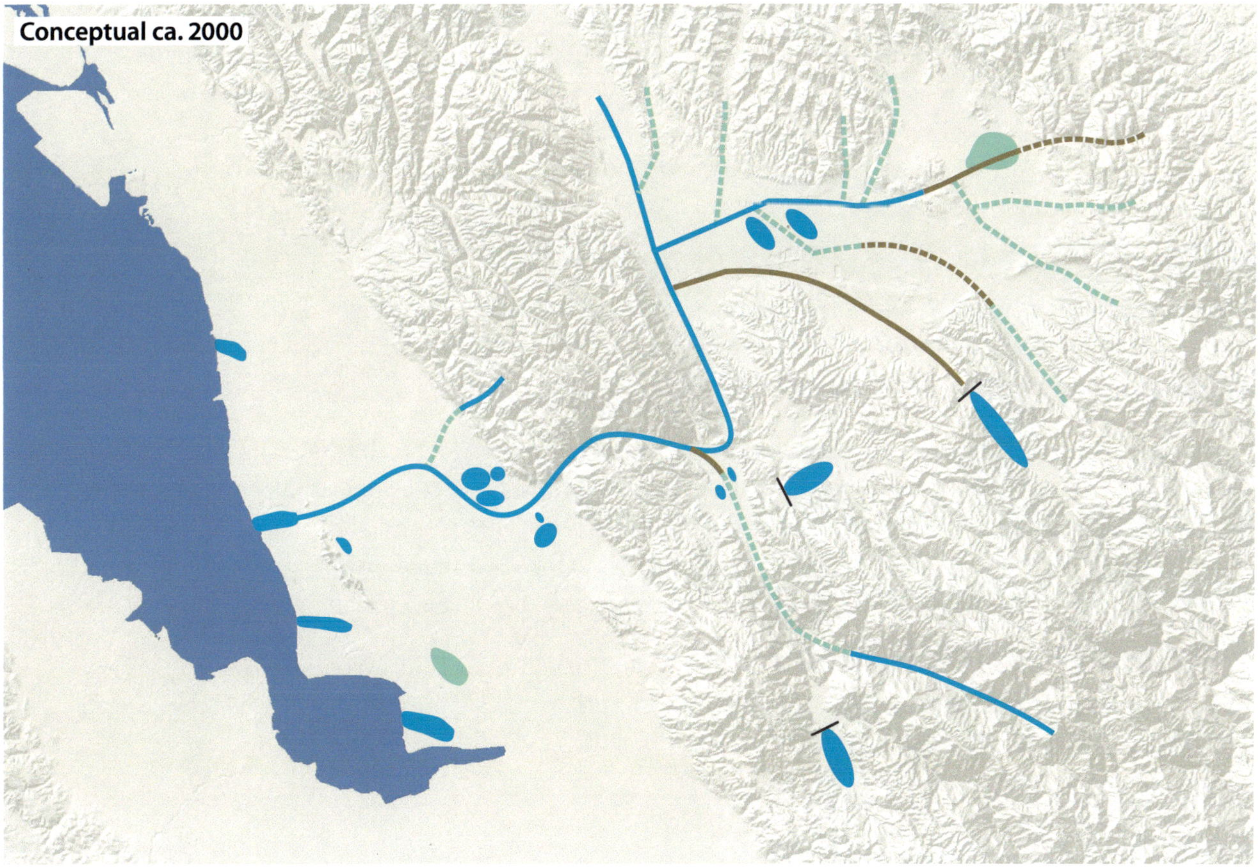
Conceptual ca. 2000

Summary concepts for wetlands

With the exception of a few fault-related wetlands, wetlands occurred in large, continuous mosaics. These mosaics of different wetland types occurred at the western end of Niles Cone, in the tidal marshes, in the Springtown sink in eastern Livermore Valley, and in the Pleasanton marsh complex in Amador Valley.

Patterns of wetland distribution have been extremely persistent over time. Although in many cases wetlands have been drained and covered, these areas are often still flood prone. For example, despite changes to the surrounding drainages, vernal pools persist in the Don Edwards National Wildlife Refuge near Newark, and parts of Pleasanton experience flooding problems, a result of the fine soils that still underlie Pleasanton, and a reminder of the historical marsh. Understanding the historical distribution of wetlands can help managers develop strategies for more effective management.

Historically, the non-tidal wetlands were largely composed of lowland seasonal wetlands. Today aquatic resources are primarily upland, artificial wetlands and open water features (reservoirs and stockponds). This shift has implications for connectivity with other habitat types, as well as the quality of habitat. These upland wetlands are no longer connected to streams, estuary margins, and seeps as they were historically.

Seasonal wetlands were the dominant wetland type across the study area. Seasonal wetlands covered almost 35,000 acres in total, and included both alkali and freshwater habitats. Surface waters covered these areas for days or weeks at a time, and they dried over the summer. Seasonal wetland types included wet meadow, alkali meadow, alkali playa complex, alkali sink scrub, and vernal pool complex. Today open water features are the dominant type. Wetland habitat has transformed from connected seasonal wetlands to isolated open water features.

Dry-season aquatic resources were historically more varied than they are today. Dry-season water was historically available in perennial marshes, ponds, in-stream pools, and stream reaches, likely totaling fewer than 1,000 acres in extent through the study area. Today more reaches have become perennial due to water inputs from the Delta, urban runoff, quarry pond pumping, and incised flood control channels. Large perennial open-water habitats exist in the form of reservoirs and gravel mining ponds. Stock ponds constructed through the rangelands of the study area add additional freshwater resources.

Most streams flowed for only a part of the year within the lowlands of southern Alameda County. Today more streams are perennial, which favors a different suite of species and may limit some xeric-adapted species. On the other hand, these historical dry-system riparian examples may be useful models for future ecological planning in the face of climate change.

Pools fed by groundwater along Alameda Creek (and likely upper Mocho and Del Valle) were able to maintain low temperatures through the summer. The historical record documents the presence of cold, perennial pools through intermittent reaches of lower Alameda Creek in particular, and recent research suggests potential prevalence of thermally stratified pools in low-flow streams (fig. 9.20; see Chapter 8). Cold water resources would have been useful for fish habitat and are largely absent today.

Natural artesian zones existed at the low points of the Niles Cone and the Livermore-Amador Valley. High pressure in groundwater aquifers created artesian springs that supported wetlands and supplied early farmers with a source of freshwater for irrigation.

Stock ponds and other artificial water bodies may provide some habitat to replace the wetland habitat that has been lost. They also trap some sediment, much as historical wetlands did. However, these artificial wetlands tend to have higher temperatures and may not support some of the native species that depended on the historical wetland types.

Re-emergent groundwater provides opportunities for wetland restoration. Along the edge of the Niles Cone, once depleted groundwater levels have recovered due to groundwater recharge efforts and may be able to support willow marshes and other wetland habitats.

Figure 9.20. Thermal stratification within in-stream pools. This figure shows conceptually how low summer flows historically allowed for thermal stratification, which created pockets of cooler water that could support fish. Higher summer flows have disrupted this pattern today. Refer to figure 8.10 for more details.

How has riparian habitat diversity changed?

The variation in channel morphology and dry season flow patterns resulted in an impressive diversity of riparian vegetation types within a small area. Stream reaches exhibited distinct patterns in both riparian width and species composition (fig. 9.21). We distinguished five riparian vegetation classes in the study area: **mixed riparian forest** along Alameda Creek through Niles Cone and confined canyons; **willow-cottonwood forest** and **willow thickets** along reaches with high groundwater such as Arroyo de la Laguna, lower Alameda Creek, and the Pleasanton marsh complex; **sycamore alluvial woodland** in the broad braided intermittent reaches of Arroyo Mocho, Arroyo del Valle, and Alameda Creek; herbaceous cover with **sparse oaks** and **sycamores** along small drainages with low flow (lower Mocho and Del Valle, small tributaries); and finally **alkali sink scrub** through the alkali reaches of Springtown. Within the study area, stream corridor width (from one edge of the riparian vegetation across the stream to the other edge) varied from extremely narrow borders of herbaceous riparian vegetation, with one occasional oak tree, to over 1,300 feet through the broad braided reaches of Arroyo del Valle and Alameda Creek.

Figure 9.21. Conceptual depiction of riparian cover. The diversity of riparian widths and types has been greatly simplified, reducing the diversity of ecological functions in the Alameda Creek watershed. In particular, the broad willow forests and floodplains are no longer present. Widths have narrowed—if we consider the Pleasanton marsh complex a riparian habitat type, the riparian corridor extended 4,000 feet wide in places. Species composition has also shifted—today, mixed riparian forest may include exotic species such as eucalyptus. In the contemporary diagram we introduced two additional classes to represent willow scrub and the sparse vegetation along flood control channels, neither of which were present historically. (For more details on historical riparian habitats see the tables at the end of each chapter.)

Different riparian habitats served distinct ecological functions, and the matrix of these habitats created an interconnected riparian network, providing important migration corridors and refugia. Riparian tree cover created shade and could support the formation of pools under undercut banks. Riparian trees also resulted in large woody debris, which could alter sediment transport and create pool habitat for fish behind natural dams. Broad willow thickets, or willow swamps, likely supported species such as the yellow-billed cuckoo (*Coccyzus americanus*) and yellow warbler (*Setophaga petechia*) and provided allochthonous input to areas downstream. Seasonally dry sycamore alluvial woodland supported an entirely distinct suite of wildlife including yellow-legged frog (*Rana boylii*), horned lizard (*Phrynosoma coronatum frontale*), and lesser nighthawk (*Chordeiles acutipennis*). Narrow bands of mixed riparian forest contained a more diverse mix of riparian tree species, including willows (*Salix*), alders (*Alnus*), cottonwoods (*Populus*), sycamores (*Platanus racemosa*), oaks (*Quercus*), and maples (*Acer*), and provided continuous corridors for whipsnake (*Masticophis lateralis euryxanthus*), migrating fish, and other animals. Riparian vegetation contributed to bank stability, while also contributing large woody debris and enabling undercut banks for fish refugia.

Along most reaches riparian vegetation has been narrowed and floodplains eliminated to make more space for housing, agriculture, and roads. Long stretches of riparian habitat have been removed to make way for flood control channels. Other changes are less obvious. Many of the reaches most well adapted to xeric conditions—sycamore alluvial woodland and sparse oaks—have been converted to wetter types. The few very wide reaches, likely important nodes for biodiversity, have been largely eliminated.

Because of the extension of the drainage network, there is more riparian forest in some places. There are also some notable remnants such as the sycamore alluvial woodlands in Livermore and Sunol valleys. A disconnected segment of the former lower Alameda Creek, bypassed by the Flood Control Channel, retains some riparian habitat.

The historical patterns represent a diverse palette to draw upon as residents and managers attempt to re-establish riparian biodiversity in contemporary stream settings. Restoration attempts should be informed by historical patterns. Along a single stream there was considerable historical variation in riparian vegetation type, controlled by differences in hydrology and morphology. The same approach should be taken today. By determining appropriate functional reaches given present-day hydrology, managers can establish riparian patterns suited to local conditions.

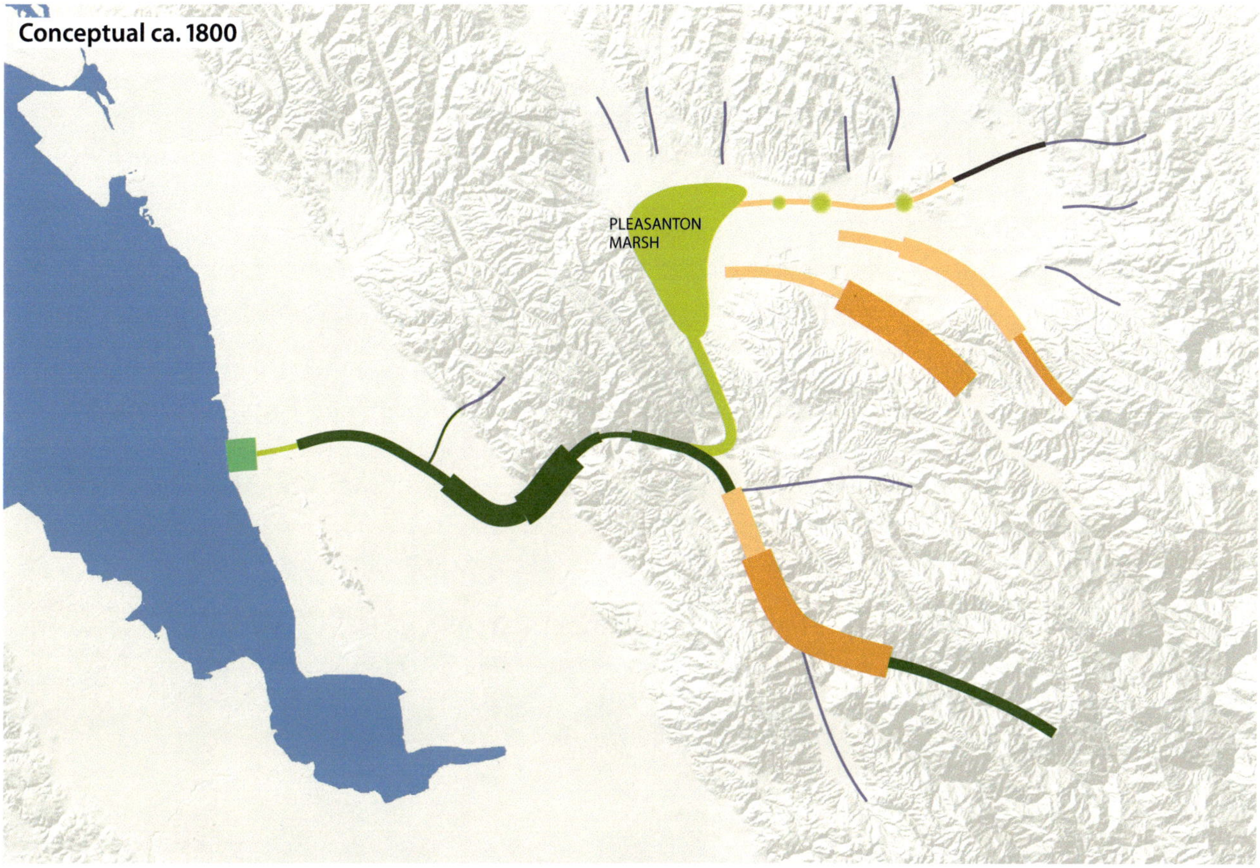
Conceptual ca. 1800
PLEASANTON MARSH

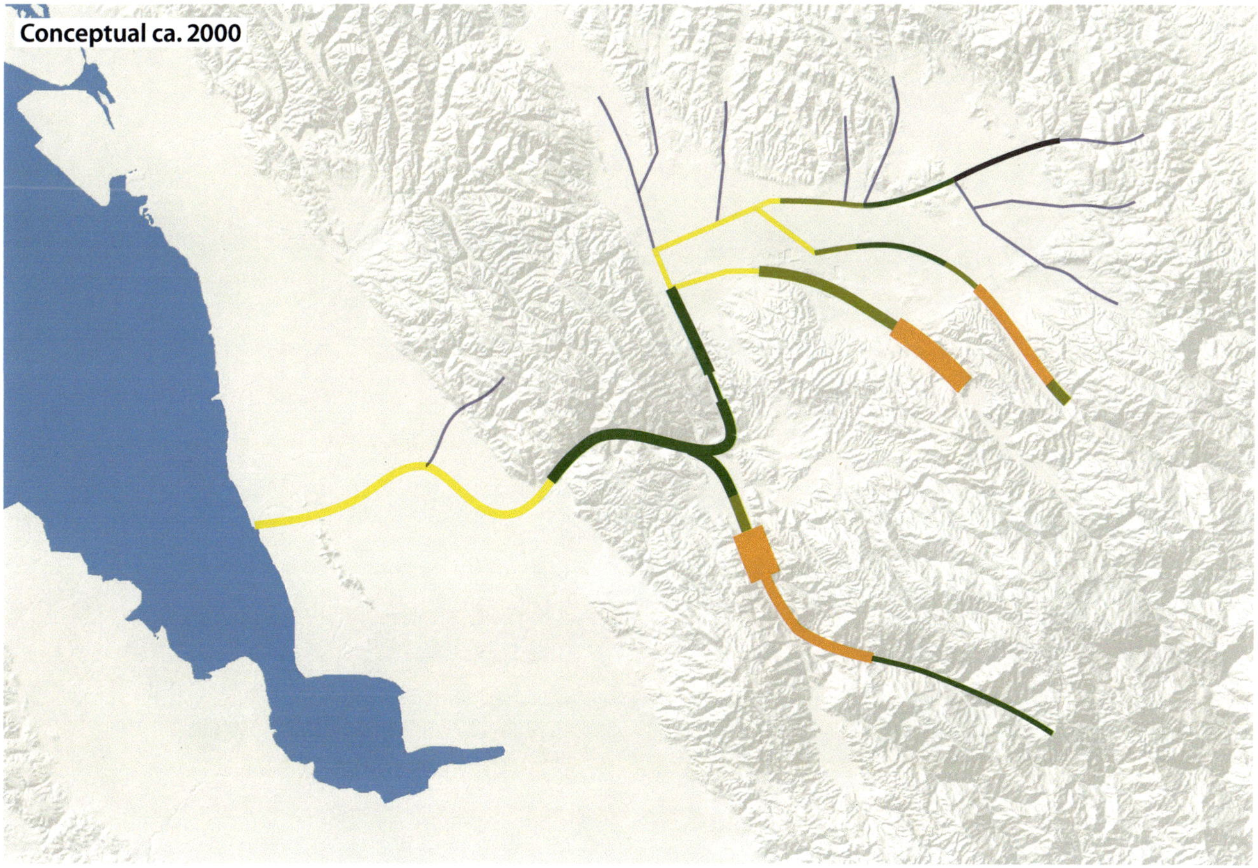
Conceptual ca. 2000

Summary concepts for riparian cover

Stream morphology and water availability both influenced riparian form, creating distinct stream reaches. Riparian vegetation was historically varied, both in species composition and in width and density of cover. Along a single stream, riparian cover could shift from mixed riparian forest to broad sycamore alluvial woodland to sparse oaks and sycamores to dense willow forest (fig. 9.22). Different restoration targets may be appropriate to different stream reaches. A wide variety of riparian habitat types existed within the project area and many reaches may have had sparse or no tree cover. Planning for riparian restoration should occur at the scale of these individual reaches.

Variation in groundwater levels and substrate can help explain morphology and vegetation patterns along Arroyo Mocho, Arroyo del Valle, and Alameda Creek. Each of these streams exited a canyon with mixed riparian forest and moved through an intermittent braided reach with sycamore alluvial woodland. Water infiltrated from the surface of the stream as it exited the canyon, and coarse gravels deposited at the top of the alluvial fan allowed this water to percolate rapidly, creating a losing intermittent reach.

Riparian diversity can be supported by maintaining natural flow patterns, including summer-dry reaches. Sycamore alluvial woodland and mixed riparian forest have different water needs, and would best be supported by different types of flow regimes.

Figure 9.22. Stream transect. This image shows patterns of groundwater, substrate and riparian vegetation along a stream transect. Refer to Chapter 1 for full text. (by Jennifer Natali)

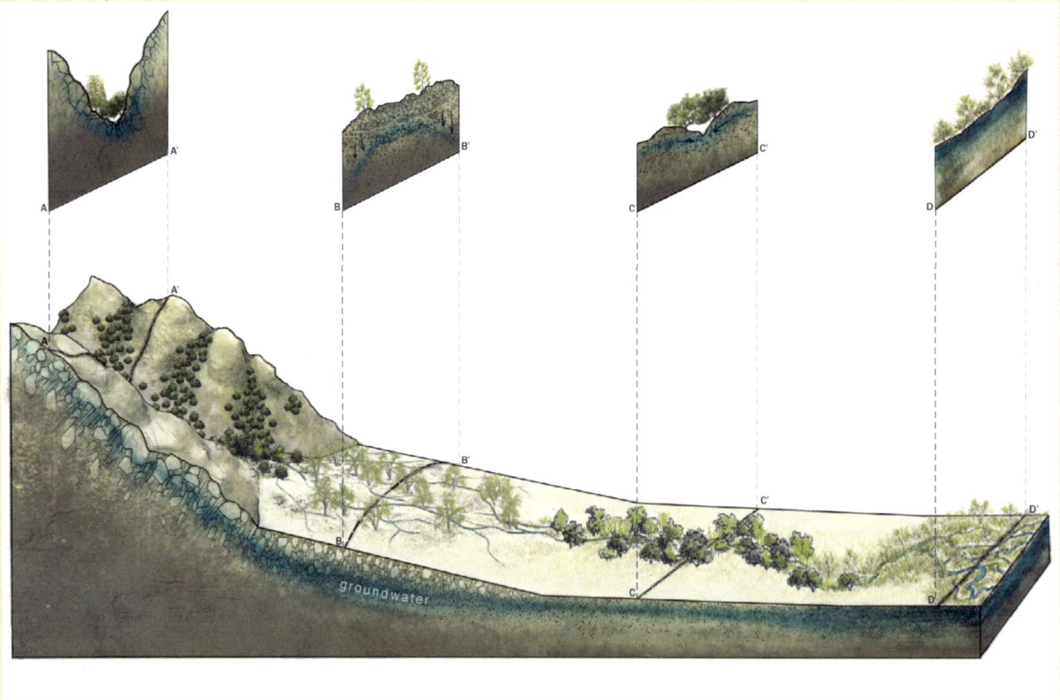

Restoration opportunities

Several potential restoration and conservation opportunities exist within the study area that merit further investigation. This study identified a number of locations within the study area where features have been lost or reduced during the past 200 years but have potential for recovery. While there are many other opportunities to recover ecological functions within the watershed, these have the potential to recover significant components of the historical system:

- **Restoration and protection of alkali wetlands near Springtown.** The Springtown alkali sink is still relatively undeveloped and supports a number of rare alkali and vernal pool species. Protection of more of both the sink and the surrounding watershed would help ensure that this area continues to provide high quality habitat in the future.

- **Restoration of fish passage up Dry Creek.** Dry Creek is positioned towards the midpoint of the Flood Control Channel, reducing the distance that fish must migrate through this exposed channel. Perennial reaches in the headwaters could potentially provide productive spawning habitat.

- **Restoration of fish habitat along Old Alameda Creek.** This channel is relatively intact, and could provide a more protected alternative to the Flood Control Channel for migrating salmonids. Feeding opportunities within the channel would help increase the size of steelhead reaching the ocean, a factor that is correlated to survival rates.

- **Restoration of wetland habitat near the Coyote Hills at the site of the historical Willows.** This area has maintained some remnant willow swamp, and the return of high groundwater here makes it a potential site for wetland restoration.

- **Re-establishment of a connection between Alameda Creek and the tidal marshlands.** The levees at the mouth of the Flood Control Channel could be set back to allow water and sediment from the creek to spread into the marshlands. This could help deliver sediment to restored tidal marshes, and provide habitat and a fresh-brackish transition for a host of species.

- **Management of flows to support sycamore alluvial woodlands.** Several stands of rare Central California sycamore alluvial woodland occur within the Alameda Creek watershed, but are regenerating very slowly. Where possible, managers should try to introduce flow patterns to support sycamore regeneration in managed streams with sycamore alluvial woodlands.

In these and other restoration efforts, watershed-scale planning will be crucial. Historically, the watershed functioned as one interconnected system, and distinct components provided different types of ecological function. Incorporation of this watershed-scale perspective will help ensure progress towards a more functional watershed.

Future Research

A number of additional topics merit further research, and would contribute to a better understanding of ecological and hydrogeomorphic pattern, process, and function in the watershed. Further information on fire ecology, the impact of invasive species introduction, early grazing impacts, and indigenous land management would contribute to a more detailed understanding. In addition, further information will certainly be discovered that can assist in the interpretation of this historical landscape and should be incorporated.

Research could also be done to add more depth to specific areas of interest. Additional, focused historical ecology studies providing data on some of the many sub-watersheds would help enhance our understanding of the historical landscape. In particular, this study did not evaluate the impact of the loss of inputs from San Antonio Creek, Arroyo Hondo, and Calaveras Creek, which would have had a substantial effect on habitat connectivity, water flows, and sediment inputs to the watershed. In addition, more detailed mapping of springs, and mapping of more riparian types on the historical habitat map, would help complete the picture of historical habitats. Expansion of the upland pilot study could help document patterns of species, vegetation, and land management before the 1930s. Upland historical ecology sources are less abundant than records in lowland areas, but further information could be gained from interviews with longtime residents and investigation of the San Jose Mission records and other historical sources. Additional techniques such as phytolith examination and dendrochronology could be used to address specific questions about tree regeneration and shifts in upland vegetation patterns over time.

Additional change analyses could clarify and quantify impacts to the watershed. Analysis of the impact of changes to sediment supply, loss of tidal prism, lowered groundwater levels, and changes in channel form (including loss of meander bends) could help interpret findings from this report. Detailed, region-specific studies could reveal changes in species composition and smaller-scale modifications such as the development of the urban forest.

Future studies could also build on this work to place ecological and biogeochemical processes within the watershed. We were not able to depict processes such as denitrification, nutrient production and storage, and carbon cycling. The historical picture would be enriched with additional focus on ecological and biogeochemical functions within the landscape and the development of additional conceptual models. Additional studies could model flooding, sediment transport, connections between groundwater and surface water, and the ranges of species of interest. In addition, the historical record could be used to develop more detailed models of stream morphology and process.

As watershed restoration and management proceed, report findings should continue to be calibrated with contemporary conditions, and with projected future climatic changes. Within this framework, the historical ecology study provides a foundation for developing a broad-based vision for the watershed—identifying the ecological functions that are needed and achievable with a spatially-specific strategy for improvement.

10 • REFERENCES

Abella R, Cook SF. 1960. *Colonial expeditions to the interior of California Central Valley, 1800-1820.* Berkeley, CA: University of California Press.

Aceituno ME, Caywood ML, Nicola SJ, et al. 1976. Occurrence of native fishes in Alameda and Coyote creeks, California. *California Department of Fish and Game* 62(3):195-206.

Adams F. 1946. *The historical background of California agriculture.* Berkeley, CA: University of California Press.

Akers TE. 1931. Mexican ranchos in the vicinity of Mission San Jose. Master of Arts, University of California, Berkeley, CA.

Alameda County Water District. 2001. *Alameda County Water District Groundwater Management Policy.*

Alameda Creek Alliance. 2012. Alameda Creek Steelhead and Salmon Documentation. Updated March 17, 2012. www.alamedacreek.org/

Alameda M. 1986. "Farming on Fremont's northern plain in the 1980s: agriculture's last stand," an oral history interview conducted by Ann Lage. *The Patterson family and ranch: southern Alameda County in transition. Vol. 1: Agriculture and farm life on Fremont's northern plain, 1890s-1980s.* Regional Oral History Office, The Bancroft Library, UC Berkeley.

Allardt GF. 1874. Official map of Alameda County, California. *Courtesy of The Bancroft Library, UC Berkeley.*

Allardt GF. 1888. Report on Alameda Creek: letter to Chas. Webb Howard, President Spring Valley Water Works. *Courtesy of San Francisco Public Utilities Commission Archives.*

Allardt GF. [1880]1907. Pleasanton and vicinity, Alameda County, Cal. Showing the original condition and extent of swamps and tules in the year 1880, subsequently drained by an extensive system of drainage canals. Traced from the original map made from surveys by the late George F. Allardt, Civil Engineer. To accompany report of Otto von Geldern, Consulting Engineer, dated San Francisco, September 1907. 1=600 *Courtesy of The Bancroft Library, UC Berkeley.*

Allen-Diaz B, Bartolome JW, McClaran MP. 1999. California oak savanna. In *Savannas, barrens, and rock outcrop plant communities of North America*, ed. RC Anderson, JS Fralish, and JM Baskin, 322-339. Cambridge, UK: Cambridge University Press.

Allen-Diaz B, Standiford RB, Jackson RD. 2007. Oak woodlands and forest, Chapter 12. *In Terrestrial Vegetation of California.* 3rd edition, ed. MG Barbour, T Keeler-Wolf, and AA Schoenherr, 313-338. Berkeley, Los Angeles, London: University of California Press.

Allen BH, Holzman BA, Evett RR. 1991. A classification-system for California hardwood rangelands. *Hilgardia* 59(2):1-45.

Alviso F. 1861. *U.S. v. Jose Dolores Pacheco, Land Case 155 ND [Santa Rita], U.S. District Court, Northern District. Courtesy of The Bancroft Library, UC Berkeley.*

Amador JM. 1855. *U.S. v. Jose Dolores Pacheco, Land Case 155 ND [Santa Rita], U.S. District Court, Northern District. Courtesy of The Bancroft Library, UC Berkeley.*

Amador JM. 1861. *U.S. v. Jose Dolores Pacheco, Land Case 155 ND [Santa Rita], U.S. District Court, Northern District. Courtesy of The Bancroft Library, UC Berkeley.*

American Ornithologists' Union. 2011. Birds of North America Online. http://www.aou.org/publications/bna/.

Anderson K. 2005. *Tending the wild: Native American knowledge and the management of California's natural resources.* Berkeley, CA: University of California Press.

Anian B. 1993. *Cultural resources inventory of Niles Canyon and archaeological survey of the right of way for two trains through the canyon.* Prepared for East Bay Regional Park District. *Courtesy of Museum of Local History, Fremont, CA.*

Anza J, Bolton HE. 1930. *Anza's California expeditions.* Berkeley, CA: University of California Press.

Atkinson K, Fuller J, Hanson C, et al. 2011. Evaluating Water Temperature and Turbidity Effects on Steelhead Life History Tactics in Alameda Creek Watershed. Unpublished report prepared for Alameda Creek Fisheries Restoration Workgroup. March 22, 2011.

Atwater BF, Conard SG, Dowden JN, et al. 1979. History, landforms, and vegetation of the estuary's tidal marshes. In *San Francisco Bay: the urbanized estuary : investigations into the natural history of San Francisco Bay and Delta with reference to the influence of man : fifty-eighth annual meeting of the Pacific Division/American Association for the Advancement of Science held at San Francisco State University, San Francisco, California, June 12-16, 1977*, ed. TJ Conomos, 493 p. San Francisco, CA: AAAS, Pacific Division.

Bailey P. 1917. *Engineer's report on Alameda Creek investigations: stream gaugings and percolation losses, 1916-17*. State Water Commission, Sacramento, CA. *Courtesy of Water Resources Center Archives.*

Bailey P. 1920. *Engineer's report on investigations on the Niles Cone 1916-1920*. State Water Commission. *Courtesy of Water Resources Center Archive.*

Bailey P. State Water Commission. 1917. Engineer's report on Alameda Creek investigations: stream gagings and percolation losses, 1916-1917. *Courtesy of Water Resources Center Archives.*

Baker JE, editor. 1914. *Past and present of Alameda County, California*. Chicago, IL: S.J. Clarke Publishing Co.

Balling RC, Wells SG. 1990. Historical rainfall patterns and arroyo activity within the Zuni River drainage basin, New Mexico. *Annals of the Association of American Geographers* 80(4):603-617.

Bancroft HH, Oak HL, Nemos W, et al. [1890]1970. *The works of Hubert Howe Bancroft. vol. XVIII-XXIV, History of California Vol. VII: 1860-1890*. San Francisco, CA: The History Company.

Bancroft HH. [1886]1966. *The works of Hubert Howe Bancroft. Volume XX, History of California vol. III: 1825-1840*. Santa Barbara, CA: Wallace Hebberd.

Barbour M, Keeler-Wolf T, Schoenherr A. 2007. *Terrestrial vegetation of California*. 3rd edition. Berkeley, Los Angeles, London: University of California Press.

Barry W. 1901. *Jane R. Clough vs. Spring Valley Water Works, Vol. 2*. MM O'Shaughnessy papers, carton 29, folder 19. *Courtesy of The Bancroft Library, UC Berkeley.*

Bartell MJ. 1914. *Report on underground water supply of Livermore Valley , California, prepared under the direction of H.E. O'Shaughnessy*. San Francisco, CA. *Courtesy of Water Resources Center Archives.*

Bartlett W. 1883. About the Bay of San Francisco. In *West of the Rocky Mountains*, ed. J Muir, 129-146. Philadelphia, PA: Running Press.

Bay Area Open Space Council. 2011. *The conservation lands network: San Francisco Bay Area upland habitat goals project report*. Berkeley, CA.

Baye PR, Faber PM, Grewell B. 2000. Tidal marsh plants of the San Francisco Estuary. In *Baylands ecosystem species and community profiles. Life histories and environmental requirements of key plants, fish and wildlife prepared by the San Francisco Bay Area Wetlands Ecosystem Goals Project*, ed., 9-32. San Francisco and Oakland, CA: U.S. Environmental Protection Agency and S.F. Bay Regional Water Quality Control Board.

Beagle JR, Eads R, Pearce S, Gilbreath AN, Gluchowski DC, and McKee LJ, 2011. *Suspended and bed sediment transport in Arroyo Mocho and Arroyo Las Positas during Water Year 2011*. A Watershed Program report of the San Francisco Estuary Institute prepared for Zone 7 Water Agency, Livermore, CA. San Francisco Estuary Institute, Oakland, CA.

Beard JL. 1901. *Jane R. Clough vs. Spring Valley Water Works, Vol. 2*. M.M. O'Shaughnessy papers, carton 29, folder 19. *Courtesy of The Bancroft Library, UC Berkeley.*

Becker GS, Reining IJ, Asbury DA, et al. 2007. *San Francisco Estuary watersheds evaluation: identifying promising locations for steelhead restoration in tributaries of the San Francisco Estuary*. Center for Ecosystem Management and Restoration, Oakland, CA.

Beechie TJ, Collins BD, Pess GR. 2001. Holocene and recent geomorphic processes, land use, and salmonid habitat in two North Puget Sound river basins. In *Geomorphic processes and riverine habitat*, ed. American Geophysical Union, 37-54. 4.

Beechie TJ, Liermann M, Pollock MM, et al. 2006. Channel pattern and river-floodplain dynamics in forested mountain river systems. *Geomorphology* 78:124-141.

Beller EE, Grossinger RM, Salomon MN, et al. 2011. *Historical ecology of the lower Santa Clara River, Ventura River, and Oxnard Plain: an analysis of terrestrial, riverine, and coastal habitats.* SFEI contribution #641. San Francisco Estuary Institute, Oakland, CA

Beller EE, Salomon MN, Grossinger RM. 2010. *Historical vegetation and drainage patterns of western Santa Clara Valley: a technical memorandum describing landscape ecology in Lower Peninsula, West Valley, and Guadalupe Watershed Management Areas.* SFEI contribution #622. San Francisco Estuary Institute, Oakland, CA.

Beller EE, Salomon MN, Grossinger RM et al. 2013. *Assessment of South Bay historical tidal-terrestrial transition zone: summary technical memorandum* (in prep). San Francisco Estuary Institute, Richmond, CA.

Benigno GM. 2011. *Invertebrate drift in neighboring perennial and seasonal tributaries of the Sacramento River.* MS Thesis, California State University , Chico, CA. 33 pp.

Bigelow P, Pearce S, McKee LJ. 2009. *Dry Creek watershed sediment source reconnaissance technical memorandum.* SFEI publication #595. San Francisco Estuary Institute, Oakland, CA.

Bigelow PE, Pearce SA, McKee LJ, et al. 2008. *A sediment budget for two reaches of Alameda Creek: analysis of natural and land use-related channel erosion and storage in Arroyo de la Laguna and upper Alameda Creek from their confluence upstream to gage locations near Verona and Welch Creek.* SFEI publication #550. San Francisco Estuary Institute and the Watersheds Science Program, Oakland, CA.

Bioletti FT. 1892. *Lasthenia glabrata* subsp. *glabrata*. Consortium of California Herbaria. http://ucjeps.berkeley.edu/consortium/.

Bjorkstedt EP, Spence BC, Garza JC, et al. 2005. *An analysis of historical population structure for evolutionary significant units of Chinook salmon, coho salmon, and steelhead in the north-central California coast recovery domain.* NOAA Technical memorandum NMFS (NOAA-TM-NMFS-SWFSC-382). October 2005. Southwest Fisheries Center, Santa Cruz, CA. 210 pp.

Blair TC, McPherson JG. 1994. Alluvial fan processes and forms. In *Geomorphology of desert environments*, ed. AD Abrahams and AJ Parsons, 354-402. London, UK: Chapman and Hall.

Bloom KJ, Bahre CJ. 2005. Aboriginal distribution of *Quercus lobata* woodland on the Kaweah River delta, California. *Madroño* 52(1).

Boardman WF. 1866. Rancho El Valle de San Jose, No. 143. County Surveyor's Office, Alameda County, CA. *Courtesy of Alameda County Department of Public Works.*

Boardman WF. 1867. Survey no. 947 for Frank Frater. County Surveyor's Office, Alameda County, CA. *Courtesy of Alameda County Department of Public Works.*

Boardman WF. 1868. Map of the town of Mission San Jose: surveyed May 1868. [ca. 1:2,400] *Courtesy of Earth Sciences and Map Library, UC Berkeley.*

Boardman WF. 1884. Map of a part of the Rancho Arroyo de la Alameda showing tracts recently sold to J. G. Clark. 10 chains to 1 inch. *Courtesy of Alameda County Department of Public Works.*

Boardman WF. ca.1870. Plat of survey: Rancho El Valle de San Jose. *Courtesy of Museum of Local History, Fremont, CA.*

Bocek BR. 1984. Ethnobotany of Costanoan Indians, California, based on collections by John P. Harrington. *Economic Botany* 38(2):240-255.

Boren JC, Engle DM, Gregory MS, et al. 1997. Landscape structure and change in a hardwood forest-tall-prairie ecotone. *Journal of Range Management* 50(3):244-249.

Borghi ER. 1987a. "Girlhood in a Patterson Ranch farm family, 1931-1948," an oral history interview conducted by Ann Lage. *The Patterson family and ranch: southern Alameda County in transition. Vol. 1: Agriculture and farm life on Fremont's northern plain, 1890s-1980s.* Regional Oral History Office, The Bancroft Library, UC Berkeley.

Borghi F. 1987b. "Dairying on the Patterson Ranch 1924-1950," an oral history interview conducted by Ann Lage. *The Patterson family and ranch: southern Alameda County in transition. Vol. 1: Agriculture and farm life on Fremont's northern plain, 1890s-1980s.* Regional Oral History Office, The Bancroft Library, UC Berkeley.

Bosqui E. 1894. Edward Bosqui letter: San Francisco, to F.O. Hopkins, Alameda County Surveyor: ALS, 1894 March 25. *Courtesy of The Bancroft Library, UC Berkeley.*

Bousman WG. 2007. *Breeding bird atlas of Santa Clara County, California*. Cupertino, CA: Santa Clara Valley Audubon Society.

Bowman JN. 1947. The area of the mission lands. *Courtesy of The Bancroft Library, UC Berkeley.*

Branner JC. 1912. *Report on the underground water conditions of the Livermore Valley and of Sunol Valley*. Joseph B. Lippincott Papers. *Courtesy of Water Resources Center Archives.*

Brewer WH. 1974. *Up and down California in 1860-1864: the journal of William H. Brewer*. New [3rd] edition. Francis Peloubet Farquhar. Berkeley, CA: University of California Press.

Brotherhood of Locomotive Engineers. 1875. Correspondence, West Oakland, October 20th, 1876. *Brotherhood of Locomotive Engineers Monthly Journal* 10(12):545-549.

Brown AK. 1960. Salt for the scraping: origin of the San Francisco Bay salt industry. *California Historical Society Quarterly* 39(2):117-120.

Brown AK. 2002. Historical oak woodland detected through *Armillaria mellea* damage in fruit orchards. *U.S. Forest Service General Technical Report* PSW-GTR-184:651-661.

Brown AK. 2005. *Reconstructing early historical landscapes in the Northern Santa Clara Valley*. Russell K. Skowronek. Santa Clara, CA: Santa Clara University.

Browne JR. 1872. *Reclamation of marsh and swamp lands, and projected canals for irrigation in California with notes on the canal systems of China and other countries*. San Francisco, CA: Alta California Printing House.

Brown R. 1986. "Observations of a ranch worker's son 1918-1950s," an oral history interview conducted by Ann Lage. *The Patterson family and ranch: southern Alameda County in transition. Vol. 1: Agriculture and farm life on Fremont's northern plain, 1890s-1980s*. Regional Oral History Office, The Bancroft Library, UC Berkeley.

Brown RE. 1901. *Jane R. Clough vs. Spring Valley Water Works. Courtesy of San Francisco Public Utilities Commission Archives.*

Brown SF. 1912. Statements of residents as to lowering of the water table, Niles Cone. Signed statement of S. F. Brown, farmer near Newark. MB051. *Courtesy of San Francisco Public Utilities Commission Archives.*

Bryant E. [1848] 1985. *What I saw in California: being the journal or a tour; by the emigrant route and south pass of the Rocky Mountains, across the continent of North America, the Great Desert basin, and through California, in the years 1846, 1847*. Lincoln, NE & London, UK: University of Nebraska Press.

Bryant HC. 1913. *Alvarado, Alameda Co., Calif. May 22*. Field Notes, Section 2, v546: "San Francisco Bay region, Southern California, Sacramento." *Archival Field Notebook. Courtesy of Museum of Vertebrate Zoology.*

Buchan LAJ, Leidy RA, Hayder MK. 1999. Aquatic resource characterization of western Mt. Hamilton stream fisheries. Prepared for The Nature Conservancy. Eisenberg, Olivieri and Associates, Oakland, CA.

Buck R. 1986. "Patterson property management, 1970s-1980s," an oral history interview conducted by Ann Lage. *The Patterson family and ranch: southern Alameda County in transition. Vol. 3: the Patterson Ranch, past and future: the family's perspective*. Regional Oral History Office, The Bancroft Library, UC Berkeley.

Bunshah. 1910. Pools in A. Mocho drying up under severe hot spell. June 4. *Courtesy of Livermore Heritage Guild.*

Buordo EA. 1956. A review of the General Land Office survey and of its use in quantitative studies of former forests. *Ecology* 37:754-768.

Burcham LT. 1956. Historical geography of the range livestock industry of California. Ph.D., Geography, University of California, Berkeley, CA.

Burcham LT. 1982. *California range land: an historico-ecological study of the range resource of California*. Davis, CA: University of California.

Burns J. 1888. A pioneer fruit region. *Overland monthly and Out West magazine* 12(67):1-14.

Burroughs, H. D. 1925. *W.E. Patterson and W.D. Patterson vs. Spring Valley Water Company. Superior Court of the state of California, in and for the county of Alameda. Courtesy of Water Resources Center Archives.*

Burtt Davy J. 1895. *Newark, Alameda Co, California USA*. Field Notes 1-3411, p.75-77. *Courtesy of Jepson Herbarium, UC Berkeley.*

Burtt Davy J. 1898. *Livermore Valley, beside the county road.* Field Notebook 13. Field Notes 4267-4289. *Courtesy of Jepson Herbarium, UC Berkeley.*

California Department of Fish and Game. 1953. Itnraoffice Correspondence, August 5, 1953, Steelhead rescue, Indian Creek, Alameda County, Yountville, 2 pp.

California Department of Transportation. 1998. *Historic property survey report for the seismic retrofit of Alameda Creek Bridge and overhead (Bridge #33-0039).* District 4-Oakland.

California Legislature Assembly. BB Redding, State Printer, 1854. *Journal of the fifth session of the legislature of the state of California, begun on the second day of January, 1854, and ended on the fifteenth day of May, 1854, at the cities of Benicia and Sacramento.*

California Natural Diversity Database (CNDDB). 2010. California Department of Fish and Game, Wildlife and Habitat Data Analysis Branch, Habitat Conservation Division.

Callaghan MG. 1915. *Spring Valley Water Company vs. City and County of San Francisco, et al. Vol. 2. Lands. District Court of the United States for the Northern District of California.*

CALVEG. McClellen, CA. USDA-Forest Service, Pacific Southwest Region, Remote Sensing Lab, 2004.

Campbell LG. 1986. "Balancing agriculture and development, family and public interests," an oral history interview conducted by Ann Lage. *The Patterson family and ranch: southern Alameda County in transition. Vol. 3: the Patterson Ranch, past and future: the family's perspective.* Regional Oral History Office, The Bancroft Library, UC Berkeley.

Capron ES. 1854. *History of California from its discovery to the present time.* Boston, MA: John P. Jewett & Company.

Carlton WH. 1874. *Field notes of the subdivision lines of township 3 and 4 south, range 2 east with lines of Rancho Valle de San Jose Mount Diablo Meridian, California.* U.S. Department of the Interior, Bureau of Land Management Rectangular Survey, California. Book 308-6. *Courtesy of Bureau of Land Management, Sacramento, CA.*

Carpenter DW, Sweeney JJ, Kasameyer PW, et al. 1984. *Geology of the Lawrence Livermore National Laboratory site and adjacent areas.* Lawrence Livermore National Laboratory.

Carpenter EJ, Cosby SW. 1939. *Soil Survey of Contra Costa County, California.* U.S. Department of Agriculture. Bureau of Soils. Series 1933. Washington, DC: Government Printing Office.

Carruthers C. 2000. *Washington Township: a pictorial history.* Fremont, CA: Washington Hospital Healthcare Foundation.

Case RM. 1978. Interstate highway road-killed animals, A data source for biologists. *Wildlife Society Bulletin* 6:8-13.

Cash. 1875. Survey no. 1624 for E. A. Rose, Murray Township. *Courtesy of Alameda County Department of Public Works.*

Cash L. County Surveyor's Office. 1870. Survey no. 1388 for A Splirala, Washington Township. Alameda Co. California. *Courtesy of Alameda County Department of Public Works.*

Castro. Alameda County Surveyor's Office. 1876. Survey no. 1718. *Courtesy of Alameda County Department of Public Works.*

Center for Conservation Biology. 1992. *Studies of Cordylanthus palmatus at the Springtown alkali sink, Livermore, California.* Stanford University, Stanford, CA. 97.

Child RE. 1913. Livermore Valley properties : Alameda County [Calif.]. [1:3,600] *Courtesy of The Bancroft Library, UC Berkeley.*

Child RE. 1914. Niles Canyon lands : Alameda County [Calif.]. [ca. 1:3,600] *Courtesy of The Bancroft Library, UC Berkeley.*

Clarke, Taylor, Beckh. 1852. *U.S. v. Jose de Jesus Vallejo, Land Case 59 ND [A tract of land near the Mission of Santa Clara], U.S. District Court, Northern District. Courtesy of The Bancroft Library, UC Berkeley.*

Clark E. 1901. *Jane R. Clough vs. Spring Valley Water Works, Vol. 2.* M.M. O'Shaughnessy papers, carton 29, folder 19. *Courtesy of The Bancroft Library, UC Berkeley.*

Clarke WC. 1952. The vegetation cover of the San Francisco Bay region in the early Spanish period. MA, Geography, University of California.

Clark WO. Department of the Interior, United States Geological Survey, 1915. *Ground-water resources of the Niles cone and adjacent areas, California; prepared in cooperation with the Department of Engineering of the State of California.* Water Supply Paper 345-H. *Courtesy of Water Resources Center Archives.*

Clark WO. Government Printing Office, 1924. *Ground water in Santa Clara valley, California.* Department of the Interior, U.S. Geological Survey (USGS): Water-supply Paper 519.

Coats R, Showers MA, Pavlik B. 1988. *A management plan for the Springtown alkali sink wetlands and the endangered plant cordylanthus palmatus.* Phillip Williams & Associates, San Francisco, CA.

Coats R, Showers MA, Pavlik B. 1993. Management plan for an alkali sink and its endangered plant *Cordylanthus palmatus. Environmental Management* 17(1):115-127.

Coffee AJ. 1862. *U.S. v. Jose Dolores Pacheco, Land Case 155 ND [Santa Rita], U.S. District Court, Northern District. Courtesy of The Bancroft Library, UC Berkeley.*

Collins BD, Montgomery DR. 2001. Importance of archival and process studies to characterizing pre-settlement riverine geomorphic processes and habitat in the Puget Lowland. *Water Science and Application* 4:227-243.

Collins JN, Grossinger RM. 2004. *Synthesis of scientific knowledge concerning estuarine landscapes and related habitats of the South Bay ecosystem. Draft final technical report of the South Bay Salt Pond Restoration Project.* San Francisco Estuary Institute, Oakland, CA. 91 pp.

Collins JN, Sutula M, Stein ED, et al. 2006. *Comparison of methods to map California riparian areas.* Final report prepared for the California Riparian Habitat Joint Venture. SFEI contribution #522. San Francisco Estuary Institute and Southern California Coastal Water Research Project, Oakland, CA.

Collins L, Leising K. 2003. Excerpt from draft Alameda Creek sediment budget and sediment source analysis: 2003 progress report. Watershed Sciences.

Collins L. 2005. Significant fluvial geomorphic alterations in the Alameda Creek watershed and preliminary sediment budget of the Alameda Creek Flood Control Channel. Powerpoint presentation.

Conerly HG. 1931. *Descriptive report. South San Francisco Bay: Charleston Slough to Calaveras and Dumbarton points. Register no. 4626.* U. S. Coast and Geodetic Survey (USCGS), Department of Commerce.

Cook SF. 1957. *The aboriginal population of Alameda and Contra Costa Counties, California.* Berkeley, London: University of California Press; Cambridge University Press.

Cook SF. 1960. *Colonial expeditions to the interior of California Central Valley, 1800-1820.* Berkeley, CA: University of California Press.

Coquhoun JA. 1893. *Illustrated album of Alameda County.*

Corbett MR. 2005. *Historical and cultural resource survey: East Alameda County.* Prepared for Alameda County Community Development Agency.

The Country Club of Washington Township. [1904]1950. *History of Washington Township, Alameda County, California.* Second edition: The Country Club of Washington Township Research Committee.

Cowardin LM, Carter V, Golet FC, et al. 1979. *Classification of wetlands and deepwater habitats of the United States.* Washington, D.C.: Fish and Wildlife Service, Biological Services Program, U.S. Department of the Interior.

Crespí J, Bolton HE. 1927. *Fray Juan Crespí, missionary explorer on the Pacific coast, 1769-1774.* Berkeley, CA: University of California Press.

Crespí J. [1772]. *Manuscript version of Juan Crespí's 1772 travel journal.* Sevilla, Spain: Excerpts from Alan K. Brown's unpublished translation. Published in Milliken et al. 1993, courtesy of Dr. Brown.

Crockett, Chittenden. 1861. *U.S. v. Jose Dolores Pacheco, Land Case 155 ND [Santa Rita], U.S. District Court, Northern District. Courtesy of The Bancroft Library, UC Berkeley.*

Cronise TF. 1868. *The natural wealth of California*. San Francisco, CA: HH Bancroft & Company.

Daily Alta California. 1860a. The Lander's wagon road expedition. May 17. *Courtesy of California Digital Newspaper Collection.*

Daily Alta California. 1860b. Notes on Amador Valley - No. 6. May 19, 1860. *Courtesy of California Digital Newspaper Collection.*

Daily Alta California. 1863. Six pages. The southeastern part of Alameda County. August 8. *Courtesy of California Digital Newspaper Collection.*

Daily Alta California. 1872. The Niles Bridge. November 9. *Courtesy of California Digital Newspaper Collection.*

Daily Alta California. 1888. The eastern shore: news from Alameda, Berkeley, Oakland and environs. September 6. *Courtesy of California Digital Newspaper Collection.*

Daily Alta California. 1889a. The eastern shore: news from Alameda, Berkeley, Oakland and environs. February 4. *Courtesy of California Digital Newspaper Collection.*

Daily Alta California. 1889b. Oakland Items. February 4, 1889. *Courtesy of the California Digital Newspaper Collection.*

Daily Alta California 1860c. Notes of a trip of the S.F.B.D. agricultural visiting committee, No. 31. September 22. *Courtesy of California Digital Newspaper Collection.*

Daniels & Osmont E. 1910. Livermore Orchard Tracts: Alameda County, California. 1:1,200.

Davis D, editor. 1976. *A pictorial history of Pleasanton*. Pleasanton, CA: Pleasanton Bicentennial Heritage Committee.

Davis FW, Borchert MI. 2006. Central Coast bioregion. In *Fire in California's ecosystems*, ed. NG Sugihara.

Davis FW, Kuhn W, Alagona P, et al. 2000. *Santa Barbara County: oak woodland inventory and monitoring program*. University of California, Santa Barbara.

Davis H. 1868. Wheat in California. *Overland Monthly and Out West Magazine* 1(5):442-452.

Davis WH. 1889. *Sixty years in California*. San Francisco, CA: A.J. Leary.

Day S. 1853. *Copy of field notes of surveys of a part of the 1st Standard Parallel south of Mount Diablo and east of the Principal Meridian, and of townships south of the base and east of the Meridian.* Book 231-7. *Courtesy of Bureau of Land Management, Sacramento, CA.*

Dedrick KG, Chu LT. 1993. Historical atlas of tidal creeks, San Francisco Bay, California. In *Coastal zone '93: proceedings of the eighth symposium on coastal and ocean management, New Orleans, Louisiana*, ed. Magoon OT, Wilson WS, Converse H, and Tobin LT: American Society of Civil Engineers.

De Mofras D. A Bertrand. 1844. Port de San Francisco dans la Haute Californie. No. 16. *Courtesy of David Rumsey Map Collection.*

DeNier FL. 1927. Robert Livermore and the development of Livermore Valley to 1860. Master of Arts, History, University of California, Berkeley, CA.

Department of Water Resources. 2006. *Livermore Valley groundwater basin*. San Francisco Hydrologic Region. California's Groundwater Bulletin 118.

Dettinger MD, Cayan DR, Diaz HF, et al. 1998. North-south precipitation patterns in western North America on interannual-to-decadal timescales. *Journal of Climate* 11(12):3095-3111.

Dillman. Spring Valley Water Co. 1911a. Map of Alameda pipe line aqueduct from station 59 to sta. 170+29.5. F-123. *Courtesy of Museum of Local History, Fremont, CA.*

Dillman. Spring Valley Water Co. 1911b. Map vicinity Niles [illegible]. F-125. *Courtesy of Museum of Local History, Fremont, CA.*

Dockweiler. 1912a. Map showing region included in and adjacent to the delta of Alameda Creek extending from San Lorenzo Creek to Coyote River and located on the easterly and southerly shores of San Francisco Bay - Alameda County, California. *Courtesy of Alameda County Water District.*

Dockweiler JH. 1912b. *Report on sources of water supply, east region of San Francisco Bay, prepared for and on behalf of the cities of Oakland and Berkeley jointly with San Francisco at the request of Percy V. Long, City Attorney.*

Dockweiler JH. 1912c. *Niles Cone. Abstract of report by J. H. Dockweiler, to Percy V. Long. Courtesy of San Francisco Public Utilities Commission Archives.*

Duerr. 1872a. Survey No 1529. *Courtesy of Alameda County Department of Public Works.*

Duerr. 1872b. Survey No. 1528. *Courtesy of Alameda County Department of Public Works.*

Duerr C. 1869. Map of the town of Pleasanton, Alameda County, California, 1869: surveyed for John W. Hollinger. [1:2,400] *Courtesy of Earth Sciences and Map Library, UC Berkeley.*

Durán FN, Cook SF. 1960. *Colonial expeditions to the interior of California Central Valley, 1800-1820.* Berkeley, CA: University of California Press.

Durán N. 1824. Plano topográfico de la mision de San José. *Courtesy of The Bancroft Library, UC Berkeley.*

Durán N. 1958. Letters from Narciso Durán, from the manuscript collections in the California Historical Society Library, translated by Francis Price. *California Historical Society Quarterly* 37(2):97-128.

Dyer EH, Hopkins JO. County Surveyor's Office. 1863. Survey no. 575 for Geo. Simpson and Geo. W. Patterson. Alameda County, CA. *Courtesy of Alameda County Department of Public Works.*

Dyer EH. 1861b. *U.S. v. Tomas Pacheco, Land Case 119 ND [Potrero de los Cerritos], U.S. District Court, Northern District. Courtesy of The Bancroft Library, UC Berkeley.*

Dyer EH. 1862a. *Field notes of the final survey of the Rancho Santa Rita. Younz, Administrator, Confirmee.* U.S. Department of the Interior, Bureau of Land Maanagement Rectangular Survey, California. Book K-38. *Courtesy of Bureau of Land Management, Sacramento, CA.*

Dyer EH. 1862b. *Field notes of the re-survey of part of the exterior boundaries of T. 3 S., R. 3 E., of the Mt. Diablo Meridian in the state of California by E. H. Dyer under his contract to subdivide said township, bearing date the 1 day of October, 1861.* U.S. Department of the Interior, Bureau of Land Management Rectangular Survey, California. Book 189-3. *Courtesy of Bureau of Land Management, Sacramento, CA.*

Dyer EH. 1862c. Plat of the Santa Rita Rancho finally confirmed to John Yountz Administrator of Estate of Jose Dolores Pacheco. E-347. 4 chs. = 1 in. *Courtesy of The Bancroft Library, UC Berkeley.*

Dyer EH. 1869. *Transcript of the field notes of the re-survey of the exterior boundaries of township 3 south, range 2 east, and the eastern boundary of the Rancho El Valle de San Jose, of the Mt. Diablo Meridian, California.* U.S. Department of the Interior, Bureau of Land Management Rectangular Survey, California. Book 271-16. *Courtesy of Bureau of Land Management, Sacramento, CA.*

Dyer EH. Alameda County Surveyor's Office. 1861a. Survey no. 402. Alameda Co. *Courtesy of Alameda County Department of Public Works.*

Dyer EH. County Surveyor's Office. 1862d. Survey no. 495 for Patterson and Stove. Alameda County. *Courtesy of Alameda County Department of Public Works.*

D'Antonio C, Bainbridge S, Kennedy C, et al. 2000. *Ecology and restoration of California grasslands with special emphasis on the influence of fire and grazing on native grassland species.* University of California, Berkeley. 99 p.

Ebersole JL, Liss WJ, Frissell CA. 2003a. Cold water patches in warm streams: physiochemical characteristics and the influence of shading. *Journal of the American Water Resources Association* (April):255-368.

Ebersole JL, Liss WJ, Frissell CA. 2003b. Thermal heterogeneity, stream channel morphology, and salmonids abundance in northeastern Oregon streams. *Canadian Journal of Fisheries and Aquatic Sciences* 60:1266-1280.

Echo. 1890. Water, as if from a large spring, rises from the ground a short distance this side of Oak Knoll & forming quite a stream, flows into the Arroyo Mocho. March 6. *Courtesy of Livermore Heritage Guild.*

Echo. 1894a. Roadmaster Beck now has 54 teams engaged in hauling gravel from Las Positas Creek onto the Collier Canyon, Doolen and Dublin roads. November 15. *Courtesy of Livermore Heritage Guild.*

Echo. 1894b. In consequence of making of a large excavation in the bed of the Arroyo Mocho... December 20. *Courtesy of Livermore Heritage Guild.*

Echo. 1894c. The arroyos Mocho and Valle were both, at times, during the late storm, unfordable... December 27. *Courtesy of Livermore Heritage Guild.*

Edmundson PE. 1858. *U.S. v. Tomas Pacheco, Land Case 119 ND [Potrero de los Cerritos], U.S. District Court, Northern District. Courtesy of The Bancroft Library, UC Berkeley.*

Edwards P. 1932. *The diary of Philip Leget Edwards; the great cattle drive from California to Oregon in 1837*. San Francisco, CA: Grabhorn Press.

Edwards SW, Thayer C. 2008. The last days of Livermore. *The Four Seasons* 13(2):1-30.

Ellsworth O. 1901. *Jane R. Clough vs. Spring Valley Water Works, Vol. 2.* M.M. O'Shaughnessy papers, carton 29, folder 19. *Courtesy of The Bancroft Library, UC Berkeley.*

[Ellsworth]. n.d. [Mouth of Niles Canyon] 15D 29A. *Courtesy of Museum of Local History, Fremont, CA.*

Elmore AJ, Manning SJ, Mustard JF, et al. 2006. Decline in alkali meadow vegetation cover in California: the effects of groundwater extraction and drought. *Journal of Applied Ecology* 43:770-779.

Enos JS. 1887. *Second biennial report of the bureau of labor statistics of the state of California, for the years 1885 and 1886.* California Bureau of Labor Statistics, Sacramento CA.

Erman DC, Leidy GR. 1975. Downstream movement of rainbow trout fry in a tributary of Sagehen Creek, under permanent and intermittent flow. *Transactions of the American Fisheries Society* 104:467-473.

Ertter B. 1997. *Annotated checklist of the East Bay flora: native and naturalized plants of Alameda and Contra Costa counties, California.* Special Publication #3. California Native Plant Society East Bay Chapter in Association with the University and Jepson Herbaria.

Espy TW. 1950. *Re - Pleasanton water rights. Memo to: Mr. A Del Thomson. MB 045. Courtesy of San Francisco Public Utilities Commission Archives.*

Evans WA. 1954. Field note, San Antonio Creek, Alameda County, September 30, 1954, California Department of Fish and Game, Yountville, 1 p.

Fages P, Bolton HE. 1911. *Expedition to San Francisco Bay in 1770: diary of Pedro Fages*. Berkeley, CA: University of California Press.

Fallon J. 1861. *U.S. v. Jose Dolores Pacheco, Land Case 155 ND [Santa Rita], U.S. District Court, Northern District. Courtesy of The Bancroft Library, UC Berkeley.*

Faulkner WB. 1866. Faulkner's hand-book and directory of Murray township, Alameda County, California. Together with interesting descriptions of the flourishing towns of Livermore, Pleasanton, Dublin, Sunol, Midway, Tracy, Bethany and Byron. The Livermore Herald Steam Printing House. *Courtesy of The Bancroft Library, UC Berkeley.*

Federal Geographic Data Committee (FGDC). 1997. *Vegetation classification standard.* United States Geological Survey, Reston, VA.

Ferriz H. 2001. Groundwater resources of Northern California: an overview. In *Engineering Geology Practice in Northern California: Association of Engineering Geologists Special Publication 12 and California Division of Mines and Geology Bulletin 210*, ed. H. Ferriz and R. Anderson.

Figuers S, Ehman KD. 2004. *Preliminary stratigraphic evaluation west side of the main basin Livermore-Amador groundwater basin.* Prepared for Zone 7 Water Agency.

Figuers S. 1998. *Groundwater study and water supply history of the East Bay Plain, Alameda and Contra Costa counties, CA, for the Friends of the San Francisco Estuary*. Norfleet Consultants, Livermore, CA.

Fisher CK. 1959. Report of barrier on Stonybrook Creek, Alameda County, July 15, 1959. California Department of Fish and Game, Yountville, 1 p.

Fisher H, Brown EG, Warne WE. 1966. *Livermore and Sunol Valleys, evaluation of ground water resources, Appendix A: Geology*. Department of Water Resources.

Fisher RB. n.d. Scene from the peak: a pictorial history of Washington Township, Alameda County: Introduction. *Courtesy of Museum of Local History, Fremont, CA.*

Follett WI. 1955a. Fish collection field notes (Field No. F55-5), Stonybrook Canyon, ca. 1.5 miles above its junction with Alameda Creek, October 14, 1955. *Courtesy of California Academy of Sciences, Department of Ichthyology, San Francisco, CA.*

Follett WI. 1955b. Fish collection field notes (Field No. F-55-8), Stonybrook Canyon, ca. 1.2 miles above its junction with Alameda Creek, October 31, 1955. *Courtesy of California Academy of Sciences, Department of Ichthyology, San Francisco, CA.*

Follett WI. 1957. Fish collection field notes (Field No. F57-4), Alameda Creek, Niles, opposite California Nursery Company, March 25, 1957. *Courtesy of California Academy of Sciences, Department of Ichthyology, San Francisco, CA.*

Follett WI. 1974. Letter from W.I. Follett, Curator Emeritus of Ichthyology, California Academy of Sciences, to S.J. Nicola, California Department of Fish and Game, on fishes of Alameda and Coyote creeks, 11 July 1974. 4 pp.

Font P, Bolton HE. 1933. *Font's complete diary: a chronicle of the founding of San Francisco*. Berkeley, CA: University of California Press.

Forbes JH. 1914. *Report on the underground waters of Niles Cone. Courtesy of San Francisco Public Utilities Commission Archives.*

Ford RS, Hills EE. State of California. Department of Water Resources, 1974. *Evaluation of Ground Water Resources: Livermore and Sunol Valleys*. Bulletin No. 118-2.

Ford WH. 1901. *Jane R. Clough vs. Spring Valley Water Works, Vol. 2*. M.M. O'Shaughnessy papers, carton 29, folder 19. *Courtesy of The Bancroft Library, UC Berkeley.*

Foster DR, Motzkin G. 2003. Interpreting and conserving the openland habitats of coastal New England: Insights from landscape history. *Forest Ecology and Management* 185:127-150.

Freeman JR. 1912. *The Hetch Hetchy water supply for San Francisco, 1912*. San Francisco, CA: The Rincon Publishing Company.

Fritts HC, Gordon GA. 1980. Annual precipitation for California since 1600 reconstructed from western Northern American tree rings. Laboratory of Tree-Ring Research, University of Arizona. Prepared for California Department of Water Resources. *Courtesy of University of California, Berkeley.*

Fuller ML. 1912. *Preliminary report on the underground waters of the Livermore and Sunol valleys. Courtesy of San Francisco Public Utilities Commission Archives.*

Furtado D. 1987. "Working for Henry Patterson 1930s-1950s," an oral history interview conducted by Bill Helfman. *The Patterson family and ranch: southern Alameda County in transition. Vol. 1: Agriculture and farm life on Fremont's northern plain, 1890s-1980s*. Regional Oral History Office, The Bancroft Library, UC Berkeley.

Gale CA. 1915. *Spring Valley Water Company vs. City and County of San Francisco, et al. Vol. 1. Lands. District Court of the United States for the Northern District of California.*

Gedan KB, Silliman BR, Bertness MD. 2009. Centuries of human-driven change in salt marsh ecosystems. *Annual Review of Marine Science*: 117-141.

Gibbes CD. 1878. Subdivisions of Plot 31 Rancho El Valle de San Jose. 1 inch=10 chains. *Courtesy of Alameda County Department of Public Works.*

Gibson RJ, Haedrich RL, Wernerheim CM. 2005. Loss of fish habitat as a consequence of inappropriately constructed stream crossings. *Fisheries* 30(1):10-17.

Gilbert GK. 1917. *Hydraulic-mining debris in the Sierra Nevada*. U.S. Geological Survey. Washington, DC: U.S. Government Printing Office.

Goals Project. 1999. *Baylands Ecosystem Habitat Goals. A report of habitat recommendations prepared by the San Francisco Bay Area Wetlands Ecosystem Goals Project.* USEPA, San Francisco, Calif./S.F. Bay Regional Water Quality Control Board, Oakland.

Gobalet KW, Schulz PD, Siefkin N. 2004. Archeological perspectives on Native American fisheries of California, with emphasis on steelhead and salmon. *Transactions of the American Fisheries Society* 133:801-833.

Gobalet KW. 1990. Native status of Sacramento perch *Archoplites interruptus* in Alameda Creek, Alameda County, California: evidence from archeological site CA-ALA-483. *California Fish and Game* 76(4):244-247.

Gobalet KW. 1992. Inland utilization of marine fishes by Native Americans along the central California coast. *Journal of California and Great Basin Anthropology* 14:72-84.

Gobalet KW. 2004. Using archaeological remains to document regional fish presence in prehistory; a central California case study. *Transactions of the Western Section of the Wildlife Society* 40:107-113.

Gobalet KW. 2006. Fish remains at archaeological site (ALA-576, aka 509-342) in Fremont, CA, near Highway 680 and Mission Blvd. Electronic mail to Robert A. Leidy, dated June 9, 2006.

Goold TL. 1986. "The Logan family farm in Alvarado," an oral history interview conducted by Ann Lage. *The Patterson family and ranch: southern Alameda County in transition. Vol. 1: Agriculture and farm life on Fremont's northern plain, 1890s-1980s.* Regional Oral History Office, The Bancroft Library, UC Berkeley.

Graymer RW, Jones DL, Brabb EE. USGS. 1996. *Preliminary geologic map emphasizing bedrock formations in Alameda County, California: a digital database.* U.S. Geological Survey Open-File Report 96-252.

Greene. 1895. *Plagiobothrys scriptus.* Consortium of California Herbaria. http://ucjeps.berkeley.edu/consortium/.

Grinnell J, Dixon JS, Linsdale JM. 1937. *Fur-bearing mammals of California: their natural history, systematic status, and relations to man.* Berkeley, CA: University of California Press.

Grinnell J, Wythe MW. 1927. *Directory to the bird-life of the San Francisco Bay region.* Berkeley, CA.

Grinnell J. 1913. *Alvarado, Alameda Co., July 28.* Field Notes, Section 8: "Journal. Alameda Co., Calif.; Solano Co., Calif." v.1230. *Courtesy of Museum of Vertebrate Zoology, Archival Field Notebook.*

Grinnell J. 1919. *Alvarado, Aug 10.* Field Notes, Section 8: "Journal. Alameda and Contra Costa cos., Calif." v.1322. *Courtesy of Museum of Vertebrate Zoology, Archival Field Notebook.*

Grossinger RM, Askevold RA, Striplen CJ, et al. 2006. *Coyote Creek watershed historical ecology study: Historical condition, landscape change, and restoration potential in the eastern Santa Clara Valley, California.* SFEI Publication #426, San Francisco Estuary Institute, Oakland, CA.

Grossinger RM, Askevold RA. 2005. *Historical analysis of California coastal landscapes: methods for the reliable acquisition, interpretation, and synthesis of archival data.* Report to the U.S. Fish and Wildlife Service San Francisco Bay Program, the Santa Clara University Environmental Studies Institute, and the Southern California Coastal Water Research Project, SFEI contribution #396. San Francisco Estuary Institute, Oakland. 48 p.

Grossinger RM, Beller EE, Salomon MN, et al. 2008. *South Santa Clara Valley historical ecology study: including Soap Lake, the upper Pajaro River, and Llagas, Uvas-Carnadero, and Pacheco creeks.* SFEI Publication #558, San Francisco Estuary Institute, Oakland, CA.

Grossinger RM, Brewster E. 2003. *A geographic history of San Lorenzo Creek watershed: landscape patterns underlying human activities in the lands of the Yrgin...* . SFEI Contribution #85, December 2003, San Francisco Estuary Institute, Oakland, CA.

Grossinger RM, Stein ED, Cayce K, et al. 2011. *Historical wetlands of the Southern California coast: an atlas of U.S. Coast Survey t-sheets, 1851-1889. SFEI contribution #586, SCCWRP technical report #589.* San Francisco Estuary Institute, Oakland, CA.

Grossinger RM, Striplen CJ, Askevold RA, et al. 2007. Historical landscape ecology of an urbanized California valley: wetlands and woodlands in the Santa Clara Valley. *Landscape Ecology* 22:103-120.

Grossinger RM. 2005. Documenting local landscape change: the San Francisco Bay Area historical ecology project. In *The historical ecology handbook: a restorationist's guide to reference ecosystems*, ed. Dave Egan and Evelyn A. Howell, 425-442. Washington, D.C.: Island Press.

Grossinger RM. 2012. *Napa Valley historical ecology atlas: Exploring a landscape of transformation and resilience.* Berkeley, CA: University of California Press.

Gudde EG, Bright W. 1998. *California place names: the origin and etymology of current geographical names.* 4th ed., rev. and enl. edition. Berkeley, CA: University of California Press.

Gutmann I. 1919. *An agricultural survey of the Pleasanton properties of the Spring Valley Water Co.,* San Francisco, CA. MB 067. *Courtesy of San Francisco Public Utilities Commission Archives.*

Haidinger TL, Keeley JE. 1993. Role of high fire frequency in destruction of mixed chaparral. *Madroño* 40(3):141-147.

Halley W. 1876. *The centennial year book of Alameda County, California.* Oakland, CA: William Halley.

Hallock C. 1877. *The sportsman's gazetteer and general guide. The game animals, birds, and fishes of North America: their habits and various methods of capture.* New York, NY: Forest and Stream Publishing Company.

Hamilton JG. 1997. Changing perceptions of pre-European grasslands in California. *Madroño* 44(4):311-333.

Hanna PT. 1951. *The dictionary of California land names.* Second edition. Los Angeles, CA: The Automobile Club of Southern California.

Hanson Environmental Inc. 2002. *Air and water temperature monitoring within Alameda Creek: 2001-2002.* Walnut Creek, CA.

Harley JB. 1989. Historical geography and cartographic illusion. *Journal of Historical Cartography* 15:80-91.

Harris JA, Hobbs RJ, Higgs E, et al. 2006. Ecological restoration and global climate change. *Restoration Ecology* 14:170-176.

Haviland & Tibbetts. 1912. *Report to Spring Valley Water Company on the artesian basin of Livermore Valley, giving an estimate of the probable yield of water. Supplementary to a report on the Pleasanton Artesian Investigation by F. H. Tibbetts of May 1907 and a Progress Report on same by Haviland & Tibbetts of June 20, 1911.* Haviland & Tibbetts Civil and Consulting Engineers, San Francisco, CA. Courtesy of Water Resources Center Archives.

Haviland PA. 1912. General route Niles Canon Road, Washington and Pleasanton Townships, Alameda Co, Cal. *Courtesy of Niles Silent Film Museum.*

Haviland PA. 1913. County road no. Niles Canyon Road being county road no. 4974 survey sta. 76 to survey sta. 86. 1 in. = 100 ft. *Courtesy of Museum of Local History, Fremont, CA.*

Haviland PA. Tribune Publishing Company. 1910. Official map of Alameda County. Oakland, CA. *Courtesy of Pleasanton's Museum on Main.*

Hazen A. 1916. *Spring Valley Water Company vs. City and County of San Francisco, et al. Vol. 7. Buildings, financial, original costs, rights of way, water rights. District Court of the United States for the Northern District of California.*

Healy C. 1863. *El Valle de San Jose.* U.S. Department of the Interior, Bureau of Land Management Rectangular Survey, California. Book 507-29. *Courtesy of Bureau of Land Management, Sacramento, CA.*

Helley EJ, Graymer RW. 1997. *Quaternary geology of Alameda County, and parts of Contra Costa, Santa Clara, San Mateo, San Francisco, Stanislaus, and San Joaquin counties, California: a digital database.* U.S. Geological Survey open-file report 97-0097.

Henion GD. 1901. *Jane R. Clough vs. Spring Valley Water Works, Vol. 2.* M.M. O'Shaughnessy papers, carton 29, folder 19. *Courtesy of The Bancroft Library, UC Berkeley.*

Higley HA. 1857b. *Field-notes of the final survey of the Rancho Agua Caliente, Fulgencio Higuera, confirmee.* U.S. Department of the Interior, Bureau of Land Management Rectangular Survey, California. *Courtesy of Bureau of Land Management, Sacramento, CA.*

Higley HA. Lith. Britton & Rey. 1857a. Official map of the County of Alameda, California [electronic resource] / surveyed and compiled by order of the Board of Supervisors. San Francisco, CA. [1:95,040] *Courtesy of Earth Sciences and Map Library, UC Berkeley.*

Higuera J. 1861. *U.S. v. Jose Dolores Pacheco, Land Case 155 ND [Santa Rita], U.S. District Court, Northern District. Courtesy of The Bancroft Library, UC Berkeley.*

Hilgard EW. 1884. *Census reports tenth census, June 1, 1880: report on cotton production in the United States; also embracing agricultural and physico-geographical descriptions of the several cotton states and of California.* Department of the Interior, Census Office, Washington, D.C.: Government Printing Office.

Hilgard EW. 1889. *Report on the hydrographic and geological features of the Alameda Creek watershed. Courtesy of San Francisco Public Utilities Commission Archives.*

Hinckley F. 1869. Map of the town of Livermore: Alameda County, California, October 1st, 1869, surveyed for Wm. M. Mendenhall. [1:2,400] 200 ft. to 1 in. *Courtesy of Earth Sciences and Map Library, UC Berkeley.*

Hobbs J. 2011. Monitoring the response of fish assemblages to restoration in the South Bay Salt Ponds. Semi-annual report (Quarter 2). Prepared for the South bay Salt Pond Restoration Project.

Hoffmann CF. Geological Survey of California. 1873. Map of the region adjacent to the Bay of San Francisco. San Francisco. 1:126,720, 2 miles to 1 in. *Courtesy of Peter J. Shields Library Map Collection, UC Davis.*

Holland RF, Placer Land Trust. 2009. Great Valley vernal pool distribution (VP_2005). Geospatial data.

Holland RF. 1978. *The geographic and edaphic distribution of vernal pools in the great Central Valley, California.* California Native Plant Society.

Holland RF. 1986. *Preliminary descriptions of the terrestrial natural communities of California. Unpublished report.* California Department of Fish and Game, Natural Heritage Division, Sacramento, CA.

Holland RF. 2009. *California's Great Valley vernal pool habitat status and loss: rephotorevised 2005.*

Holmes LC, Nelson JW. 1917. *Reconnaissance soil survey of the San Francisco Bay region, California [report].* Government Printing Office, Washington, D.C.

Holmes LC, Nelson JW. Government Printing Office. 1914. Reconnaissance soil survey of the San Francisco Bay region, California [map]. Washington, D.C.

Holmes P. 1997. *Two centuries at Mission San Jose 1797-1997.* Fremont, CA: Fremont Museum of Local History.

Holmes TH. 1990. Botanical trends in Northern California oak woodland. *Rangelands* 12(1):3-7.

Holstein G. 2000. Plant communities ecotonal to the Baylands. In *Baylands ecosystem species and community profiles. Life histories and environmental requirements of key plants, fish and wildlife. Prepared by the San Francisco Bay Area Wetlands Ecosystem Goals Project,* ed. PR Olofson, 49-68. San Francisco and Oakland, CA: U.S. Environmental Protection Agency and San Francisco Bay Regional Water Quality Control Board.

Holstein G. 2001. Pre-agricultural grassland in Central California. *Madroño* 48(4):253-264.

Holzman BA, Allen-Diaz BH. 1991. *Vegetation Change in Blue Oak Woodlands.* United States Department of Agriculture, U.S. Forest Service, Pacific Southwest Research Station, Berkeley, CA.

Hopkinson P, Huntsinger L. 2005. Are East Bay hills grasslands a historical artifact? Phytolith evidence and a potential candidate for the true East Bay vegetation type. *Grasslands* 15(1):7-9.

Hopkirk JD. 1973. Endemism in fishes of the Clear Lake region. *University of California Publications in Zoology* 96:160.

Howard AD. 1979. *Geologic history of middle California.* Berkeley, CA: University of California Press.

Howe R. 1851. *Field notes on a meridian line south of Mount Diablo in the state of California, under contract dated October 1, 1851.* U.S. Department of the Interior, Bureau of Land Management

Rectangular Survey, California. Book 231-3. *Courtesy of Bureau of Land Management, Sacramento, CA.*

ICF International. 2010a. *East Alameda conservation strategy, final draft, prepared for the East Alameda County Conservation Strategy Steering Committee.* ICF International, San Jose, CA.

ICF International. 2010b. *Alameda watershed habitat conservation plan: working draft.* Produced for the San Francisco Public Utilities Commission.

Jepson WL. 1891. *Plagiobothrys canescens.* Consortium of California Herbaria. http://ucjeps.berkeley.edu/consortium/.

Jepson WL. 1896. *Berkeley to San Jose, May 26.* Field Book of W. L. Jepson. Volume 1, p. 77. *Courtesy of Jepson Herbarium, UC Berkeley.*

Jepson WL. 1905. *Berkeley to Mission San Jose: Niles, April 8.* Field Book of W. L. Jepson. Volume 12, p.67-8. *Courtesy of Jepson Herbarium, UC Berkeley.*

Jepson WL. 1909. *A flora of California.* San Francisco: Cunningham, Curtiss & Welch.

Jepson WL. 1920. *Berkeley to the Salinas Valley, April 5.* Field Book of W. L. Jepson. Volume 37. *Courtesy of Jepson Herbarium, UC Berkeley.*

Jokerst J. 1986. Valley Sink Scrub. California Natural Diversity Database (CNDDB). California Department of Fish and Game, Wildlife and Habitat Data Analysis Branch, Habitat Conservation Division.

Keeler-Wolf T, Elam DR, Lewis K, et al. State of California Department of Fish and Game, 1998. *California vernal pool assessment preliminary report.*

Keeler-Wolf T, Lewis K, Roye C. 1996. *The definition and location of sycamore alluvial woodland in California.* State of California, The Resources Agency, Department of Fish and Game, Sacramento, CA.

Keeley JE. 2002. Native American impacts on fire regimes of the California coastal ranges. *Journal of Biogeography* 29:303-320.

Keeley JE. 2004. American Indian influence on fire regimes in California's coastal ranges. *Fire management today* 64(3).

Keeley JE. 2005. Fire history of the San Francisco East Bay region and implications for landscape patterns. *International Journal of Wildland Fire* 14(3):285-296.

Keith AJ. 1995. Effects of Disturbance on the Trophic Structure of an Ephemeral California Stream. MS Thesis, July 1995. San Francisco State University. 105 pp.

Kellogg. 1901. *Jane R. Clough vs. Spring Valley Water Works, Vol. 2.* M.M. O'Shaughnessy papers, carton 29, folder 19. *Courtesy of The Bancroft Library, UC Berkeley.*

Kellogg L. 1909. *Numerous short trips afield: near Newark, Alameda Co., Calif.* Field Notes, Section 2: "Journal and catalog. Alameda, Contra Costa, Marin, Solano, and Sonoma Cos., Calif." v.1384. p.25-26. *Courtesy of Museum of Vertebrate Zoology, Archival Field Notebook.*

Kern RE. ca. 1983. *The Vallejos of Mission San Jose. Courtesy of California State Library.*

Kerr D. U. S. Coast Survey (USCS). 1857a. San Francisco Bay, California, register no. 634. 1:10,000.

Kerr D. U. S. Coast Survey (USCS). 1857b. San Francisco Bay, California, register no. 635. 1:10,000.

King MG. Oakland Tribune Publishing Company. 1880. Oakland Daily & Weekly Tribune map of Alameda County: Compiled from the most reliable surveys, and corrected to date. Oakland, CA. 1:95,040 *Courtesy of The Bancroft Library, UC Berkeley.*

Kirwan ML, Murray AB, Donnelly JP, et al. 2011. Rapid wetland expansion during European settlement and its implication for marsh survival under modern sediment delivery rates. *Geological Society of America* 39(5):507-510.

Knighton D. 1998. *Fluvial forms and processes: A new perspective.* New York, NY: Hodder-Arnold.

Knudsen KL, Sowers JM, Witter RC, et al. 2000. *Preliminary maps of quaternary deposits and liquefaction susceptibility, nine-county San Francisco Bay region, California: a digital database.* U.S. Geological Survey. U.S.G.S. Open File Report 2006-1037.

Kohlmann S, Alvarez J, Clark C, et al. 2008. *A baseline assessment of environmental and biological conditions at Five Pillars Farms, Alameda County, California*. Tierra Resource Management, Castro Valley, CA.

Koltermann CE, Gorelick SM. 1992. Paleoclimatic signature in terrestrial flood deposits. *Science* 256(5065):1775-1782.

Kroeber AL. [1925]1976. *Handbook of the Indians of California*. New York, NY: Dover Publications, Inc.

La Croze J. 1860. *Field Notes of the obsolete survey of the Rancho Santa Rita, Yontz, Administrator, Confirmee*. U.S. Department of the Interior, Bureau of Land Management Rectangular Survey, California. Book Obs Santa Rita 37. *Courtesy of Bureau of Land Management, Sacramento, CA*.

La Croze J. 1862a. Plat of the Santa Rita rancho, finally confirmed to John Yontz administrator of estate of Jose Dolores Pacheco E-346. 40 chains = 1 in. *Courtesy of The Bancroft Library, UC Berkeley*.

La Croze J. 1862b. *U.S. v. Jose Dolores Pacheco, Land Case 155 ND [Santa Rita], U.S. District Court, Northern District. Courtesy of The Bancroft Library, UC Berkeley*.

Lajoie KR. 2003. Alameda Creek alluvial fan. 1 in. = 1000 ft. Unpublished map.

LaMarche VC. 1974. Paleoclimatic inferences from long tree-ring records: intersite comparison shows climatic anomalies that may be linked to features of the general circulation. *Science* 183(4129):1043-1048.

Lawrence WB. 1920. *Niles Water District. MB 042. Courtesy of San Francisco Public Utilities Commission Archives*.

Lee CH. 1912. *Appendix I. Evaporation and transpiration from the wet lands in Livermore Valley*. San Francisco. In *The future water supply of San Francisco*, Spring Valley Water Company, pp. 482-493.

Lee CH. 1916a. *Report of damage to Spring Valley Water Co. lands in Livermore Valley, California, resulting from pumping operations of the company at Pleasanton wells*. San Francisco, CA. *Courtesy of Water Resources Center Archives*.

Lee CH. 1916b. *Spring Valley Water Company vs. City and County of San Francisco, et al. Vol. 8. Depreciation, financial, going value, lands, overhead rating base, reservoir values, water rights. District Court of the United States for the Northern District of California*.

Leidy RA, Becker G, Harvey BN. 2005c. Historical status of coho salmon in streams of the urbanized San Francisco Estuary, California. *California Fish and Game* 91(4):1-36.

Leidy RA, Becker GS, Harvey BN. 2005a. *Historical distribution and current status of steelhead/rainbow trout* (Oncorhynchus mykiss) *in streams of the San Francisco Estuary, California*. Center for Ecosystem Management and Restoration, Oakland, CA.

Leidy RA, Becker GS, Harvey BN. 2005b. *Historical distribution and current status of steelhead* (Oncorhynchus mykiss), *coho* (O. kisutch), *and Chinook salmon* (O. tshawytscha) *in streams of the San Francisco Estuary, California*. Center for Ecosystem Management and Restoration, Oakland, CA.

Leidy RA, Cervantes-Yoshida K, Carlson SM. 2011. Persistence of native fishes in small streams of the urbanized San Francisco Estuary, California: acknowledging the role of urban streams in native fish conservation. *Aquatic Conservation: Marine and Freshwater Ecosystems* 21:472-483.

Leidy RA. 2007. Ecology, assemblage structure, distribution, and status of fishes in streams tributary to the San Francisco Estuary, California. SFEI contribution #530, April 2007, San Francisco Estuary Institute, Oakland, CA.

Leopold LB, Wolman MG, Miller JP. 1964. *Fluvial Processes in Geomorphology*. San Francisco, CA: W.H. Freeman.

Leopold LB. 1951. Rainfall frequency: an aspect of climatic variation. *Transactions, American Geophysical Union* 32(3).

Lewis HT. 1973. *Patterns of Indian burning in California: ecology and ethnohistory*. Ramona, CA: Ballena Press.

Lewis HT. 1985. Why Indians burned: specific versus general reasons. In *Proceedings, symposium and workshop on wilderness fire*, ed. JE Lotan, BM Kilgore, WC Fischer, and RW Mutch, 75-86. Missoula, MT: Intermountain Forest and Range Experiment Station, Forest Service, U.S. Department of Agriculture.

Lewis WJ. 1850a. Plan of a part of the San Jose Mission land: surveyed for John M. Hormer, Esq. on the 11th, 12th, Dec. 1850 [1:4,752] *Courtesy of Earth Sciences and Map Library, UC Berkeley.*

Lewis WJ. 1850b. Survey no. 29: plat of land purchased by Earl Marsh of J. M. Horner & Co. *Courtesy of Santa Clara County Surveyors Office.*

Lewis WJ. 1852. Map of land surveyed for Jones and Strode, J. M. Horner, E. L. Beard and J. C. Smith, by Wm. J. Lewis. *Courtesy of Santa Clara County Surveyors Office.*

Lewis WJ. 1860. Plat of the rancho Potrero de los Cerritos, finally confirmed to Thomas Pacheco and Augustin Alviso, E-262. 40 chs = 1 in. *Courtesy of The Bancroft Library, UC Berkeley.*

Lewis WJ. 1861. *U.S. v. Jose Dolores Pacheco, Land Case 155 ND [Santa Rita], U.S. District Court, Northern District. Courtesy of The Bancroft Library, UC Berkeley.*

Lewis WJ. 1862. *U.S. v. Tomas Pacheco, Land Case 119 ND [Potrero de los Cerritos], U.S. District Court, Northern District. Courtesy of The Bancroft Library, UC Berkeley.*

Lewis WJ. ca.1850. Plat of the rancho Potrero de los Cerritos, D-264. Alameda County. *Courtesy of The Bancroft Library, UC Berkeley.*

Lightfoot KG, Luby EM. 2002. Late Holocene in the San Francisco Bay Area: temporal trends in the use and abandonment of shell mounds in the East Bay. In *Catalysts to complexity: late Holocene societies of the California coast*, ed. JM Erlandson and TL Jones: Cotsen Institute of Archaeology, University of California, Los Angeles.

Likins MJW. 1874. *Six years experience as a book agent in California, including my trip from New York to San Francisco via Nicaragua.* 168 pp. Available through California As I Saw It: First-Person Narratives of California's Early Years, 1849-1900, Library of Congress. Women's union book and job printing office, San Francisco, CA.

Limm MP, Marchetti MP. 2009. Juvenile Chinook salmon (*Oncorhynchus tshawytscha*) growth in off-channel and main-channel habitats on the Sacramento River, CA, using otolith increment widths. *Environmetal Biology of Fishes* 85: 141-151.

Lindenmayer D, Hobbs RJ, Montague-Drake R, et al. 2008. A checklist for ecological management of landscapes for conservation *Ecology Letters* 11:78-91.

Lyman CS. 1848. [Journals of Chester Lyman, No. 12 and 13]. *Courtesy of California Historical Society, San Francisco, CA.*

Magney DL. 1992. Descriptions of three new Southern California vegetation types: Southern Cactus Scrub, Southern Coastal Needlegrass Grassland, and Scalebroom Scrub. *Crossosoma* 18(1):1-9.

Malamud-Roam FP, Dettinger M, Ingram BL, et al. 2007. Holocene climates and connections between the San Francisco Bay estuary and its watershed: A review. *San Francisco Estuary and Watershed Science* 5(1):Article 3.

Malamud-Roam FP, Ingram BL, Hughes MK, et al. 2006. Holocene paleoclimate records from a large California estuarine system and its watershed region: Linking watershed climate and bay conditions. *Quaternary Science Reviews* 25(13-14):1570-1598.

Manies KL. 1997. Evaluation of General Land Office survey records for analysis of the northern Great Lakes hemlock-hardwood forests. University of Wisconsin, Madison.

Marinacci B, Marinacci R. 1997. *California's Spanish place names: what they mean and how they got there.* Houston, TX: Gulf Publishing Company.

Martinez D. 2010. Fire, grass, and chaparral in prehistoric southwestern California (in prep). *Ecological Restoration.*

Mayfield DW. 1978. Ecology of the pre-Spanish San Francisco Bay Area. M.A., San Francisco State University, San Francisco, CA.

McBain S, Trush W. 2008. *Alameda creek population recovery strategies and instream flow assessment for steelhead trout: Final study plan.* McBain & Trush, Inc., Arcata, CA.

McCann WE, Hinkel EJ. 1937. *History of rural Alameda County in two volumes.* Oakland, CA.

McCarthy FF. 1958. *The history of Mission San Jose, California: 1797-1835.* Fresno, CA: Academy Library Guild.

McEwan D, Jackson TA. 1996. *Steelhead restoration and management plan for California.* California Department of Fish and Game, Sacramento, CA.

McEwan D. 2001. Central Valley steelhead. In *Contributions to the biology of Central Valley salmonids*, ed. RL Brown: California Department of Fish and Game, Fish Bulletin 179.

McGown JE. 1902. *Progressive Pleasanton.* The Pleasanton Times. *Courtesy of Pleasanton's Museum on Main.*

McKeown N. 1975a. "A neighboring farmer recalls the early days," an oral history interview conducted by Donald Patterson. *The Patterson family and ranch: southern Alameda County in transition. Vol. 1: Agriculture and farm life on Fremont's northern plain, 1890s-1980s.* Regional Oral History Office, The Bancroft Library, UC Berkeley.

McKeown W. 1975b. "A neighboring farmer recalls the early days," an oral history interview conducted by Donald Patterson. *The Patterson family and ranch: southern Alameda County in transition. Vol. 1: Agriculture and farm life on Fremont's northern plain, 1890s-1980s.* Regional Oral History Office, The Bancroft Library, UC Berkeley.

Mensing S, Byrne R. 1999. Invasion of Mediterranean Weeds into California before 1769. *Fremontia* 27(3):6-9.

Metsker CF. n.d. Metsker's map of Alameda County. San Francisco, CA. 1/2 in. = 1 mile *Courtesy of Livermore Heritage Guild.*

Mikesell LR, Weissmann GS, Karachewski JA. 2010. Stream capture and piracy recorded by provenance in fluvial fan strata. *Geomorphology* 115:267-277.

Milliken R, Shoup LH, Ortiz BR. 2009. *Ohlone/Costanoan Indians of the San Francisco Peninsula and their neighbors, yesterday and today.* Technical report prepared by Archaeological and Historical Consultants, Oakland, California for the National Park Service, Golden Gate Recreation Area, Fort Mason, San Francisco, California.

Milliken R. 1995. *A time of little choice: the disintegration of tribal culture in the San Francisco Bay area, 1769-1810.* Menlo Park, CA: Ballena Press.

Milliken R. 2008. *Native Americans at Mission San Jose.* Banning, CA: Malki-Ballena Press.

Minnich RA. 2008. *California's fading wildflowers: lost legacy and biological invasions.* Berkeley and Los Angeles, CA: University of California Press.

Montgomery DR. 1999. Erosional processes at an abrupt channel head: implications for channel entrenchment and discontinuous gully formation. In *Incised river channels: processes, forms, engineering and management*, ed. SE Darby and A Simon. Chichester, UK: John Wiley & Sons.

Morenhout JA. [1849]1935. The inside story of the gold rush. AP Nasatir, editor. California as I saw it: first-person narratives of California's early years, 1849-1900. California Historical Society, San Francisco, CA.

The Morning Call. 1892. Camping at Niles: parties in the canyon, notwithstanding Spring Valley's protests. June 23. *Courtesy of Chronicling America.*

The Morning Call. 1895. Alameda Roads: they are in bad condition and need attention. January 21. *Courtesy of Chronicling America.*

The Morning Oregonian. 1880. [Flood]. April 22. *Courtesy of Union City Historical Museum.*

Morris JM, Thronson RE, Nicklen RR. Department of Water Resources, Division of Resource Planning, 1960. *Intrusion of salt water into ground water basins of southern Alameda County, Bulletin no. 81.*

Morse F, Westdahl F. U. S. Coast and Geodetic Survey (USCGS). 1896-7. Re-survey of San Francisco Bay, California, southern extremity of Bay, Alviso to Warm Springs, Register no. 2313. 1:10,000.

Morse F, Westdahl F. U.S. Coast and Geodetic Survey (USCGS). 1896. Pacific Coast Resurvey of San Francisco Bay, California, Dumbarton Point to Head of Bay, Register No. 2258. Washington, D.C. 1:10000. *Courtesy of the National Ocean Service, Rockville, MD.*

Morse HN, San Francisco Chronicle. 1898. Training cattle thieves. October 2. *Courtesy of ProQuest Historical Newspapers.*

Mortimer WJ. 1915. *Spring Valley Water Company vs. City and County of San Francisco, et al. Vol 1. Lands. District Court of the United States for the Northern District of California.*

Mosier P, Mosier D. 1986. *Alameda County place names.* Fremont, CA: Mines Road Books.

Moyle PB, Crain PK, Whitener K. 2007. Patterns in the use of a restored California floodplain by native and alien fishes. *San Francisco Estuary and Watershed Science* 5(3): Article 1.

Moyle PB, Hobbs J, O'Rear T. 2012. Tidal marsh fishes of the San Francisco Estuary. In *Ecology, conservation and restoration of tidal marshes: The San Francisco Estuary,* ed. A. Palaima. Berkeley, CA: University of California Press.

Moyle PB. 2002 *Inland fishes of California.* Revised and expanded. Berkeley, CA: University of California Press.

Mudd S. 2011. The life and death of salt marshes in response to anthropogenic disturbance of sediment supply. *Geological Society of America* 39(5):511-512.

Mulholland W, Lippincott. 1912. [Resources of the Spring Valley Water Company in the Livermore Valley: Supplement in response to questions]. *Courtesy of Water Resources Center Archives.*

Munyon E. 1861. *U.S. v. Tomas Pacheco, Land Case 119 ND [Potrero de los Cerritos], U.S. District Court, Northern District. Courtesy of The Bancroft Library, UC Berkeley.*

Nabhan GP, Martinez D. 2012. Traditional ecological knowledge and endangered species recovery: is ethnobiology for the birds? *Journal of Ethnobiology* 32(1):1-5.

National Marine Fisheries Service (NMFS). 2007. Federal recovery outline for the distinct population segment of Central California Coast steelhead. Memorandum dated May 31, 2007, Long Beach, CA.

Nielsen JL, Lisle TE, Ozaki V. 1994. Thermally stratified pools and their use by steelhead in northern California streams. *Transactions of the American Fisheries Society* 123:613-626.

Nusbaumer. 1885. Washington Township part 2. *Courtesy of Museum of Local History, Fremont, CA.*

Nusbaumer GL, Boardman WF. Tribune Publishing Company. 1889. Official map of Alameda County, California. Oakland, CA. *Courtesy of Museum of Local History, Fremont, CA.*

Nusbaumer GL. 1889b. Survey No. 2529 for Spring Valley Water Co, Washington Tp. *Courtesy of Alameda County Department of Public Works.*

Nusbaumer GL. ca. 1900. Map showing the property of Geo. L. Nusbaumer in Murray Tp Alameda Co. Cal. 8 chs = 1 in. *Courtesy of Alameda County Department of Public Works.*

Nusbaumer GL. Tribune Publishing Company. 1889a. Official Map of Alameda County. Oakland, CA. *Courtesy of Museum of Local History, Fremont, CA.*

Oakland Tribune. 1875. City of Oakland eyes Alvarado water. September 16. *Courtesy of Union City Historical Museum.*

Oakland Tribune. 1876a. [Flood]. January 8. *Courtesy of Union City Historical Museum.*

Oakland Tribune. 1876b. [Another flood]. March 10. *Courtesy of Union City Historical Museum.*

Oakland Tribune. 1898. *Alameda County illustrated: the Eden of the Pacific; the flower garden of California.* Oakland, CA: The Tribune Publishing Company.

Oakland Tribune. 1909. [Flood]. January 24. *Courtesy of Union City Historical Museum.*

Oakland Tribune. 1915. Alvarado Clippings. May 29, 1915. *Courtesy of the Union City Historical Museum.*

Oakland Tribune. 1921. [Flood]. December 29. *Courtesy of Union City Historical Museum.*

Olson B. 1991. Sycamore Alluvial Woodland. California Natural Diversity Database (CNDDB). California Department of Fish and Game, Wildlife and Habitat Data Analysis Branch, Habitat Conservation Division.

Overacker H. 1901. *Jane R. Clough vs. Spring Valley Water Works, Vol. 2.* M.M. O'Shaughnessy papers, carton 29, folder 19. *Courtesy of The Bancroft Library, UC Berkeley.*

Patterson D. 1977. "Family lore: the Pattersons and their land since the 1850s," an oral history interview conducted by Stanley Bry. *The Patterson family and ranch: southern Alameda County in transition. Vol. 3: the Patterson Ranch, past and future: the family's perspective.* Regional Oral History Office, The Bancroft Library, UC Berkeley.

Patterson G. 1987. "Recalling the Pattersons' past: the family, land, and historic homes," an oral history interview conducted by Ann Lage. *The Patterson family and ranch: southern Alameda County in transition. Vol. 3: the Patterson Ranch, past and future: the family's perspective.* Regional Oral History Office, The Bancroft Library, UC Berkeley.

Patterson WD. 1955. "The Alameda County Water District, 1914-1955," an oral history interview conducted by John Caswell. *The Patterson family and ranch: southern Alameda County in transition. Vol. 2: Water, development, and preservation in southern Alameda County.* Regional Oral History Office, The Bancroft Library, UC Berkeley.

Pierce JW. 1948. Salt water infiltration into the Alameda County Water District. Engineer. Department of Civil Engineering, Stanford University, Stanford, CA.

Poathon [?] EC. County Surveyor's Office. 1895. Survey no. 2799, proposed public road, Washington Township. *Courtesy of Alameda County Department of Public Works.*

Porter RP. Government Printing Office, 1884. *Report on valuation, taxation and public indebtedness in the United States as returned at the tenth census (June 1, 1880).*

Raab L, Jones T. 2004. *Prehistoric California: Archaeology and the Myth of Paradise.* Salt Lake City, UT: University of Utah Press.

Radeloff VC, Mladenoff DJ, He HS, et al. 1999. Forest landscape change in the northwestern Wisconsin Pine Barrens from pre-European settlement to the present. *Canadian Journal of Forest Research* 29:1649-1659.

Radeloff VC, Mladenoff DJ, Manies KL, et al. 1998. Analyzing forest landscape restoration potential: presettlement and current distribution of oak in the northwest Wisconsin Pine Barrens. *Transcriptions of the Wisconsin Academy of Sciences Arts and Letters* 86:189-205.

Rasmussen P. 1915. *Spring Valley Water Company vs. City and County of San Francisco, et al. Vol. 2. Lands. District Court of the United States for the Northern District of California.*

Rensch HE, Rensch EGH, Hoover MB, et al. 1966. *Historic spots in California.* 3d edition. Stanford, CA: Stanford University Press.

Richards I. 1901. *Jane R. Clough vs. Spring Valley Water Works, Vol. 2.* M.M. O'Shaughnessy papers, carton 29, folder 19. *Courtesy of The Bancroft Library, UC Berkeley.*

Robinson A, Becker G, Pearce S, et al. 2012. *Alameda Creek Flood Control Channel Wildlife Support Functions.* Draft Report, May 23, 2012. Prepared by the San Francisco Estuary Institute, Richmond, CA, and Center for Ecosystem Management and Restoration, Oakland, CA. 24 pp.

Robinson A. [1846]1947. *Life in California: a historical account of the origin, customs, and traditions of the Indians of Alta-California.* California Centennial Edition, Number IX edition. Biobooks: Oakland, CA.

Rodgers A, Kerr D. U. S. Coast Survey (USCS). 1857. San Francisco Bay, California, Register no. 676. 1:10,000 *Courtesy of NOAA.*

Rogers P, *Silicon Valley Mercury News.* 2010. Just add water. August 17.

Rosenthal JS, Praetzellis A, Meyer J. 1998. *An archaeological investigation of artifacts and human remains from CA-CCO-637, Los Vaqueros Project area, Contra Costa County, California.* Rohnert Park, Calif.: Anthropological Studies Center, Sonoma State University Academic Foundation.

Russell WH, McBride J. 2003. Landscape scale vegetation-type conversion and fire hazard in the San Francisco Bay Area open spaces. *Landscape and Urban Planning* 64:201-208.

Ryan WR. 1850. Personal adventures in Upper and Lower California, in 1848-9; with the author's experience at the mines. Illustrated by twenty-three drawings taken on the spot. California as I saw it: first person narratives of California's early years, 1849-1900. William Shoberl, London.

Safford H, North M, et al. 2012. Climate change and the relevance of historical forest conditions. Managing Sierra Nevada forests. General Technical Report PSW-GTR-237. M. North. Albany, CA, U.S. Department of Agriculture, Forest Service, Pacific Southwest Research Station: 23-45.

Sal H. 1795. Document 11: Report which Lieutenant Don. Hermenegildo Sal. takes to the senior colonel and governor Don Diego Borica of the places which have been examined in La Alameda in conformity with the superior order of the ninth of the present month prepared by his honor accompanied by the revered Father Fran Antonio Danti, Minister of the Mission of San Francisco [translated by Museum of Local History]. *Courtesy of Museum of Local History, Fremont, CA.*

Sanborn Mapping. 2007. Alameda County LiDAR data. Alameda County, CA.

Sanchez N. 1930. *Spanish and Indian place names of California.* San Francisco, CA: A.M. Robertson.

Sandercock PJ, Hooke JM, Mant JM. 2007. Vegetation in dryland river channels and its interaction with fluvial processes. *Progress in Physical Geography* 31(2):107-129.

Sanderson EW, Huron A. 2011. Conservation in the city. *Conservation Biology* 25(3):421-423.

Sandoval J. n.d. *A century of Mission San Jose in the East Bay.* Hayward, CA. 48.

Sandoval JS. 1985. *The history of Washington Township.* Hayward, CA: Mt. Eden Historical Publishers.

San Francisco Call. 1860. Jumps from a wheel to help drowning boy. Charles Mitchell saves a life in the waters of Alameda Creek. July 6. *Courtesy of California Digital Newspaper Collection.*

San Francisco Call. 1865. Life in Livermore. May 7. *Courtesy of California Digital Newspaper Collection.*

San Francisco Call. 1900. Just as Matthew Whitfield was sinking. July 6. *Courtesy of California Digital Newspaper Collection.*

San Francisco Call. 1903a. Spring Valley must battle to retain Niles watershed. January 11. *Courtesy of California Digital Newspaper Collection.*

San Francisco Call. 1903b. Throw victim from a bridge. July 17. *Courtesy of California Digital Newspaper Collection.*

San Francisco Chronicle. 1870. Across the Bay. April 8, 1870. *Courtesy of Proquest Historical Newspapers.*

San Francisco Chronicle. 1885. Southern Alameda: Rambles by a beautiful stream. May 25, 1885. *Courtesy of Proquest Historical Newspapers.*

San Francisco Chronicle. 1889a. Industrial Topics. January 4, 1889. *Courtesy of Proquest Historical Newspapers.*

San Francisco Chronicle. 1889b. Oakland news: Damage by the heavy rains. December 12. *Courtesy of Proquest Historical Newspapers.*

San Francisco Chronicle. 1890. Alameda County. July 13. *Courtesy of Proquest Historical Newspapers.*

San Francisco Chronicle. 1892. Oakland news: Sunol Glen campers. January 24. *Courtesy of Proquest Historical Newspapers.*

San Francisco Chronicle. 1893. California names: a translation of Spanish terms. Jan 17, 1893. *Courtesy of Proquest Historical Newspapers.*

San Francisco Chronicle. 1895. Southern Alameda. Rambles by a beautiful stream. May 25. *Courtesy of Proquest Historical Newspapers.*

San Francisco Chronicle. 1899a. Oakland news. A women killed by a local train. No more camping in Sunol glen. January 24. *Courtesy of Proquest Historical Newspapers.*

San Francisco Chronicle. 1899b. More water for San Francisco. Spring Valley to build big filter. It will purify Alameda Creek water in the rainy season. February 21. *Courtesy of Proquest Historical Newspapers.*

San Francisco Estuary Institute (SFEI). 1998. Bay Area EcoAtlas: Geographic information system of wetland habitats past and present. http://www.sfei.org/ecoatlas/.

San Francisco Estuary Institute (SFEI). 2011. Bay Area aquatic resources inventory. SFEI. http://www.californiawetlands.net/tracker/ba/map. http://www.californiawetlands.net/tracker/ba/map.

San Francisco Water and Power. 2005. *A history of the municipal water department and Hetch Hetchy system.* SFPUC Communications.

Satterthwaite WH, Beakes MP, Collins EM, et al. 2009a. Steelhead life history on California's Central Coast: insights from a state-dependent model. *Transactions of the American Fisheries Society* 138: 532-548.

Satterthwaite WH, Beakes MP, Collins EM, et al. 2009b. State-dependent life history models in a changing (and regulated) environment: steelhead in the California Central Valley. *Evolutionary Applications* (2010): 221-243.

Sawyer JO, Keeler-Wolf T, Evens J. 2009. *A manual of California vegetation, second edition.* Sacramento, CA: California Native Plant Society.

Sawyer JO, Keeler-Wolf T. 1995. *A manual of California vegetation.* Sacramento, CA: California Native Plant Society.

Sawyer TL, Unruh JR. 2004. Characterization of late Quaternary deformation of the Mt. Diablo fold-and-thrust best, eastern San Francisco Bay Area, California. Powerpoint Presentation AGU meeting.

Sawyer TL. 1999. *Assessment of contractional deformation rates of the Mt. Diablo fold and thrust belt, eastern San Francisco Bay region, Northern California.* U.S. Geological Survey National Earthquake Hazards Reduction Program.

Saylor CF. Government Printing Office, 1906. *Progress of the beet-sugar industry in the United States in 1905, report no. 82.*

Schulz PD. 1986. *Fish remains from test excavations at CA-ALA-453, Alameda County, California.* Unpublished report by Brienes, West and Schulz, Davis, CA. 8.

Schussler H. 1901. *Jane R. Clough vs. Spring Valley Water Works, Vol. 2.* M.M. O'Shaughnessy papers, carton 29, folder 19. *Courtesy of The Bancroft Library, UC Berkeley.*

Schussler H. 1911. *Report of the resources of the Livermore Valley. Courtesy of San Francisco Public Utilities Commission Archives.*

Schussler H. 1911. *Report of the Resources of the Livermore Valley. Spring Valley Water Company. Map 6.* MB 068. 1,000: 1 inch. *Courtesy of San Francisco Public Utilities Commission Archives.*

Schween HC. 1915. *Spring Valley Water Company vs. City and County of San Francisco, et al. Vol. 1. Lands. District Court of the United States for the Northern District of California.*

Scoppettone GG, Smith JJ. 1978. Additional records on the distribution and status of native fishes in Alameda and Coyote creeks, California. *California Fish and Game* 64(1):62-65.

Scott J. 1871. *Information concerning the terminus of the railroad system of the pacific coast.* Oakland, CA: Daily Transcript Book and Printing Office.

Seale A. 1934. Aquarium fishes from a California stream. *The Aquarium Journal* 7:150-152.

Sedell JR, Reeves GH, Hauer FR, et al. 1990. Role of refugia in recovery from disturbances: modern fragmented and disconnected river systems. *Environmental Management* 14(5):711-724.

Shapovalov, L. 1938b. California Department of Fish and Game, stream and river survey files, Calaveras Creek field notes, Alameda County, May 19, 1938, Yountville, 2 pp.

Shapovalov L. 1938a. California Department of Fish and Game, stream and river survey files, Alameda Creek field notes, Alameda County, May 19, 1938, Yountville, 4 pp.

Shapovalov L. 1938c. California Department of Fish and Game, stream and river survey files, La Costa Creek field notes, Alameda County, May 19, 1938, Yountville, 2 pp.

Shapovalov L. 1944. CDFG, stream and river files, Arroyo Mocho, Alameda County, 18 November 1944. 1 p.

Shapovalov L. 1998. CDFG, stream and river files, Alameda Creek, Alameda County, 19 May 1938. 2 pp.

Sharsmith HK. 1945. Flora of the Mount Hamilton Range of California (a taxonomic study and floristic analysis of the vascular plants). *American Midland Naturalist* 34(2):289-367.

Shinn CH. 1901. *Jane R. Clough vs. Spring Valley Water Works, Vol. 2.* M.M. O'Shaughnessy papers, carton 29, folder 19. *Courtesy of The Bancroft Library, UC Berkeley.*

Shinn CH. 1907. The Alvarado Squatters' League. *Out West magazine.* Archaeological Institute of America. Southwest Society.

Shinn CH. [1889] 1991. *Historical sketches of southern Alameda County, first published in the Oakland Enquirer as a series of articles June 8-November 18,1889.* Oakland, CA: Alameda County Historical Society.

Shinn MW. 1883. Summer Cañons. In *The Overland Monthly: Devoted to the development of the country*, ed. Brett Hart, 205-210. San Francisco, CA: Samuel Carson, Publisher. II.-Second Series, July-December 1883.

Shuford WD, Gardali T. 2008. *California bird species of special concern: a ranked assessment of species, subspecies, and distinct populations of birds of immediate conservation concern in California. Studies of Western Birds 1.* Western Field Ornithologists, and California Department of Fish and Game.

Shumm SA. 1973. Geomorphic thresholds and complex response of drainage systems. In *Fluvial geomorphology*, ed. M Morisawa, 299-310. Binghampton, NY: State University of New York.

Sibrian Y. 1861. *U.S. v. Jose Dolores Pacheco, Land Case 155 ND [Santa Rita], U.S. District Court, Northern District. Courtesy of The Bancroft Library, UC Berkeley.*

Sickley TA, Mladenoff DJ, Radeloff VC, et al. 2000. A Pre-European Settlement Vegetation Database for Wisconsin. http://gis.esri.com/library/userconf/proc00/professional/papers/PAP576/p576.htm.

Skinner JE. 1962. *An historical review of the fish and wildlife resources of the San Francisco Bay Area. Sacramento, CA*: California. Dept. of Fish and Game. Water Projects Branch.

Sloan D. 2006. *Geology of the San Francisco Bay region.* Berkeley, CA: University of California Press.

Smith JJ. 1998. Steelhead and other fish resources of western Mt. Hamilton streams. Unpublished report. Department of Biological Sciences, San Jose State University, San Jose, CA.

Snyder JO. 1905. Notes on fishes of the streams flowing into the San Francisco Bay. *Report of the Bureau of Fisheries Appropriations* 5:327-338.

Sommer TR, Harrell WC, Nobriga ML. 2005. Habitat use and stranding risk of juvenile Chinook salmon on a seasonal floodplain. *North American Journal of Fisheries Management* 25: 1493-1504.

Sommer TR, Nobriga ML, Harrell WC, et al. 2001. Floodplain rearing of juvenile Chinook salmon: evidence of enhanced growth and survival. *Canadian Journal of Aquatic Sciences* 58:325-333.

Sowers JM, Richard CM. Oakland Museum of California. 2003. Creek and watershed map of the Pleasanton and Dublin area. Oakland, CA.

Sowers JM. Oakland Museum of California. 1999. Creek and watershed map of Fremont and vicinity. Oakland, CA.

Spina AP. 2007. Thermal ecology of juvenile steelhead in a warm-water environment. *Environmental Biology of Fishes* 80: 23-34.

Spring Valley Water Company (SVWC). 1910. Property owners in the vicinity of Pleasanton. *Courtesy of Earth Sciences and Map Library, UC Berkeley.*

Spring Valley Water Company (SVWC).1912a. *Spring Valley Water Company : hydrographic investigation and water rights. Courtesy of Water Resources Center Archives.*

Spring Valley Water Company (SVWC). 1912b. *The future water supply of San Francisco; a report to the Honorable the secretary of the interior and the advisory board of engineers of the United States army.*

Spring Valley Water Company (SVWC). 1915. *Recent floods, a drainage district needed.* Alameda County Water District, 1914-1916. *Courtesy of The Bancroft Library, UC Berkeley.*

Spring Valley Water Company (SVWC). 1916. *Silt in the flood waters of Alameda Creek: January and February of 1916. Courtesy of San Francisco Public Utilities Commission Archives.*

Spring Valley Water Company (SVWC). [191-?]. Property owners in the vicinity of Pleasanton. [1:16,800] *Courtesy of Earth Sciences and Map Library, UC Berkeley.*

Spring Valley Water Company (SVWC). ca.1900-1929. *Dams and watersheds of the Spring Valley Water Company*. San Francisco Bay region views. *Courtesy of The Bancroft Library, UC Berkeley.*

Spring Valley Water Company (SVWC). n.d. *The water supply of San Francisco*. San Francisco: The Blair-Murdock Company.

Stahle D, Therrell M, Cleaveland M. 2001. Ancient Blue Oaks Reveal Human Impact on San Francisco Bay Salinity. *Eos* 82(12):141, 144-145.

Stanford B, Grossinger RM, Askevold RA, et al. 2011. *East Contra Costa County historical ecology study*. Prepared for Contra Costa County and the Contra Costa Watershed Forum. A report of SFEI's Historical Ecology program, SFEI Publication #648. San Francisco Estuary Institute, Oakland, CA.

Stanger FM, Brown AK. 1969. *Who discovered the Golden Gate? The explorers' own accounts, how they discovered a hidden harbor and at last found its entrance*. San Mateo, CA: San Mateo County Historical Association.

Steere JT, Schaefer N. 2001. *Restoring the estuary: implementation strategy of the San Francisco Bay Joint Venture*. San Francisco Bay Joint Venture, Oakland, CA. 124.

Stevenson, ML, TC Goddard, LM Kiguchi, PJ Kinney. 1987. South Bay Dischargers Authority Water Quality Monitoring Program. Final Monitoring Report, December 1981-November 1986. Prepared by Kinnetic Laboratories, Inc. in joint venture with Larry Walker Associates, Inc.

Stewart OC, Lewis HT, Anderson K. 2002. *Forgotten fires: Native Americans and the transient wilderness*. Norman, OK: University of Oklahoma Press.

Stockton CW, Meko DM. 1975. A long-term history of drought occurrence in western United States as inferred from tree rings. *Weatherwise* 28(6):244-249.

Stokle JG. 1968. Mission San Jose and the Livermore Valley 1798-1842: demography and ecology of a mission hinterland. Master of Arts, Geography, University of California, Berkeley.

Stoner DS, Jamison RH, Nunes EC. 1915. Investigation of the water supply of the Niles Cone, California. Topographic and hydrographic map of Alameda County Water District and some adjacent territory. Location of contours, streams, wells, highways, railroads, towns, etc. established by actual surveys made under direction of Cyril Williams, Jr. *Courtesy of Alameda County Water District.*

Stratton JT. 1862. *U.S. v. Tomas Pacheco, Land Case 119 ND [Potrero de los Cerritos], U.S. District Court, Northern District. Courtesy of The Bancroft Library, UC Berkeley.*

Sullivan MS. 1934. The travels of Jedediah Smith: a documentary outline including the journal of the great American pathfinder. *The Fine Arts Press.*

Superior Court. 1920. *H. H. Patterson and W. D. Patterson vs. Spring Valley Water Company. Superior Court of the State of California, in and for the county of Alameda. Courtesy of San Francisco Public Utilities Commission Archives.*

Supreme Court of California. 1929. *H. H. Patterson et al., v. Spring Valley Water Company, S.F. No.12451.*

Taylor B. 1850. *Eldorado, or, adventures in the path of empire: comprising a voyage to California, via Panama; life in San Francisco and Monterey; pictures of the gold region, and experiences of Mexican travel*. 2nd edition. New York, NY: G.P. Putnam.

Tays G, editor. 1938. *Historical sites and landmarks of Alameda County, California*. Oakland, CA: Alameda County Library.

Thompson, West. 1878. *Official and historical atlas map of Alameda County, California* Oakland, CA: Thompson & West.

Thompson GH. HM Newhall & Co. 1875. Map of the lands of the Newark Land Co. in Alameda County, Cal. *Courtesy of Fisher Scrapbook Collection, Museum of Local History, Fremont, CA.*

Tibbetts FH. 1907a. *Report on artesian basin of Livermore Valley. Courtesy of Water Resources Center Archives.*

Tibbetts FH. 1907b. Topographic map of the lower end of Livermore Valley, Alameda Co., Cal. [1:12,000] *Courtesy of Earth Sciences and Map Library, UC Berkeley.*

Tibbetts H. 1912. *Supplementary report to Spring Valley Water Company on the artesian basin of Livermore Valley*. Haviland & Tibbetts Civil and Consulting Engineers, San Francisco. *Courtesy of Water Resources Center Archives.*

Tiegs SD, O'Leary JF, Pohl MM, et al. 2005. Flood disturbance and riparian species diversity on the Colorado River Delta. *Biodiversity and Conservation* 14:1175-1194.

Tooth S. 2000. Process, form and change in dryland rivers: a review of recent research. *Earth-Science Reviews* 51(1-4):67-107.

Township Register. 1910. Advancement edition - the Township Register. How the county was named. *Courtesy of Museum of Local History, Fremont, CA.*

Tracy. 1852. No. 110. Anders Pico et al. Mission San Jose, Map H.H. No.1. annexed to the deposition of Charles P. Tracy before Coms. Hiland Hall. E-866. *Courtesy of The Bancroft Library, UC Berkeley.*

Tracy CC. 1852a. *Field notes of the survey, under contract bearing date September 29, 1852, of the exterior township lines of the townships between the base line and the 1st standard parallel south, and the principal meridian and the bay of San Francisco, under contract dated September 29, 1852.* U.S. Department of the Interior, Bureau of Land Management Rectangular Survey, California. Book 231-4. *Courtesy of Bureau of Land Management, Sacramento, CA.*

Tracy CC. 1852b. *Field notes of the survey of the standard parallel, commencing at the corner of townships 4 and 5, south of ranges 1 east and 1 west, and running east and west from the standard meridian in the state of California.* U.S. Department of the Interior, Bureau of Land Management Rectangular Survey, California. Book 231-5. *Courtesy of Bureau of Land Management, Sacramento, CA.*

Tracy CC. 1853. *Under contract dated September 28, 1852. Survey of township south of first standard south, and west of the Meridian.* U.S. Department of the Interior, Bureau of Land Management Rectangular Survey, California. Book 254-20. *Courtesy of Bureau of Land Management, Sacramento, CA.*

Tracy CC. 1857. *Field-notes of the final survey of the Rancho Arroyo de la Alameda, Jose de Jesus Vallejo confirmee.* U.S. Department of the Interior, Bureau of Land Management Rectangular Survey, California. Book I-8. *Courtesy of Bureau of Land Management, Sacramento, CA.*

Trask JB. 1854. *Document No. 9: Report on the geology of the coast mountains, and part of the Sierra Nevada: embracing their industrial resources in agriculture and mining.* Journal of the fifth session of the legislature of the state of California, begun on the second day of January, 1854, and ended on the fifteenth day of May, 1854, at the cities of Benicia and Sacramento.

Treutlein TE, Fages P. 1972. Fages as Explorer, 1769-1772. *California Historical Society* 51(4):338-356.

Tyson J. 1901. *Jane R. Clough vs. Spring Valley Water Works, Vol. 2.* M.M. O'Shaughnessy papers, carton 29, folder 19. *Courtesy of The Bancroft Library, UC Berkeley.*

U. S. Army Corps of Engineers (USACE). 1961. *Review report for flood control and allied purposes: Alameda Creek, Alameda County, California. Courtesy of Water Resources Center Archives.*

U. S. Census Bureau. 2010. 2010 Census Data. 2010.census.gov/2010census.

U. S. Department of Agriculture (USDA), Western Division Laboratory. 1939-40. [Aerial photos of Alameda County]. Scale: 1:20,000. Agricultural Adjustment Administration (AAA). *Courtesy of Earth Sciences & Map Library, UC Berkeley, and the Alameda County Resource Conservation District (ACRCD) and National Resources Conservation Service (NRCS).*

U. S. Department of Agriculture (USDA). 2005. [Natural color aerial photos of Alameda County]. Ground resolution: 1m. National Agriculture Imagery Program (NAIP). Washington, D.C.

U. S. Department of Agriculture (USDA). 2009. [Natural color aerial photos of Alameda County]. Ground resolution: 1m. National Agriculture Imagery Program (NAIP). Washington, D.C.

U. S. Department of Energy. 2004. *Draft site-wide environmental impact statement for continued operation of Lawrence Livermore National Laboratory and supplemental stockpile stewardship and management programmatic environmental impact statement, Appendix H: Seismicity.* National Nuclear Security Administration, Livermore, CA.

U. S. District Court (USDC), Northern District. 1861. Potrero de los Cerritos, Tomas Pacheco et al., clmts. 119 ND. Land case map D-263. *Courtesy of The Bancroft Library, UC Berkeley.*

U. S. District Court (USDC), Northern District. ca. 1840a. Santa Rita, Jose Dolores Pacheco, clmt. 155 ND. Land case map B-343. *Courtesy of The Bancroft Library, UC Berkeley.*

U. S. District Court (USDC), Northern District. ca. 1840b. [El Valle de San Jose] Antonio Sunol, et al. clmt. 102 ND. Land case map D-238. *Courtesy of The Bancroft Library, UC Berkeley.*

U. S. District Court (USDC), Northern District. ca. 1840c. San Lorenzo "Diseno" Barbara Soto et al. Exhibit to deposition Nov 15 1854. 300 N.D. Land case map B-595. *Courtesy of The Bancroft Library, UC Berkeley.*

U. S. District Court (USDC), Northern District. ca. 1840d. [Potrero de los Cerritos]. Tomas Pacheco et al. clmts. 119 ND. Land case map B-258. *Courtesy of The Bancroft Library, UC Berkeley.*

U. S. District Court (USDC), Northern District. ca. 1840e. Half "San Ramon" or "La Juntas" Contra Costa County, Domingo Peralta, clmt. 322 N.D. Land case map D-652. *Courtesy of The Bancroft Library, UC Berkeley.*

U. S. District Court (USDC), Northern District. ca. 1840f. Santa Rita, Jose Dolores Pacheco, clmt. 155 ND. Land case map D-342. *Courtesy of The Bancroft Library, UC Berkeley.*

U. S. District Court (USDC), Northern District. ca. 1840g. Arroyo del Alameda, Jose de Jesus Vallejo, clmt. 59 ND. Land case map B-133. *Courtesy of The Bancroft Library, UC Berkeley.*

U.S. District Court (USDC), Northern District. ca. 1840h. San Ramon, Jose Maria Amador, clmt. 144 ND. Land case map D-329. *Courtesy of The Bancroft Library, UC Berkeley.*

U. S. Geological Survey (USGS). 1899a. Hayward Quadrangle, California : 15-minute series (Topographic). 1:62,500.

U. S. Geological Survey (USGS). 1899b. San Jose Quadrangle, California : 15-minute series (Topographic). 1:62,500.

U. S. Geological Survey (USGS). 1906. Livermore Quadrangle, California : 15-minute series (Topographic) 1:62,500.

U. S. Geological Survey (USGS). 1907. Altamont Quadrangle, California : 15-minute series (Topographic). 1:62,500.

U. S. Geological Survey (USGS). 1962. Floods at Fremont. 1:24,000.

Unknown. 1850. Rancho el Valle de San Jose. *Courtesy of Livermore Heritage Guild.*

Unknown. 1865. Plat 560 private claims volume Q. *Courtesy of Bureau of Land Management, Sacramento, CA.*

Unknown. 1913. Map of survey 54 & portions of surveys 'e' & 5[?] of the ex-mission lands and a portion of plot 63 of the Sunol portion of the Rancho El Valle de San Jose, Alameda Co., Calif. For Spring Valley Water Company. 1"=300'. *Courtesy of Museum of Local History, Fremont, CA.*

Unknown. 1915. *Alameda Creek system hydrography season 1914-1915. PWC 008. Courtesy of San Francisco Public Utilities Commission Archives.*

Unknown. 1938. *Agreement no. 872, SFPUC v. Pattersons HH and WD.*

Unknown. 1942. Patterson and Crandall Sloughs. Compiled from county surveys and aerial photos. 1": 1,000' *Courtesy of Alameda County Water District.*

Unknown. ca. 1890. Map of Sunol Valley, Alameda Co. 300 ft = one inch. *Courtesy of San Francisco Public Utilities Commission.*

Unruh J, Sundermann S. 2006. *Digital compilation of thrust and reverse fault data for the northern California map database: collaborative research with William Lettis & Associates, Inc., and the U.S. Geological Survey.* U.S. Geological Survey National Earthquake Hazards Reduction Program.

Unruh JR, Sawyer TL. 1997. *Assessment of blind seismogenic sources, Livermore Valley, eastern San Francisco Bay region.* U.S. Geological Survey National Earthquake Reduction Program.

Vaider J, Cook SF. 1960. *Colonial expeditions to the interior of California Central Valley, 1800-1820.* Berkeley, CA: University of California Press.

Vallejo J. 1840. *U.S. v. Jose de Jesus Vallejo, Land Case 59 ND [A tract of land near the Mission of Santa Clarra], U.S. District Court, Northern District.* Courtesy of The Bancroft Library, UC Berkeley.

Vickery FP. 1925. The structural dynamics of the Livermore region. *The Journal of Geology* 33(6):608-628.

von Geldern O. 1912. *Report on Alameda Creek yield. Courtesy of San Francisco Public Utilities Commission Archives.*

Warren M, Pardew M. 1998. Road crossings as barriers to small-stream fish movement. *Transactions of the American Fisheries Society* 127:637-644.

Welch LE, Huff RC, Dierking RA, et al. 1966. *Soil survey of the Alameda area, California*. United States Department of Agriculture Soil Conservation Service. Series 1961, No. 41.

Welch LE. 1981. *Soil survey of Alameda County, California, western part*. U. S. Department of Agriculture, Soil Conservation Service.

Welch WR. 1931. Game reminiscences of yesteryear. *California Fish and Game* 17(3):255-263.

Wemple B, Jones J, Grant G. 1996. Channel network extension by logging roads in two basins, Western Cascades, Oregon. *Water Resources Bulletin* American Water Resources Association(32):6.

West CH. 1937. *Groundwater resources of the Niles cone, Alameda County, California*. Alameda County Water District.

Westdahl F. 1896c. *Descriptive report to accompany topographic sheet entitled U. S. Coast and Geodetic Survey. W. W. Duffield, Superintendent. Re-survey of San Francisco Bay Roberts' Landing to Alameda Creek, California. Topographic survey under the direction of assistant Aug. F. Rodgers. Register no. 2252*. U. S. Coast and Geodetic Survey (USCGS).

Westdahl F. 1896d. *Descriptive report to accompany topographic sheet entitled U.S. Coast and Geodetic Survey, W.W. Duffield, Superintendent, Pacific Coast. Re-survey of San Francisco Bay, Alameda Creek to Beards Creek, California. Topographic survey under the direction of Assistant Aug. F. Rodgers. Register no. 2353*. U.S. Coast and Geodetic Survey (USCGS).

Westdahl F. U. S. Coast and Geodetic Survey (USCGS). 1896a. Re-survey of San Francisco Bay Roberts Landing to Alameda Creek, California, Register No. 2252. Washington, D.C. 1:10,000. *Courtesy of National Oceanic and Atmospheric Administration (NOAA).*

Westdahl F. U. S. Coast and Geodetic Survey (USCGS). 1896b. Re-survey of San Francisco Bay, Alameda Creek to Beard's Creek, California, Register no. 2353. 1:10,000. *Courtesy of National Oceanic and Atmospheric Administration (NOAA).*

Westover HL, van Duyne C. Government Printing Office, 1911. *Soil survey of the Livermore area, California.*

Westover HL, van Duyne C. U. S. Department of Agriculture, Bureau of Soils. 1910. Livermore Soil Map. 1:625,000.

Whipple AA, Grossinger RM, Davis FW. 2011. Shifting baselines in a California oak savanna: nineteenth century data to inform restoration scenarios. *Restoration Ecology* 19(101):88-101.

Whipple JC. 1901. *Jane R. Clough vs. Spring Valley Water Works, Vol. 2*. M.M. O'Shaughnessy papers, carton 29, folder 19. *Courtesy of The Bancroft Library, UC Berkeley.*

White CA. 1991. *A history of the rectangular survey system*. Washington, D.C.: U. S. General Printing Office.

White MD, Greer KA. 2006. The effects of watershed urbanization on the stream hydrology and riparian vegetation of Los Penasquitos Creek, California. *Landscape and Urban Planning* 74(2):125-138.

Whitfield MP. 1986. "General manager of the Alameda County Water District, 1953-1977," an oral history interview conducted by Ann Lage. *The Patterson family and ranch: southern Alameda County in transition. Vol. 2: Water, development, and preservation in southern Alameda County.* Regional Oral History Office, The Bancroft Library, UC Berkeley.

Whitney JD. State Geological Survey. 1873. Map of the region adjacent to the Bay of San Francisco. Sacramento, CA.

Wickwon EJ. 1915. *Spring Valley Water Company vs. City and County of San Francisco et al. Vol. 3. Lands. District Court of the United States for the Northern District of California.*

Wiens, JA, GD Hayward, HD Safford, et al., editors. 2012. *Historical Environmental Variation in Conservation and Natural Resource Management*. West Sussex, UK: Wiley-Blackwell.

Wieslander AE, Yates HS, Jensen HA, et al. 1933. *Manual of field instruction for Vegetation Type Map of California.*

Wieslander AE. 1935. A vegetation type map of California. *Madroño* 3(3):140-144.

Williams C, Jr. 1912. *Report on the water supply of the Alameda Creek watershed with particular reference to the Livermore Valley underground supply, prepared at the request of Percy V. Long*. San Francisco, CA.

Williams G. 1986. "The L. S. Williams Company: farming in southern Alameda County, 1930s-1983," an oral history interview conducted by Ann Lage. *The Patterson family and ranch: southern Alameda County in transition. Vol. 1: Agriculture and farm life on Fremont's northern plain, 1890s-1980s*. Regional Oral History Office, The Bancroft Library, UC Berkeley.

Williams HT, Hayden FV, King C, et al. 1878. *The Pacific tourist: Williams' illustrated trans-continental guide of travel, from the Atlantic to the Pacific Ocean*. Henry T. Williams.

Williams M, Collins J, Pearce S, et al. 2012. *Napa River watershed profile: past and present characteristics with implications for future management of the changing Napa River Valley*. SFEI contribution #615. San Francisco Estuary Institute (SFEI), Richmond, CA.

Wilson AM. 1861. *U.S. v. Jose Dolores Pacheco, Land Case 155 ND [Santa Rita], U.S. District Court, Northern District. Courtesy of The Bancroft Library, UC Berkeley*.

Witter RC, Knudsen KL, Sowers JM, et al. 2006. *Maps of Quaternary deposits and liquefaction susceptibility in the central San Francisco Bay region, California*. U.S. Geological Survey Open-File Report 2006-1037.

Wondzell SM. 2012. Hyporheic zones in mountain streams: physical processes and ecosystem functions. Stream Notes: January-April 2012. *Stream Systems Technology Center,* Rocky Mountain Research Station, Fort Collins, CO.

Wood MW. 1883. *History of Alameda County, California: including its geology, topography, soil, and productions: together with a full and particular record of the Spanish grants*. Oakland, CA: Pacific Press.

Zone 7 Water Agency. 2006. *Stream management master plan*. Section 3: needs and opportunities. Livermore, CA.

Zone 7 Water Agency. 2011. *Annual report for the groundwater management program. 2010 water year. Livermore Valley groundwater basin*. Livermore, CA.

PERSONAL COMMUNICATIONS

Banke, Paul. 2011. Land use interview. May 17. Alameda County Resource Conservation District Office. Livermore, CA.

Fletcher, Rich. 2011. Land use interview. May 17. Alameda County Resource Conservation District Office. Livermore, CA.

Hopkirk, J.D. 2000.

Koopman, Tim. 2011. Land use interview. May 24. Alameda County Resource Conservation District Office. Livermore, CA.

Leonardson, Rebecca. 2012. March 20. SFPUC

Sacchau, Dan. 2011. Land use interview. May 24. Alameda County Resource Conservation District Office. Livermore, CA.

Trush, B. 2012.

Made in United States
Troutdale, OR
11/11/2026

57125324R00228